T0215113

Scheduling and Control of Queueing Networks

Applications of queueing network models have multiplied in the last generation, including scheduling of large manufacturing systems, control of patient flow in health systems, load balancing in cloud computing, and matching in ride sharing. These problems are too large and complex for exact solution, but their scale allows approximation.

This book is the first comprehensive treatment of fluid scaling, diffusion scaling, and many-server scaling in a single text presented at a level suitable for graduate students. Fluid scaling is used to verify stability, in particular treating max weight policies, and to study optimal control of transient queueing networks. Diffusion scaling is used to control systems in balanced heavy traffic, by solving for optimal scheduling, admission control, and routing in Brownian networks. Many-server scaling is studied in the quality and efficiency driven Halfin–Whitt regime and applied to load balancing in the supermarket model and to bipartite matching in ride-sharing applications.

GIDEON WEISS is Professor Emeritus in the Department of Statistics at the University of Haifa, Israel. He has previously held tenured positions at Tel Aviv University and at Georgia Tech Industrial and Systems Engineering and visiting positions at Berkeley, MIT, Stanford, NYU, and NUS. He is author of some 90 research papers and served on the editorial boards of leading journals on operations research and applied probability. His work includes significant contributions to the fields of time series, stochastic scheduling, bandit problems, fluid analysis of queueing networks, continuous linear programming, and matching problems.

INSTITUTE OF MATHEMATICAL STATISTICS
TEXTBOOKS

Editorial Board
Nancy Reid (University of Toronto)
Arnaud Doucet (University of Oxford)
Xuming He (University of Michigan)
Ramon van Handel (Princeton University)

ISBA Editorial Representative
Peter Müller (University of Texas at Austin)

IMS Textbooks give introductory accounts of topics of current concern suitable for advanced courses at master's level, for doctoral students and for individual study. They are typically shorter than a fully developed textbook, often arising from material created for a topical course. Lengths of 100–290 pages are envisaged. The books typically contain exercises.

In collaboration with the International Society for Bayesian Analysis (ISBA), selected volumes in the IMS Textbooks series carry the "with ISBA" designation at the recommendation of the ISBA editorial representative.

Other Books in the Series (*with ISBA)

1. *Probability on Graphs*, by Geoffrey Grimmett
2. *Stochastic Networks*, by Frank Kelly and Elena Yudovina
3. *Bayesian Filtering and Smoothing*, by Simo Särkkä
4. *The Surprising Mathematics of Longest Increasing Subsequences*, by Dan Romik
5. *Noise Sensitivity of Boolean Functions and Percolation*, by Christophe Garban and Jeffrey E. Steif
6. *Core Statistics*, by Simon N. Wood
7. *Lectures on the Poisson Process*, by Günter Last and Mathew Penrose
8. *Probability on Graphs (Second Edition)*, by Geoffrey Grimmett
9. *Introduction to Malliavin Calculus*, by David Nualart and Eulàlia Nualart
10. *Applied Stochastic Differential Equations*, by Simo Särkkä and Arno Solin
11. **Computational Bayesian Statistics*, by M. Antónia Amaral Turkman, Carlos Daniel Paulino, and Peter Müller
12. *Statistical Modelling by Exponential Families*, by Rolf Sundberg
13. *Two-Dimensional Random Walk: From Path Counting to Random Interlacements*, by Serguei Popov

Scheduling and Control of Queueing Networks

GIDEON WEISS

University of Haifa, Israel

CAMBRIDGE
UNIVERSITY PRESS

CAMBRIDGE
UNIVERSITY PRESS

University Printing House, Cambridge CB2 8BS, United Kingdom

One Liberty Plaza, 20th Floor, New York, NY 10006, USA

477 Williamstown Road, Port Melbourne, VIC 3207, Australia

314–321, 3rd Floor, Plot 3, Splendor Forum, Jasola District Centre,
New Delhi – 110025, India

103 Penang Road, #05–06/07, Visioncrest Commercial, Singapore 238467

Cambridge University Press is part of the University of Cambridge.

It furthers the University's mission by disseminating knowledge in the pursuit of
education, learning, and research at the highest international levels of excellence.

www.cambridge.org
Information on this title: www.cambridge.org/9781108415323
DOI: 10.1017/9781108233217

© Gideon Weiss 2022

This publication is in copyright. Subject to statutory exception
and to the provisions of relevant collective licensing agreements,
no reproduction of any part may take place without the written
permission of Cambridge University Press.

First published 2022

A catalogue record for this publication is available from the British Library.

Library of Congress Cataloging-in-Publication Data
Names: Weiss, Gideon, author.
Title: Scheduling and control of queueing networks / Gideon Weiss,
The University of Haifa.
Description: First edition. | New York : Cambridge University Press, [2021] |
Series: Institute of mathematical statistics textbooks | Includes
bibliographical references and index.
Identifiers: LCCN 2021029970 | ISBN 9781108415323 (hardback)
Subjects: LCSH: Queuing theory. | System analysis–Mathematics. |
Scheduling–Mathematics. | BISAC: MATHEMATICS / Probability & Statistics / General
Classification: LCC QA274.8 .W43 2021 | DDC 519.8/2–dc23
LC record available at https://lccn.loc.gov/2021029970

ISBN 978-1-108-41532-3 Hardback
ISBN 978-1-108-40117-3 Paperback

Additional resources for this publication at www.cambridge.org/9781108415323

Cambridge University Press has no responsibility for the persistence or accuracy of
URLs for external or third-party internet websites referred to in this publication
and does not guarantee that any content on such websites is, or will remain,
accurate or appropriate.

Contents

Notation   xi

Introduction   xiii

Part I   The Single Queue   1

1   Queues and Their Simulations, Birth and Death Queues   3
1.1   The Single Queue   3
1.2   Simulation of a Queue   6
1.3   Birth and Death Queues   7
1.4   Historical Notes, Sources and Extensions   13
       Exercises   13

2   The M/G/1 Queue   16
2.1   Little's Law   16
2.2   Work Conservation   21
2.3   Some Renewal Theory   23
2.4   Length Biasing   25
2.5   Stationary Point Processes and Palm Measure   27
2.6   PASTA — Poisson Arrivals See Time Averages   28
2.7   M/G/1 Average Waiting Times   31
2.8   Busy Periods   33
2.9   Supplementary Material: Embedded Markov Chains   34
2.10   Sources   39
       Exercises   39

3   Scheduling   42
3.1   Batch Scheduling   43
3.2   Value of Information   45
3.3   Stream Scheduling   46
3.4   Optimality of SRPT   47

3.5   Mean and Variance for Non-Preemptive, Non-Predictive
       Scheduling   49
3.6   M/G/1 Priority Scheduling   50
3.7   M/G/1 under Preemptive LCFS and under PS   52
3.8   Smith's Rule, Klimov's Control of M/G/1, and the $c\mu$ Rule.   53
3.9   Bandit Processes and the Gittins Index   53
3.10   Sources   55
       Exercises   56

Part II   Approximations of the Single Queue   59

4   The G/G/1 Queue   61
4.1   Loynes Construction for Stability of G/G/1 Queues   61
4.2   G/G/1 and the Random Walk   64
4.3   Bounds for G/G/1 Waiting Times   67
4.4   Sources   68
       Exercises   69

5   The Basic Probability Functional Limit Theorems   71
5.1   Convergence of Stochastic Processes   71
5.2   Functional Limit Theorems for Random Walks   73
5.3   Functional Limit Theorems for Renewal Processes   79
5.4   Sources   81
       Exercises   81

6   Scaling of G/G/1 and G/G/∞   83
6.1   Primitives and Dynamics of G/G/1   83
6.2   Scaling Time, Space, and Initial Conditions   85
6.3   Skorohod's Reflection Mapping for the Single-Server Queue   86
6.4   Fluid Scaling and Fluid Limits of G/G/1   88
6.5   Diffusion Scale Limits for Underloaded and Overloaded G/G/1   90
6.6   Diffusion Scaling of the G/G/1 Queue in Heavy Traffic   93
6.7   Approximations of the G/G/1 Queue   95
6.8   Another Approach: Strong Approximation   97
6.9   Diffusion Limits for G/G/∞ in Heavy Traffic   98
6.10   Sources   100
       Exercises   101

7   Diffusions and Brownian Processes   102
7.1   Diffusion Processes, Brownian Motion, and Reflected Brown-
       ian Motion   102
7.2   Diffusions and Birth and Death Processes   105

7.3   Approximation of the G/G/1 Queue   107
7.4   Two Sided Regulation of Brownian Motion   107
7.5   Optimal Control of a Manufacturing System   109
7.6   Oblique Reflection and Multivariate RBM   111
7.7   Supplement: Calculations for Brownian Motion and Derived
      Diffusions   112
7.8   Sources   120
      Exercises   120

Part III   Queueing Networks   123

8   Product-Form Queueing Networks   125
8.1   The Classic Jackson Network   125
8.2   Reversibility and Detailed Balance Equations   127
8.3   Partial Balance and Stationary Distribution of Jackson Networks 129
8.4   Time Reversal and the Arrival Theorem   131
8.5   Sojourn Times in Jackson Networks   133
8.6   Closed Jackson Networks   134
8.7   Kelly Networks   135
8.8   Symmetric Queues and Insensitivity   138
8.9   Multi-Class Kelly-Type Networks   140
8.10 Sources   140
      Exercises   141

9   Generalized Jackson Networks   143
9.1   Primitives and Dynamics   143
9.2   Traffic Equations and Stability   144
9.3   Centering and Skorohod Oblique Reflection   146
9.4   Fluid Limits   147
9.5   Diffusion Limits   149
9.6   Stability of Generalized Jackson Networks   153
9.7   Sources   154
      Exercises   154

**Part IV   Fluid Models of Multi-Class Queueing
Networks**   157

**10   Multi-Class Queueing Networks, Instability, and Markov Rep-
         resentations**   159
10.1 Multi-Class Queueing Networks   159
10.2 Some Unstable MCQN   161

10.3   Policy Driven Markovian Description of MCQN   165
10.4   Stability and Ergodicity of Uncountable Markov Processes   168
10.5   Sources   173
        Exercises   174

11   Stability of MCQN via Fluid Limits   **175**
11.1   System Equations, Fluid Limit Model, Fluid Equations and
        Fluid Solutions   175
11.2   Stability via Fluid Models   181
11.3   Some Fluid Stability Proofs   185
11.4   Piecewise Linear Lyapunov Functions and Global Stability   191
11.5   Sources   194
        Exercises   194

12   Processing Networks and Maximum Pressure Policies   **197**
12.1   A More General Processing System   198
12.2   Maximum Pressure Policies   201
12.3   Rate Stability Proof via the Fluid Model   204
12.4   Further Stability Results under Maximum Pressure Policy   206
12.5   Applications   209
12.6   Sources   215
        Exercises   215

13   Processing Networks with Infinite Virtual Queues   **219**
13.1   Introduction, Motivation, and Dynamics   219
13.2   Some Countable Markovian MCQN-IVQ   221
13.3   Fluid Models of Processing Networks with IVQs   226
13.4   Static Production Planning Problem and Maximum Pressure
        Policies   228
13.5   An Illustrative Example: The Push-Pull System   230
13.6   Sources   233
        Exercises   234

14   Optimal Control of Transient Networks   **236**
14.1   The Semiconductor Wafer Fabrication Industry   237
14.2   The Finite Horizon Problem Formulation   238
14.3   Formulation of the Fluid Optimization Problem   240
14.4   Brief Summary of Properties of SCLP and its Solution   242
14.5   Examination of the Optimal Fluid Solution   245
14.6   Modeling Deviations from the Fluid as Networks with IVQs   247
14.7   Asymptotic Optimality of the Two-Phase Procedure   250
14.8   An Illustrative Example   252
14.9   Implementation and Model Predictive Control   253

14.10   Sources   255
        Exercises   255

        Part V   Diffusion Scaled Balanced Heavy Traffic   257

15   **Join the Shortest Queue in Parallel Servers**   259
15.1   Exact Analysis of Markovian Join the Shortest Queue   259
15.2   Variability and Resource Pooling   262
15.3   Diffusion Approximation and State Space Collapse   264
15.4   Threshold Policies for Routing to Parallel Servers   266
15.5   A Note about Diffusion Limits for MCQN   266
15.6   Sources   267
        Exercises   268

16   **Control in Balanced Heavy Traffic**   270
16.1   MCQN in Balanced Heavy Traffic   270
16.2   Brownian Control Problems   273
16.3   The Criss-Cross Network   276
16.4   Sequencing for a Two-Station Closed Queueing Network   281
16.5   Admission Control and Sequencing for a Two-Station MCQN   287
16.6   Admission Control and Sequencing in Multi-Station MCQN   295
16.7   Asymptotic Optimality of MCQN Controls   300
16.8   Sources   301
        Exercises   302

17   **MCQN with Discretionary Routing**   305
17.1   The General Balanced Heavy Traffic Control Problem   305
17.2   A Simple Network with Routing and Sequencing   306
17.3   The Network of Laws and Louth   309
17.4   Further Examples of Pathwise Minimization   312
17.5   Routing and Sequencing with General Cuts   315
17.6   Sources   318
        Exercises   319

        Part VI   Many-Server Systems   321

18   **Infinite Servers Revisited**   323
18.1   Sequential Empirical Processes and the Kiefer Process   323
18.2   Stochastic System Equations for Infinite Servers   325
18.3   Fluid Approximation of Infinite Server Queues   326
18.4   Diffusion Scale Approximation of Infinite Server Queues   327

18.5	Sources	329
	Exercises	329
19	**Asymptotics under Halfin–Whitt Regime**	330
19.1	Three Heavy Traffic Limits	330
19.2	M/M/s in Halfin–Whitt Regime	331
19.3	Fluid and Diffusion Limits for G/G/s under Halfin–Whitt Regime	336
19.4	sources	345
	Exercises	345
20	**Many Servers with Abandonment**	346
20.1	Fluid Approximation of G/G/n+G	346
20.2	The M/M/n+M System under Halfin–Whitt Regime	350
20.3	The M/M/n+G System	352
20.4	Sources	358
	Exercises	358
21	**Load Balancing in the Supermarket Model**	360
21.1	Join Shortest of d Policy	361
21.2	Join the Shortest Queue under Halfin–Whitt Regime	373
21.3	Approaching JSQ: Shortest of $d(n)$ and Join Idle Queue	381
21.4	Sources	382
	Exercises	383
22	**Parallel Servers with Skill-Based Routing**	385
22.1	Parallel Skill-Based Service under FCFS	386
22.2	Infinite Bipartite Matching under FCFS	390
22.3	A FCFS Ride-Sharing Model	401
22.4	A Design Heuristic for General Parallel Skill-Based Service	404
22.5	Queue and Idleness Ratio Routing	406
22.6	Sources	409
	Exercises	409
References		413
Index		427

Notation

$\mathbb{P}$ probability

$\mathbb{E}$ expectation

$\mathbb{1}$ characteristic function

$\mathbb{P}_x$ probability from initial state x

$\mathbb{E}_x$ expectation from initial state x

$\mathbb{N}$ natural numbers, $0, 1, \ldots$

$\mathbb{Z}$ integers

$\mathbb{R}$ the real line

$\mathbb{C}$ space of continuous functions

$\mathbb{D}$ space of functions right continuous with left limits

$\mathbf{1}$ vector of 1's

π stationary probabilities

$\mathcal{A}(t)$ arrival process

$\mathcal{D}(t)$ departure process

$Q(t)$ queue length

$\mathcal{W}(t)$ workload process

$\mathcal{S}(t)$ service process

$\mathcal{T}(t)$ busy time

$\mathcal{I}(t)$ idle time

$\mathcal{J}(t)$ free time

a_ℓ arrival time of customer ℓ

T_ℓ interarrival time, $T_\ell = a_\ell - a_{\ell-1}$ (Chapters 1–4)

u_ℓ interarrival time, $u_\ell = a_\ell - a_{\ell-1}$ (Chapters 5–20)

X_ℓ service requirement of customer ℓ (Chapters 1–4)

υ_ℓ service requirement of customer ℓ (Chapters 5–20)

F distribution of interarrival times

G distribution of service times

H distribution of patience times

V_ℓ waiting time of customer ℓ

$\bar{V}$ mean waiting time

W_ℓ sojourn time of customer ℓ

$\bar{W}$ mean sojourn time

$\bar{\mathcal{W}}$ mean workload

ρ offered load or traffic intensity

α_i exogenous arrival rates

λ_i total arrival rates

μ_i service rates

$p_{i,j}$ routing probabilities

ν_k nominal allocation

c_a coefficient of variation of interarrival times

c_s coefficient of variation of service time

C constituency matrix

C_i constituency of server i

A resource consumption matrix

R input-output matrix

R^{-1} work requirement matrix

B_κ a compact neighborhood of the origin

$\mathscr{C}$ set or subset of customer types (Chapter 22)

$\mathscr{S}$ set or subset of server types (Chapter 22)

$\mathcal{U}$ set or subset of customer types unique to some servers (Chapter 22)

$\mathcal{G}$ compatibility graph (Chapter 22)

$\mathcal{P}(J)$ permutations of the set J

$\emptyset$ the empty set

r.h.s. right-hand side

i.i.d. independent identically distributed

u.o.c. uniformly on compacts

a.s. almost surely

c.o.v. coefficient of variation

pdf probability density function

cdf cumulative distribution function

BM Brownian motion

RBM reflected Brownian motion

RCLL right continuous with left limits

MCQN multi-class queueing networks

PASTA Poisson arrivals see time averages

BP busy period

EFSBP exceptional first service busy period

HOL head of the line policy

FCFS first come first served

LCFS last come first served, pre-emptive

PS processor sharing

FBFS first buffer first served, in re-entrant line

LBFS last buffer first served, in re-entrant line

SPT shortest processing time

SEPT shortest expected processing time

SRPT shortest remaining processing time

IVQ infinite virtual queue

BCP Brownian control problem

ED efficiency driven service

QD quality driven service

QED quality and efficiency driven service

CRP complete resource pooling

PSBS parallel skilled based service

ALIS assign longest idle server

SD server dependent service rates

QIR queue and idleness ratio policy

Introduction

Queueing networks are all pervasive; they occur in service, manufacturing, communication, computing, internet and transportation. Much of queueing theory is aimed at performance evaluation of stochastic systems. Extending the methods of deterministic optimization to stochastic models so as to achieve both performance evaluation and control is an important and notoriously hard area of research. In this book our aim is to familiarize the reader with recent techniques for scheduling and control of queueing networks, with emphasis on both evaluation and optimization.

Queueing networks of interest are discrete, stochastic dynamical systems, often of very large size, and exact analysis is usually out of the question. Furthermore, the data necessary for exact analysis is rarely available. Thus, to obtain useful results we are led to use approximations. In this book, our emphasis is on approximations obtained from scaled versions of the systems, and analyzing the limiting behavior when the scale tends to infinity. We will be studying three types of scaling: fluid, diffusion, and many server.

Fluid scaling (Part IV): We count time in units of n and space, expressed by number of items, in units of n. This will be a reasonable model to ask what happens to a system with n items in a time span in which n items are processed. Under fluid scaling, a discrete stochastic system may converge to a deterministic continuous process, its fluid model. Fluid models are used in two ways: first, to answer the question of stability – is the system capable of recovering from extreme situations, in which case it may converge to a stationary behavior. Second, perhaps more exciting, we can use fluid scaling to obtain asymptotically optimal control of transient systems over finite time horizons.

Diffusion scaling (Part V): Space is scaled by units of n, and time by units of n^2. On this scale a stable system may reach stationary behavior, and reveal the congested elements of the network in balanced heavy traffic. Approximation of these by stochastic diffusion processes can be used to evaluate performance measures. Furthermore, on the diffusion scale we may formulate and solve Brownian control problems and derive efficient policies.

Many-server scaling (Part VI): Increasingly, recent applications involve systems with many servers and a large volume of traffic. For such systems time is not scaled, but the number of servers and the arrival rates are scaled. These models preserve not just first moment parameters of the fluid scaling and second moment parameters of the diffusion scaling, but the full service time distribution of individual items moving through the system. Many-server models are used to answer staffing-level questions, and to achieve quality of service goals.

The first three parts of the book cover more conventional material, as well as introducing some of the techniques used later. *Part I* covers birth and death queues and the M/G/1 queue, and a chapter on scheduling. *Part II* deals with approximations to G/G/1, introducing fluid and diffusion scaling and many-server G/G/∞. Two chapters, Chapter 5 and Chapter 7 survey some of the essential probability theory background, at a semi-precise level. *Part III* of the book introduces Jackson networks and related queueing networks with product-form stationary distributions, and generalized Jackson networks.

The book is aimed at graduate students in the areas of Operations Research, Operations Management, Computer Science, Electrical Engineering, and Mathematics, with some background in applied probability. It can be taught as a two part course, using the first three parts of the book as a basic queueing course, or it can be taught to students already familiar with queueing theory, where the first three parts are skimmed and the emphasis is on the last three parts. I have taught this material three times, at three different schools, as a PhD-level course in a single semester, though it was somewhat tight to include all of the material of the second half of the book. I tried to make each chapter as self-contained as possible, to enable more flexibility in teaching a course, and to be more useful for practitioners.

Each chapter in the book is followed by a list of sources, and by exercises. Some of the exercises lead to substantial extensions of the material, and I provide references for those. A few problems that require further study and much more effort are in addition marked by (∗). A solution manual will accompany the book, and be placed on the book website, www.cambridge.org/9781108415323 .

I conclude this introduction with six examples of potential applications that the techniques developed in this book are designed for:

Semiconductor wafer fabrication plant: Given the current state of the plant, how to schedule production for the next six weeks. While this appears at first to be a deterministic job-shop scheduling problem, optimal schedules never work due to unexpected stochastic interference. In Chapter 14 we formulate this as a control problem of a discrete stochastic transient queueing network. We use fluid scaling to obtain and solve a deterministic continuous control problem, and then track the optimal fluid solution using decentralized control.

Input queued crossbar switches: Scheduling the traffic through ultra-high speed communication switches so as to achieve maximum throughput is solved by a maximum pressure policy as described in Chapter 12.

Joint management of operating theaters in a hospital: To control the flow of patients, surgeons, equipment, and theaters on a long-term basis, this can be formulated as a multi-class queueing network, which is operating in stationary balanced heavy traffic. In Chapters 16 and 17 we use diffusion scaling to formulate and solve such problems as stochastic Brownian control problems, and the optimal solution of the Brownian problem is used to determine policies that use admission limits, choice of routes, scheduling priorities, and thresholds.

Control of a call center: This is modeled as a parallel service system, where types of customers are routed to pools of compatible skilled servers. Here design of the compatibility graph, balancing staffing levels, and maintaining acceptable levels of abandonment need to be determined, based on many-server scaling in Chapters 19 and 20.

Cloud computing and web searching: Balancing the utilization of the servers and controlling the lengths of queues at many servers needs to be achieved with a minimum amount of communication. Asymptotic optimality here is achieved by routing tasks to the shorter of several randomly chosen servers. This is modeled by the so-called supermarket problem, as studied in Chapter 21.

Ride sharing: Drivers as well as passengers become available in a random arrival stream, and have limited patience waiting for a match to determine a confirmed trip. Matching available drivers to passengers according to their compatibility, and dispatching on first come first served is analyzed and used to design regimes of operation in Chapter 22.

In writing this book I benefitted from the help of many colleagues, foremost I wish to thank to Ivo Adan, Onno Boxma, Asaf Cohen, Liron Ravner, Shuangchi He, Rhonda Righter, Dick Serfozo, and Hanqin Zhang for their many useful comments and suggestions, and to my students who kept me in check.

Part I

The Single Queue

In the first part of this book we introduce the single queue, a system in which arriving customers require a single service from the system, and this service is provided by one or more servers. We study properties of some special queueing systems that are amenable to exact analysis.

In Chapter 1 we define the single queue, introduce notations and some relations and properties, and present the most tractable examples of queues, so-called birth and death queues . We also discuss simulation of queues.

In Chapter 2 we study a queueing system with memoryless Poisson arrivals and generally distributed processing times, the so-called M/G/1 system. Performance measures of this system can be derived exactly, using the principle of work conservation, and the property of PASTA (Poisson arrivals see time averages).

Chapter 3 considers the scheduling of batches of jobs, and of stationary streams of jobs. We discuss priority queues and other service policies.

1

Queues and Their Simulations, Birth and Death Queues

In this chapter we start our exploration of queues. We describe a single queue consisting of a single stream of customers where each of them requires a single service operation and this service is provided by a single service station, manned by one or more servers, operating under some service policy. We introduce notation to describe such queues and derive two basic relationships: the first presents the queue length as arrivals minus departures, the second, Lindley's equation, provides a recursive calculation of waiting times of successive customers. We discuss simulation of queues using the Lindley's equation recursion. Next we study examples of single queues with Poisson arrivals and exponential service times, which are modeled by Markovian birth and death processes, and derive the stationary distribution of the queue length, using detailed balance equations.

1.1 The Single Queue

A queueing system consists of two parts: on the demand side there are the streams of customers, each with its service requests; on the service side there are one or more service stations, with one or more servers in each. We start our exploration of queueing theory by considering a single stream of customers, each requiring a single service operation, and a single service station that provides the service. We refer to the sequence of arrivals and services as the primitives of the system. We model the customer arrivals by a stochastic point process $\mathcal{A}(t)$, $t > 0$, which counts the number of arrivals in the time interval $(0, t]$. We let A_n, $n = 1, 2, \ldots$ be the arrival times of the customers, and $T_1 = A_1$, $T_n = A_n - A_{n-1}$, $n = 2, 3, \ldots$ be the interarrival times. We have that $\mathcal{A}(t) = \max\{n : A_n \leq t\}$. We will frequently assume that T_n are independent identically distributed random variables with distribution F and finite expectation $\mathbb{E}(T_1) = 1/\lambda$, so that $\mathcal{A}(t)$ is a renewal process, with arrival rate λ. In particular, if interarrivals are exponentially distributed, then $\mathcal{A}(t)$ is a Poisson process and we say that arrivals are Poisson with rate λ.

Customer n requires service for a duration X_n, $n = 1, 2, \ldots$. We will always assume that the sequence of service durations is independent of the arrival times, and service durations are independent identically distributed with distribution G and finite expectation $m = 1/\mu$.

The customers are served by a single service station, which may have one or more servers that provide the service. Typically, an arriving customer will join a queue and wait, and will then move to a server and be served.

A common notation introduced by D.G. Kendall describes the single queue by a three-field mnemonic: the first describes the arrival process, the second the service distribution, and the third the service station. Thus M/M/1 denotes a queue with Poisson arrivals, exponential service times, and a single server, where M stands for memoryless. D/G/s denotes a queue with deterministic arrivals, generally distributed independent service times, and s servers. G/G/∞ is a queue with independent, generally distributed interarrival times, general independent service times, and an infinite number of servers, which means that arriving customers start service immediately and there is no waiting. G/·/· will denote a queue that has a general stationary sequence of interarrival times. If the system can only contain a limited number of customers at any time, this limit is sometimes added as a fourth field. Thus, M/M/K/K is a queueing system with Poisson arrivals, exponential service times, K servers, and a total space for K customers. This system is the famous Erlang loss system, which Erlang has used to model a telephone exchange with K lines. In this system, when all the lines are busy, arriving customers are lost.

The interaction between the customers described by $\mathcal{A}(t)$, $t > 0$ and X_n, $n = 1, 2, \ldots$ on the one hand and the service station on the other hand creates waiting and queues. To describe this interaction we need to specify also the service policy. We list a few commonly used service policies: FCFS – first come first served (also known as FIFO – first in first out) in which customers enter service in order of arrival; LCFS – last come first served (sometimes called LIFO – last in first out) in which whenever a customer arrives it enters service immediately, sometimes preempting the service of an earlier customer; PS – processor sharing, the station divides its service capacity equally between all the customers in the system.

The following two very simple relationships are the basis of much of queueing theory. The first describes the dynamics of customers. We denote the queue length by $Q(t)$, which is the number of customers in the system at time t, including both those waiting for service and those being served. We also denote by $\mathcal{D}(t)$ the number of customers that have left the system in the time interval $(0, t]$, which we call the departure process. Then we have the obvious relation

$$Q(t) = Q(0) + \mathcal{A}(t) - \mathcal{D}(t), \tag{1.1}$$

i.e. what is in the system at time t is what was there initially at time 0, plus all arrivals, minus all departures. The queueing dynamics are illustrated in Figure 1.1.

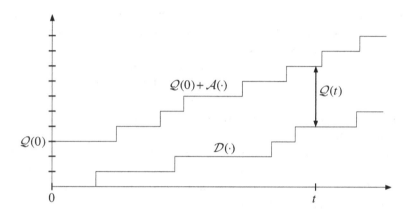

Figure 1.1 Queue length is arrivals minus departures.

The second relation calculates the waiting time of a customer under FCFS, and is known as Lindley's equation. We denote by V_n the waiting time of the nth arriving customer, from his arrival time to the start of service. For a single server operating under FCFS, we then have:

$$V_{n+1} = (V_n + X_n - T_{n+1})^+, \tag{1.2}$$

where $(x)^+ = \max(0, x)$ is the positive part of x. We explain this relation: Customer n departs the system $V_n + X_n$ time units after his arrival, while customer $n + 1$ arrives T_{n+1} time units after the arrival of customer n. If T_{n+1} exceeds $V_n + X_n$, then customer $n + 1$ will enter service immediately and not wait. If T_{n+1} is less than $V_n + X_n$, then customer $n + 1$ will wait $V_n + X_n - T_{n+1}$. This proves (1.2). Lindley's equation is illustrated in Figure 1.2.

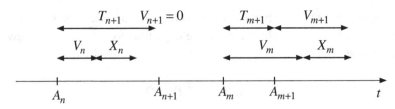

Figure 1.2 Waiting time calculation using Lindley's equation.

1.2 Simulation of a Queue

Simulation is a powerful tool for studying queueing systems. The detailed analysis of many queueing systems is intractable, but various performance measures of the queues can be estimated by simulation. There is a rich theory of how to use simulation, which we will not cover in this text; however, we suggest some books in the sources in Section 1.4. Here we indicate only the elementary method that can be used for simple explorations by the reader. Thinking of simulation is also a way of getting a different view from what can be obtained from theorems and equations.

Recursive relations such as Lindley's equation can be used to simulate the queues. In the case of a single queue with a single server, operating under FCFS, the simulation will work as follows: Initialize the system with the time at which the server will be available after serving all the customers present at time zero. Thereafter, generate successive interarrival and service times for successive customers, and use these to obtain arrival time, service start time, and departure time of successive customers. From this, obtain the waiting time and sojourn time (we define sojourn time as waiting plus service, or more generally, time from arrival to departure) of each customer. Furthermore, by counting arrivals minus departures, the queue lengths at any time can be obtained.

Example 1.1 The following tables illustrate part of a simulation of a queue. In the first table we start from the 17th customer who arrives at time 39.1, and the server is available to start serving this customer at time 42.8. We denote by S_n the start of service of customer n, by D_n his departure, by W_n his sojourn time, and by V_n his waiting time before being served. The simulation then proceeds as follows. Successive interarrivals T_n and service requirements X_n are generated pseudorandomly from the interarrival and service time distributions, and we then calculate recursively: $A_n = A_{n-1} + T_n$, $S_n = \max(A_n, D_{n-1})$, $V_n = S_n - A_n$, $D_n = S_n + X_n$, $W_n = V_n + X_n$. The round-off numbers are:

Customer	T_n	A_n	X_n	Start	Depart	Sojourn	Wait
17		39.1	2.2	42.8	45.1	6.0	3.8
18	2.8	41.9	2.7	45.1	47.7	5.9	3.2
19	4.3	46.1	1.0	47.7	48.7	2.6	1.6
20	2.5	48.6	0.3	48.7	49.0	0.4	0.1
21	4.1	52.8	1.5	52.8	54.2	1.5	0.0
22	4.8	57.6	2.0	57.6	59.5	2.0	0.0
23	1.3	58.9	3.9	59.5	63.5	4.6	0.6

The second table calculates the queue length. Order all the arrival and departure times, attaching 1 to each arrival and −1 to each departure. The queue length is then obtained, by adding the initial queue length and all the +1's and −1's up to each time t. The table lists times of arrival and departure, the identity of the customer that arrives or departs (with positive sign for arrival and negative sign for departure), and the queue length at the time of this event. The queue lengths are plotted in Figure 1.3.

Time	39.1	41.9	42.1	42.8	45.1	46.1	47.7	48.6	48.7
Customer	17	18	-15	-16	-17	19	-18	20	-19
Queue	3	4	3	2	1	2	1	2	1

Time	49.0	52.8	54.2	57.6	58.9	59.5	59.8	61.6	63.5
Customerr	-20	21	-21	22	23	-22	24	25	-23
Queue	0	1	0	1	2	1	2	3	2

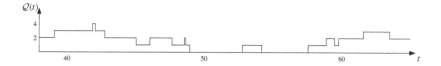

Figure 1.3 Simulation of queue length.

1.3 Birth and Death Queues

We now consider queues with Poisson arrivals and exponential service times. For such a queue, at any time t, the remaining time to the next arrival is exponentially distributed, as are also the remaining processing times of all the customers present in the system, either waiting or currently in process. All these times are independent of anything that happened prior to t. As a result, the queue length process $Q(t)$ is a continuous time Markov chain, with states $m = 0, 1, 2, \ldots$. Furthermore, when there are m customers in the system, $Q(t) = m$, the next event to happen will, with probability 1, be either a single arrival or a single service completion and departure, so the state will change by ± 1. Such queues are called birth and death queues. They are fully described by the transition rates

$$q(m, m + 1) = \lambda_m, \qquad q(m, m - 1) = \mu_m. \qquad (1.3)$$

Here λ_m, the birth rate, is the rate at which arrivals occur when there are m customers in the system, and μ_m, the death rate, is the rate at which departures occur when there are m customers in the system. Note, for Poisson arrivals $\lambda_n = \lambda$, but letting the birth rate depend on the state allows for more general models, some of which we will encounter soon.

A major descriptor of the queueing process $Q(t)$ is its stationary distribution,

$$\pi(m) = \lim_{t \to \infty} \mathbb{P}(Q(t) = m), \quad m \geq 0, \qquad (1.4)$$

provided these limits exist. This is sometimes called the limiting or long-run distribution, since it describes the state of the process after a time at which it no longer depends on the initial state. We will define stability of the queueing system if such a stationary distribution exists.

Continuous time birth and death processes are time reversible, and their stationary probabilities, $\pi(m) = \lim_{t \to \infty} \mathbb{P}(Q(t) = m)$, satisfy the detailed balance equations:

$$\pi(m)q(m, m+1) = \pi(m+1)q(m+1, m), \text{ i.e. } \pi(m+1) = \pi(m)\frac{\lambda_m}{\mu_{m+1}}. \quad (1.5)$$

We will discuss reversibility and balance equations in greater detail in Section 8.3. To interpret (1.5), note that it equates the rate at which transitions from m to $m + 1$ occur, to the rate at which transitions back from $m + 1$ to m occur. These rates are sometimes referred to as flux, borrowing a term from electricity networks. The detailed balance equations say that at stationarity these must be equal. From the detailed balance equations (1.5) we obtain the stationary distribution of a general birth and death queue:

$$\pi(m) = \pi(0)\frac{\lambda_0\lambda_1\cdots\lambda_{m-1}}{\mu_1\mu_2\cdots\mu_m}, \qquad (1.6)$$

where $\pi(0)$ is obtained as the normalizing constant. The necessary and sufficient condition for ergodicity (irreducibility and positive recurrence of the Markov chain, see later Definition 2.4) is that the normalizing constant is > 0, i.e. that the sum of the terms on the r.h.s. of (1.6) converges. In that case, we say that the queue is stable.

We now describe several important models of birth and death queues. We use (1.6), to derive their stationary distributions. It is customary to denote by ρ the offered load of the system, which is the average amount of work that arrives at the service station per unit of time: if arrivals are Poisson with rate λ and service is exponential with rate μ then average service time is $m = 1/\mu$, and the average amount of work arriving per unit time is $\rho = \lambda/\mu$.

Example 1.2 (The M/M/1 queue)   The M/M/1 queue is the simplest queueing model, for which almost every property or performance measure can be expressed by a closed form formula. Arrivals are Poisson at rate λ, service is exponential at rate μ, and there is a single server. Figure 1.4 illustrates the states and transition rates of $Q(t)$.

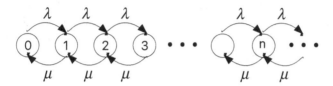

Figure 1.4 The M/M/1 queue, states and transition rates.

From (1.6) we have immediately:

Stationary distribution of the M/M/1 queue:
$$\pi(n) = (1 - \rho)\rho^n, \quad n = 0, 1, 2, \ldots, \quad \rho < 1. \quad (1.7)$$

The queue is stable if and only if $\rho < 1$, and the stationary distribution of the queue length is geometric with parameter $1 - \rho$ (denoted $\sim \text{Geom}_0(1 - \rho)$), and mean $\frac{\rho}{1-\rho}$.

Theorem 1.3 *The sojourn time of a customer in the stationary M/M/1 queue under FCFS is exponentially distributed with rate $\mu - \lambda$, $W_n \sim Exp(\mu - \lambda)$, with mean $\frac{1}{\mu-\lambda}$.*

Proof If a customer arrives and there are j customers in the queue then his waiting time will be the sum of j i.i.d. exponential rate μ random variables, and his sojourn time will be the sum of $j + 1$; this has an Erlang distribution. The probability that there are j customers in the queue is $(1 - \rho)\rho^j$. So the pdf of his sojourn time $f_W(t)$ is

$$f_W(t) = \sum_{j=0}^{\infty} (1 - \rho)\rho^j \frac{\mu^{j+1}t^j}{j!} e^{-\mu t} = \mu(1 - \rho)e^{-\mu(1-\rho)t},$$

where $\frac{\mu^{j+1}t^j}{j!}e^{-\mu t}$ is the density of the Erlang distribution with parameters $j+1$ and μ, denoted ~Erlang$(j+1, \mu)$. Hence, W_n is distributed exponentially with parameter $\mu(1-\rho) = \mu - \lambda$. Here we assume that the number of customers in the system at the time of an arrival is distributed as the stationary distribution of $Q(t)$. We justify this assumption, by proving that Poisson arrivals see time averages (PASTA) in Section 2.6.   □

Remark (Resource pooling)   There is an important lesson to be learned here: We note that the queue length, described by (1.7), depends only on $\rho = \lambda/\mu$, and does not depend directly on λ or μ. If we speed up the server, and speed up the arrival rate, say by a factor s, the number of customers in the system will remain the same. However, the expected sojourn time will decrease by a factor of s: $\frac{1}{s\mu - s\lambda}$. In other words, suppose we had s single-server M/M/1 queues to process s streams of customers, and were able to use instead an s time faster service rate and pool them all into one queue. In that case we would see the same length of queue at the pooled single queue as we saw in each of the s queues, but customers would move at a speed increased by a factor of s. This is the phenomena of *resource pooling*. We will encounter it later throughout the text.

Example 1.4 (The M/M/∞ queue)   Arrivals are Poisson at rate λ, service time is exponential with rate μ, and there is an unlimited number of servers, so that customers enter service immediately on arrival and there is no waiting. The queue length process $Q(t)$ is now the number of customers in service, which is also the number of busy servers. Figure 1.5 illustrates the states and transition rates of $Q(t)$.

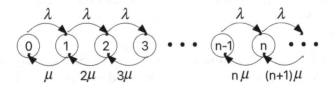

Figure 1.5 The M/M/∞ queue, states and transition rates.

The M/M/∞ queue is always stable, and the stationary distribution of the queue length is, by (1.6), Poisson with parameter ρ, with mean and variance ρ:

Stationary distribution of the M/M/∞ queue:

$$\pi(n) = \frac{\rho^n}{n!}e^{-\rho}, \quad n = 0, 1, 2, \ldots . \tag{1.8}$$

Example 1.5 (The M/M/K/K Erlang loss system)   Arrivals are Poisson at rate λ, service is exponential with rate μ, and there are K servers wth room for only K customers. This was Erlang's model for a telephone exchange with K lines. Customers that arrive when all the servers are busy (all the lines are engaged) do not enter the system and are not served; they are lost. Figure 1.6 illustrates the states and transition rates of $Q(t)$.

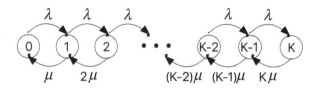

Figure 1.6 The M/M/K/K queue, states and transition rates.

The stationary distribution of the number of customers in the system, which is also the number of busy servers, is

Stationary distribution of the M/M/K/K queue:

$$\pi(n) = \frac{\rho^n}{n!} \Big/ \sum_{m=0}^{K} \frac{\rho^m}{m!}, \quad n = 0, 1, \ldots, K. \tag{1.9}$$

Two performance measures are of interest here: the fraction of customers lost and the average number of busy servers. The fraction of customers lost is the probability that an arriving customer to the stationary queue will find it in state K with all servers busy:

The M/M/K/K loss probability: $\qquad \pi(K) = \frac{\rho^K}{K!} \Big/ \sum_{m=0}^{K} \frac{\rho^m}{m!}. \tag{1.10}$

The average number of busy servers is $\rho(1 - \pi(K))$ (see Exercise 1.4 as well as Section 2.1).

Example 1.6 (The M/M/s queue)   Arrivals are Poisson at rate λ, service is exponential with rate μ, and there are s servers. Figure 1.7 illustrates the states and transition rates of $Q(t)$. The condition for stability is $\rho < s$, and the stationary distribution is:

Stationary distribution of M/M/s queue:

$$\pi(n) = \pi(0) \begin{cases} \dfrac{\rho^n}{n!}, & n \le s, \\[2mm] \dfrac{\rho^s}{s!}(\rho/s)^{n-s}, & n > s, \end{cases} \qquad \rho < s. \tag{1.11}$$

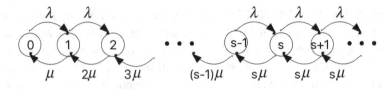

Figure 1.7 The M/M/s queue, states and transition rates.

Note that despite its simplicity, this is not easy to evaluate. We shall meet approximations to this formula, and important applications in Chapter 19, when we talk about the Halfin–Whitt regime.

1.3.1 Simulation of Markovian Queues

For the simulation of birth and death queues, as well as of more complex queueing systems that are described by a continuous-time countable state Markov chain, such as some queueing networks, one can use a uniformization approach: One simulates a Poisson process of events at rate θ which is the sum of all arrival and all the service rates. Then, for each event, one generates the type of event, arrival or service completion. One then generates accordingly the type of arrival, or the type of service completion. Finally, one performs the transition of adding or removing customers. An event is a dummy event with no change in state if it is a service completion at an empty queue.

Example 1.7 Consider a system with Poisson arrivals of rate λ and two servers in tandem, with respective exponential service times of rates μ_1, μ_2. In this system, each customer first queues up for server 1, and when he completes service at server 1, he moves to queue up for server 2, and departs when he completes service at server 2. We take $\theta = \lambda + \mu_1 + \mu_2$, and when an event of the Poisson process of rate θ occurs, we generate its type and change the state of the system as follows:

- Arrival with probability λ/θ, when we add a customer to the queue of the first server.
- Departure from the first server and arrival to the second server with probability μ_1/θ, if the first server is busy, and if it is idle this is a dummy event and the state of the system does not change.
- Departure from the second server with probability μ_2/θ, or a dummy event if the second queue is empty.

1.4 Historical Notes, Sources and Extensions

Queueing theory was invented by Agner Krarup Erlang to model telephone exchanges, while working for the Copenhagen Telephone Company, a subsidiary of Bell Telephone. His original two papers were Erlang (1909, 1917). The Kendall notation was suggested in Kendall (1953). Lindley's equation appeared in Lindley (1952). For an introduction to birth and death processes, see the popular book, *Stochastic Models*, by Sheldon Ross (2009).

The stationary distribution of the queue length, which is the limiting distribution of the queue if it has been running for a long time, allows the calculation of several system performance measures. However, sometimes one is also interested in the distribution of the queue length at time t, when the queue started empty at time 0 or started at some other given state. This requires the transient distribution of the queue length, which can in principle be obtained for birth and death queues from Kolmogorov's forward or backward equation, but is not straightforward; see Baccelli and Massey (1989). In reality, queues are never as well behaved as the models discussed here. In particular, arrival rates and service rates may be time varying, $\lambda(t)$, $\mu(t)$, which also requires analysis of transient systems. A thorough book on simulation by two top researchers is *Stochastic Simulation: Algorithms and Analysis*, Asmussen and Glynn (2007).

Exercises

1.1 Use Excel to simulate the following two queueing systems, using the same pattern as in Example 1.1:
   (i) A single queue with two servers. Arrivals are Poisson rate 0.2, service is exponential with mean $m = 8$.
   (ii) Two servers in tandem. Arrivals are Poisson rate 0.25, each arrival visits server 1 and then server 2. Service requirements are exponential, with average $m = 3$.
1.2 Find an analog to Lindley's equation for s servers, under FCFS service.
1.3 Derive the waiting and sojourn times for customers in a stationary M/M/s queue.
1.4 Calculate the average number of busy servers in an M/M/K/K queue.
1.5 A taxi stand is modeled as follows :
   - The stand has space for waiting taxi cabs and for waiting passengers.
   - Maximal number of waiting taxi cabs: 2.
   - Maximal number of waiting passengers: 3.
   - Taxis arrive in a Poisson stream of rate $\lambda = 1/6$ cabs per minute.
   - Passengers arrive in a Poisson stream of rate $\mu = 1/8$ passengers per minute.

- A passenger that arrives when there are cabs in the stand leaves immediately with a cab. If there is no cab, he joins the queue of waiting customers if there is space for waiting; otherwise, he leaves without service.
- A cab that arrives when there are passengers at the stand, leaves immediately with a passenger. If there are no passengers the cab joins the queue of other cabs, if there is space in the stand; otherwise, it leaves without a passenger.

Suggest a birth and death description of this system.

   (i) Make a diagram of states and of the transition rates.

  (ii) Calculate the stationary distribution of the state of the system – in particular, the probabilities that the stand is empty, has 1 or 2 or 3 passengers, or has 1 or 2 cabs.

 (iii) What fraction of the passengers depart with a cab immediately?

 (iv) What fraction of the passengers do not receive service?

  (v) What is the distribution of the waiting time of passengers that receive service?

1.6   *A K machines, M repairmen queueing system:* A workshop has a total of K machines, with an operating time between failures (TBF) exponential with rate λ. When a machine fails it needs service by a repairman for an exponential repair time with rate μ. There are $M \leq K$ repairmen. Describe this system as a birth and death queue, illustrate its state and transition rates diagram, and derive its stationary behavior. Performance measures for this system are the fraction of time that machines are operational, and the fraction of time that each repairman is busy. For $\mu/\lambda = 4$, prepare a table of these two performance measures for $K = 3, \ldots, 7$ and $M = 1, \ldots, K$.

1.7   A gas station is modeled as follows:

- There are two pumps, and space for a total of 4 cars.
- Cars arrive in a Poisson stream of rate $\lambda = 20$ cars per hour.
- Time to fill up at the pump is exponential with mean 10 minutes.
- A car that finds the station full leaves without service.

Suggest a birth and death description of this system.

   (i) Make a diagram of states and of the transition rates.

  (ii) Calculate the stationary distribution of the state of the system.

 (iii) What fraction of the cars is lost?

 (iv) What is the distribution of the sojourn time of cars that receive service, and what is its average?

1.8   For the gas station of the previous problem, simulate the system (you can use Excel or any other code) under the following alternative sets of conditions:

   (i) The fill-up time is exponential with mean 10.

  (ii) The fill-up time is distributed uniformly $\sim U(8, 12)$.

In either case, simulate the system for 1100 cars, discard the first 100, and present the following analysis of the results:

   (i) What is the fraction of cars that abandon without service?

(ii) What is the mean and standard deviation of the sojourn time of cars that get service?

(iii) Plot a histogram of the sojourn times.

1.9 *A queue with reneging (abandonments):* In a single-server queue, customers arrive at rate λ, service is at rate μ, but customers have patience with mean $1/\theta$ and leave the system without service (renege, or abandon) if their waiting time exceeds their patience. Assume Poisson arrivals, and exponential service and patience times. Present this as a Markovian birth and death queue, and derive its stationary distribution, and the average waiting time of customers that get served.

1.10 *A queue with balking:* In a single-server queue, customers arrive at rate λ, service is at rate μ, but a customer that sees a queue of n customers in the system only joins the queue with probability $p(n)$ (otherwise he balks). Assume Poisson arrivals, and exponential service times, and assume that $p(n) = p^n$. Present this as a Markovian birth and death queue, and derive its stationary distribution, the fraction of customers that balk, and the average waiting time of customers that join the queue.

1.11 *The Israeli queue:* This describes a typical scenario in a queue for cinema tickets in Israel on a weekend in the 1950s. In a small country, any two people know each other with probability p. When a person arrives at an M/M/1 queue, he scans the customers waiting in queue from first to last and joins with the longest waiting customer who is an acquaintance, to be served jointly with him (at no extra service time), or if he does not find an acquaintance, he joins the end of the queue. Assume Poisson arrivals at rate λ, exponential service at rate μ. Present this as a Markovian birth and death queue, and derive the stationary distribution of the queue length, the percentage of customers that find an acquaintance to join, and the distribution of the waiting time [Perel and Yechiali (2014)].

1.12 (∗) *A queue with vacations:* In an M/M/1 queue, the server goes on a vacation of duration $\sim \mathrm{Exp}(\theta)$ whenever the queue becomes empty. Describe the queue length for this system as a Markov chain and derive its stationary distribution [Servi and Finn (2002)].

2

The M/G/1 Queue

Poisson arrivals are often close to reality, in particular when there are many users who approach the system independently of each other. It is also realistic to assume that the service time of a customer is independent of the arrival times, and independent of service times of other customers. This makes the M/G/1 queue a realistic model. It has Poisson rate λ arrivals, and service times are i.i.d. with distribution G, service rate μ, and mean service time $m = 1/\mu$.

In this chapter we study the M/G/1 queue, and derive the famous Pollaczek--Khinchine formula for the mean waiting time, as well as the expected length of the M/G/1 busy period. To derive these we first need to introduce Little's law, point processes, work conservation, some results in renewal theory, length biasing, and PASTA. All of these are topics of major interest in themselves, and will occupy Sections 2.1–2.6. The reader may at this point skip some of the proofs in these sections.

We also add a more detailed analysis of queues that can be analyzed via embedded Markov chains, including in addition to the M/G/1 queue also the G/M/1 queue.

An important additional parameter for the M/G/1 queue is the coefficient of variation of the service time defined by c.o.v. $= c_s = \sigma(X)/\mathbb{E}(X)$. For the exponential distribution, c.o.v. $= 1$, which is very high. Most realistic service time distributions are much less variable, though in the internet one also encounters much higher c.o.v. due to the phenomenon of mice and elephants: messages, queries and files are a mixture of items whose sizes differ by several orders of magnitude.

2.1 Little's Law

Consider the total time spent in a system by n items that arrive and leave in the time interval $[0, T]$. It can be calculated in two ways: by adding the sojourn times W_ℓ over the n items, or by integrating the number of items in

16

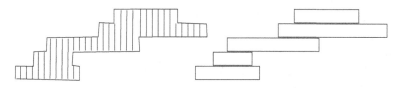

Figure 2.1 Two ways to add up all the sojourn times.

the system $Q(t)$ over $[0, T]$. Both ways are equal (see Figure 2.1), so

$$\int_0^T Q(t)dt = \sum_{\ell=1}^{n} W_\ell.$$

Multiplying and dividing by n and dividing by T, we get

$$\frac{1}{T} \int_0^T Q(t)dt = \frac{n}{T} \times \frac{\sum_{\ell=1}^{n} W_\ell}{n}. \tag{2.1}$$

So the *time average of number in the system* is equal to the *arrivals per unit time* $\times$ *the customer average sojourn time*, all this for the set of the n customers served within $(0, T)$.

The relation expressed by (2.1) is an example of a relationship between *time average* and *customer average*. We shall encounter more examples of useful relations of this type.

Observing a system with general arrivals and departures over $(0, T)$, we have

$$\frac{1}{T} \int_0^T Q(t)dt \approx \frac{\mathcal{A}(T)}{T} \times \frac{\sum_{\ell=1}^{\mathcal{A}(T)} W_\ell}{\mathcal{A}(T)}.$$

This is only an approximation, because the left-hand side includes all the remaining service times of the $Q(0)$ customers that are in the system initially, and the right-hand side includes all the remaining service times after T of the $Q(T)$ customers that are still in the system at time T.

Letting $T \to \infty$ we may get

$$\lim_{T \to \infty} \frac{1}{T} \int_0^T Q(t)dt = \lim_{T \to \infty} \frac{\mathcal{A}(T)}{T} \frac{\sum_{j=1}^{\mathcal{A}(T)} W_j}{\mathcal{A}(T)}. \tag{2.2}$$

Theorem 2.1 (Little's law) *Little's law states that the long-term average number of items in a system, denoted by* L, *equals the long-term average arrival rate, denoted by* λ, *times the long-term average sojourn time, denoted by* $\overline{W}$:

$$\textbf{Little's law} \qquad L = \lambda \overline{W}. \tag{2.3}$$

A sufficient condition for the validity of Little's law is that the following three hold:

(i) $\lim_{T\to\infty} \mathcal{A}(T) = \infty$,

(ii) $\max_{1\le\ell\le n} W_\ell / \sum_{\ell=1}^n W_\ell = o(n)$,

(iii) $\lim_{T\to\infty} Q(T)/T = 0$.

Proof From (i) we have that $Q(0)$ is negligible relative to $\mathcal{A}(t)$, and (ii) guarantees that the effect of the first $Q(0)$ remaining processing times at 0 is negligible. From (iii) the same argument also holds for the $Q(T)$ remaining processing times at time T. Hence, the limiting expression (2.2) is valid almost surely (a.s.). □

Definition 2.2 We say that $Q(t)$ is rate stable if

$$\textbf{Rate stability} \qquad \lim_{T\to\infty} \frac{Q(T)}{T} = 0, \text{ a.s.} \qquad (2.4)$$

2.1.1 Stationarity and Ergodicity

Definition 2.3 A stochastic process $X(\cdot)$ is stationary if for all $t_1,\ldots,t_n$ and τ, the joint distributions of $X(t_1),\ldots,X(t_n)$ and of $X(t_1+\tau),\ldots,X(t_n+\tau)$ are the same. The definition for a discrete time process is analogous.

Ergodicity is defined in general for any probability space and transformation. For our purpose, we need the following definition of ergodicity for stationary processes:

Definition 2.4 A stationary process $X(\cdot)$ is ergodic if for any function $f(x_1,\ldots,x_n)$, such that $\mathbb{E}(f(X(t_1),\ldots,X(t_n)))$ exists, the following holds:

$$\lim_{T\to\infty} \frac{1}{T} \int_0^T f(X(t_1+\tau),\ldots,X(t_n+\tau))d\tau = \mathbb{E}(f(X(t_1),\ldots,X(t_n))) \text{ a.s.}$$

In words, almost surely every sample path contains all the information about the process. So, averaging a performance measure over a single infinite sample path equals averaging over the sample space of the stationaray process at time 0 (or at some arbitrary time t, since the process is stationary).

The following property of Markov chains is well known:

Theorem 2.5 *A continuous time discrete space Markov chain that is irreducible and positive recurrent has a stationary version that is ergodic.*

Furthermore, starting from any initial state $X(0)$ of the Markov chain, the distribution of $X(t)$ approaches the stationary distribution as $t \to \infty$.

Such a Markov chain is called ergodic. A discrete time Markov chain is ergodic if, in addition, it is non-periodic.

To illustrate how a positive recurrent Markov chain may not be ergodic, consider two situations: If a discrete time positive recurrent Markov chain X_n has period 2, then it will possess a stationary distribution, but the average of a sample path at all even times will not yield the expected value of the stationary distribution. If a continuous time Markov chain is reducible, then averaging over a single sample path will give a different result for sample paths in different non-communicating parts of the sample space. Such Markov chains can be decomposed to chains on subsets of the sample space that are ergodic.

We note that for an M/G/1 system, the queue length process $Q(t)$ is not a Markov process: When the number of customers is $Q(T) = n > 0$, the future depends also on the amount of time since the customer currently in service has started service. For a general single queue that is not Markovian, we can still have stationarity and ergodicity.

We give a somewhat cryptic definition of stability and ergodicity of a queueing network. It will be made more precise and definite as we continue to develop the theory step-by-step in later chapters.

Definition 2.6 We say that a queueing system is stable if there exists a stationary version of the process $Q(\cdot)$. In particular for this stationary version, the marginal distribution of $Q(t)$ will be the same for all t.

We say that a stable queueing system is ergodic if, starting from any state at time 0, $Q(t)$ converges in distribution to this stationary distribution as $t \to \infty$, equivalently, the system having started at the distant past will have stationary $Q(t)$, $-\infty < t < \infty$.

Our formulation of Little's law was derived by considering long-term averages over a sample path. By the definition of ergodicity, we have:

Proposition 2.7 *If the process $Q(t)$ in (2.2)–(2.3) is stationary and ergodic, then this implies that Little's law (2.3) holds for the expectations of the stationary distribution.*

2.1.2 Applications of Little's Law

We now give two examples of the use of Little's law. We will first define two concepts that will accompany us throughout the book.

Definition 2.8 In a queueing system:
- We say that the service is work conserving if servers are non-idling: a server does not idle if there is a customer in the system that he can serve.
- We say that service is non-preemptive if service of a customer once started is not interrupted before it is complete.

Example 2.9 Consider a single-server station with stationary arrival process $\mathcal{A}(t)$, a stationary sequence of service times X_n, and work-conserving non-preemptive service. Assume further that the queue length $Q(t)$, including waiting customers and customers in service, has a stationary distribution and is ergodic. Then Little's law holds for the whole system. It tells us that for the stationary process, $\mathbb{E}(Q(t)) = \lambda \, \mathbb{E}(W_n)$. We can verify this for the M/M/1 queue:

$$\mathbb{E}(Q(t)) = \frac{\rho}{1-\rho}, \qquad \lambda \, \mathbb{E}(W_n) = \lambda \frac{1}{\mu - \lambda} = \frac{\rho}{1-\rho}.$$

Furthermore, we can also consider the two subsystems of the queue, the "waiting room" of customers waiting for service, and the server, see Figure 2.2. For the waiting room subsystem we have as before $\mathbb{E}(\text{number waiting}) = \lambda \, \mathbb{E}(V_n)$.

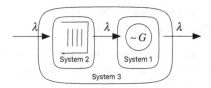

Figure 2.2 Subsystems of the single-server queue.

More interesting is the subsystem consisting of the server, which has customers arriving and departing at rate λ (since $Q(T)/T \to 0$ as $T \to \infty$). Each customer sojourns at the server for exactly his service duration X_n. The number of customers at the server is either 1 if the server is busy, or it is 0, if the server is idle. Hence, Little's law applied to the server subsystem tells us that:

$$L = \mathbb{E}(\text{Number of customers at the server}) \tag{2.5}$$

$$= \mathbb{P}(\text{Server is busy}) = \lambda \, \mathbb{E}(X_n) = \lambda \frac{1}{\mu} = \rho,$$

i.e. the server is busy a fraction ρ of the time, and idle a fraction $1 - \rho$ of the time. By ergodicity, $1 - \rho$ is also the probability that the stationary queue is empty.

Example 2.10 Consider the Erlang loss system M/M/K/K. We would like to know the expected number of busy servers (calculated in Exercise 1.4). We have that customers arrive to the system at rate λ, but only a fraction $1 - \mathbb{P}(\text{Loss})$ enter into service (see Figure 2.3). They stay in service for a

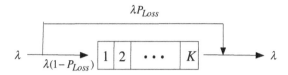

Figure 2.3 Little's law for the Erlang loss system.

time X_n. Hence, for the stationary system, the expected number of busy servers is:

$$\mathbb{E}(Q(t)) = \lambda\,(1 - \mathbb{P}(\text{Loss}))\,\mathbb{E}(X_n) = \rho \sum_{m=0}^{K-1} \frac{\rho^m}{m!} \Big/ \sum_{m=0}^{K} \frac{\rho^m}{m!}. \qquad (2.6)$$

The quantity $\mathbb{E}(Q(t))/K$ measures *the utilization of the servers*, i.e. the fraction of time each server is busy.

2.2 Work Conservation

We define the workload process $\mathcal{W}(t)$ to be the total amount of work in the system, consisting of all the remaining processing times of all the customers in the system at time t. In a single-server queue, the workload $\mathcal{W}(t)$ behaves as follows: At arrival time A_ℓ of a customer ℓ, it increases by the service time X_ℓ. At all other times, when the queue is not empty, it decreases at the constant rate 1, and it remains equal to 0 when the system is empty; see Figure 2.4. We state the result that follows directly:

Theorem 2.11 (Work conservation) *In a single-server system with work conserving service, $\mathcal{W}(t)$ is independent of the service policy.*

Assume now that service is FCFS. We again relate time average and customer average as in Section 2.1. We calculate the total inventory over time in two ways: as an integral over time of $\mathcal{W}(t)$, and as a sum of the contributions of the individual customers. For each customer, his contribution to the total inventory consists of two parts: contribution during his waiting time V_n, where his workload inventory is X_n, which amounts to $V_n X_n$, and contribution during service, when the workload reduces linearly from X_n to

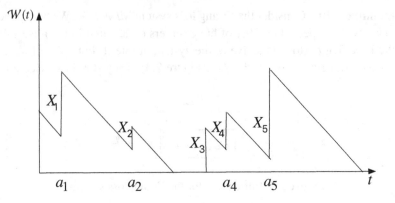

Figure 2.4 The evolution of workload.

zero, which amounts to $X_n^2/2$. This is illustrated in Figure 2.5. Equating the two ways of calculating the workload for n customers served within $(0, T)$, we have

$$\int_0^T \mathcal{W}(t)dt = \sum_{n=1}^{\mathcal{A}(T)} (V_n X_n + X_n^2/2).$$

Theorem 2.12 *The expected workload in the system for a stationary M/G/1 queue is given by* $\mathbb{E}(\mathcal{W}) = \lambda \left(\mathbb{E}(V)\mathbb{E}(X) + \mathbb{E}(X^2)/2 \right).$

Proof We proceed as in the derivation of Little's law. Averaging over the

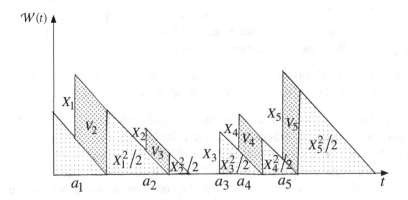

Figure 2.5 Workload inventories of the customers, FCFS.

long-run, we get

$$\mathbb{E}(\mathcal{W}) = \lim_{T \to \infty} \frac{1}{T} \int_0^T \mathcal{W}(t)dt = \lim_{T \to \infty} \frac{\mathcal{A}(T)}{T} \frac{\sum_{n=1}^{\mathcal{A}(T)}(V_n X_n + X_n^2/2)}{\mathcal{A}(T)}$$

$$= \lambda \left(\mathbb{E}(V)\mathbb{E}(X) + \mathbb{E}(X^2)/2 \right). \tag{2.7}$$

Note that V_n depends only on previous customers and on the arrival time A_n of customer n, and it is independent of the service time X_n; hence, we get $\mathbb{E}(V_n X_n) = \mathbb{E}(V_n)\mathbb{E}(X_n)$. In words, the long-run average work in the system equals the arrival rate multiplied by the sum of two terms: the long-term average waiting time multiplied by the expected processing time, and half the expected value of the squared processing time. □

2.3 Some Renewal Theory

Renewal processes appear in many queueing models. In this section, we provide a brief summary of some of their properties. The reader may already be familiar with those, and so we leave out most of the proofs but provide guidance to those in the exercises. Let T_j be a sequence of i.i.d. non-negative, not identically 0, random variables distributed with distribution function F, with finite or infinite mean $1/\mu$. Define $A_0 = 0$, $A_n = \sum_{j=1}^n T_j$, and $\mathcal{A}(t) = \max\{n : A_n \leq t\}$. Then $\mathcal{A}(t)$ is a renewal process, and A_n are the renewal epochs of $\mathcal{A}(t)$. We define $m(t) = \mathbb{E}(\mathcal{A}(t))$ to be *the renewal function* of $\mathcal{A}(t)$.

We note first that $m(t) < \infty$ for all $t > 0$, and that $m(t)$ as well as $\mathcal{A}(t)$ converge to ∞ as $t \to \infty$; see Exercise 2.2.

Theorem 2.13 (Law of large numbers and elementary renewal theorem)

$$\lim_{t \to \infty} \frac{\mathcal{A}(t)}{t} = \mu \text{ a.s.,} \qquad \lim_{t \to \infty} \frac{m(t)}{t} = \mu. \tag{2.8}$$

Proof of the law of large numbers for renewal processes The first part of the theorem is the law of large numbers for renewal processes, and it follows directly from the strong law of large numbers (SLLN). We have that $A_{\mathcal{A}(t)} \leq t < A_{\mathcal{A}(t)+1}$; hence,

$$\frac{\sum_{n=1}^{\mathcal{A}(t)} T_n}{\mathcal{A}(t)} = \frac{A_{\mathcal{A}(t)}}{\mathcal{A}(t)} \leq \frac{t}{\mathcal{A}(t)} < \frac{A_{\mathcal{A}(t)+1}}{\mathcal{A}(t)} = \frac{\sum_{n=1}^{\mathcal{A}(t)+1} T_n}{\mathcal{A}(t)},$$

but, as $t \to \infty$ also $\mathcal{A}(t) \to \infty$ and therefore by the strong law of large numbers all three expressions converge almost surely to $\mathbb{E}(T_1)$, and the result follows. □

The second part of the theorem is the Elementary Renewal Theorem, $\frac{m(t)}{t} \to \mu$. It can be proved in two ways: using exchange of limit and expectation by showing uniform integrability of $\mathcal{A}(t)$, or by using Wald's equation to bound $\mathbb{E}(\mathcal{A}(t))$. Both these approaches involve concepts that are very useful. We provide the definitions and results necessary for both approaches to the proof. We outline the proofs themselves in Exercises 2.3 and 2.4.

It is not in general true that for a sequence of random variables, $X_n \to X$ almost surely implies also that $\mathbb{E}(X_n) \to \mathbb{E}(X)$. Uniform integrability is the concept that allows such interchange of limits.

Definition 2.14 (Uniform integrability) We say a collection of random variables X_t, $t \in T$ is uniformly integrable if for every $\epsilon > 0$ there exists κ such that $\mathbb{P}(X_t > \kappa) \le \epsilon$, for all $t \in T$.

Theorem 2.15 (Interchange of limit and expectation) *For any collection of random variables X_t, $t \in T$, the interchange $\mathbb{E}(\lim_t X_t) = \lim_t \mathbb{E}(X_t)$ holds if and only if X_t, $t \in T$ are uniformly integrable.*

For a sequence of i.i.d. random variables, one often needs to calculate $\mathbb{E}(X_1 + \cdots + X_N)$ where N is a random variable. In such cases, one hopes to have $\mathbb{E}(X_1 + \cdots + X_N) = \mathbb{E}(N)\mathbb{E}(X_1)$, but that is not always true. Wald's equation provides a sufficient condition, namely, that N is a stopping time.

Definition 2.16 (Stopping time) A random time τ with values in $T = (0, \infty)$ or $T = 0, 1, 2, \ldots$ is a stopping time for the stochastic process $X(t)$, $t \in T$, if the event $\{\tau = t\}$ is independent of $\{X(s),\ s > t\}$.

Two important special cases of a stopping time are if τ is independent of all $X(s)$, $s \in T$, or if τ is determined by $\{X(s),\ s \le t\}$.

Theorem 2.17 (Wald's equation) *Let N be a stopping time for a sequence of i.i.d. random variables T_n. Then:*

$$\mathbb{E}\left(\sum_{n=1}^{N} T_n \right) = \mathbb{E}(T_n)\,\mathbb{E}(N).$$

Exercises 2.3 and 2.4 use these concepts to prove the elementary renewal theorem, the second part of Theorem 2.13.

A *renewal reward process* is defined as follows: For a renewal process $\mathcal{A}(t)$, a reward of C_n is obtained for every renewal interval between A_{n-1} and A_n, where T_n, C_n are jointly distributed i.i.d., and C_n has a finite mean. Define $C(t)$ as the cumulative reward over $(0, t]$, and $c(t) = \mathbb{E}(C(t))$.

Theorem 2.18 (Renewal reward theorem)

$$\lim_{t \to \infty} \frac{C(t)}{t} = \mu \mathbb{E}(C_n) = \frac{\mathbb{E}(C_n)}{\mathbb{E}(T_n)} \quad a.s., \qquad \lim_{t \to \infty} \frac{c(t)}{t} = \mu \mathbb{E}(C_n) = \frac{\mathbb{E}(C_n)}{\mathbb{E}(T_n)}.$$

For the proof of the renewal reward theorem see Exercise 2.5.

This theorem is very useful in deriving results for regenerative processes, and will already be used extensively in the next section. While the proof is quite simple, the result itself is rather surprising, since at first glance one might expect the reward to be an average of C_n/T_n, and of course $\mathbb{E}(C_n/T_n) \neq \mathbb{E}(C_n)/\mathbb{E}(T_n)$.

2.4 Length Biasing

Consider observing a rate 1 Poisson process at an arbitrary time t. The distance to the next event is exponentially distributed with rate 1, by the memoryless property, and by reversibility of the Poisson process so is the distance to the previous event. Furthermore, by the memoryless property the two are independent. Hence, observing the Poisson process at arbitrary time t, we are in an interval of length equal to the sum of two independent exponentials, $\sim$Erlang$(2, 1)$. This is an example of length biasing: our arbitrary t is more likely to fall in a long interval than in a short interval.

For a general renewal process with i.i.d. intervals $T_n \sim F$, we observe the process at time t and record the *length of the current interval* $T_{Int}(t)$, the time to the next event, *forward recurrence or excess time* $T_{Fwd}(t)$, and the time to the previous event, *the backwards recurrence time or age* $T_{Bwd}(t)$.

The processes $T_{Int}(t) = \text{Length}(t)$, $T_{Bwd}(t) = \text{Age}(t)$, and $T_{Fwd}(t) = \text{Excess}(t)$ that give the values at time t for the interval in which t falls, are stochastic processes. In fact, they are continuous time continuous state space piecewise deterministic Markov processes. They are illustrated in Figure 2.6. We wish to calculate for each of these processes the long-run fraction of times t at which they are greater than x. If $\mathbb{E}(T)$ is finite, these processes are in fact ergodic, so these time averages are equal to the sample averages, i.e. they are the probabilities that length, excess, and age are greater than x when the stationary process is observed at time t. Recall the definitions of stationarity and ergodicity in Section 2.1.1; we will have a further discussion of stationarity in Section 2.5.

We start with the excess process. To calculate $\mathbb{P}(T_{Fwd}(t) > x)$, we use the renewal reward theorem. The reward per renewal interval is $(T_j - x)^+$, the

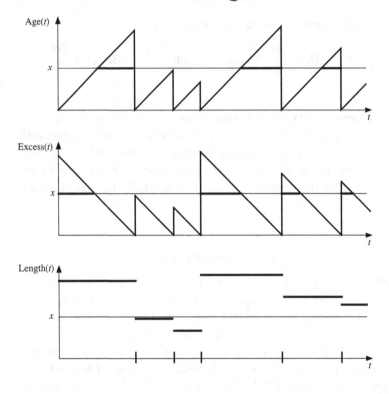

Figure 2.6 Age, excess, and length processes.

length of time during the interval that the forward recurrence time exceeded x, which is 0 if the interval is shorter than x:

$$\mathbb{P}(T_{Fwd}(t) > x) = \lim_{t \to \infty} \frac{\text{measure of } r \in (0, t) \text{ with forward recurrence} > x}{t}$$

$$= \frac{\mathbb{E}(T - x)^+}{\mathbb{E}(T)} = \mu \int_x^\infty (y - x)dF(y) = \mu \int_x^\infty (1 - F(y))dy.$$

Here the first equality follows by ergodicity, the second by the renewal reward theorem, and the last from integration by parts.

Hence, forward recurrence time has a density $f_{Fwd}(x) = \mu(1 - F(x))$. This is the density of the so-called *equilibrium distribution*, which is also denoted F_{eq}. The same calculation shows that $f_{Fwd}(x) = f_{Bwd}(x)$, as we might expect by time reversibility.

Remark The age and excess are in general not independent; see Exercises 2.6– 2.8.

To calculate $\mathbb{P}(T_{Int}(t) > x)$, we use a reward which is T if $T > x$ and 0 if $T \leq x$, then:

$$\mathbb{P}(T_{Int}(t) > x) = \lim_{t \to \infty} \frac{\text{measure of } r \in (0, t) \text{ that are in interval} > x}{t}$$

$$= \frac{\mathbb{P}(T > x)\mathbb{E}(T|T > x)}{\mathbb{E}(T)} = \mu \int_x^\infty y dF(y).$$

Hence, the length biased interval length for an observer at arbitrary time t has density $f_{Int}(x) = \mu x f(x)$ (assuming the density of F exists). Note that this density at x is the original density at x, multiplied by a quantity proportional to x. So the length biasing is exactly linear in the length.

2.5 Stationary Point Processes and Palm Measure

Arrival times are an example of a point process. We define a point process as a sequence of the times of ordered events on the real line $-\infty < t < \infty$, $\cdots < A_{-m} < \cdots < A_0 \leq 0 < A_1 < \cdots < A_n < \cdots$. It has a sequence of intervals $T_n = A_n - A_{n-1}$, $n \in \mathbb{Z}$, and a counting process $\mathcal{A}(t)$, where $\mathcal{A}(t) - \mathcal{A}(s)$ counts the number of events in $(s, t]$ for any $s < t$, and we set $\mathcal{A}(0) = 0$.

We say that the point process is stationary if the increments of the counting process are stationary, that is if for any $s_1 < t_1 < \cdots < s_n < t_n$, and any Δ, the joint distributions of $(A(t_j) - A(s_j))_{j=1}^n$ and of $(A(t_j + \Delta) - A(s_j + \Delta))_{j=1}^n$ are the same.

If the point process is stationary, then by length biasing the sequence $(T_n)_{n \in \mathbb{Z}}$ is not stationary, because already the interval around 0 will not be distributed like its neighboring intervals, by length biasing.

We can alternatively take a stationary sequence of intervals, $(T_n)_{n \in \mathbb{Z}}$, and create a point process by fixing $A_0 = 0$. Now the sequence of intervals is stationary, but the counting process $\mathcal{A}(t)$ will not be stationary; the number of events in $(0, t]$ will be distributed differently from the number of events in $(\tau, \tau + t]$.

The stationary point process and the process of stationary intervals with an event at time 0, are connected by the Palm calculus. For a stationary point process with probability measure $\mathfrak{P}$, there is an associated Palm probability measure of the process, $\mathfrak{P}^0$, which is the conditional measure of $\mathfrak{P}$ conditional on an event at 0. There is a one-to-one correspondence between the two, and simple calculus that is used to go from one to the other. The stationary point process measure $\mathfrak{P}$ can be obtained from a point process constructed by a stationary sequence of intervals with an event at 0 that has measure $\mathfrak{P}^0$.

As far as renewal processes are concerned, the standard version has an event at time 0 (which we do not count, event 0), and this is the Palm version. The stationary version is obtained by taking the time of the first event, A_1 distributed with the excess time distribution, i.e. $T_1 \sim F_{Fwd}$, and $T_n \sim F$, $n > 1$. This is called the delayed (or equilibrium) renewal process; it has $m(t) = \mu t$, and we denote the delay distribution by F_{eq} (a synonym for F_{Fwd}).

2.6 PASTA — Poisson Arrivals See Time Averages

In many stochastic models, particularly in queueing theory, Poisson arrivals both observe (see) a stochastic process and interact with it. It is a remarkable property that the fraction of Poisson arrivals that see the process in some state is equal to the fraction of time that the process is in that state. This property is called PASTA (Poisson arrivals see time averages), and it holds whenever the process that is observed does not anticipate the events of the Poisson process. In this section, we will give a precise formulation of this property and present a proof. The proof is somewhat technical, but is included because of the important role of PASTA in queueing theory. That PASTA is remarkable is illustrated by the following example, and by Example 2.21.

Example 2.19 Consider a D/D/1 queue (deterministic arrivals, deterministic service, single server), where customers arrive at times A_n, $n = 1, 2, 3, \ldots$ and require service $X_n = 0.9$. Then the number of customers in the system is 1 for 0.9 fraction of the time. But every arriving customer will find 0 customers in the system.

We now give a precise formulation of PASTA: Consider a stochastic process $X(t)$, a Poisson process $\Lambda(t)$, and a set of states B, and let $U(t)$ be the indicator of $X(t) \in B$. Let

$$\bar{U}(T) = \frac{1}{T} \int_0^T U(t)dt, \qquad \mathcal{A}(T) = \int_0^T U(t)d\Lambda(t),$$

so that $\bar{U}(T)$ is the fraction of time out of $(0, T)$ that $X(t)$ spends in states B, and the $\mathcal{A}(T)$ is the count of events of $\Lambda(t)$ in $[0, T]$ that find $X(t)$ in B.

Make the following assumptions:

(i) Assume that $U(t)$ is left continuous with right limits almost surely. The left continuity implies that an event of $\Lambda(\cdot)$ at t, has no effect on $U(t)$ (because $U(t) = U(t-)$).

(ii) Assume lack of anticipation, which is defined by

$$\{\Lambda(t + u) - \Lambda(t) : u \geq 0\} \text{ is independent of } \{U(s) : 0 \leq s \leq t\}.$$

Theorem 2.20 (PASTA) *Under assumptions (i) and (ii),* $\lim_{T \to \infty} \bar{U}(T) = \bar{U}(\infty)$ *almost surely if and only if* $\lim_{T \to \infty} \mathcal{A}(T)/\Lambda(T) = \bar{U}(\infty)$ *almost surely, i.e. if one of these limits exists almost surely, then so does the other, and they are equal.*

Proof We assume as always that sample paths of $X(t)$ belong to $\mathbb{D}$, the space of right continuous with left limits (RCLL) functions. Such sample paths can have only a finite number of discontinuities in any finite interval. As a result, the same holds for $U(t)$.

The main idea of the proof is to approximate $\mathcal{A}(t)$ by a sequence of finite sums:

$$\mathcal{A}_n(t) = \sum_{k=1}^{n} U(kt/n)(\mathcal{A}((k + 1)t/n) - \mathcal{A}(kt/n)).$$

$\mathcal{A}_n(t)$ differs from $\mathcal{A}(t)$ only if for some k, $U(\frac{kt}{n}) = 0$, there is an event of Λ at $s \in (\frac{kt}{n}, \frac{(k+1)t}{n})$, and $U(s) = 1$. Note that in that case there is a discontinuity of U in the interval $(\frac{kt}{n}, s)$. We show that this cannot happen when n is large enough. Almost surely, for every sample path of U and Λ there will be a finite number of discontinuities of U, and events of Λ. By the right continuity and lack of anticipation, with probability 1, none of the events of Λ coincides with a discontinuity of U. So for every sample path we can find n_0 large enough that for all $n > n_0$, none of the intervals $(\frac{kt}{n}, \frac{(k+1)t}{n})$ contains both a discontinuity and an event, and for such n, $\mathcal{A}_n(t) = \mathcal{A}(t)$. We have shown that

$$\mathcal{A}_n(t) \text{ converges almost surely to } \mathcal{A}(t) \text{ as } n \to \infty.$$

Likewise,

$$\frac{1}{n} \sum_{k=1}^{n} U(kt/n) \text{ converges almost surely to } \bar{U}(t) \text{ as } n \to \infty.$$

By lack of anticipation, we have:

$$\mathbb{E}(\mathcal{A}_n(t)) = \lambda t \, \mathbb{E}\Big(\sum_{k=1}^{n} U(kt/n) \Big/ n \Big).$$

Since $\mathcal{A}_n(t) \leq \mathcal{A}(t)$, and $\sum_{k=1}^n U(kt/n)\big/n \leq 1$, by bounded convergence we can change the order of limit and expectation and obtain

$$\mathbb{E}(\mathcal{A}(t)) = \lambda t\, \mathbb{E}(\bar{U}(t)) = \lambda\, \mathbb{E}\left(\int_0^t U(s)ds \right). \tag{2.9}$$

What we wish to show is that $\mathcal{R}(t) = \mathcal{A}(t) - \lambda t\, \bar{U}(t)$ converges to 0 almost surely as $t \to \infty$. We now look at U, Λ from t onward, conditioning on the past. Clearly, (i) and (ii) still hold, and we can therefore also write:

$$\mathbb{E}(\mathcal{A}(t + h) - \mathcal{A}(t) \,|\, \mathfrak{F}_t) = \lambda\, \mathbb{E}\left(\int_t^{t+h} U(s)ds \,|\, \mathfrak{F}_t \right)$$

and we see that $\mathbb{E}(\mathcal{R}(t + h) - \mathcal{R}(t) \,|\, \mathfrak{F}_t)=0$ (this means that $\mathcal{R}(t)$ is a martingale). We can also bound second moments of $\mathcal{R}(t)$:

$$\mathbb{E}(\mathcal{R}^2(t)) \leq \mathbb{E}(\mathcal{A}(t)^2) + \lambda^2 t^2 \mathbb{E}(\bar{U}(t)^2) \leq \mathbb{E}(\Lambda(t)^2) + \lambda^2 t^2 = \lambda t + 2\lambda^2 t^2 := k(t). \tag{2.10}$$

We now look at increments of $\mathcal{R}(t)$, defining:

$$X_n = \mathcal{R}(nh) - \mathcal{R}((n-1)h), \quad n = 1, 2, \ldots.$$

Then by the above, $\mathbb{E}(X_n \,|\, \mathfrak{F}_{nh}) = 0$, and, similar to (2.10),

$$\mathbb{E}(X_n^2) \leq k(h), \quad \text{and therefore} \quad \sum_{n=1}^{\infty} \mathbb{E}(X_n^2)/n^2 < \infty.$$

Theorem 3, p. 243, of Feller (1971) states that for a sequence of random variables (which may be dependent) if $\mathbb{E}(X_n \,|\, \mathfrak{F}_{nh}) = 0$ and $\sum_{n=1}^{\infty} \mathbb{E}(X_n^2)/n^2 < \infty$, then as $n \to \infty$, $\sum_{k=1}^n X_k/n \to 0$ almost surely. Therefore:

$$\mathcal{R}(nh)/n = \sum_{k=1}^n \mathbb{E}(X_k)/n \to 0 \quad \text{almost surely as } n \to \infty,$$

but

$$\mathcal{R}(nh) - \lambda h \leq \mathcal{R}(t) \leq \mathcal{R}((n+1)h) + \lambda h \quad \text{for } nh \leq t \leq (n+1)h,$$

and we have shown that $\mathcal{R}(t)/t \to 0$ a.s. as $t \to \infty$.

Finally, the theorem follows from

$$\mathcal{R}(t)/t = (\mathcal{A}(t)/\Lambda(t))(\Lambda(t)/t) - \lambda\bar{U}(t)$$

since $\Lambda(t)/t \to \lambda$ a.s. $\square$

Note, all it says in the theorem is that if the limits exist, they are equal. The natural application then is to assume more about $X(t)$ and B: we get PASTA if the process is stationary and ergodic. In particular, then, any measurable property of the process at time t will also be what is observed at Poisson times. The following example highlights how remarkable this property is:

Example 2.21 Consider an M/G/1 queue with $\rho < 1$, so that the queue length and the workload are stable (this will be discussed later). Consider the workload process $\mathcal{W}(t)$, and let $\bar{\mathcal{W}}$ be the longtime average workload in the system. Now sample the workload just before each arrival $\mathcal{W}(A_n-)$, just after each arrival $\mathcal{W}(A_n+)$, and in the middle of the interval between arrivals $[\mathcal{W}(A_n+) + \mathcal{W}(A_{n+1}-)]/2$, and take an averages of each of these quantities over many intervals. Intuitively, one would think that the pre-event average is less than $\bar{\mathcal{W}}$, the post-event average is bigger than $\bar{\mathcal{W}}$, and the middle of the interval average should be closest to $\bar{\mathcal{W}}$. In fact, for general arrivals one cannot make any deductions on which is closest to $\bar{\mathcal{W}}$. However, in the case of Poisson arrivals, $\bar{\mathcal{W}}$ is equal, by PASTA, to the pre-event average. It is the arrivals that observe, at the time of their arrival, the time average $\bar{\mathcal{W}}$.

Note We have used PASTA in our calculation of the M/M/1 sojourn time, in Section 1.3 Theorem 1.3.

2.7 M/G/1 Average Waiting Times

At this point, we have prepared all the necessary results to calculate the average waiting times for the M/G/1 queue. Consider an M/G/1 queue with Poisson arrivals of rate λ and service times X_n i.i.d. distributed as G, with mean $m = 1/\mu$ and variance σ^2, and c.o.v. (coefficient of variation) $c_s = \frac{\sigma}{m} = \mu\sigma$. Assume $\rho = \lambda/\mu < 1$, which, as we shall see in Sections 2.9 and 4.1, guarantees stability of the queue.

Theorem 2.22 *The average delay or waiting time per customer in an M/G/1 queue under FCFS is given by the Pollaczek–Khinchine formula:*

$$\bar{V} = \frac{\lambda\mathbb{E}(X^2)}{2(1-\rho)} = m\frac{\rho}{1-\rho}\frac{1+c_s^2}{2}. \tag{2.11}$$

Proof We have seen that by work conservation, for non-preemptive service, long-term average workload can be expressed in terms of long-term average waiting time and average squared processing time,

$$\bar{\mathcal{W}} = \rho\bar{V} + \tfrac{1}{2}\lambda\mathbb{E}(X^2).$$

By PASTA, this is equal to the average workload that an arriving customer will see. But under FCFS this will be his waiting time, i.e.

$$\bar{W} = \bar{V}.$$

Combining these two equations, we obtain (2.11).   □

Remark Note that the stationary waiting time has a finite expectation only if the service time has a finite second moment. We can have a stable queue, with guaranteed finite stationary waiting time distribution, and infinite expected waiting time.

It is worth noting the expression for $\bar{V}$ in (2.11), and interpreting its various parts: The average waiting time is proportional to the average processing time. It is then multiplied by the term $\frac{1+c_s^2}{2}$ that is the effect of the variability in the processing times. Finally, the term $\frac{\rho}{1-\rho}$ expresses the drastic effect of the workload (the traffic intensity). We get a similar interpretation for a bound on the average waiting time for the G/G/1 queue in (4.11).

Similar to the M/M/1 queue, the M/G/1 queue also has the resource pooling property, see Exercise 2.18.

We have used the fact that service is FCFS to derive (2.11). However, the Pollaczek–Khinchine formula holds under most policies.

Theorem 2.23 $\bar{V}$ *is also the average delay under any non-preemptive, non-predictive (i.e. it does not use values of future service times) policy.*

Proof Assume that all the jobs are identical and require the same amount of work, but the actual service times of customers are still random because the server takes a random time X_ℓ to perform the ℓth job that he serves, where X_ℓ are i.i.d. Clearly, the departure times of customers are now independent of the policy. But in this new model, customers still are served for times X_n which are i.i.d., and equal to those which customer n would have under FCFS. So the departure process $\mathcal{D}(t)$ is independent of the policy, and hence the queue length process $Q(t)$ is independent of the policy, and by Little's law the average waiting time is independent of the policy.   □

Remark The distribution of the queue length $Q(t)$ is independent of the policy (assuming non-preemptive, work conserving, non-predictive). This implies, as we have seen, that average waiting time is independent of the policy (under the same assumptions). However, this is not the case for the distribution of the waiting time, which is very much dependent on the policy, as we will see in Section 3.5, Theorem 3.7.

2.8 Busy Periods

The server of a single-server queue is busy for a fraction ρ of the time (by Little's law). If we assume Poisson arrivals, then by the memoryless property of Poisson arrivals the time from the end of a busy period to the next arrival has $\mathrm{Exp}(\lambda)$ distribution, independent of all else, and therefore the intervals between busy periods have expected duration $1/\lambda$ (note that this assumption cannot be made under other arrival streams). Simple balancing using the renewal reward theorem then shows that the average busy period satisfies $\overline{BP}/\frac{1}{\lambda} = \rho/(1 - \rho)$.

Theorem 2.24 *The expected duration of the busy period of the M/G/1 queue is*

$$\overline{BP} = \frac{1}{\lambda}\frac{\rho}{1 - \rho} = \frac{m}{1 - \rho}. \tag{2.12}$$

Remark For an M/M/1 queue, the mean length of a busy period is equal to the mean sojourn time of a single customer. This is another illustration of the great variability that results from exponential service times. See also comment in Section 3.7 which discusses sojourn times in the M/G/1 queue under LCFS or PS.

Remark The expression $m/(1 - \rho)$ is misleading. This is another illustration of length biasing. Starting from an arbitrary time (starting from steady state), the time until the end of the busy period is $\sim m/(1 - \rho)^2$ (see Exercises 2.12–2.15). This indicates that the variability of the length of a busy period is huge.

Remark For an M/M/1 queue the expression $m/(1 - \rho)$ is the expected sojourn time for an arriving customer, i.e. it is the time it takes for the waiting time to decrease to 0. On the other hand, $m/(1 - \rho)^2$ is the time that it takes from the arrival of a customer until the work in the stationary queue decreases to zero. These are hugely different in heavy traffic. This illustrates the so-called *snapshot principle*: In heavy traffic, during the presence of a single customer in the queue, the queue itself will hardly change at all. This is significant in networks, where customers move through a network much faster than any changes will occur in the state of the network.

Assume now that a busy period starts with a service of length x. During the service duration x, there will be $N \sim \mathrm{Poisson}(\lambda x)$ arrivals. The busy period will consist of the service of these N arrivals, and the service of arrivals during their service times, etc. until the system is empty. This entire period can therefore be broken up into N busy periods, each starting with one of the N arrivals and serving it and all later arrivals to exhaustion. Note that this would in fact be the order of processing under non-preemptive LCFS. By Wald's equation and the formula (2.12) for the busy period of the M/M/1 queue, the entire busy period starting with x will have expected duration

$$\overline{\text{EFSBP}} = x + \lambda x \frac{m}{1 - \rho} = \frac{x}{1 - \rho}, \qquad (2.13)$$

where EFSBP stands for *exceptional first service busy period*. Thus, the busy period generated by a service of length x has expected length proportional to x. The multiplier $1/(1 - \rho)$ appears in many queueing formulas, and expresses the slowdown due to congestion.

2.9 Supplementary Material: Embedded Markov Chains

For most queueing models, $Q(t)$ is not a continuous time Markov chain. It is sometimes possible to find *regeneration times*, which are defined as follows:

Definition 2.25   A sequence of times $\tau_0 < \tau_1 < \cdots < \tau_n < \cdots$ are regeneration times for the process $X(t)$ if for $n = 1, 2, \ldots$:
  (i) $X(\tau_n + t)$, $t > 0$ have the same distribution as $X(\tau_0 + t)$, $t > 0$,
  (ii) $X(\tau_n + t)$, $t > 0$ is independent of $X(s)$, $s < \tau_n$.

In a process with regeneration times, many quantities can be calculated from the renewal reward theorem. An example of a regeneration time for $Q(t)$ in a G/G/s system is the beginning of a busy period. The end of a busy period is not a regeneration time, since the length of time to the next arrival depends on the length of the busy period.

Alternatively, we may find a sequence of times t_n such that while $Q(t)$ is not Markov, the sequence $Q_n = Q(t_n)$ is a discrete time Markov chain. We call Q_n an embedded Markov chain.

Finally, if processing times and interarrival times have *phase-type distributions*, i.e. they can be expressed as the time to absorption in a continuous time finite state Markov chain, then one may be able to study the resulting queueing system as a *quasi birth and death queue* (QBD). We will not discuss QBDs in this text.

We next study two examples of processes with embedded Markov chains, the M/G/1 queue and the G/M/1 queue.

2.9.1 The M/G/1 Queue — Embedded Markov Chain at Service Completions

We consider an M/G/1 queue under a non-preemptive service discipline. Let D_n be the departure time of customer n, and define the sequence $Q_n = Q(D_n)$ for $n = 1, 2, \dots$. Recall that $Q(t)$ is taken as right continuous with left limits, so Q_n is the number of customers in the queue immediately after a departure. At D_n all customers in the queue have remaining processing times of length distributed as G, and the time to the next arrival is exponentially distributed, with parameter λ, so that Q_n is an embedded Markov chain. We use Q_n to derive the stationary distribution of $Q(t)$.

The dynamics of Q_n are described by:

$$Q_{n+1} = \begin{cases} Q_n + K_{n+1} - 1, & Q_n \geq 1, \\ K_{n+1}, & Q_n = 0, \end{cases} \tag{2.14}$$

where K_n is the number of arrivals during the nth service period. Letting $k_j = \mathbb{P}(K_n = j)$, the transition matrix of Q_n is

$$\mathbf{P} = [p_{i,j}] = \begin{bmatrix} k_0 & k_1 & k_2 & \dots \\ k_0 & k_1 & k_2 & \dots \\ 0 & k_0 & k_1 & \dots \\ 0 & 0 & k_0 & \dots \\ \vdots & \vdots & \vdots & \dots \end{bmatrix}, \tag{2.15}$$

and writing the global balance equations $\pi = \pi \mathbf{P}$ we get:

$$\pi_i = \pi_0 k_i + \sum_{j=1}^{i+1} \pi_j k_{i-j+1}, \qquad i = 0, 1, 2, \dots,$$

where π is the stationary distribution of Q_n.

Let $\Pi(z) = \sum_{i=0}^{\infty} \pi_i z^i$, $K(z) = \sum_{i=0}^{\infty} k_i z^i$ be the generating functions of Q_n and K_n. Then we get (after some manipulations, see Exercise 2.16):

$$\Pi(z) = \frac{\pi_0(1-z)K(z)}{K(z) - z} = \frac{(1-\rho)(1-z)K(z)}{K(z) - z}, \tag{2.16}$$

where we use $\Pi(1) = K(1) = 1$, $K'(1) = \mathbb{E}(K) = \rho$, and L'Hopital's rule, to get $\pi_0 = 1 - \rho$. Then π_0 is the stationary probability that $Q_n = 0$, and it follows that Q_n is ergodic if $\rho < 1$.

The number of arrivals during a service period of length x is $\sim\text{Poisson}(x)$, and so:

$$K(z) = \mathbb{E}\Big(\sum_{i=0}^{\infty} \frac{(\lambda X z)^i}{i!} e^{-\lambda X} \Big) = \mathbb{E}(e^{-\lambda(1-z)X}) = G^*(\lambda(1-z)),$$

where G^* is the Laplace transform of G defined as $G^*(s) = \mathbb{E}(e^{-sX})$, so we get the Pollaczek–Khinchine formula for the generating function of queue length:

$$\Pi(z) = \frac{(1-\rho)(1-z)G^*(\lambda(1-z))}{G^*(\lambda(1-z)) - z}. \tag{2.17}$$

We found an expression for the stationary distribution of Q_n, the Markov chain of number of customers in the system left after a departure.

Proposition 2.26 *In a stable single-server queue, the stationary distribution of the number of customers left after departure equals the stationary distribution of the number of customers in the system found at arrivals.*

Proof The number of times in interval $(0, T)$ that departures leave n customers is the number of down-crossings from $n + 1$ to n, while the number of times in interval $(0, T)$ that arrivals find n customers is the number of up-crossings from n to $n + 1$, so they differ by at most 1. In the long-run, their ratio tends to 1. The proposition follows. □

But in an M/G/1 queue, arrivals are Poisson, and thus by PASTA, π_i are also the stationary probabilities for $Q(t)$.

2.9.2 The Distributional Form of Little's Law

Theorem 2.27 (The distributional form of Little's law, Keilson and Servi (1988)) *A stream of items that enter in a Poisson stream of rate λ, stay for a while without affecting either future arrivals or the time spent in the system by previous items, and leave in the same order as they arrived, satisfies the following relation between steady-state number N of items in the system, and the time W spent in the system by each item:*

$$\mathbb{E}(z^N) = \mathbb{E}(e^{-\lambda W(1-z)}). \tag{2.18}$$

Proof Consider the system at an arrival time and at a departure time, when arrivals are Poisson. Then the stationary queue length, the number found on arrival, and the number found at departure have the same distribution, say π_i. Consider then the number at an arrival of a customer and the number at the departure of that customer; they should have the same distribution. But at the departure, all the customers that were there at the arrival will have departed, and in their place will be all the customers that arrived during the sojourn of the customer. Conditional on sojourn t the number of arrivals is Poisson(λt). Hence,

$$\pi_k = \int_0^\infty \frac{(\lambda t)^k}{k!} e^{-\lambda t} dW(t),$$

and therefore,

$$\mathbb{E}(z^N) = \sum_{k=0}^\infty z^k \int_0^\infty \frac{(\lambda t)^k}{k!} e^{-\lambda t} dW(t)$$

$$= \int_0^\infty e^{-\lambda(1-z)t} dW(t) = \mathbb{E}(e^{-\lambda W(1-z)}).$$

□

Using this, we get that the Laplace transform of the stationary sojourn time and waiting time in M/G/1 is:

$$W^*(s) = \frac{(1-\rho)sG^*(s)}{s - \lambda(1 - G^*(s))}, \qquad V^*(s) = \frac{(1-\rho)s}{s - \lambda(1 - G^*(s))}. \qquad (2.19)$$

It is easy to check from $V^*(s)$ that the mean delay is as in (2.11).

2.9.3 The G/M/1 Queue — Embedded Markov Chain at Arrival Times

At arrival times, the time to the next arrival is an interarrival time, $T \sim F$, while the remaining processing times of all remaining customers, including the one in service, are exponentially distributed by the memoryless property. Denoting by A_n the arrival time of customer n, $Q_n = Q(A_n-)$, the number of customers in system seen by the arriving customer, is an embedded Markov chain. We use Q_n to derive the stationary distribution of $Q(t)$.

The dynamics of Q_n are described by

$$Q_{n+1} = Q_n + 1 - B_n, \qquad B_n \le Q_n + 1, \qquad (2.20)$$

where B_n is the number of customers that are served during the interarrival time between the n and $n + 1$ arrival, T_{n+1}. Letting $b_j = \mathbb{P}(B_n = j)$, the transition matrix of Q_n is:

$$\mathbf{P} = [p_{i,j}] = \begin{bmatrix} 1 - b_0 & b_0 & 0 & 0 & 0 & \cdots \\ 1 - \sum_{j=0}^1 b_j & b_1 & b_0 & 0 & 0 & \cdots \\ 1 - \sum_{j=0}^2 b_j & b_2 & b_1 & b_0 & 0 & \cdots \\ \vdots & \vdots & \vdots & \vdots & \vdots & \cdots \end{bmatrix}. \qquad (2.21)$$

From the global balance equations, the solution is of the form

$$\pi_i = (1 - \alpha)\alpha^i, \qquad i = 0, 1, 2, \dots. \qquad (2.22)$$

To see this, recall that for an ergodic Markov chain, π_j/π_i equals the average number of visits to j between visits to i, so we need to show that π_{i+1}/π_i is independent of i. Consider then the times $t_1 < t_2 < t_3$ such that: t_1 is an up-crossing by $Q(t)$ from i to $i+1$, t_2 is the next down-crossing from i to $i-1$, and t_3 is the next up-crossing from i to $i+1$. Then $t_1 t_3$ are two successive times at which the Markov chain Q_n visits i (arrival finds i items in the queue). In the interval $t \in [t_2, t_3)$, we have $Q(t) \leq i$, so there are no arrivals that find $i+1$ in that interval. Hence, all arrivals that find $i+1$ customers on arrival happen in (t_1, t_2). But during $t \in [t_1, t_2)$, we have $Q(t) \geq i$. In the period $t \in [t_1, t_2)$ the process $Q(t) - i$ is distributed exactly like $Q(t)$ in a single busy period. So the number of visit of Q_n to the state $i+1$ between two visits to state i are distributed exactly like the number of visits to the state 1 (arrivals finds one item in queue), in a busy period of $Q(t)$, and it is independent of i. Hence, $\pi_{i+1}/\pi_i = \alpha$, and we must have $\alpha < 1$ so that (2.22) will converge.

We now obtain an equation for α. From the balance equations for π_1, we have

$$\pi_1 = (1 - \alpha)\alpha = \sum_{i=0}^{\infty}(1 - \alpha)\alpha^i b_i,$$

or in terms of the generating function of B_n:

$$\alpha = \mathbb{E}(\alpha^{B_n}).$$

This equation will have a unique solution $0 < \alpha < 1$, if and only if $\mathbb{E}(B_n) = \mu/\lambda = 1/\rho > 1$ (see Exercise 2.17). This proves that $\rho < 1$ is a necessary and sufficient condition for stability.

Conditional on $T_{n+1} = t$, $B_n \sim$ Poisson(μt), hence,

$$\mathbb{E}(\alpha^{B_n}) = \int_0^{\infty} \sum_{j=0}^{\infty} \alpha^j \frac{(\mu t)^j}{j!} e^{-\mu t} dF(t)$$

$$= \int_0^{\infty} e^{-\mu(1-\alpha)t} dF(t) = F^*(\mu(1 - \alpha)),$$

where $F^*(s) = \mathbb{E}(e^{-sT})$ is the Laplace transform of F, and α is the unique solution of:

$$\alpha = F^*(\mu(1 - \alpha)). \tag{2.23}$$

Thus, π_i are the stationary distribution of the queue length at arrival times. However, we would like to have the stationary distribution of the continuous time process $Q(t)$, which we denote p_j. As we already argued in the discussion of the embedded Markov chain for the M/G/1 queue, π_i

is also the stationary distribution of the queue length at departure times. But the departure times occur at times of a Poisson process of rate μ, with dummy departures if the queue is empty. By PASTA, departures from state j occur at rate μp_j. But this rate equals the rate of arrivals to state $j-1$, which are $\lambda \pi_{i-1}$. Hence, $\lambda \pi_{j-1} = \mu p_j$. We obtain that the stationary distribution of $Q(t)$ is:

$$p_j = \rho \pi_{j-1} = \begin{cases} \rho(1-\alpha)\alpha^{j-1}, & j = 1, 2, \ldots \\ 1 - \rho, & j = 0. \end{cases} \tag{2.24}$$

2.10 Sources

The topics in this chapter are contained in all standard textbooks on queues. The textbook *Stochastic Models*, Ross (2009) has a very clear elementary treatment of renewal processes, and the textbook *Stochastic Modeling and the Theory of Queues*, Wolff (1989), has a very good treatment of PASTA. A recent very nice book is *Queues: A Course in Queueing Theory*, Haviv (2013). The proof of PASTA is taken from Wolff (1982). An early introduction of renewal theory is Smith (1958) and it is studied in depth by Feller (1971). Piecewise deterministic Markov chains were introduced by Davis (1984). Palm calculus is studied in Baccelli and Brémaud (2013). Phase type distributions were introduced by Neuts (1974), who developed the rich theory in his monograph *Matrix-Geometric Solutions in Stochastic Models: an Algorithmic Approach*, Neuts (1981), which enabled exact analysis of many models of quasi birth and death processes, and matrices of M/G/1 type. Proceedings on this area of research are included in Latouche and Taylor (2000, 2002); Latouche et al. (2012).

Exercises

2.1 Take Figure 2.4 and, similar to Figure 2.5, plot the contributions to the workload of each job under preemptive LCFS, and under PS (for the latter, assume that $Q(0) = 2$, with equal processing times for both).

2.2 Prove that the renewal function satisfies $m(t) < \infty$ for all $t > 0$, that $\mathcal{A}(t) < \infty$ for all $t > 0$, and that $m(t) \to \infty$, as well as $\mathcal{A}(t) \to \infty$ almost surely, as $t \to \infty$.

2.3 Derive the following steps toward the proof of the elementary renewal theorem, by the use of Wald's equation:

    (i) Show that $\mathcal{A}(t) + 1$ is a stopping time for the sequence T_n, $n = 1, 2, \ldots$, while $\mathcal{A}(t)$ is not.

    (ii) Use Wald's equation to calculate $\mathbb{E}(A_{\mathscr{A}(t)+1})$ and obtain a lower bound on $m(t)/t$.

    (iii) Obtain an upper bound for bounded T_n, and prove the theorem for bounded T_n.

    (iv) Consider the renewal process with truncated inter-event times $\tilde{T}_n = \min(T_n, a)$, to upper bound $m(t)/t$, and complete the proof.

2.4   Derive the following steps toward the proof of the elementary renewal theorem, by showing that $\mathscr{A}(t)/t$, $t \geq 0$ are uniformly integrable:

    (i) Obtain a random variable $\tilde{T}_n$ with values 0 or a that is less or equal to T_n.

    (ii) Calculate the renewal function of the renewal process $\tilde{\mathscr{A}}(t)$ of $\tilde{T}_n$, to show it is bounded above by $c_1 t + c_2 t^2$, for some constant c_1, c_2.

    (iii) Use Markov's (Chebyshev's) inequality to show that for all $t > 1$, $\mathbb{P}(\tilde{\mathscr{A}}(t)/t > x) \leq \frac{c_1+c_2}{x^2}$ and therefore $\mathscr{A}(t)/t$, $t \geq 0$ are uniformly integrable to complete the proof.

2.5   Prove the renewal reward Theorem.

2.6   Obtain the conditional probability $\mathbb{P}(\text{Excess} > y \mid \text{Age} = x)$, and use it to derive the joint distribution of age and excess.

2.7   Derive the joint distribution of age and excess directly from the renewal reward theorem, by considering the length of time that excess $> x$ and also age $> y$, within an interval of length T.

2.8   Prove that when you observe a stationary renewal process at an arbitrary time t, t is uniformly distributed along the interval that includes t.

2.9   A bus arrives at your station according to a stationary renewal process, with interarrival time distributions $T \sim F$. You arrive at some arbitrary time. Calculate the distributions of: (1) the length of the interval in which you arrived, (2) the time since the last bus arrived, (3) your waiting time for the next bus, in the following cases:

    (i) T is uniform, $T \sim U(a, b)$, i.e. $f_T(t) = \frac{1}{b-a}$, $\quad a < t < b$.

    (ii) T is exponential, $T \sim \exp(\lambda)$, i.e. $f_T(t) = \lambda e^{-\lambda t}$, $\quad t > 0$.

    (iii) T is Erlang 2, $T \sim \gamma(2, \lambda)$, i.e. $f_T(t) = \lambda^2 t e^{-\lambda t}$, $\quad t > 0$.

2.10 Use (2.11) to obtain the average number of customers in the queue, the average number of customers waiting for service, and the average number of customers at the server, for an M/G/1 queue.

2.11 Calculate the average waiting time for an M/G/1 system with given ρ, when the service time is distributed as (i) Exp(1), (ii) Erlang(k, k), (iii) Deterministic $= 1$, (iv) $f_X(x) = \frac{1}{1+a}\frac{1}{a}e^{-x/a} + \frac{a}{1+a}ae^{-ax}$.
Note: the last distribution is called a hyper-exponential distribution and has a large c.o.v.

2.12 Consider a stationary M/M/1 queue at an arbitrary time t, and calculate the remaining length of the busy period. This will be the waiting time of a so-called *standby customer*, who only starts service when the queue is empty.

2.13 In an M/G/1 queue, let $G^*(s)$ be the Laplace transforms of the service time distribution, and $BP^*(s)$ be the Laplace transform of the distribution of the busy period length. Prove the equation:

$$BP^*(s) = G^*(s + \lambda - \lambda BP^*(s)).$$

2.14 Derive the variance and the second moment of the length of an M/G/1 busy period [Cohen (2012), Section II-2.2].

2.15 Derive the Laplace transform of the length of a busy period for an M/M/1 queue. Use it to obtain the pdf of the length of a busy period of an M/M/1 queue.

2.16 Verify the derivation for the generating function of the M/G/1 embedded Markov chain, given by equation (2.16)

2.17 Prove that the equation $\alpha = \mathbb{E}(\alpha^{B_n})$ has a unique solution $0 < \alpha < 1$ if and only if $\mathbb{E}(B_n) > 1$.

2.18 Use equations (2.17), (2.19), to show that the M/G/1 queue has the resource pooling property: when service is speeded up s-fold, and arrivals are speeded up s-fold, the queue length distribution remains exactly the same, while waiting and sojourn times are improved by a factor of s.

2.19 Explain why the sojourn time of a job under preemptive LCFS equals the length of a busy period. In particular, use this to explain why in the M/M/1 queue, the expected busy period equals the expected sojourn time of a customer under any non-predictive work-conserving policy.

2.20 Derive the expected length of a busy period for an M/G/1 queue, in which the server goes on vacation at the end of each busy period, where the vacation time has distribution H.

3

Scheduling

To improve the performance of a queueing system, we may be able to control some aspects of the process. One way of doing this is scheduling, deciding when and in what order to perform the processing of jobs in the system. In this chapter, we discuss scheduling for the single queue.

The well-established theory of scheduling deals with scheduling of batches of jobs of known sizes on various configurations of machines with various objective functions. It falls under the subject of combinatorial optimization. The purpose is to find optimal schedules. Because the data is deterministic, there is no question of evaluating the performance of the optimal schedule, as it changes from instance to instance. Solution of most scheduling problems is NP-hard, and approximations to optimal schedules are sought. Comparisons of various policies and evaluations of approximations for the optimal solution are done on the basis of worst case analysis, or under the even stricter criterion of competitive ratio analysis. This may not give a good idea about *typical* instances, because there is no concept of typical, and worst case certainly does not capture it.

Robust optimization is one approach to remedy this: Instead of the single instance, a family of scenarios rich enough to model the neighborhood of an instance is considered, and the best solution for all scenarios is found. This approach provides a more realistic picture on performance, by assessing the possible values obtained over the various scenarios. Another alternative approach, somewhere in between robust optimization and worst case analysis, is to introduce a probability model of the problem: With this, one can obtain the distribution of the objective, and search for a policy that optimizes its expected value.

In the context of queueing theory, we are in a strong position to evaluate the performance of scheduling rules, since we assume a probabilistic model for our scheduling problems. This allows us to perform an average rather than a worst case analysis. We stress the advantages of this approach in the current chapter and in further analyses developed in later chapters.

We start this chapter with the scheduling of a deterministic and a stochastic batch of jobs. In the deterministic case, we know all the processing times precisely, whereas in the stochastic case we know the distributions of the various processing times. We focus on the minimization of the flowtime, the sum of completion times, or equivalently the average sojourn time.

We then discuss the scheduling of a stream of jobs that arrive according to a stationary arrival process. For preemptive, fully informed scheduling we prove the optimality of the shortest remaining processing time (SRPT) rule, which is pathwise optimal almost surely.

In the remaining sections of this chapter we evaluate the performance of several scheduling policies for M/G/1 queues. We derive average waiting times for priority scheduling with K priority classes. We evaluate expected waiting times for the non-preemptive shortest processing time first policy and for SRPT. We discuss both preemptive and non-preemptive last come first served (LCFS) and processor sharing (PS). In these sections of the chapter, we leave several of the proofs as exercises. Finally, we briefly describe control of the M/G/1 queue using the Gittins index priority rule.

3.1 Batch Scheduling

A set of N jobs is available for scheduling, all of them available at time 0. Job j requires service time X_j. If we schedule the jobs in the order $1, 2, \ldots, N$ then job n will complete at time $D_n = \sum_{j=1}^{n} X_j$ (see Figure 3.1).

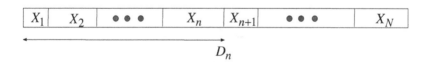

Figure 3.1 Schedule of N jobs, with flowtime of job n.

Theorem 3.1 *The flowtime of the jobs, $\sum_{n=1}^{N} D_n$, is minimized by the shortest processing time first (SPT) policy.*

Proof We present three proofs, illustrating important features of the problem.

(i) *Pairwise interchange:* Assume that $X_n > X_{n+1}$. Changing their order will not affect D_k, $k \neq n$, but D_n will be earlier by $X_n - X_{n+1} > 0$. Hence, any non-SPT order can be improved by a sequence of pairwise interchanges, until SPT is reached (see Figure 3.2).

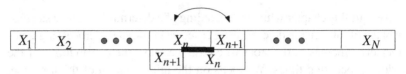

Figure 3.2 First optimality proof: pairwise interchange.

(ii) *Euler summation:* Rewrite the flowtime as

$$\sum_{n=1}^{N} D_n = \sum_{n=1}^{N} \sum_{k=1}^{n} X_k = \sum_{k=1}^{N} (N - k + 1) X_k ; \qquad (3.1)$$

then X_k is multiplied by $N - k + 1$, which indeed says that job k is delaying itself and the $N - k$ jobs scheduled after it. By the Hardy–Littlewood–Polya inequality, equation (3.1) is minimized by pairing $N - k + 1$ with the kth shortest job (see Figure 3.3).

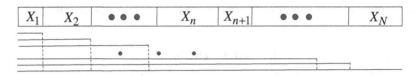

Figure 3.3 Second optimality proof: Euler summation.

(iii) *Adding the delays for each pair of jobs:* Rewrite the flowtime as

$$\sum_{n=1}^{N} D_n = \sum_{n=1}^{N} X_n + \sum_{n=1}^{N} \sum_{k \neq n} \delta_{n,k} X_k, \qquad (3.2)$$

where $\delta_{n,k} = 1$ if job k precedes job n and $\delta_{n,k} = 0$ if job k is after job n. This is minimized for every pair if we schedule the shorter job before the longer job, which is SPT (see Figure 3.4).

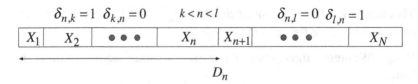

Figure 3.4 Third optimality proof: adding pairwise delays.

□

Next we consider a stochastic version of this scheduling problem. The processing time of job j is a random variable X_j, and all the scheduler knows is the distribution of each X_j, for $j = 1, \ldots, N$.

Corollary 3.2 *For jobs with random processing times X_j the expected flowtime is minimized by the shortest expected processing time first (SEPT) policy.*

Proof This follows directly from Theorem 3.1 and the linearity of expectation, since flowtime is linear in the X_j. □

3.2 Value of Information

In this section, we quantify the advantage of using information for scheduling a batch of jobs. We assume there is a population of jobs, of varying lengths, distributed according to a distribution G. We let m denote the expected value of the distribution G, and we let $m_{(i:n)}$ denote the expected value of the ith order statistic (the i smallest) in a sample of size n from G. We need to schedule a batch of N jobs, chosen randomly from G. We ask how much better can we do if we know in advance the length of each job, compared to knowing nothing.

Under these scenarios, if we have no information about the size of the individual jobs in the batch, then we cannot do better than to schedule them in random order. On the other hand, if we know the sizes of all the jobs, we can schedule them optimally using SPT. The difference in expected flowtime under the two scenarios measures the advantage of SPT over random order, and the value of full information to the scheduler. We formulate this as a theorem:

Theorem 3.3 *For a batch of N jobs with i.i.d. processing times, the difference in expected flowtime, between scheduling with no information, and with full information is:*

$$\textbf{Value of information} = \frac{N(N-1)}{2}(m - m_{(1:2)}).$$

Proof The expected flowtime under a random order is $\frac{N(N+1)}{2}m$.

The expected flowtime under SPT is, by taking expectation over the expression (3.2), $Nm + \frac{N(N-1)}{2}m_{(1:2)}$

The theorem follows. □

By taking expectations over expression (3.1) and comparing it to the expectation over (3.2), we obtain an interesting formula relating the expectations of order statistics:

$$Nm + \frac{N(N-1)}{2}m_{(1:2)} = \sum_{k=1}^{N}(N-k+1)m_{(k:N)}. \qquad (3.3)$$

For a distribution G, we define the SPT-advantage dimensionless parameter $d_G = \dfrac{m - m_{(1:2)}}{\sigma}$. For some commonly used distributions we obtain (see Exercise 3.3):

$$d_{\text{Exponential}} = \frac{1}{2} = 0.5, \quad d_{\text{Normal}} = \sqrt{\frac{1}{\pi}} \approx 0.56, \quad d_{\text{Uniform}} = \sqrt{\frac{1}{3}} \approx 0.58.$$

For processing time distribution G, with standard deviation σ_G and coefficient of variation c_G, the value of information for a batch of N jobs is $\approx \sigma_G d_G N^2/2$, and the performance ratio $\dfrac{\text{Random Order Flowtime}}{\text{SPT Flowtime}}$ is $\approx \frac{m_{1:2}}{m} = 1 - c_G d_G$.

3.3 Stream Scheduling

Stream scheduling differs from batch scheduling in that there is a stream of a potentially infinite number of jobs, which only become available for processing at their arrival times. In contrast to batch scheduling, in stream scheduling we define the flowtime as the sum of the sojourn times, i.e. we take the sum of the departure times minus the sum of the arrival times. Usually we do not control the arrival times of jobs, so minimizing the sum of sojourn times or minimizing the sum of departure times is equivalent. However, if N jobs arrive in regular intervals, the sum of sojourns is of order N, while the sum of departure times is of order N^2.

Scheduling a batch of N jobs with given arrival times A_n and service times X_n, without preemptions, to minimize flowtime, is an NP-hard combinatorial optimization problem. The following two examples show that (i) SPT is not necessarily optimal, (ii) work conservation is not necessarily optimal, i.e. it may be optimal to insert idle time in the schedule:

Example 3.4   There are five jobs, with $A_1 = A_2 = 0, A_3 = A_4 = A_5 = 3$, $X_1 = 2, X_2 = 3, X_3 = X_4 = X_5 = 1$. It is optimal to schedule job 2 before job 1. This is illustrated in Figure 3.5, where the left side shows job sizes and arrival times data, and on the right schedule A is SPT, while schedule B is optimal.

Example 3.5   There are four jobs, with $A_1 = 0, A_2 = A_3 = A_4 = 1$,

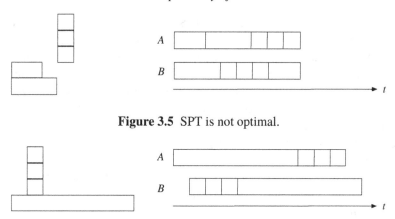

Figure 3.5 SPT is not optimal.

Figure 3.6 Optimal to insert idle time.

$X_1 = 8, X_2 = X_3 = X_4 = 1$. It is optimal to insert idle time until A_2, and schedule jobs 2 and 3 and 4, and then 1. This is illustrated in Figure 3.6, where the left shows the job sizes and arrival times, and on the right schedule A is work conserving (no idling), while schedule B is optimal.

This emphasizes the difficulty of finding optimal scheduling rules that minimize flowtime in queues. We shall next prove two very general results: The first is that if preemptions are allowed and the required processing time of a job becomes known at its moment of arrival, then an optimal schedule exists and is easy to implement. In the second, if we have no information on the processing times of jobs, so that all appear the same to the scheduler, and if preemptions are not allowed, then all work-conserving schedules will have the same expected flowtime, though FCFS will minimize the variance of the flowtime.

3.4 Optimality of SRPT

When preemption is allowed and service times are known for every job at its moment of arrival, then one can implement the policy of SRPT: at any moment in time the server should work on the job with the shortest remaining processing time. This can be used to schedule a batch of jobs with known arrival and service times, as well as to schedule a stream of jobs in a queueing system.

For a stream of jobs, SRPT means that once processing of a job is started, it may only be interrupted by the arrival of a new job. So processing is non-preemptive between arrivals, and so it is SPT between arrivals, but at every arrival the arriving job may preempt the current job if it is shorter than the current job's remaining processing time. While preemptions will be used by SRPT, there will always be only one job in process, with shortest remaining processing time, which will then become even shorter. Even if we start with two jobs of equal length, once we start working on one it will stay shorter than the other from that time onward. Therefore, splitting the processing capacity to work on more than one job simultaneously is never needed.

Theorem 3.6 *Preemptive shortest remaining processing time first (SRPT) minimizes the number of customers in the system at all times, for every sample path.*

Proof Consider a single realization of arrival times $A_1 < A_2 < \ldots$ and of service requirements $X_1, X_2, \ldots$.

Under SRPT, define the process $l_k(t)$ as the sum of the k longest remaining processing times at time t, and if $Q(t) \le k$ define it as $\mathcal{W}(t)$, the total workload. Define $l'_k(t)$ in the same way for some arbitrary work-conserving (non-idling) policy.

We will show that $l_k(t)$ is maximal among all work-conserving policies. We show $l_k(t) \ge l'_k(t)$ by induction on k. For $k = 0$, $l_0(t) = l'_0(t) = 0$. By definition, both $l_k(t)$ and $l'_k(t)$ are nonincreasing between arrivals. By work conservation, the amount of work in the system, $\mathcal{W}(t)$, is the same under SRPT or under any other work-conserving policy, and it is an upper bound on both $l_k(t)$ and $l'_k(t)$. Assume at some time t_0, $l_k(t_0) \ge l'_k(t_0)$, and let t_1 be the first arrival after t_0. In the interval $t \in [t_0, t_1)$, $l_k(t)$ will remain unchanged for as long as $Q(t) > k$, since we use SRPT, and once $Q(t) \le k$ is reached, $l_k(t)$ will remain equal to $\mathcal{W}(t)$ at least until the next arrival. Hence, $l_k(t) = \min(l_k(t_0), \mathcal{W}(t))$, which is the upper bound, and is therefore $\ge l'_k(t)$.

At the arrival time t_1, let X be the processing requirement of the arriving job. Consider now $l_k(t_1)$ and $l'_k(t_1)$. If X is included in either of these sums, it will replace the smallest of the k largest remaining processing times. Hence, $l_k(t_1) = \max(l_k(t_1-), l_{k-1}(t_1-) + X)$ and $l'_k(t_1) = \max(l'_k(t_1-), l'_{k-1}(t_1-) + X)$. By what we showed, $l_k(t_1-) \ge l'_k(t_1-)$, and by induction $l_{k-1}(t_1-) \ge l'_{k-1}(t_1-)$, which completes the induction proof.

Starting at $t_0 = 0$, we can proceed from one arrival time to the next, so we have shown:

$$l_k(t) \ge l'_k(t) \qquad \text{for all } k \text{ and all } t.$$

However, by work conservation,

$$\mathcal{W}(t) = l_{Q(t)}(t) = l'_{Q'(t)}(t) = \mathcal{W}'(t).$$

Hence, for all t,

$$Q(t) \le Q'(t).$$

It is also clear that inserting idle time will not be optimal: assume job j is available at time t and the server is idle until $t + h$, then just working on job j in the interval $(t, t + h)$ for some time δ, and then later idling for the same amount of time δ when job j is finished, will not change completion of any job except j, which will complete a time δ earlier. Note that job j can still be preempted at exactly the same time points as before. $\square$

This is a very strong optimization result: SRPT minimizes the queue for every sample path (that is, for every realization of the values of A_n, X_n when they are chosen randomly). Furthermore, the minimization of $Q(t)$ is pathwise, i.e. $Q(t)$ is minimized jointly for all t, which is much stronger than minimization of the sum of the sojourn times. Even more, there are no assumptions at all on the arrival and service times. By Little's law, this will certainly imply the much weaker conclusion that SRPT minimizes the average sojourn time.

3.5 Mean and Variance for Non-Preemptive, Non-Predictive Scheduling

We saw in Theorem 2.23 that if jobs with i.i.d. processing time are scheduled without prediction of the processing time of each job, and non-preemptively, then all work conserving (non-idling) schedules achieve the same expected waiting times. Recall that this is because the distribution of the departure process is independent of the order in which jobs are scheduled: When we decide to put a job in service, among all jobs present, its processing time has the same distribution as all the other jobs, so the distribution of the next departure is not affected by which job is chosen. This result holds for any sequence of arrival times. Consider now the variance of the sojourn times.

Theorem 3.7 *In a G/G/1 stable queue, scheduling jobs in FCFS order on a single machine will minimize the variance of the sojourn and waiting times, among all non-preemptive work conserving non-predictive policies. Scheduling in non-preemptive LCFS order will maximize it.*

Proof Consider a set of n jobs, starting from an empty system. Let $A_1 <$

$\cdots < A_n$ be their arrival times, and $D_1, \ldots, D_n$ their departure times, with A_j the arrival and D_j the departure of job j, and $W_j = D_j - A_j$ the sojourn time of job j. Let $X_1, \ldots, X_n$ be the processing times of the n jobs, ordered by the order of processing, so that X_j is the processing time of the j'th to be processed by the server. Then the set of departure times $\{D_1, \ldots, D_n\}$ is independent of the order in which we schedule the jobs to be processed, and the actual values for jobs $1, \ldots, n$ (ordered by their arrival times), will be a permutation of these departure times.

We now calculate $\sum_{j=1}^{n} W_j^2 = \sum_{j=1}^{n} D_j^2 + \sum_{j=1}^{n} A_j^2 - 2\sum_{j=1}^{n} A_j D_j$. The first two terms are independent of the order in which the jobs are scheduled. Hence, we only need to consider $\sum_{j=1}^{n} A_j D_j$. Under FCFS, we will have $D_1 < \cdots < D_n$. By the Hardy–Littlewood–Polya rearrangement inequality, this permutation will maximize $\sum_{j=1}^{n} A_j D_j$, and hence minimize $\sum_{j=1}^{n} W_j^2$.

We have shown that FCFS will minimize $\sum_{j=1}^{n} W_j^2$ for any sequence of jobs starting from empty. Hence, if the system is stable it will minimize the long-term average, and if the system is ergodic it will minimize the expected value of the squared sojourn time. We have already shown in Theorem 2.23 that the expected sojourn time is independent of the schedule. Hence FCFS minimizes the variance of the sojourn times. Since in a non-preemptive schedule the sojourn is the sum of waiting and processing times, which are independent, FCFS also minimizes the variance of the waiting time.

The proof for LCFS non-preemptive is similar and left as Exercise 3.5.

□

3.6 M/G/1 Priority Scheduling

We now consider an M/G/1 system with several customer types, $k = 1, \ldots, K$. Arrivals of the different types are independent Poisson processes, with arrival rates λ_k. Processing times of customers of type k are distributed G_k with rate μ_k. The workloads are $\rho_k = \lambda_k/\mu_k$, with total workload $\rho = \sum_{k=1}^{K} \rho_k < 1$. We now consider this system operated under a priority scheduling policy. For $j < k$, jobs of type j have non-preemptive priority over jobs of type k, while jobs of the same type are scheduled FCFS (this is known as head of the line (HOL) within types).

Consider the waiting time of a job of type k: When the job arrives at the system he sees, by PASTA, a time average stationary state of the work in the system, $\mathcal{W}(t)$. The expected work at the server is $\lambda \mathbb{E}(X^2)/2$ and the expected work of waiting customers of type j is $\rho_j \overline{V}_j$ (see Exercise 3.6). He need not wait for customers of type $j > k$, so the expected work that concerns the customer of type k in the system when he arrives is:

$$\sum_{j \leq k} \rho_j \overline{V}_j + \lambda \mathbb{E}(X^2)/2. \tag{3.4}$$

However, while this work is processed, new arrivals of types $j < k$ will arrive with higher priority, and he needs to wait for all of them. This amounts to a busy period of an M/G/1 queue with types $j < k$, and a special initial period that equals the delay that he finds on arrival. The total expected delay is then, from equation (2.13)

$$\overline{V}_k = \Big(\sum_{j \leq k} \rho_j \overline{V}_j + \lambda \mathbb{E}(X^2)/2\Big)\Big/\Big(1 - \sum_{j < k} \rho_j\Big).$$

By induction, one then has (see Exercise 3.7):

Theorem 3.8 *The expected waiting time of a customer of type k in a multi-type M/G/1 queue under a non-preemptive priority scheduling policy is*

$$\overline{V}_k = \frac{\lambda \mathbb{E}(X^2)/2}{(1 - \sum_{j<k} \rho_j)(1 - \sum_{j \leq k} \rho_j)}. \tag{3.5}$$

In particular, we can assign priorities according to SEPT.

Corollary 3.9 *SEPT minimizes the overall expected waiting time of customers in a multi-type M/G/1 queue, among all non-preemptive priority scheduling policies.*

See Exercise 3.8 for the proof.

Next we onsider an M/G/1 queue where a customer's service time becomes known at the moment of arrival. In that case, one can schedule jobs according to non-preemptive shortest processing time (SPT). We denote $\rho(x) = \lambda \int_0^x y\,dG(y)$; it is the workload of customers with processing time $\leq x$. We then have

Corollary 3.10 *Consider an M/G/1 queue, operated under non-preemptive SPT. Then the expected waiting time of a customer with processing time x is*

$$\overline{V}(x) = \frac{\lambda \mathbb{E}(X^2)/2}{(1 - \rho(x-))(1 - \rho(x))}. \tag{3.6}$$

The expected waiting time over all customers is

$$\overline{V}[\text{non-preemptive SPT}] = \frac{\lambda \mathbb{E}(X^2)}{2} \int_0^\infty \frac{dG(x)}{(1 - \rho(x-))(1 - \rho(x))}. \tag{3.7}$$

See Exercise 3.9 for the proof.

A similar calculation is also possible for SRPT (see Exercise 3.11).

3.7 M/G/1 under Preemptive LCFS and under PS

Under the policy of preemptive last come first served (LCFS), the server always works on the customer that is the last to have arrived in the system. That means that every time a new customer arrives, the server preempts service of the customer he is serving. Under the policy of processor sharing (PS), the server is dividing his service capacity equally between all the customers present in the system. These two policies have the remarkable property that the expected sojourn time of a customer in the system, from his arrival to his departure, is proportional to the amount of processing that he requires. This may be viewed as fair treatment of customers.

This property is a result of the even more remarkable property of insensitivity, which we will discuss at greater length when we look at networks of queues, in Section 8.8. By insensitivity, we mean that the stationary distribution for the number of customers depends on the service time distribution only through its mean. We shall see in Section 8.8 that LCFS and PS are symmetric policies which, by Theorem 8.17 implies insensitivity. The following theorem is a direct outcome of Theorem 8.17.

Theorem 3.11 *For the M/G/1 queue, under PS and under preemptive LCFS policies, the stationary distribution of $Q(t)$, the number of customers in the system, is the same as for the M/M/1 queue. Furthermore, at time t the ages of all customers are i.i.d. with density $g_{eq}(x) = (1 - G(x))/\mathbb{E}(X)$ (the forward recurrence or equilibrium distribution), independent of the number in the system.*

Theorem 3.12 *The expected sojourn time under PS and under preemptive LCFS for a job of length x is $\frac{x}{1-\rho}$, and the average sojourn time is $\frac{m}{1-\rho} = \frac{1}{\mu - \lambda}$.*

Proof (i) Consider first preemptive LCFS. The proof does not depend on Theorem 3.11. Each time a job is preempted, it will need to wait expected time $\mathbb{E}(X)/(1 - \rho)$ for the busy period started by the interrupting job to end. A job of length x will have a Poisson number of interruptions with expectation λx. By Wald's formula, the expectation of its delay is $\rho x/(1-\rho)$, and its total expected sojourn is therefore $x/(1 - \rho)$.

(ii) Consider PS. We use Theorem 3.11. When observed at an arbitrary time, by Theorem 3.11, the stationary distribution of job sizes is

$xdG(x)/\mathbb{E}(X)$, independent of the number of jobs in the system, and the expected number of jobs in the system, which is the same as for the M/M/1 queue, is $\frac{\rho}{1-\rho}$. Hence, the expected number of jobs of size x in the system is $xdG(x)\lambda/(1 - \rho)$. The arrival rate of jobs of size x is $\lambda dG(x)$. By Little's formula: $xdG(x)\lambda/(1 - \rho) = \lambda dG(x) \times \overline{W}(x)$. Hence, $\overline{W}(x)$, the expected sojourn time of jobs of length x, equals $x/(1 - \rho)$.

Averaging over all job sizes, we get the average sojourn time, which equals that of the M/M/1 queue.

Note, proof (ii) works for any symmetric policy, as defined in Section 8.8 (including preemptive LCFS). □

3.8 Smith's Rule, Klimov's Control of M/G/1, and the $c\mu$ Rule.

Consider now a different objective function. We wish to minimize weighted flow time, where job j will have a cost C_j per unit of sojourn time (weight of the job). For a batch of N jobs with processing times X_j and weights C_j, on a single machine the objective is to minimize $\sum_{j=1}^{N} C_j D_j$. The optimal schedule for this is given by Smith's rule: schedule jobs with smaller X_j/C_j first. All three proofs for the optimality of SPT for flowtime, work also for optimality of Smith's rule for weighted flowtime (see Exercise 3.13). If the processing times and the weights are stochastic, the optimal policy is Smith's rule for the expectation: Schedule jobs with smaller $\mathbb{E}(X_j)/\mathbb{E}(C_j)$ first (see Exercise 3.14).

For the M/G/1 queue, Smith's rule is the optimal static priority policy, as can be seen from (3.5) (see Exercise 3.15).

This leaves open the question of whether a dynamic rule, which takes into account the contents of the queues at time t, can do better. Klimov has shown that for the M/G/1 queue, when job holding costs and processing times become known upon arrival, or more generally, if jobs are of several types, with expected processing time $1/\mu_j$ and expected holding costs c_j, priority given to jobs with largest $c_j\mu_j$ (smallest $\mathbb{E}(X_j)/\mathbb{E}(C_j)$) is optimal. This version of Smith's rule is called the $c\mu$ rule. Klimov's model is a precursor to the Gittins index priority rule for control of multi-armed bandits, which we discuss next.

3.9 Bandit Processes and the Gittins Index

Consider a system consisting of a family of K alternative so-called bandit processes, where at time t the state of the system is given by the vector

$\mathbf{Z}(t) = (Z_1(t), \ldots, Z_K(t))$ of the states $Z_k(t)$ of the bandit processes $k = 1, \ldots, K$. We assume that these bandits move on countable state spaces E_k, so $Z_k(t) \in E_k, k = 1, \ldots, K$.

At any point in time, $t = 0, 1, 2, \ldots$, we need to take one of K possible actions, namely choose to activate one of the bandit processes, which will then yield a reward and undergo a Markovian state transition, while all the other bandit processes are passive – they yield no reward, and their states remain frozen. More precisely, if we choose at time t action $k(t) = k$, then bandit k in state $Z_k(t) = i$ will be activated. This action will yield a reward $R_k(i)$, where R_k is the reward function for bandit k, and bandit k will undergo a transition, from state i to state j according to $p_k(i, j) = \mathbb{P}(Z_k(t + 1) = j | Z_k(t) = i)$. For all other bandits, $l \neq k(t)$, there will be no change in state, so $Z_l(t + 1) = Z_l(t)$, and no reward, so the reward for period t will be given by $\tilde{R}(t) = R_{k(t)}(Z_{k(t)}(t)) = R_k(i)$.

We will assume that $|R_k(i)| \leq C$ uniformly for all states and bandits. The objective is to choose a policy π for activating the bandits so as to maximize total expected discounted reward

$$V_\pi(\mathbf{i}) = \mathbb{E}_\pi \Big\{ \sum_{t=0}^{\infty} \alpha^t \tilde{R}(t) | \mathbf{Z}(0) = \mathbf{i} \Big\},$$

where $\mathbf{Z}$ and $\mathbf{i}$ denote the state vector, and $0 < \alpha < 1$ is the discount factor.

This problem, introduced by Richard Bellman (1957) as the *multiarmed bandit problem*, is clearly a dynamic programming problem, with a countable state space, a finite action space, bounded rewards, and discounted infinite horizon objective. As such, by the theory of dynamic programming (cf. Ross (1983)), it has an optimal solution given by a stationary policy, which can be calculated using various general schemes. However, such a direct approach to the problem is impractical due to the high dimensionality of the state space.

What makes the problem tractable is Gittins' discovery that the problem is solved by a priority policy – one needs only to calculate a priority index for each of the bandits, based on its own current state, and independent of all the other bandits, and then activate the bandit with the highest index. Formally

Theorem 3.13 (Gittins, 1976)   *There exist functions, $G_k(Z_k(t)), k = 1, \ldots, K$, such that a policy π^*, which will in state $\mathbf{Z}(t)$ activate a bandit process (arm) $k(t) \in \arg\max_{1 \leq k \leq K} G_k(Z_k(t))$ is optimal. The function $G_k(\cdot)$ is calculated from the dynamics of process k alone.*

Gittins defined his Dynamic Allocation Index, now known as the Gittins

Index for a single-armed bandit process $Z(t)$, as follows:

$$v(i) = \sup_{\sigma > 0} v(i, \sigma) = \sup_{\sigma > 0} \frac{\mathbb{E}\left\{\sum_{t=0}^{\sigma-1} \alpha^t R(Z(t)) | Z(0) = i\right\}}{\mathbb{E}\left\{\sum_{t=0}^{\sigma-1} \alpha^t | Z(0) = i\right\}}. \qquad (3.8)$$

Here, $v(i, \sigma)$ is the expected discounted reward per expected unit of discounted time, when the arm is operated from initial state i, for a duration σ, and σ is a $Z(t)$ positive stopping time. The value $v(i)$ is the supremum of $v(i, \sigma)$ over all positive stopping times. By the boundedness of the rewards, $v(i)$ is well defined for all i, and bounded.

Starting from state i, we define the stopping time ($\leq \infty$):

$$\tau(i) = \min\left\{t : v(Z(t)) < v(i)\right\}. \qquad (3.9)$$

An important property of the Gittins index is that the supremum (3.8) is achieved, and in fact it is achieved by $\tau(i)$.

It can be shown that the problem of optimal non-preemptive scheduling of the M/G/1 queue to minimize expected average number of customers in the system, or equivalently average sojourn times, is solved by a Gittins index priority policy. The index is given by the $c\mu$ rule. In particular, this proves that the SEPT policy for the M/G/1 queue with K job types is not only the optimal policy among all priority policies, but it is optimal among all non-preemptive policies.

3.10 Sources

The standard textbook on scheduling is Pinedo (2012), *Scheduling: Theory, Algorithms, and Systems*. It contains many more results on the scheduling of batches of jobs, both deterministic and stochastic. Mor Harchol-Balter (2013) examines queueing applications for computer systems and has a whole section dealing with scheduling of M/G/1 queues. A remarkable early book in two volumes with a thorough engineering approach is Kleinrock (1975, 1976). The value of information comes from Matloff (1989). Priority scheduling for M/G/1 queues was originally derived by Cox and Smith (1961). Priority scheduling and further calculations of sojourn times in M/G/1 queues under various policies are extensively studied in Wolff (1989). The optimality of SRPT was shown by Linus Schrage (1968), and calculation of sojourn time for M/G/1 queues under SRPT is derived by Schrage and Miller (1966). Optimality of the $c\mu$ rule was proved by Klimov (1975). It is a special case of the Gittins index, first presented by Gittins and Jones (1974). Four alternative proofs of the optimality of the Gittins index are surveyed by Frostig and Weiss (2016). A complete treatment of the Gittins index is found in the book by Gittins, Glazebrook and Weber (2011).

Exercises

3.1 Prove that SPT minimizes the flowtime of a batch of N jobs scheduled on M parallel identical machines [Lawler et al. (1993)].

3.2 What is the non-preemptive schedule that will minimize the flowtime for a batch of N jobs, on parallel machines that are working at different speeds, $s_1, \ldots, s_M$ so that job j with processing requirement X_j, performed on machine i will need processing for a duration X_j/s_i [Lawler et al. (1993)].

3.3 The SEPT-savings parameter of a distribution measures the amount by which knowing the values of the job processing times and using SPT is better than scheduling them in random order. For jobs with processing time distribution G, it is defined as $d_G = \frac{m - m_{1:2}}{\sigma}$, where m is the mean, σ the standard deviation, and $m_{1:2}$ the mean of the smaller in a sample of 2, drawn from G. Calculate d_G for the following distributions:

    (i) Exponential.

    (ii) Normal.

    (iii) Uniform.

3.4 Consider the policy of preemptive LRPT (longest remaining processing time), for the G/G/1 queue. Describe how this policy works when you allow processor splitting, and show that it maximizes the number of customers in the system at all times.

3.5 Show that in a G/G/1 queue LCFS non-preemptive maximizes the variance of the sojourn and of the waiting time among all non-predictive, non-preemptive, work-conserving policies.

3.6 In a multi-type M/G/1 queue under priority scheduling policy, show that the long-term average amount of relevant work that a customer of type k finds on arrival, is given by (3.4).

3.7 Prove by induction the formula (3.5) for the waiting time of customers of type k under a priority scheduling policy.

3.8 Prove that among all priority scheduling policies for M/G/1 queues, SEPT minimizes the average flowtime.

3.9 Prove the formulas (3.6) and (3.7) for expected waiting times in the M/G/1 queue under SPT [Wolff (1989)].

3.10 Jobs arrive at a single-server station in a Poisson stream of rate $\lambda = 0.1$ jobs per unit time. The processing times of the jobs are i.i.d. uniformly distributed $\sim U(6, 12)$.

    (i) If no information about the processing time of the jobs is available to the server, calculate the average waiting time per job (in steady state).

(ii) If each arriving job can be classified into one of three types, $\sim U(6, 8)$, $\sim U(8, 10)$, or $\sim U(10, 12)$, and jobs are served according to non-preemptive SEPT, calculate the average waiting time of each type, and the overall average waiting time.

(iii) If the processing time of each job becomes known upon arrival, calculate the average waiting time of each job, and the overall average waiting time for non-preemptive SPT.

3.11 (∗) Derive the expected sojourn time of a customer with processing time x in a stationary M/G/1 queue under the SRPT policy, and the expected sojourn time over all customers [Schrage and Miller (1966)].

3.12 In an M/G/1 queue, under what conditions should one use preemptions or processor splitting to shorten the expected sojourn time?

3.13 Prove the optimality of Smith's rule for minimizing flow time of a batch of N deterministic jobs, using the analogs of the three proofs for optimality of SPT for flowtime.

3.14 Show that Smith's rule minimizes expected weighted flowtime for stochastic jobs, where priority is given to jobs with the smallest value of $\mathbb{E}(X_j)/\mathbb{E}(C_j)$ where (X_j, C_j), $j = 1, \ldots, N$ are N independent, two-dimensional random variables.

3.15 Show that the $c\mu$ rule minimizes expected weighted flowtime among all static priority policies for the M/G/1 queue with K customer types.

Part II

Approximations of the Single Queue

The single queue models of Part I could all be analyzed via discrete state Markov chains. This enabled us to obtain explicit expressions for the stationary distribution of the queue lengths, from which many performance measures could be calculated and the Markovian transition probabilities allowed evaluation of probabilities of transient events.

This is no longer the case for more general single queues. Without the assumption that the arrival process or the processing times are memoryless, questions of stability of the queue and the exact calculation of any performance measures become much harder. In this part, we consider the G/G/1 queue, discuss its stability, and describe methods to approximate its behavior.

In Chapter 4, we present the ingenious scheme devised by Loynes to show that the G/G/1 queue with stationary arrival and service processes is stable when the traffic intensity $\rho < 1$, and transient if $\rho > 1$. Under the stronger *renewal* assumption, that interarrivals as well as services are i.i.d. we explore the connection of the G/G/1 queue with the general random walk, and obtain an insightful upper bound on waiting times.

In Chapter 6, we discuss fluid and diffusion approximations to the G/G/1 queue, by scaling time and space. We also introduce the G/G/∞ queueing system, and study it under many-server scaling. The three types of scaling, fluid, diffusion, and many server, will form the backbone for Parts IV, V and VI, where we apply them to networks of queues.

Chapters 5 and 7 provide the probability tools used in the approximations, and will serve us throughout the subsequent text. In Chapter 5, we discuss weak convergence, the functional strong law of large numbers, and the central limit theorem. In Chapter 7, we introduce Brownian motion and related processes.

Part II

Approximations of the Single Dot

4

The G/G/1 Queue

In this chapter, we consider more general single-server queues. At the highest level of generality, when all we know is that arrivals and services are stationary, we can only hope to show that if traffic intensity is less than 1, then the queue is stable. This is shown by the beautiful Loynes scheme. For the more specific case of renewal arrivals and i.i.d. service times, there is a close connection between queues and random walks, which makes analysis possible, but the main conclusion is that all performance measures of the queue depend on the full shape of service and interarrival distributions. Some bounds on waiting time, which become exact approximations in heavy traffic, are obtained by a fresh look at Lindley's equation due to Kingman.

4.1 Loynes Construction for Stability of G/G/1 Queues

So far, we have seen that in all the single-server models that we considered (M/M/1, M/G/1, G/M/1), traffic intensity $\rho < 1$ implies stability, while $\rho > 1$ implies transience. The following construction due to Loynes shows that this is true for the most general possible model of G/G/1 queues, where all that is assumed is that the joint sequence of interarrival and service times is stationary.

This ingenious Loynes scheme, which involves forward and backward coupling, has since been used for several more complex models (cf. Borovkov and Foss (1992), Baccelli and Brémaud (2013)). We will return to it in Section 22.2. We describe the Loynes construction now.

Theorem 4.1 *Let* $\{T_n, X_n\}_{-\infty < n < \infty}$ *be a stationary sequence of non-negative random variables, with finite means* $a = \mathbb{E}(T_1)$, *and* $b = \mathbb{E}(X_1)$, *which satisfies the strong law of large numbers (SLLN), i.e.*

$$\lim_{n-m \to \infty} \frac{1}{n-m} \sum_{i=m}^{n} T_i = a, \qquad \lim_{n-m \to \infty} \frac{1}{n-m} \sum_{i=m}^{n} X_i = b, \quad a.s. \quad (4.1)$$

If $a < b$, then there exists a stationary sequence $\{V_n\}_{-\infty<n<\infty}$ that satisfies Lindley's equation:

$$V_{n+1} = \max(0, V_n + X_n - T_n). \tag{4.2}$$

Proof We can of course think of this as modeling a single-server queue with service time of customer n given by X_n, and interarrival time from the arrival of customer n to the arrival of customer $n+1$ given by T_n, and then V_n is the waiting time of customer n. Assume, to start with, that customer 0 arrives to an empty queue. Then $V_0 = 0$, and using Lindley's equation (4.2) we calculate recursively:

$$
\begin{aligned}
V_n &= \max(0, X_{n-1} - T_{n-1} + V_{n-1}) \\
&= \max(0, X_{n-1} - T_{n-1}, X_{n-1} - T_{n-1} + X_{n-2} - T_{n-2} + V_{n-2}) \\
&= \max(0, X_{n-1} - T_{n-1}, X_{n-1} - T_{n-1} + X_{n-2} - T_{n-2}, \\
&\qquad X_{n-1} - T_{n-1} + X_{n-2} - T_{n-2} + X_{n-3} - T_{n-3} + V_{n-3}) \\
&= \ldots,
\end{aligned}
$$

to get:

$$V_n = \max_{0\le j\le n} \sum_{i=j}^{n-1}(X_i - T_i). \tag{4.3}$$

Note that V_n coincides in distribution with

$$V_n \circ \theta^{-n} = \max_{-n\le j\le 0} \sum_{i=j}^{-1}(X_i - T_i), \tag{4.4}$$

which is the waiting time of customer 0, if customer $-n$ started with an empty queue. Here θ is the time shift operator, which preserves the distribution by the stationarity of the sequence $\{T_n, X_n\}_{-\infty<n<\infty}$.

The sequence $(V_n \circ \theta^{-n})_{n=0,1,\ldots}$ is non-decreasing with n, so it must converge to a finite limit or to infinity. By the SLLN and the assumption that $a < b$, the sums in the sequence of $V_n \circ \theta^{-n}$ will a.s. be negative from some point onward, and so the max will reach a finite value and stay there a.s., i.e. V_n will a.s. couple with a finite limit,

$$V^0 = \sup_{j\le 0} \sum_{i=j}^{-1}(X_i - T_i).$$

One can similarly define, for any $-\infty < m < \infty$,

$$V^m = V^0 \circ \theta^m = \sup_{j\le m} \sum_{i=j}^{m-1}(X_i - T_i).$$

Then $\{V^m\}_{-\infty < m < \infty}$ is a stationary sequence of random variables, and

$$V^{m+1} = \max(0, V^m + X_m - T_m).$$

Thus, we have constructed a stationary solution to the recursive Lindley's equation. □

The beauty of the Loynes construction is that while the forward expression for V_n in (4.3) varies wildly with n, the backward expression for V^n in (4.4) is monotone in n.

We can show a stronger property of the stationary solution constructed above:

Theorem 4.2 *The sequence V^m constructed by the Loynes scheme in Theorem 4.1 is almost surely the unique stationary solution to Lindley's equation. Furthermore, starting from $v_0 = y \geq 0$, the sequence of waiting times v_n obtained from Lindley's equation will couple with the Loynes construction solution after a finite time.*

Proof First we show that if y^n is a sequence that satisfies Lindley's equation, then it must be a.s. $\geq V^n$. To show this, note that because $y^0 \geq 0 = V_0$, we have by monotonicity of Lindley's equation that $y^m > V_m$, for all $m > 0$. Using the same monotonicity when starting from $y^{-m} \geq 0$, and the fact that the y^n satisfy Lindley's equation, we have for all $m > 0$:

$$y^0 = \max\left(y^{-m} + \sum_{i=-m}^{-1}(X_i - T_i), \max_{-m+1 \leq j \leq 0} \sum_{i=j}^{-1}(X_i - T_i)\right)$$

$$\geq \max_{-m \leq j \leq 0} \sum_{i=j}^{-1}(X_i - T_i),$$

and letting $m \to \infty$, we get

$$y^0 \geq \max_{j \leq 0} \sum_{i=j}^{-1}(X_i - T_i) = V^0.$$

By the same argument, $y^n \geq V^n$ for all n.

Next one can see that if y^n is a sequence that satisfies Lindley's equation then it must be zero for some times m. Otherwise, if $y^k > 0$, $k = 1, \ldots, m$, then $y^k = y^0 + \sum_{j=1}^{k}(X_j - T_j) > 0$, $k = 1, \ldots, m$, and by $a > b$ this cannot be true for more than a finite m.

But $y^m = 0$ means that $V^m = y^m$, and by Lindley's equation they couple and remain equal for all times after m. This argument holds for some $m > n$

where n can be arbitrary. Let $n \to -\infty$ and we have that y^m coincides with V^m a.s.

Next, we show that starting with arbitrary V_0, the waiting times V_n will couple with V^n, for almost all paths of $\{T_n, X_n\}_{-\infty < n < \infty}$. This is because almost surely for some finite m, $V_m = 0$, and so at that time $V^m \geq V_m$, and then V^n and V_n will couple at the first time $n > m$ for which $V^n = 0$. In particular, this implies that $V_m \to V^m$ in distribution, as $m \to \infty$.   □

Note that the Loynes construction is a pathwise construction: For every $\omega \in \Omega$ such that $\frac{1}{n-m} \sum_{i=m}^{n} T_i(\omega) \to a$, $\frac{1}{n-m} \sum_{i=m}^{n} X_i(\omega) \to b$, the construction will give a unique sequence $V^m(\omega)$ that satisfies Lindley's equation, and any sequence starting from $v_0 = y$ will couple with this sequence.

We list more properties of the G/G/1 queue under the assumptions of Theorem 4.1. With the Loynes construction, we also obtain the process $\mathcal{W}(t)$ of the stationary workload. Furthermore, an arrival finds an empty queue infinitely often, and one can define the sequence of times B_n when an arrival finds an empty queue. These are the ends of idle periods. We then have a sequence of times C_n between B_n and B_{n+1} at which the workload reaches 0, when the queue becomes empty. These are the ends of busy periods. With $B_0 \leq 0 < B_1$, this defines a stationary sequence of busy times, idle times, and cycle times. Finally, the queue length process $Q(t)$ of number of customers in the system also has a unique stationary version if $a > b$.

If $a < b$, then the construction yields $V^0 = \infty$ a.s., and so there is no stationary solution. This general model does not provide enough information to decide on the behavior when $a = b$. One can construct examples where a stationary solution exists, though for most models that we will consider, no stationary solution exists.

Recall the definition of stability of a queue, Definition 2.6. We state:

Corollary 4.3   *The G/G/1 queue with stationary interarrivals and service times with rates λ, μ is stable if $\rho = \lambda/\mu < 1$ and is not stable if $\rho > 1$.*

4.2 G/G/1 and the Random Walk

We now consider the G/G/1 queue, where we have more information about the primitives of the system. Unless stated differently, we will from now on assume that in the G/G/1 queue interarrivals between customers $n-1$ and n are i.i.d. $T_n \sim F$ with rate λ, services for customer n are i.i.d. $X_n \sim G$ with rate μ, and services and arrivals are independent. Let $Y_n = X_n - T_{n+1}$, $n =$

$1, 2, \ldots$; it is an i.i.d. sequence. We again use Lindley's equation to derive the waiting times, starting from an empty system at $t = 0$:

$$V_{n+1} = \max\{0, Y_n + V_n\}$$
$$= \max\{0, Y_n, Y_n + Y_{n-1}, \ldots, Y_n + \cdots + Y_1\}.$$

With the additional assumptions of G/G/1 we now have immediately:

Proposition 4.4 (Duality result) *From $Y_1, \ldots, Y_n$ i.i.d. it follows that $(Y_1, Y_2, \ldots, Y_n)$ and $(Y_n, Y_{n-1}, \ldots, Y_1)$ have the same joint distribution. This implies that*

$$\mathbb{P}(V_{n+1} > x) = \mathbb{P}(\max\{0, Y_n, Y_n + Y_{n-1}, \ldots, Y_n + \cdots + Y_1\} > x)$$
$$= \mathbb{P}(\max\{0, Y_1, Y_1 + Y_2, \ldots, Y_1 + \cdots + Y_n\} > x). \quad (4.5)$$

Note that we again use a forward and a backward construction of V_n, as in the Loynes construction.

At this point we introduce the concept of a random walk:

Definition 4.5 Let $Y_1, Y_2, \ldots$, be a sequence of i.i.d. random variables. The sequence $S_0 = 0$, $S_n = S_{n-1} + Y_n$, $n = 1, 2, \ldots$ is called the random walk of the sequence Y_n.

What we see from (4.5), and using the results of the Loynes construction (Section 4.1), is that the stationary distribution of the waiting time in a G/G/1 queueing system is the same as the distribution of the maximum of the random walk, if that exists:

Theorem 4.6 *For the G/G/1 queue starting empty at $t = 0$, with $Y_n = X_n - T_{n+1}$, the waiting time V_n of customer n converges in distribution as $n \to \infty$ to the distribution of the maximum of the random walk of the Y_n, if that exists.*

There is a rich theory of random walks; the following theorem is well known:

Theorem 4.7 (One-dimensional random walk) *For a random walk, one of the following four possibilities holds: (i) $S_n = 0$ for all n a.s., (ii) $S_n \to \infty$ as $n \to \infty$ a.s., (iii) $S_n \to -\infty$ as $n \to \infty$ a.s., (iv) $\limsup S_n = \infty$ and $\liminf S_n = -\infty$ a.s.*
Specifically, if Y_1 has a finite expectation then:
- if $\mathbb{E}(Y_1) > 0$ then (ii), $S_n \to \infty$ a.s.,
- if $\mathbb{E}(Y_1) < 0$ then (iii), $S_n \to \infty$ a.s.,

- *if* $\mathbb{E}(Y_1) = 0$ *and* $\mathbb{P}(Y_1 = 0) < 1$ *then (iv),* $\limsup S_n = \infty$ *and* $\liminf S_n = -\infty$ *a.s.*

In particular, in case (iii), when $S_n \to -\infty$ *a.s., the maximum of the random walk* $M = \max\{S_n, n = 0, 1, 2, \ldots\}$ *is well defined and* $\mathbb{P}(M < \infty) = 1$.

Remark   Clearly, in case (iv), which is called the null recurrent case, the random walk will switch from positive to negative infinitely often. It can be shown that in case (iv), the expected length of time that it stays positive (negative) is infinite. See Exercise 4.6 for a simple case.

Theorem 4.8   *The G/G/1 queue is stable if* $\mu > \lambda$. *If* $\mu < \lambda$, *then the queue will diverge, i.e. a.s. for every sample path, for any* $x > 0$ *there will exist* n_x *such that* $V_n > x$ *for all* $n > n_x$. *If* $\mu = \lambda$, *then the queue will not be stable, but it will be empty infinitely often.*

Proof   The first and second parts hold for the more general G/G/1 queue as we saw in Section 4.1. If $\mu = \lambda$, the random walk will change infinitely often between positive and negative. Let $N = \min\{n : n > 0, S_n \leq 0\}$, then $\mathbb{P}(N < \infty) = 1$. But then, for the queue starting from empty, N is exactly the random number of customers until the queue is empty. So the queue returns to empty after a finite time.   □

For the stable G/G/1 queue, by the independence of V_n and Y_n, we obtain from $V_{n+1} = (Y_n + V_n)^+$ the following relation for the distribution functions:

$$F_{V_{n+1}}(x) = \int_{0-}^{\infty} F_{Y_n}(x - y) dF_{V_n}(y),$$

(here the lower bound $0-$ is for $\mathbb{P}(V_n = 0) > 0$, i.e. the distribution has an atom at 0). When we let $n \to \infty$, we get *Lindley's integral equation*:

$$F_V(x) = \int_{0-}^{\infty} F_Y(x - y) dF_V(y), \tag{4.6}$$

where F_V, F_Y are the distributions of the stationary waiting time V_∞, and of $Y_n = X_n - T_{n+1}$. This looks almost like a convolution, except that the integral is not over $(-\infty, \infty)$ but only over $(-0, \infty)$, and $F_V(x) = 0$ for $x < 0$.

Solution of the equation (4.6) for the distribution of the maximum of a random walk with negative drift is done by Wiener–Hopf decomposition, but is far from simple. In the queueing context, it is somewhat easier since $Y_n = X_n - T_{n+1}$ is the difference of two independent non-negative random variables. As a result, if the service time distribution has a rational Laplace transform, one can get expressions for the Laplace transform of the waiting time distribution.

The stationary work in the system is given by:

$$W(\infty) \overset{D}{=} (V_\infty + X_n - T_{eq})^+, \tag{4.7}$$

where the three variables are independent, and T_{eq} has the equilibrium distribution of the interarrival times (see Exericse 4.7). Note, this is the long time average, and because PASTA does not hold, it is not what a customer sees on arrival, or leaves behind at departure.

4.3 Bounds for G/G/1 Waiting Times

The results of the previous two sections were derived under the assumption that interarrival times and service times have finite means, but no further moments were required for the existence of stationary distributions of the queue length and the waiting time. However, as we have already seen for the M/G/1 queue (Section 2.7), existence of finite mean waiting time and finite mean queue length required the existence of the second moment of the service time. We will in this section assume, for the G/G/1 queue, that both interarrival times and service times have finite second moments. We denote the variance of the interarrival times by σ_a^2, and the variance of the service times by σ_s^2, with coefficients of variation $c_a = \lambda \sigma_a$, $c_s = \mu \sigma_s$.

As we stated in the previous section, the expected waiting time in a G/G/1 queue is hard to calculate. Furthermore, to calculate it one needs full information on the distribution of $Y_n = X_n - T_{n+1}$, which is usually not available in practice. The following bound, which uses only first and second moment information on X_n and T_n, was derived by Kingman, who took a fresh look at Lindley's equation.

Consider the stationary G/G/1 queue (with $\rho < 1$). As we saw (Lindley's equation), $V_{n+1} = (V_n + Y_n)^+$. Define

$$I_n = (V_n + Y_n)^- = -\min(0, V_n + Y_n). \tag{4.8}$$

Then I_n is the idle period generated between the n'th customer's departure, and the $n+1$'st customer's arrival (which takes the value 0, if customer $n+1$ needs to wait). Note that the process regenerates at the $n+1$'st arrival if $I_n > 0$. We have, by their definitions,

$$V_{n+1} - I_n = V_n + Y_n \quad \text{and} \quad V_{n+1} I_n = 0. \tag{4.9}$$

Taking expectations of the two sides of the first equality in (4.9), for the stationary queue

$$\mathbb{E}(I_n) = -\mathbb{E}(Y_n) = \frac{1}{\lambda} - \frac{1}{\mu} = \frac{1}{\lambda}(1 - \rho).$$

Squaring both sides of the first equality of (4.9), taking expectations, and using the second part of (4.9), we get after some manipulations,

$$\overline{V} = \mathbb{E}(V_n) = \frac{\mathbb{E}(Y_n^2) - \mathbb{E}(I_n^2)}{-2\mathbb{E}(Y_n)} = \frac{\lambda}{2(1 - \rho)}(\sigma_a^2 + \sigma_s^2 - \mathbb{V}\mathrm{ar}(I_n)).$$

The unknown part here is $\mathbb{V}\mathrm{ar}(I_n)$. It can possibly be estimated directly from the data, with greater ease than $\overline{V}$. In any case, it is positive, so we have an upper bound:

$$\overline{V} \le \frac{\lambda}{2(1 - \rho)}(\sigma_a^2 + \sigma_s^2). \tag{4.10}$$

A lower bound that is half the size of the upper bound can also be derived, and the bounds can be improved in several ways. The lower bound may be close to reality in light traffic. In heavy traffic, since I_n, the idle period between the n'th and $n + 1$'st arrival, is mostly 0, the variance of I_n is negligible, and so, in heavy traffic, i.e. when $\rho \approx 1$ (but still $\rho < 1$),

$$\overline{V} \approx \frac{\lambda}{2(1 - \rho)}(\sigma_a^2 + \sigma_s^2) \approx \frac{1}{1 - \rho}m_s\frac{c_a^2 + c_s^2}{2}, \tag{4.11}$$

where m_s, c_s^2, c_a^2 are the mean service time and the squared coefficients of variation of service time and interarrival time. Note that this expression looks like a generalization of the Pollaczek–Khinchine formula, with c_a^2 replacing the value 1, which is the squared coefficient of variation for Poisson interarrivals; usually we would expect to have $c_a^2 < 1$.

By Little's formula, we also obtain a heavy traffic approximation for the average number of customers in the queue in heavy traffic:

$$\overline{Q} \approx \frac{1}{1 - \rho}\frac{c_a^2 + c_s^2}{2}. \tag{4.12}$$

4.4 Sources

This chapter is based on papers by three of the giants of the early age of applied probability. Loynes construction is in his paper Loynes (1962). The solution of the distribution of waiting times for G/G/1, by the Wiener–Hopf decomposition, when service times have rational Laplace transforms, is given by Smith (1953), and the Kingman bound is derived in his paper Kingman (1962). At this point, the reader may want to consult some more advanced queueing texts, two of which are *Applied Probability and Queues,*

Asmussen (2003) and *Elements of Queueing Theory: Palm Martingale Calculus and Stochastic Recurrences,* Baccelli and Brémaud (2013). Finally, *The Single Server Queue,* J.W. Cohen (2012) is a perfect guide to a host of derivations by analytic methods.

Exercises

4.1 Explain the difference between the sequence $v_n = \max_{0 \le j \le n} \sum_{i=j}^{n-1} (X_i - T_i)$ and the sequence: $v_n \circ \theta^{-n} = \max_{-n \le j \le 0} \sum_{i=j}^{-1} (X_i - T_i), n = 0, 1, 2, \ldots ,$ which is used in the Loynes construction.

4.2 Explain why the sequence V^n is stationary, and verify that it satisfies Lindley's equation.

4.3 (∗) Use a Loynes-type construction to show that the waiting time V_n process for a G/G/*s* queueing system is stable if $\rho = \lambda/s\mu < 1$ and unstable if $\rho > 1$. As before, assume that (T_n, X_n) are a stationary sequence satisfying SLLN. Use the recursion relation analogous to Lindley's equation for V_n (see Exercise 1.2), [Loynes (1962); Kiefer and Wolfowitz (1955); Brandt et al. (1990) page 165].

4.4 Show that the sequence of waiting time vectors for G/G/*s*, obtained by the Loynes construction, is minimal in some sense. What else is needed to show that it is unique?

4.5 Consider the following sequences: $a_n = 1, b_{2n} = 2, b_{2n+1} = 3/2, -\infty < n < \infty$. Let a G/G/2 system have interarrivals $\{T_n = a_n\}_{-\infty<n<\infty}$, and sequence of service times with probability $\frac{1}{2}$ given by $\{X_n = b_n\}_{-\infty<n<\infty}$, and with probability $\frac{1}{2}$ given by $\{X_n = b_{n+1}\}_{-\infty<n<\infty}$. Show that the stationary sequence of waiting times for this system is not unique. Explain why.

4.6 Let $Y_j = 1$ or $Y_j = -1$ with equal probabilities, and $S_0 = 0, S_n = Y_1 + \cdots + Y_n$. S_n is the simple symmetric random walk. The path of the random walk is at (t, k) if it is in position k at time t, i.e. $S_t = k$.

    (i) Let $N_{t,k}$ be the number of paths that reach k in t steps, and $p_{t,k} = \mathbb{P}(S_t = k)$. Calculate $N_{t,k}$ and $p_{t,k}$.

    (ii) Prove the reflection principle: For $(t_0, k_0), (t_1, k_1)$, with $k_0 > 0, k_1 > 0$, the number of paths from (t_0, k_0) to (t_1, k_1) that cross the time axis, equals the total number of paths from $(t_0, -k_0)$ to (t_1, k_1).

    (iii) Prove the ballot theorem: For $a > b$, of all the paths with a positive and b negative steps, show that the probability of paths where the number of positives exceeds negatives at all times is $\frac{a-b}{a+b}$.

    (iv) Calculate $u_{2n} = \mathbb{P}(S_{2n} = 0)$, the probability to return to 0 at time $2n$.

    (v) Show that the probability to return to 0 for the first time at time $2n$ is:

$$f_{2n} = \mathbb{P}(S_{2n} = 0, S_j \ne 0, j = 1, \ldots, 2n - 1)$$

$$= u_{2n-2} - u_{2n} = \frac{1}{2n-1} u_{2n} = \frac{1}{2n-1} \binom{2n}{n} \frac{1}{2^{2n}}.$$

   (vi) Prove that the simple symmetric random walk returns to 0 infinitely often.

   (vii) Show that the expected time to return to 0 is infinite (use Stirling's formula).

4.7   Prove the formula (4.7) for the distribution of the stationary workload.

4.8   Find the average waiting time for a D/M/1 queue and compare it to the Kingman bound.

4.9   Find the average waiting time for a G/M/1 queue and compare it to the Kingman bound.

5

The Basic Probability Functional Limit Theorems

To study approximations of the queueing process, we need some background on theorems that govern probability for sample paths of stochastic processes, i.e. probability in function spaces. We start with the definition of convergence of stochastic processes. We then discuss functional strong law of large numbers and functional central limit theorem for random walks. Finally we use these to derive the functional strong law of large numbers and the functional central limit theorem for renewal processes. Recall that arrivals and service in G/G/1 are often assumed to be renewal processes, so this is relevant for our purpose. We also discuss the alternative approach of strong approximation.

5.1 Convergence of Stochastic Processes

We consider stochastic processes with sample paths almost surely in the space $\mathbb{D}$ of right continuous with left limit (RCLL) functions, or in the space $\mathbb{C}$ of continuous functions. In our context, sample paths of the arrival, departure, queue length, and workload processes are all in $\mathbb{D}$. Limits of scaled versions of these processes however, are often in $\mathbb{C}$.

In $\mathbb{C}$ the topology is of uniform convergence on compacts (u.o.c.):

$$x_n \to x \text{ u.o.c.} \quad \text{if} \quad \sup_{0 \le s \le t} |x_n(s) - x(s)| \to 0, \text{ for all } t > 0.$$

In $\mathbb{D}$ the topology is determined by the Skorohod metric, which essentially says that two functions x and y are "close" not only if $x(t) - y(t)$ is small but also if $x(t) - y(\zeta(t))$ is small, where ζ is a "small" deformation of time. For instance, under the Skorohod topology,

$$y_n = \begin{cases} 0 & x < \frac{1}{n} \\ 1 & x \ge \frac{1}{n} \end{cases} \quad \to_d \quad y = \begin{cases} 0 & x < 0 \\ 1 & x \ge 0 \end{cases},$$

where $\to_d$ is convergence in the Skorohod metric. In the widely occurring special case that $x_n \in \mathbb{D}$ and $x \in \mathbb{C}$, $x_n \to_d x$ is equivalent to $x_n \to x$ u.o.c.

71

Having defined convergence in the function spaces $\mathbb{C}$, $\mathbb{D}$, we now define two types of convergence for stochastic processes that have paths that are a.s. in $\mathbb{D}$.

Definition 5.1   We say that $Z_n \to Z$ almost surely if $Z_n(t, \omega) \to_d Z(t, \omega)$ for almost all sample paths (i.e all ω except a set of measure 0). If paths of Z are a.s. continuous, it is enough to check that $Z_n(t, \omega) \to Z(t, \omega)$ u.o.c.

Definition 5.2   We say that Z_n converges weakly (converges in distribution) to Z, $Z_n \to_w Z$, if we can construct a probability space that contains copies (with the same distribution) Z_n' of each Z_n, and a copy Z' of Z, such that $Z_n' \to_{a.s.} Z'$.

We interpret this condition a little later. It is difficult to construct these copies in a way that one can show that they converge. Such constructions are usually not necessary since equivalent conditions exist.

Theorem 5.3 $Z_n \to_w Z$ *if and only if for all bounded continuous functionals* h *(mapping* $\mathbb{D}$ *to the real line),* $\mathbb{E}(h(Z_n)) \to \mathbb{E}(h(Z))$.

The difficulty here is to characterize all continuous functionals.

Theorem 5.4 $Z_n \to_w Z$ *if and only if all the finite-dimensional joint distributions of* Z_n *converge in distribution to the finite-dimensional distributions of* Z, *and* Z_n *is tight.* Z_n *is tight if for all* $\epsilon > 0$ *there exists a compact (in the Skorohod topology) set* $K_\epsilon \subseteq \mathbb{D}$ *such that* $P\{Z_n \in K_\epsilon\} > 1 - \epsilon$ *for all* n.

Theorem 5.4 is the more useful way of verifying convergence: The convergence of finite-dimensional distributions takes us back to standard probability tools for random variables. The harder part is verifying tightness. It is done by considering the continuity properties of the sample paths.

We will also make use of the following ideas, which enable us to infer convergence of a sequence of processes from the known convergence of simpler sequences:

Theorem 5.5 (Continuous mapping principle)   *If* h *is a mapping,* $h : \mathbb{D} \to \mathbb{D}$, *which is continuous (in the Skorohod metric), then* $Z_n \to_w Z$ *implies* $h(Z_n) \to_w h(Z)$.

Theorem 5.6 (Time change lemma)   *Let* $Z_n(\omega)$, $\Phi_n(\omega) \in \mathbb{D}$, Φ_n *non-decreasing, and assume* $(Z_n, \Phi_n) \to_w (Z, \Phi)$, *in* $\mathbb{D} \times \mathbb{D}$. *Then* $Z_n \circ \Phi_n \to_w Z \circ \Phi$, *where* $Z \circ \Phi(t, \omega) = Z(\Phi(t, \omega), \omega)$.

Theorem 5.7 (Convergence together)   *If $Z_n \to_w Z$, and $Z_n - Z'_n \to_w 0$, then $Z'_n \to_w Z$.*

5.1.1 Discussion: The Difference between Weak and a.s. Convergence

For $Z_n \to Z$ a.s., all Z_n and Z are defined on an original probability space, with whatever relations exist between them. In particular, the joint distribution of $Z, Z_{n_1}, \ldots, Z_{n_k}$, for any subset $(n_1, \ldots, n_k)$, is prescribed, and determines jointly the sample paths for all n and for the limit. For $Z_n \to_w Z$ we make copies Z' of Z, and Z'_n of Z_n, with the same distributions, but we do not preserve the joint distribution of $Z, Z_{n_1}, \ldots, Z_{n_k}$ in the joint distribution of $Z', Z'_{n_1}, \ldots, Z'_{n_k}$. So sometimes we can make this construction such that $Z'_n(\omega) \to Z'(\omega)$ a.s. in this new probability space, but $Z_n \to Z$ a.s. does not hold. The actual construction of copies of a sequence of stochastic processes and their weak limit in a new probability space is done by defining several sequences of independent stochastic processes, using the so-called Prohorov construction.

Analogous to the Prohorov construction for sequences of stochastic processes is the following construction for a sequence of random variables. A sequence of random variables X_n with distributions F_n is said to converge in distribution (weak convergence) to X with distribution F if $\lim_{n \to \infty} F_n(t) = F(t)$ at all continuity points of F. The construction is to take a single uniform random variable U, and to construct copies of X_n, X by defining $X'_n = F_n^{-1}(U)$, $X' = F^{-1}(U)$. Then $F_n(t) \to F(t)$ at all continuity points of F if and only if $X'_n \to X'$ a.s. The Prohorov construction is based on a similar idea; however, it is more complicated.

5.2 Functional Limit Theorems for Random Walks

Let X_j, $j = 1, 2, \ldots$, be i.i.d. random variables with mean m and variance σ^2 (when they exist). Let $S_0 = 0$, $S_n = \sum_{j=1}^n X_j$ be the random walk defined by X_j. Define the continuous time stochastic process:

$$S(t) = S_{\lfloor t \rfloor} = \sum_{j=1}^{\lfloor t \rfloor} X_j, \, t \geq 0,$$

where $S(t)$ is simply the piecewise constant process in $\mathbb{D}$ (right continuous with left limits) that coincides with S_n at $t = n$.

The two basic limit laws of probability theory, the strong law of large numbers and the central limit theorem which are formulated for S_n, can be formulated as functional limit theorems for $S(t)$. We consider first the sequence of scaled processes:

$$\bar{S}^n(t) = \frac{1}{n}S(nt) = \frac{1}{n}\sum_{j=1}^{\lfloor nt \rfloor} X_j.$$

Theorem 5.8 (The functional strong law of large numbers – FSLLN)   *If* $\mathbb{E}(X_1) = m < \infty$, *then*

$$\bar{S}^n(t) = \frac{1}{n}S(nt) = \frac{S_{\lfloor nt \rfloor}}{n} \xrightarrow[u.o.c.]{a.s.} mt.$$

The FSLLN is illustrated in Figure 5.1. Here $S(t)$ is piecewise constant, and is changing by random jumps distributed like X_j at time intervals of 1, while $\bar{S}^n(t)$ is piecewise constant, and is changing by random jumps distributed like X_j/n a total of n times in every time unit. Thus, $\bar{S}^n(t) = \frac{1}{n}S(nt)$ is the original process $S^1(t)$ scaled by a factor of n both in time and in space. The figure shows the beginning of $S(t)$ and of $\frac{1}{4}S(4t)$, the latter getting closer to mt, which is the almost sure *fluid limit*. The proof follows directly from the strong law of large numbers for a sequence of random variables.

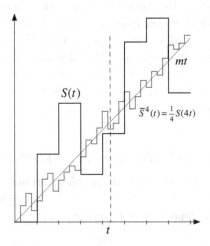

Figure 5.1 FSLLN convergence for random walk.

Proof of the FSLLN   By the SLLN for the sequence X_j for almost every sample path, $\frac{1}{n}\sum_{j=1}^{n}X_n \to m$. Consider a single sample path for which convergence holds. Then for any $\epsilon > 0$ there exists n_0 so that $\left|\frac{1}{n}\sum_{j=1}^{n}(X_j - m)\right| < \epsilon$ for all $n > n_0$. We then have:

$$\left|\frac{1}{n}\sum_{j=1}^{\lfloor nt \rfloor}(X_j - m)\right| \leq \begin{cases} \epsilon t, & \text{if } nt > n_0, \\ \dfrac{n_0}{n}max_{j=1,...,n_0}|X_j - m|, & \text{if } nt \leq n_0. \end{cases}$$

For a compact interval $0 < t < T$, the first expression can be made arbitrarily small by choosing ϵ and n_0. Clearly, $max_{j=1,...,n_0}|X_j - m|/n \to 0$ as $n \to \infty$ a.s. for every sample path. So the second expression can be made arbitrarily small for all $n > n_1$ by choosing n_1/n_0 large enough. This proves the convergence to mt u.o.c. a.s.   □

Next we consider the sequence:

$$\hat{S}^n(t) = \sqrt{n}\left(\bar{S}^n(t) - mt\right).$$

Here we consider the deviations of $\bar{S}^n(t)$ from mt, and we rescale them by multiplying with $\sqrt{n}$, so time is still measured in units of n, but the size of the deviations is measured in units of $\sqrt{n}$.

Theorem 5.9 (The functional central limit theorem – FCLT, Donsker's theorem)   *If* $\mathbb{V}ar(X_1) = \sigma^2 < \infty$, *then:*

$$\hat{S}^n(t) = \sqrt{n}(\bar{S}^n(t) - mt) \to_w \sigma BM(t),$$

where BM is a (standard) Brownian motion.

Figure 5.2 illustrates Donsker's theorem: We will discuss Brownian motion in Section 7.1. We note here just that $BM(t)$ is a Gaussian random

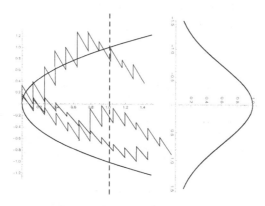

Figure 5.2 FCLT convergence of random walk paths.

variable with mean 0 and variance t. In Figure 5.2 we see some sample paths of $\hat{S}^n(t)$, and the curve of $\pm\sigma\sqrt{t}$, which is the standard deviation of $\hat{S}^n(t)$. For large enough n, at time t, the distribution of $\hat{S}^n(t)$ is close to the plotted Gaussian distribution. The theorem says that for large enough n the distribution of these sample paths is close to the distribution of paths of $BM(t)$.

In practical terms, convergence in distributions is all we need to approximate a random walk observed over a long period, since the expected value of any performance measure, by Theorem 5.3, will be close to that of the Brownian motion process.

5.2.1 Discussion: Why Is CLT Only Weak Convergence?

We discuss SLLN and CLT for a sequence of i.i.d. random variables, normalized to have mean 0 and variance 1. SLLN states that $S_n/n \to_p 0$ as well as $S_n/n \to_{a.s.} 0$, while *CLT* says that $S_n/\sqrt{n} \to_w Z$, $Z \sim N(0,1)$ (convergence in distribution, or weak convergence). The law of the iterated logarithm (see Figure 5.3) is somewhat in between, it states

$$\limsup_{n\to\infty} \frac{S_n}{\sqrt{n\log\log n}} = \sqrt{2}, \ a.s.,$$

and by considering $-X_j$, also $\liminf_{n\to\infty} \frac{S_n}{\sqrt{n\log\log n}} =_{a.s.} -\sqrt{2}$. This implies that

$$\limsup_{n\to\infty} S_n/\sqrt{n} = \infty, \qquad \liminf_{n\to\infty} S_n/\sqrt{n} = -\infty, \qquad a.s.$$

So almost surely $S_n/\sqrt{n}$ does not converge (for almost every ω it diverges).

We can also see that it does not converge in probability: Assume to the contrary that for some random variable Z, $\mathbb{P}(|\frac{S_n}{\sqrt{n}} - Z| > \epsilon) \to 0$ as $n \to \infty$. Then it follows that $\mathbb{P}(|\frac{S_n}{\sqrt{n}} - \frac{S_{2n}}{\sqrt{2n}}| > \epsilon) \to 0$. But:

$$\frac{S_{2n}}{\sqrt{2n}} = \frac{1}{\sqrt{2}}\left(\frac{S_{2n}-S_n}{\sqrt{n}} + \frac{S_n}{\sqrt{n}}\right),$$

and $\frac{S_{2n}-S_n}{\sqrt{n}}$ and $\frac{S_n}{\sqrt{n}}$ are two sequences of i.i.d. random variables that are also pairwise independent. Each of these sequences converges in distribution to $N(0,1)$, which contradicts that some linear combination of them converges to 0 in probability.

Figure 5.3, by Dean P. Foster, illustrates the SLLN and the law of the iterated logarithm (LIL). It traces a single sample path of sample averages, $\bar{X}_n = S_n/n$, for a sequence of X_j that take values $+1$ and -1 with probability

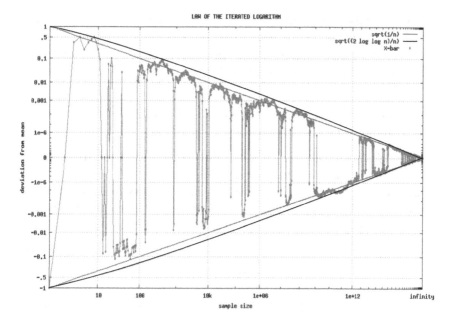

Figure 5.3 Simulation to illustrate the law of iterated logarithms, reproduced from Wikipedia, original be Dean P. Foster.

$1/2$ each (that is, S_n is the simple symmetric random walk). Also shown are $\sqrt{1/n}$, which is the standard deviation of S_n/n, and $\sqrt{(2\log\log n)/n}$, which, as seen in the figure, is where the supremum and infimum visit alternatingly.

The law of the iterated logarithm provides the right scaling to distinguish between the three modes of convergence: Clearly $\frac{S_n}{\sqrt{n\log\log n}} \not\to_{a.s} 0$, but it can be shown that $\frac{S_n}{\sqrt{n\log\log n}} \to_p 0$ holds.

5.2.2 On the Proof of Donsker's Theorem and Skorohod Embedding

While the FSLLN follows directly from the SLLN, Donsker's theorem does not follow directly from the CLT. The standard proof of Donsker's theorem is based on Theorem 5.4: (i) Show that the finite-dimensional distributions of the random walk converge to those of BM, which follows directly from CLT. (ii) Show tightness, which is done by considering the modulus of continuity of the process $\hat{S}^n(t) = \frac{S(nt)-mnt}{\sqrt{n}}$.

A completely different approach is to use Skorohod embedding. The following theorem is by Skorohod:

Theorem 5.10 (Skorohod embedding)   *Let X be a random variable with $\mathbb{E}(X) = 0$, $\mathbb{V}ar(X) < \infty$. Let $BM(t)$ be a standard Brownian motion. Then there exists a stopping time T such that: (i) $BM(T)$ is distributed as X, (ii) $\mathbb{E}(T) = \mathbb{V}ar(X)$, and (iii) if $\mathbb{E}(X^4) < \infty$, then one can also have $\mathbb{V}ar(T) < \infty$.*

We discuss this theorem in greater detail in Exercises 7.12–7.19. We now use it to prove Donsker's theorem.

Sketch of proof of Donsker's theorem   Let X be distributed like the X_j, and assume mean 0 and variance 1. Assume also that $E(X^4) < \infty$. Let T be the corresponding stopping time for a Brownian motion as stated in the Skorohod embedding theorem. For a given Brownian motion $BM(t)$, we construct a sequence of i.i.d. stopping times τ_j distributed like T so that

$$\mathbb{P}(S_n \leq x) = \mathbb{P}\Big(BM\Big(\sum_{j=1}^{n} \tau_j\Big) \leq x\Big), \qquad n = 1, 2, \dots .$$

Let $S'_n = BM\left(\sum_{j=1}^{n} \tau_j\right)$ and $t'_n = BM(n)$; then, by $\mathbb{E}(\tau_j) = 1$, $\mathbb{V}ar(\tau_j) < \infty$, and by CLT, $\sum_{j=1}^{n} \tau_j - n = O(n^{1/2})$, and so:

$$|S'_n - t'_n| = \Big|BM\Big(\sum_{j=1}^{n} \tau_j\Big) - BM(n)\Big| = O(n^{1/4}),$$

because $BM\left(\sum_{j=1}^{n} \tau_j\right) - BM(n)$ is a Gaussian random variable with mean 0, and conditional variance $\sum_{j=1}^{n} \tau_j - n = O(n^{1/2})$, and standard deviation $O(n^{1/4})$.

When time is scaled by a factor of n and space by a factor of $\sqrt{n}$, the convergence follows.

In fact, $\mathbb{V}ar(X) < \infty$ is all that is needed, and the proof can be strengthened to hold under this weaker condition.   □

5.2.3 Strong Approximations

The FSLLN for random walks holds whenever X_j have a finite expectation. The FCLT for random walks holds whenever X_j have a finite variance. If we require the slightly more stringent condition that $\mathbb{E}(X_j^r)$ exists and is finite for some $r > 2$, then a more informative limit theorem, called *strong approximation*, holds.

Komlos, Major, and Tusnady construct an embedding that is closer than the Skorohod embedding, and show the following:

Theorem 5.11 (Strong approximation theorem)   *Assume X_j, $j = 1, 2, \ldots$ are i.i.d. with mean m and variance σ^2, and assume in addition that $\mathbb{E}(|X_j|^r) < \infty$, for some $r > 2$. Then one can construct a probability space with a copy $S'(t)$ of $S(t) = \sum_{j=1}^{\lfloor t \rfloor} X_j$ and a $BM(t)$ so that:*

$$\sup_{0 \le t \le T} |S'(t) - mt - \sigma\, BM(t)| =_{a.s} o(T^{1/r}) \quad \text{as } T \to \infty. \tag{5.1}$$

If $\mathbb{E}(e^{aX_j})$ also exists for all a in a neighborhood of zero (a condition that implies but is in fact stronger than the existence of moments of all orders), then

$$\sup_{0 \le t \le T} |S'(t) - mt - \sigma BM(t)| =_{a.s} O(\log(T)) \quad \text{as } T \to \infty.$$

It is seen immediately that Theorem 5.11 implies $\bar{S}^n(t) \to_{u.o.c.}^{a.s.} mt$ and $\hat{S}^n(t) = \sqrt{n}(\bar{S}^n(t) - mt) \to_w \sigma BM(t)$, i.e. the FSLLN and FCLT for random walks (see Exercise 5.4). It does more, however, in that it quantifies the rate of convergence.

Strong approximation can be used to obtain convergence rates for expectations of continuous bounded functionals.

Theorem 5.12   *Let $S(t)$ be the random walk process for X_j i.i.d. with mean m, variance σ^2, and $\mathbb{E}(|X_1|^r) < \infty$ for some $r > 2$, and let $\hat{S}^n(t) = (S(nt) - nmt)/\sqrt{n}$. Then, for any continuous bounded functional h,*

$$\mathbb{E}(h(\hat{S}^n(\cdot))) = \mathbb{E}(h(m\mathbf{e}(\cdot) + \sigma BM(\cdot))) + o(t^{1/r - 1/2}),$$

where $\mathbf{e}(t) = t$ is the identity function.
If also $\mathbb{E}(e^{aX_1}) < \infty$ for a in a neighborhood of 0, then:

$$\mathbb{E}(h(\hat{S}^n(\cdot))) = \mathbb{E}(h(m\mathbf{e}(\cdot) + \sigma BM(\cdot))) + O(\log(t)/t^{\frac{1}{2}}).$$

5.3 Functional Limit Theorems for Renewal Processes

Let X_n be non-negative i.i.d., with finite mean $m = 1/\mu$ and finite variance σ^2, let also $c = \sigma/m = \mu\sigma$ denote the coefficient of variation of X_n. We consider the *random walk* $S_0 = 0$, $S_n = X_1 + \cdots + X_n$, and the *renewal process* $\mathcal{A}(t)$

$$S_n = \sum_{j=1}^{n} X_j, \qquad \mathcal{A}(t) = \max\{n : S_n \le t\}.$$

We see that $\mathcal{A}(t)$ is in some sense the inverse of S_n, since $\mathcal{A}(S_n) = n$, and $S_{\mathcal{A}(t)} \approx t$, or more precisely, $S_{\mathcal{A}(t)} \leq t < S_{\mathcal{A}(t)+1}$, which sandwiches t in an interval of length X_j, with $j = \mathcal{A}(t) + 1$.

We have already discussed some properties of renewal processes in Section 2.3. In this section, we derive functional limit laws for the renewal process $\mathcal{A}(t)$. These follow from the functional limit laws for the random walk S_n.

Theorem 5.13 (FSLLN for renewal processes)

$$\frac{1}{n}\mathcal{A}(nt) \to_{a.s.} \mu t \quad u.o.c.$$

Proof   This is an extension of Theorem 2.13. The only thing to show is that convergence is a.s. uniform on compacts. We use the fact that $\frac{1}{n}\mathcal{A}(nt) \to_{a.s.}$ μt pointwise, and follow the steps as in the proof of the FSLLN for random walks, Theorem 5.8; see Exercise 5.5.   □

Theorem 5.14 (FCLT for renewal processes)

$$\sqrt{n}\left(\frac{1}{n}\mathcal{A}(nt) - \mu t\right) \to_w \sigma\mu^{3/2}BM(t) = \mu^{1/2}cBM(t).$$

Proof   Combining the above FSLLN and Donsker's theorem for S_n we have:

$$\left(\sqrt{n}(\frac{1}{n}S_{\lfloor nt \rfloor} - \frac{t}{\mu}), \frac{1}{n}\mathcal{A}(nt)\right) \to_w (\sigma BM(t), \mu t).$$

Here the convergence is for the two-dimensional process, and it holds for the joint distribution because the limit of the second component, $\frac{1}{n}\mathcal{A}(nt)$, is a constant function μt (see Exercise 5.6).

Using the time change $t \mapsto \frac{1}{n}\mathcal{A}(nt)$ (Theorem 5.6), we get

$$\sqrt{n}(\frac{1}{n}S_{\mathcal{A}(nt)} - \frac{1}{\mu}\frac{1}{n}\mathcal{A}(nt)) \to_w \sigma BM(\mu t).$$

Multiply by μ, use the fact that $BM(t) =_w -BM(t)$ and $BM(at) =_w a^{1/2}BM(t)$ (see Section 7.1.1) to get

$$\sqrt{n}(\frac{1}{n}\mathcal{A}(nt) - \mu\frac{1}{n}S_{\mathcal{A}(nt)}) \to_w \sigma\mu^{3/2}BM(t).$$

Finally, by $S_{\mathcal{A}(nt)} \leq nt < S_{\mathcal{A}(nt)+1}$, we have:

$$\frac{1}{n}S_{\mathcal{A}(nt)} - t \to_{a.s.} 0,$$

and by Theorem 5.7 (convergence together), the proof is complete.   □

Theorem 5.15 (strong approximation for renewal processes) *If X_n have finite moments of order $r > 2$, we can construct a copy $\mathcal{A}'(t)$ of $\mathcal{A}(t)$, for which*

$$\sup_{0 \le t \le T} |\mathcal{A}'(t) - \mu t - \mu^{1/2} c_a BM(t)| =_{a.s} o(T^{1/r'}) \quad as \ T \to \infty,$$

where $r' = \min(r, 4 - \delta)$ for arbitrary $\delta > 0$.

5.4 Sources

The presentation in this chapter follows in part a survey paper *Diffusion Approximations* by Peter Glynn (1990). The classic book on weak convergence is *Convergence of Probability Measures*, Billingsley (1999). Donsker's theorem has a long history including Donsker (1951), and a current proof can be found in Billingsley's book. A survey on Skorohod embedding is Obłój (2004), and a particularly elegant version is by Dubins (1968), which is described beautifully by Meyer (1971). Strong approximation was developed in a series of papers by Komlós et al. (1975, 1976) and Major (1976). Figure 5.3 is by Dean P. Foster, taken from Wikipedia (`commons.wikimedia.org/wiki/File:Law_of_large_numbers.gif`). In this figure, the scaling of the Y-axis is $y^{0.12}$, and of the X-axis is $x^{-0.06}$, which is close to logarithmic.

Exercises

5.1 A sequence of stochastic processes Z_n converges in probability to a stochastic process Z (written $Z_n \to_p Z$), if for any $\epsilon > 0$

$$\mathbb{P}(d(Z_n, Z) > \epsilon) \to 0, \text{ as } n \to \infty.$$

Here $d(\cdot, \cdot)$ is the u.o.c. distance if $Z_n \in \mathbb{D}$ and $Z \in \mathbb{C}$, or it is the $J1$ topology distance if $Z_n \in \mathbb{D}$ and $Z \in \mathbb{D}$.
Show that $Z_n \to Z$ a.s. implies $Z_n \to_p Z$.

5.2 For a seqeunce of random variables X_n, and a constant c, show that if $X_n \to_w c$, then $X_n \to_p c$.

5.3 For a sequence of stochastic processes Z_n with paths in $\mathbb{D}$, and a deterministic continuous function z, show that if $Z_n \to_w z$, then $Z_n \to_p z$, i.e. weak convergence of a sequence of stochastic processes to a continuous deterministic function implies convergence in probability.

5.4 Show that the Strong approximation theorem 5.11, implies that $\bar{S}^n(t) = \frac{1}{n} S(nt) \to_p mt$ and $\hat{S}^n(t) = \sqrt{n}(\bar{S}^n(t) - mt) \to_w \sigma BM(t)$, i.e. the FSLLN (in probability but not a.s.) and the FCLT for random walks.

5.5   Complete the proof of the FSLLN for renewal processes.

5.6   Prove the following result:

If $Z_n(t) \rightarrow_w Z(t)$, and $Y_n(t) \rightarrow_w y(t)$ where $y(t)$ is deterministic, then the jointly distributed sequence $(Z_n(t), Y_n(t)) \rightarrow_w (Z(t), y(t))$.

Some further exercises relevant to this chapter will be posed in Chapter 7, after the definition and discussion of the properties of Brownian motion.

6

Scaling of G/G/1 and G/G/∞

As we saw, once we go beyond Poisson arrivals and exponential service, exact analysis becomes much harder. For G/G/1 even the calculation of expected queue length requires the knowledge of the full distribution of the service times, and there are no explicit expressions to gain insights. Therefore one is led to look for approximations, with the hope that these will provide more insight.

In this chapter, we take our first steps in this direction. The single-server queue is, as we saw before, driven by the sequence of arrival times and service times. If we assume renewal arrivals and i.i.d. service times, then by scaling time and space we saw that we get FSLLN and FCLT for these two processes. These can be used to approximate the renewal processes, using just means and variances. Our analysis now will be to express the queue length and workload processes in terms of the arrival and service sequences, and then translate the FSLLN and FCLT of those to limit laws for the scaled queue and workload processes. The limiting processes will be used as approximations. The big advantage of these approximations is that they give a tractable description of the whole process, not just the stationary distribution at one time point. In addition to G/G/1 we also study approximations of the G/G/∞ queue.

6.1 Primitives and Dynamics of G/G/1

Our primitives for the single-server G/G/1 queue are the renewal stream of arrivals $\mathcal{A}(t)$ with interarrivals u_j with mean $1/\lambda$, variance σ_a^2, and coefficient of variation $c_a = \lambda \sigma_a$, and the renewal stream of service times, $\mathcal{S}(t)$ with individual service times v_j, with mean $1/\mu$, variance σ_s^2, and coefficient of variation $c_s = \mu \sigma_s$. Here $\mathcal{S}(t)$ counts service completions over a period of continuous processing of length t.

On a point of notation: This $\mathcal{S}(t)$ is a continuous time process that has integer values, not to be confused with earlier defined random walk process

$S(t) = S_{\lfloor t \rfloor}$ that had real valued jumps at integer times. Also, we have now changed our notations for interarrivals and service times from T_j, X_j to u_j, v_j, which we will retain for the rest of the book.

Our policy to control the system is non-preemptive and work conserving. Note that $S(t)$ is defined by the ordered sequence of i.i.d. processing times, and does not depend on the scheduling rule that determines the identities of the jobs that are scheduled. For concreteness, we assume the policy is FCFS, also referred to as head of the line (HOL). These primitives, together with the initial state and the policy, determine the queue and workload.

The dynamics of the queue length (including customer in service) are given by:

$$Q(t) = Q(0) + \mathcal{A}(t) - \mathcal{D}(t) = Q(0) + \mathcal{A}(t) - S(\mathcal{T}(t)). \qquad (6.1)$$

Here the *busy time process* $\mathcal{T}(t)$ is the total actual accumulated time that the server was working on jobs (this is the control, determined by the policy), and we express the departure process $\mathcal{D}(t)$ as the number of service completions during $\mathcal{T}(t)$. This is correct because service is non-preemptive. To complement the processing time, we have the idle time process, $\mathcal{I}(t)$, with $\mathcal{T}(t) + \mathcal{I}(t) = t$. Both processes, $\mathcal{T}(t)$ and $\mathcal{I}(t)$, start at zero and are non-decreasing. Naturally, we require that the queue length is non-negative. Under work conservation (non-idling), we also have that $\mathcal{I}(t)$ increases only when the system is empty. We then have in addition to (6.1):

$$Q(t) \geq 0, \qquad \int_0^t Q(s)d\mathcal{I}(s) = 0. \qquad (6.2)$$

We shall sometimes write $Q(t)d\mathcal{I}(t) = 0$ as shorthand for the work conservation condition.

Remark   Since both $\mathcal{T}(t), \mathcal{I}(t)$ are non-decreasing, and sum up to t, we have $\mathcal{T}(t) - \mathcal{T}(s) \leq t - s$ for $s < t$, so that $\mathcal{T}(t)$ is Lipschitz continuous, and similarly $\mathcal{I}(t)$ is Lipschitz continuous. This implies that they are absolutely continuous, they have derivatives almost everywhere, and they are integrals of these derivatives. We shall make use of this property repeatedly throughout the book.

Given the primitives and using a work-conserving non-preemptive policy, equations (6.1), (6.2) determine the evolution of the system. The equations are implicit (recursive), but a simple induction argument shows that, for given $Q(0), \mathcal{A}(t), S(t)$, they determine $Q(t), \mathcal{T}(t), \mathcal{I}(t)$ for all t (see Exercise 6.1).

We can also write down equations for the workload:

$$\mathcal{W}(t) = \mathcal{W}(0) + S_{\mathcal{A}(t)} - \mathcal{T}(t) = \mathcal{W}(0) + S_{\mathcal{A}(t)} - t + I(t), \qquad (6.3)$$

where $S_n = \sum_{j=1}^{n} v_j$ is the random walk of the non-negative processing times, and again, work is non-negative, and idle time increases only when there is no work in the system,

$$\mathcal{W}(t) \geq 0, \qquad \int_0^t \mathcal{W}(s)dI(s) = 0. \qquad (6.4)$$

We see from (6.3) that the workload decreases at a rate $d\mathcal{W}(t)/dt = -1$ whenever $\mathcal{W}(t) > 0$, and increases in jumps at arrival times.

6.2 Scaling Time, Space, and Initial Conditions

We saw (Section 4.2) by considering the random walk generated by $y_j = v_j - u_j$ that for the G/G/1 queue, the waiting times of customers $(V_n)_{n=0}^{\infty}$ are stable (in the sense of possessing a stationary distribution) for $\rho < 1$, transient for $\rho > 1$, and null recurrent for $\rho = 1$, and the same holds for $Q(t)$. In this analysis we considered the queue in its natural scale, where $Q(t)$ is piecewise constant, with jumps of ± 1 up and down, that are controlled by rates λ and μ. To obtain approximations to the behavior of the system, we introduce two additional scales.

For a function $z(t), z \in \mathbb{D}$, we introduce the fluid scaling $\bar{z}^n(t) = z(nt)/n$. This looks at the process in time units of n, so time is scaled by a factor of n, and it counts the material in the system (customers and workload) in units of n, so space is scaled by a factor of n. If $\bar{z}^n(t)$ converges u.o.c. to a continuous deterministic function, we call the limit $\bar{z}(t)$ the fluid limit of $z(t)$. Fluid scaling is the correct scaling to implement the FSLLN: We have, for the renewal stream of arrivals $\mathcal{A}(t, \omega)$, by Theorem 5.13, that $\bar{\mathcal{A}}^n(t, \omega) = \mathcal{A}(nt, \omega)/n \rightarrow \lambda t$ as $n \rightarrow \infty$, u.o.c. a.s. Similarly, $\bar{S}^n(t) \rightarrow \mu t$ u.o.c. a.s. So $\lambda t, \mu t$ are first-order, rough approximations to the behavior of the arrival and service processes, their fluid approximations.

For a function $z(t), z \in \mathbb{D}$ with fluid scaling $\bar{z}^n(t)$ and fluid limit $\bar{z}(t)$, we define the diffusion scaling as $\hat{z}^n(t) = \sqrt{n}(\bar{z}^n(t) - \bar{z}(t))$, so we center around the fluid limit, and then scale the deviations from the fluid limit, with the time measured in units of n, i.e. scaled by a factor of n, and space (values) measured in units of $\sqrt{n}$, i.e. scaled by a factor of $\sqrt{n}$. It is the correct scaling to implement the FCLT: For the arrivals $\mathcal{A}(t)$, by Theorem 5.14, $\hat{\mathcal{A}}^n(t) = \sqrt{n}(\bar{\mathcal{A}}^n(t) - \lambda t) \rightarrow_w \lambda^{1/2} c_a BM(t)$ as $n \rightarrow \infty$, and similarly, $\hat{S}^n(t) \rightarrow_w \mu^{1/2} c_s BM(t)$. So these limits are a second-order approximation, they approximate the behavior of the random deviations of the arrival and service processes from their fluid approximations.

We wish to study the behavior of $\bar{Q}(t)$ and $\hat{Q}(t)$, which raises two problems: First, we note that for any fixed finite deterministic or random variable $Q(0)$, a.s. $Q(0)/n \to 0$ and similarly, $Q(0)/\sqrt{n} \to 0$. But we would like to have fluid and diffusion paths starting from $\bar{Q}(0) > 0$ and $\hat{Q}(0) \neq 0$, because we are interested in transient properties of the scaled sample paths. Second, we would like to alleviate the sharp discontinuity between stable and transient systems that occurs at $\rho = 1$, by considering $\rho \to 1$ at an appropriate rate. The standard way to achieve these goals is to consider a sequence of systems, indexed by n. These systems will differ in the initial queue length, $Q^n(0)$, and they may have different arrival and service rates λ^n, μ^n, but they all share the same sequence of random variables that determines the interarrival and service times. We will let $u_j^n = u_j/\lambda^n$, $v_j^n = v_j/\mu^n$, $n = 1, 2, \ldots$, where u_j, v_j, $j = 1, 2, \ldots$ are two independent sequences of i.i.d. random variables with mean 1 and c.o.v.s c_a, c_s. They may have any general distributions F, G, which will not be relevant to the limits of the scaled processes.

In this framework, we study the processes derived from the primitives, namely the queue lengths and workloads. We look for fluid limits in which for almost all ω the sequence $\bar{Q}^n(t)$ with the sequence of initial values $\bar{Q}^n(0)$ converges u.o.c. to a fluid limit $\bar{Q}(t)$, and we look for diffusion limits, where the sequence $\hat{Q}^n(t)$, with the sequence of initial values $\hat{Q}^n(0)$, and a sequence of rates λ^n, μ^n, centered around $\bar{Q}(t)$, converges weakly (i.e. in distrubution) to a stochastic diffusion process (diffusions will be discussed in Chapter 7), with similar results for workloads. These fluid and diffusion limits may then serve as approximations for the actual process of queue length and workload.

6.3 Skorohod's Reflection Mapping for the Single-Server Queue

Let $x \in \mathbb{D}$, $x(0) \geq 0$. The *reflection mapping* in $\mathbb{D}$ is a pair of operators φ, ψ, which starting from the function x, yield two functions $y = \varphi(x)$, $z = \psi(x)$, both in $\mathbb{D}$ that satisfy
(i) $z(t) = x(t) + y(t) \geq 0$,
(ii) $y(0) = 0$, and $y(t)$ is non-decreasing,
(iii) $y(t)$ is increasing only when $z(t) = 0$.
An equivalent alternative to (iii) is the requirement:
(iii') $y(t)$ is the minimal function satisfying (i) and (ii).

An explicit expression for $y(t)$ is (see Exercise 6.2):

$$y(t) = -\min(0, \inf_{0 \leq s \leq t} x(s)) = \sup_{0 \leq s \leq t} x(s)^-.$$

What this means is that $y(t)$ is added to $x(t)$ in order to prevent it from becoming negative, and the result is non-negative $z(t)$. Since $x(0) \geq 0$, we have $y(0) = 0$. The fact that $y(t)$ is non-decreasing means that the difference between $x(t)$ and $z(t)$ is non-decreasing. In other words, there is no backlogging of the negative values of $x(t)$. On the other hand conditions (iii) and (iii') say that regulation is applied only when $z(t) = 0$ and is about to go negative, and as a result it is minimal. We say that y is the regulator and z is the reflection of x.

Skorohod has shown the following

Theorem 6.1 *The reflection mapping is well defined and unique for every $x \in \mathbb{D}$, $x(0) \geq 0$, and it is continuous on $\mathbb{D}$, that is, if $x_n \to x$ (in the Skorohod metric) then $\varphi(x_n) \to \varphi(x)$, $\psi(x_n) \to \psi(x)$, in $\mathbb{D}$.*

For the proof of existence, uniqueness, and minimality see Exercises 6.2–6.4. The proof of continuity is harder and we do not give it here.

The Skorohod reflection is used to describe the dynamics of the G/G/1 queue. We rewrite the equation (6.1) of the queue dynamics, by centering and separating it into two components:

$$Q(t) = \Big[Q(0) + (\lambda - \mu)t + [\mathcal{A}(t) - \lambda t] - [\mathcal{S}(\mathcal{T}(t)) - \mu\mathcal{T}(t)]\Big] + \Big[\mu\mathcal{I}(t)\Big]$$
$$= \mathcal{X}(t) + \mathcal{Y}(t). \tag{6.5}$$

In this decomposition of $Q(t)$, we centered $\mathcal{A}(t)$ around its mean (its fluid approximation λt), and $\mathcal{S}(\mathcal{T}(t))$ around its mean (its fluid approximation $\mu\mathcal{T}(t)$), and to compensate we added $(\lambda - \mu)t$ to obtain $\mathcal{X}(t)$, with a further compensation of $\mu(t - \mathcal{T}(t)) = \mu\mathcal{I}(t)$ to obtain $\mathcal{Y}(t)$. The process $\mathcal{Y}(t)$ is the cumulative expected lost output, at rate μ, accumulated during $\mathcal{I}(t)$, due to idling. We call $\mathcal{X}(t)$ the netput of the queue, and $\mathcal{Y}(t)$ the regulator of the queue. Note that the netput rate of change equals the rate of change of the queue length as long as $Q(t) > 0$, whereas in periods when $Q(t) = 0$, the netput rate of change is equal to minus the rate of change of the regulator $\mathcal{Y}(t)$ (see Figure 6.1).

We see that, for work-conserving policies, by (6.2), the conditions (i)–(iii) for the Skorohod decomposition hold, and so

$$Q(\cdot) = \psi(\mathcal{X}(\cdot)), \qquad \mathcal{Y}(t) = \varphi(\mathcal{X}(\cdot)),$$

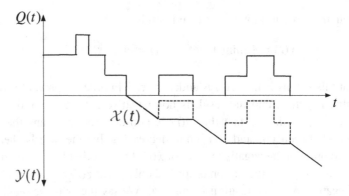

Figure 6.1 Skorohod reflection mapping of the G/G/1 queue.

where the explicit expression for $\mathcal{Y}(t)$ is

$$\mathcal{Y}(t) = - \inf_{0 \le s \le t} (X(s), 0) = \sup_{0 \le s \le t} X(s)^- .$$

6.4 Fluid Scaling and Fluid Limits of G/G/1

We now look at a sequence of systems, with different initial queue lengths $Q^n(0)$ but with shared interarrival and service time sequences, with rates λ and μ. We assume that $Q^n(0)/n \to \bar{Q}(0)$, and use fluid scaling, replacing t with nt and dividing by n. We rewrite the decomposed dynamics (6.5) of the fluid scaled queue lengths of the sequence:

$$\bar{Q}^n(t) = \left[\bar{Q}^n(0) + (\lambda - \mu)t + [\bar{\mathcal{A}}^n(t) - \lambda t] - [\bar{S}^n(\bar{\mathcal{T}}^n(t)) - \mu \bar{\mathcal{T}}^n(t)] \right] \\ + \left[\mu \bar{I}(t) \right] = \bar{X}^n(t) + \bar{\mathcal{Y}}^n(t). \tag{6.6}$$

Note that fluid scaled λt is λt, and we also see, for the nth system, that $S^n(T^n(nt))/n = S^n(n(T^n(nt)/n))/n = S^n(n\bar{T}^n(t))/n = \bar{S}^n(\bar{T}^n(t))$. The conditions for Skorohod's decomposition are met here, as they were for the unscaled system. We have $\bar{Q}^n(t) = \psi(\bar{X}^n(t))$, $\bar{\mathcal{Y}}^n(t) = \varphi(\bar{X}^n(t))$.

We now let $n \to \infty$ and obtain the following limit for the sequence of scaled netputs $\bar{X}^n(t)$. By the FSLLN for renewal processes, $\bar{\mathcal{A}}^n(t) - \lambda t \to 0$ u.o.c. a.s. as $n \to \infty$. Similarly, $\bar{S}^n(t) - \mu t \to 0$ u.o.c. a.s., that is, almost surely $\sup_{0 \le s \le t} (\bar{S}^n(s) - \mu s) \to 0$. By $\mathcal{T}^n(t) \le t$, we also have $\bar{\mathcal{T}}^n(t) \le t$, and therefore $\bar{S}^n(\bar{\mathcal{T}}^n(t)) - \mu \bar{\mathcal{T}}^n(t) \to 0$ u.o.c. a.s. as $n \to \infty$. Therefore,

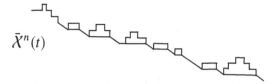

$$\bar{X}^n(t)$$

Figure 6.2 Fluid scaling of the netput of the G/G/1 queue.

by the FSLLN (see Figure 6.2):

$$\bar{X}^n(t) \to_{u.o.c.}^{a.s.} \bar{Q}(0) + (\lambda - \mu)t = \bar{X}(t).$$

Hence, by the continuity of the Skorohod reflection mapping,

$$\bar{Q}^n(\cdot) \to_{u.o.c.}^{a.s.} \bar{Q}(\cdot) = \psi(\bar{X}(\cdot)), \qquad \bar{\mathcal{Y}}^n(\cdot) \to_{u.o.c.}^{a.s.} \bar{\mathcal{Y}}(\cdot) = \varphi(\bar{X}(\cdot)).$$

Since the fluid $\bar{X}(t)$ is a linear function, it is easy to see that the following fluid limits are, depending on the value of ρ (see Figure 6.3):

$$\bar{Q}(t) = \begin{cases} \bar{Q}(0) + (\lambda - \mu)t, & \rho > 1, \\ \bar{Q}(0), & \rho = 1, \\ \begin{cases} \bar{Q}(0) - (\mu - \lambda)t, & t < t_0, \\ 0, & t \geq t_0, \end{cases} & \rho < 1, \end{cases}$$

$$t_0 = \frac{\bar{Q}(0)}{\mu - \lambda},$$

$$\bar{\mathcal{T}}(t) = \begin{cases} t, & \rho > 1, \\ t, & \rho = 1, \\ \begin{cases} t, & t < t_0, \\ t_0 + \rho(t - t_0), & t \geq t_0, \end{cases} & \rho < 1. \end{cases}$$

$$(6.7)$$

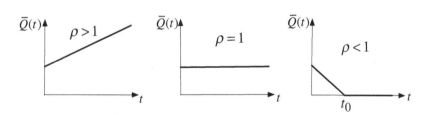

Figure 6.3 Fluid limits for the G/G/1 queue.

For the workload, we obtain

$$\bar{W}(t) = \bar{Q}(t)/\mu. \tag{6.8}$$

This follows directly from the FSLLN (see Exercise 6.7), or by scaling and Skorohod reflection on (6.3)–(6.4) (see Exercise 6.8).

6.5 Diffusion Scale Limits for Underloaded and Overloaded G/G/1

We now look at the G/G/1 under diffusion scaling, which will allow us to study the finer scale deviations from the fluid limit. In this section, we assume that ρ is fixed away from 0, and consider the cases $\rho > 1$ and $\rho < 1$, we will consider $\rho \approx 1$ in the next section. We scale time by n, center all quantities around the fluid limit, and measure the deviations in units of $\sqrt{n}$. Thus the centered and diffusion scaled arrival and service processes are

$$\hat{\mathcal{A}}^n(t) = \sqrt{n}(\frac{1}{n}\mathcal{A}(nt) - \lambda t), \qquad \hat{S}^n(t) = \sqrt{n}(\frac{1}{n}S(nt) - \mu t).$$

We consider again a sequence of systems that have initial states $Q^n(0)$, and we assume that

$$\bar{Q}^n(0) = \frac{1}{n}Q^n(0) \to \bar{Q}(0), \qquad \hat{Q}^n(0) = \sqrt{n}(\frac{1}{n}Q^n(0) - \bar{Q}(0)) \to \hat{Q}(0).$$

The Skorohod decomposition for $\sqrt{n}\bar{Q}^n(nt)$ is, similar to (6.6),

$$\sqrt{n}\bar{Q}^n(t) = \sqrt{n}\big[\bar{Q}^n(0) + (\lambda - \mu)t + [\hat{\mathcal{A}}^n(t) - \lambda t]$$
$$-[\hat{S}^n(\bar{T}^n(t)) - \mu\bar{T}^n(t)]\big] + \sqrt{n}\big[\mu(t - \bar{T}^n(t))\big] \quad (6.9)$$
$$= \sqrt{n}\bar{X}^n(t) + \sqrt{n}\bar{Y}^n(t),$$

and we have the Skorohod mapping relation $\sqrt{n}\bar{Q}^n(t) = \psi(\sqrt{n}\bar{X}^n(t))$ and $\sqrt{n}\bar{Y}^n(t) = \varphi(\sqrt{n}\bar{X}^n(t))$.

Centering and scaling the queue, $\hat{Q}^n(t) = \sqrt{n}(\bar{Q}^n(t) - \bar{Q}(t))$, by (6.7), we obtain:

$$\hat{Q}^n(t) = \sqrt{n}\big[\bar{X}^n(t) - ((\bar{Q}(0) + (\lambda - \mu)t) \wedge 0)\big] + \varphi(\sqrt{n}\bar{X}^n(t)). \quad (6.10)$$

We consider first the case of $\rho > 1$.

Theorem 6.2 *If $\rho > 1$, then $\hat{Q}^n(t)$ converges weakly for all $t > 0$, to a Brownian motion with initial state $\hat{Q}(0)$, drift $\lambda - \mu$, and diffusion coefficient $\lambda c_a^2 + \mu c_s^2$. The same holds if $\rho < 1$, for $0 < t < \frac{\bar{Q}(0)}{\mu - \lambda}$.*

Proof We notice that for large enough n, if $\rho > 1$, then $\bar{X}^n(t) > 0$ for all $t > 0$, and if $\rho < 1$, then $\bar{X}^n(t) > 0$ for $0 < t < \frac{\bar{Q}(0)}{\mu - \lambda}$. Hence, under those conditions $\bar{\mathcal{Y}}^n(t) = 0$, and:

$$\hat{Q}^n(t) = \sqrt{n}(\bar{Q}^n(t) - \bar{Q}(t)) = \sqrt{n}\bar{X}^n(t) - \sqrt{n}\left[\bar{Q}(0) + (\lambda - \mu)t\right]$$

$$= \sqrt{n}\left[\bar{Q}^n(0) - \bar{Q}(0) + [\bar{\mathcal{A}}^n(t) - \lambda t] - [\bar{S}^n(\bar{\mathcal{T}}^n(t)) - \mu\bar{\mathcal{T}}^n(t)]\right].$$

We then have as $n \to \infty$ the following limits:

(i) $\sqrt{n}(\bar{Q}^n(0) - \bar{Q}(0)) \to \hat{Q}(0)$ by our assumption,

(ii) $\sqrt{n}(\bar{\mathcal{A}}^n(t) - \lambda t) \to_w (\lambda c_a^2)^{1/2} BM(t)$ by the FCLT for renewal processes,

(iii) $\bar{\mathcal{T}}^n(t) = t$ as long as $\bar{X}^n(t) > 0$, so $\bar{\mathcal{T}}^n(t) \to_w t$ u.o.c. for $0 < t < \frac{\bar{Q}(0)}{\mu-\lambda}$ in the case of $\lambda < \mu$ and for all $t > 0$ in the case of $\lambda > \mu$. Also $\bar{\mathcal{T}}^n(t)$ is non-decreasing for all t.

(iv) We can use time change to see that $\sqrt{n}(\bar{S}^n(\bar{\mathcal{T}}^n(t)) - \mu\bar{\mathcal{T}}^n(t))$ has the same limit as $\sqrt{n}(\bar{S}^n(t) - \mu t)$, and by the FCLT for renewal processes, $\sqrt{n}(\bar{S}^n(t) - \mu t) \to_w (\mu c_s^2)^{1/2} BM(t)$.

(v) Furthermore, $A^n(t)$ and $S^n(t)$ are independent.

We obtain:

$$\hat{Q}^n(t) \to_w \hat{Q}(0) + (\lambda - \mu)t + (\lambda c_a^2 + \mu c_s^2)^{1/2} BM(t), \quad \begin{cases} \rho > 1, \ t > 0 \\ \rho < 1, \ 0 < t < \frac{\bar{Q}(0)}{\mu-\lambda}. \end{cases}$$

$\square$

We now look at $\rho < 1$. We already obtained its diffusion limit over $0 < t < \frac{\bar{Q}(0)}{\mu-\lambda}$. For simplicity, we take $\lim_{n \to \infty} \frac{1}{\sqrt{n}} Q^n(0) = \hat{Q}(0)$, so $\bar{Q}(0) = 0$. In the case that $\rho < 1$, we know that G/G/1 is stable, i.e. the queue length $Q(t)$ has a stationary distribution, and converges to this stationary distribution as $t \to \infty$ from any initial queue length. The following theorem is therefore not surprising, but its proof is both delicate and enlightening:

Theorem 6.3 *For $\rho < 1$, with $\bar{Q}(0) = 0$, we have for all $t > 0$ that $\hat{Q}^n(t) \to_w 0$.*

Proof We note first that with $\bar{Q}(0) = 0$ and $\lambda < \mu$, $\bar{Q}(t) = 0$, so $\hat{Q}^n(t) = \sqrt{n}\bar{Q}^n(t) = Q(nt)/\sqrt{n}$.

We will show that $\hat{Q}^n(t) \to_p 0$ u.o.c., which is equivalent to $\hat{Q}^n(t) \to_w 0$, (in general, convergence in probability implies convergence in distribution, and if the limit is constant the two are equivalent; see Exercise 5.2). We wish to show for $t > 0$ and $\epsilon > 0$:

$$\mathbb{P}(\sup_{0 < s \leq t} \hat{Q}^n(s) > \epsilon) \to 0, \quad \text{as } n \to \infty.$$

Assume first that $\hat{Q}(0) = 0$. Then for n large enough, $\hat{Q}^n(0) < \epsilon/2$. Consider the event $\sup_{0<s\le t} \hat{Q}^n(s) > \epsilon$. If $\hat{Q}^n(0) < \epsilon/2$ and $\sup_{0<s\le t} \hat{Q}^n(s) > \epsilon$, then we can define $0 \le \tau^{*n} < \tau^n \le t$ such that:

$$\tau^n = \inf\{s : \hat{Q}^n(s) > \epsilon\}, \qquad \tau^{*n} = \sup\{s : 0 \le s < \tau^n, \hat{Q}^n(s) \le \epsilon/2\}.$$

In the interval (τ^{*n}, τ^n), $\hat{Q}^n$ is positive all the time, and it increases by at least $\epsilon/2$. We then have:

$$\epsilon/2 \le \hat{Q}^n(\tau^n) - \hat{Q}^n(\tau^{*n})$$
$$= \sqrt{n}\left[\left(\bar{\mathcal{A}}^n(\tau^n) - \bar{\mathcal{A}}^n(\tau^{*n})\right) - \left(\bar{S}^n(\bar{\mathcal{T}}^n(\tau^n)) - \bar{S}^n(\bar{\mathcal{T}}^n(\tau^{*n}))\right)\right]$$
$$= \sqrt{n}\left[\left(\bar{\mathcal{A}}^n(\tau^n) - \bar{\mathcal{A}}^n(\tau^{*n})\right) \right.$$
$$\left. - \left(\bar{S}^n\left(\bar{\mathcal{T}}^n(\tau^{*n}) + (\tau^n - \tau^{*n})\right) - \bar{S}^n(\bar{\mathcal{T}}^n(\tau^{*n}))\right)\right]$$
$$\le \sup_{0\le r < s \le t} \sqrt{n}\left[\left(\bar{\mathcal{A}}^n(s) - \bar{\mathcal{A}}^n(r)\right) - \left(\bar{S}^n(\bar{\mathcal{T}}^n(r) + s - r) - \bar{S}^n(\bar{\mathcal{T}}^n(r))\right)\right],$$

where the second equality holds, because in the interval $t \in (\tau^{*n}, \tau^n)$, $Q^n(t) > 0$, so that $d\mathcal{T}^n(t)/dt = 1$.

We have therefore shown that:

$$\mathbb{P}\left(\sup_{0<s\le t} \hat{Q}^n(s) > \epsilon\right) \le \mathbb{P}\left(\sup_{0\le r < s \le t} \sqrt{n}\left[\left(\bar{\mathcal{A}}^n(s) - \bar{\mathcal{A}}^n(r)\right)\right.\right.$$
$$\left.\left. - (\bar{S}^n(\bar{\mathcal{T}}^n(r) + s - r) - \bar{S}^n(\bar{\mathcal{T}}^n(r)))\right] > \epsilon/2\right).$$

Now we have for any $0 \le r < s \le t$:

$$\sqrt{n}\left[\left(\bar{\mathcal{A}}^n(s) - \bar{\mathcal{A}}^n(r)\right) - \left(\bar{S}^n(\bar{\mathcal{T}}^n(r) + s - r) - \bar{S}^n(\bar{\mathcal{T}}^n(r))\right)\right]$$
$$= \sqrt{n}(\lambda - \mu)(s - r) + \sqrt{n}\left[\left(\bar{\mathcal{A}}^n(s) - \bar{\mathcal{A}}^n(r) - \lambda(s - r)\right)\right.$$
$$\left. - \left(\bar{S}^n(\bar{\mathcal{T}}^n(r) + s - r) - \bar{S}^n(\bar{\mathcal{T}}^n(r)) - \mu(s - r)\right)\right]$$
$$:= \sqrt{n}(\lambda - \mu)(s - r) + \mathcal{Z}^n(r, s),$$

where

$$\mathcal{Z}^n(r, s) \to_w (\lambda c_a^2 + \mu c_s^2)^{-1/2} BM(s - r), \quad \text{as } n \to \infty,$$

and we need to evaluate:

$$\mathbb{P}\left(\sup_{0\le r < s \le t}\left[\sqrt{n}(\lambda - \mu)(s - r) + \mathcal{Z}^n(r, s)\right] > \epsilon/2\right),$$

and show it converges to 0 as $n \to \infty$.

We can choose δ small enough so that for $(s - r) < \delta$, we can make $\mathcal{Z}^n(r, s)$ arbitrarily small when n is large, while for $s - r > \delta$, $\sqrt{n}(\lambda - \mu)(s - r) + \mathcal{Z}^n(r, s)$ is negative when n is large. This completes the proof for $\hat{Q}(0) = 0$.

We have assumed that $\hat{Q}(0) = 0$. In the case that $\hat{Q}(0) < 0$, the proof needs no modification. In the case that $\hat{Q}(0) > 0$, one can easily see that $\inf\{t : \hat{Q}(t) = 0\} = 0$, and the rest of the proof is the same. □

Turning to the workload we can also obtain fluid and diffusion scaling limits directly from (6.3), and we can show (see Exercises 6.11 and 6.12):

Theorem 6.4 *The diffusion scaled workload and queue length are proportional in the limit,*

$$\frac{\mu \hat{W}^n(t)}{\hat{Q}^n(t)} \to_w 1, \tag{6.11}$$

where we let $0/0 = 1$.

6.6 Diffusion Scaling of the G/G/1 Queue in Heavy Traffic

We now consider G/G/1 in heavy traffic, i.e. in the neighborhood of $\rho = 1$. We use a sequence of systems with scaled initial condition and scaled rates, with a fixed sequence of unscaled interarrivals and services, as described in Section 6.2. Our sequence of systems has λ^n, μ^n, $Q^n(0)$ defined so that as $n \to \infty$

$$\lambda^n \to \lambda, \quad \sqrt{n}(\lambda^n - \mu^n) \to \theta, \quad Q^n(0)/\sqrt{n} \to \hat{Q}(0). \tag{6.12}$$

The important feature here is that the traffic intensity $\rho^n \to 1$, so the systems are in heavy traffic, but we are exploring the neighborhood of $\rho = 1$. As we shall see, the behavior of such systems does not exhibit a discontinuity at $\rho = 1$.

We have first:

Proposition 6.5 *The fluid limit in the heavy traffic case satisfies: If* $Q^n(0)/n \to 0$, $\lambda^n \to \lambda$, *and* $\lambda^n - \mu^n \to 0$, *then* $\bar{Q}(t) = 0$ *and* $\bar{T}(t) = t$.

Proof See Exercise 6.9. □

Because the fluid limit is $\bar{Q}(t) = 0$, we define $\hat{Q}^n(t) = Q^n(nt)/\sqrt{n}$.

Proposition 6.6 *The diffusion limits for the arrival and service processes satisfy: If* $Q^n(0)/n \to 0$, $\lambda^n \to \lambda$, *and* $\lambda^n - \mu^n \to 0$ *then:* $\frac{1}{\sqrt{n}}(\mathcal{A}^n(nt) - \lambda nt) \to_w (\lambda c_a^2)^{1/2} BM(t)$ *and* $\frac{1}{\sqrt{n}}(S^n(\mathcal{T}(nt)) - \lambda nt) \to_w (\lambda c_s^2)^{1/2} BM(t)$, *and the joint distribution of the diffusion scaled arrival and service processes converges weakly jointly to the joint distribution of two independent Brownian motions.*

Proof   See Exercise 6.10.   □

We now write:

$$\hat{Q}^n(t) = \frac{1}{\sqrt{n}}\Big\{\big[Q^n(0) + (\lambda^n - \mu^n)nt + (\mathcal{A}^n(nt) - \lambda^n nt) - (S^n(T^n(nt))$$

$$-\mu^n T^n(nt))\big] + \big[(\mu^n nt - \mu^n \mathcal{T}^n(nt)\big]\Big\} = \hat{X}^n(t) + \hat{Y}^n(t),$$

and we still have that $\hat{Q}^n(t) \geq 0$, and $\hat{Y}^n(t)$ only increases when $\hat{Q}^n(t) = 0$, as required for the Skorohod reflection.

From Propositions 6.5 and 6.6 we have $\frac{1}{\sqrt{n}}(\mathcal{A}^n(nt) - \lambda^n nt) \to_w (\lambda c_a^2)^{1/2}$ $BM_1(t)$, $\bar{T}^n(t) \to t$ u.o.c. a.s., and, by time change, $\frac{1}{\sqrt{n}}(S^n(T^n(nt)) - \mu^n T^n(nt)) \to_w (\lambda c_s^2)^{1/2} BM_2(t)$, where the two Brownian motions are independent. Hence, we obtain

$$\hat{X}^n(t) \to_w \hat{X}(t) = \hat{Q}(0) + \theta t + (\lambda(c_a^2 + c_s^2))^{1/2} BM(t).$$

It is important to note the difference between the case of $\rho^n \to 1$ and $\rho^n = \rho \neq 1$ fixed for all n. As we saw, for fixed $\rho > 1$ we can for every $\epsilon > 0$ find n_0 so that for all $n > n_0$, $\bar{X}^n(t) > 0$ for all $t > \epsilon$. As a result $\bar{Y}^n(t) = 0$, $\hat{Y}^n(t) = 0$, and $\hat{Q}^n(t) = \hat{X}^n(t)$, so the fluid scaled queue is positive linear, and the diffusion scaled deviations from it converge to Brownian motion. For fixed $\rho < 1$, we saw that $\hat{Q}^n(t) \to_w 0$, and $\hat{Y}^n(t) \approx (\mu - \lambda)t$. In contrast, for $\rho^n \to 1$ at the rates given by (6.12), the behavior is that $\bar{X}^n(t) \to 0$, $\bar{Q}^n(t) \to 0$, and now $\hat{X}^n(t)$ behaves close to: $\hat{X}(t) =_D \hat{Q}(0) + \theta t + (\lambda(c_a^2 + c_s^2))^{1/2} BM(t)$. Here the mean and the standard deviation of $\hat{X}(t)$ will both be of order t, and the process will fluctuate between positive and negative, so that $\mathcal{Y}^n(t)$ is neither always 0 (as it is for $\rho > 1$), nor always $(\mu - \lambda)t$ (as it is for $\rho < 1$).

As a result we have proved, using Skorohod reflection:

Theorem 6.7   *In the heavy traffic case, with $\sqrt{n}(\lambda^n - \mu^n) \to \theta$, the diffusion limits as $n \to \infty$ are given by*

$$\hat{X}^n(t) \to_w \hat{Q}(0) + \theta t + (\lambda(c_a^2 + c_s^2))^{1/2} BM(t), \tag{6.13}$$

$$\hat{Q}^n(t) \to_w \hat{Q}(t) = \psi\big(\hat{Q}(0) + \theta t + (\lambda(c_a^2 + c_s^2))^{1/2} BM(t)\big), \tag{6.14}$$

$$\mu^n \hat{I}^n(t) \to_w \hat{Y}(t) = \varphi\big(\hat{Q}(0) + \theta t + (\lambda(c_a^2 + c_s^2))^{1/2} BM(t)\big), \tag{6.15}$$

$$\hat{W}^n(t) \to_w \frac{1}{\mu}\hat{Q}(t), \text{ where in the limit } \mu = \lambda. \tag{6.16}$$

The process $\hat{Q}(t) = \psi\big(\hat{Q}(0) + \theta t + (\lambda(c_a^2 + c_s^2))^{1/2} BM(t)\big)$ is a *reflected*

Brownian motion. We discuss reflected Brownian motion in Section 7.1. See Exercise 6.12 for proof of (6.16).

6.7 Approximations of the G/G/1 Queue

We have proven some limit theorems for the G/G/1 queue under fluid scaling and under diffusion scaling. We would now like to use those for approximations of an unscaled queue. We assume renewal arrivals and services, with rates λ, μ and coefficients of variation c_a, c_s. We are interested in the process starting at initial queue length N, and our time horizon is T.

Consider first the case of $\rho = \frac{\lambda}{\mu} > 1$. In that case, the fluid approximation to the queue is $Q(t) \approx Q(0) + (\lambda - \mu)t$, i.e. the queue is approximated by a deterministic linear function. The approximation is tighter for longer periods, where by the law of the iterated logarithm $\sup_{0 < t < T} |Q(t) - Q(0) + (\lambda - \mu)t| = O(\sqrt{T \log \log T})$. The deviations from the linear fluid approximation are approximated by a Brownian motion, $(\lambda c_a^2 + \mu c_s^2)^{1/2} BM(t)$, where the approximation becomes better for large $t \in [0, T]$; in particular, for large T we have approximately that $Q(T) - Q(0) - (\lambda - \mu)T \sim N(0, (\lambda c_a^2 + \mu c_s^2)T)$. This enables us to estimate where the queue will be in a future time T where T is large.

For the case of $\rho = \frac{\lambda}{\mu} < 1$, the fluid approximation is $Q(t) \approx (Q(0) + (\lambda - \mu)t)^+$. It is useful to describe the transient behavior of the queue when it starts from $Q(0) = N$ where N is large: Then the queue will decrease approximately at the linear rate of $\mu - \lambda$ until it reaches 0, at the time $t_0 = \frac{N}{\mu} \frac{1}{1-\rho}$, where the first term is the initial workload, which is also the waiting time of an arrival at time 0, and the second term is the ratio between the waiting time of an arrival at time 0 and the time for the queue to empty. For $t > t_0$, the fluid approximation is 0, which is uninformative. The diffusion approximation in this case is also identically 0 – this looks like a contradiction, but we note that for diffusion scaling we scale time by a factor n and space by a factor $\sqrt{n}$, or, equivalently, space by a factor of N and time by a factor of N^2. So if the fluid starting from N reaches 0 after $t_0 = O(N)$, on the diffusion scale $t_0 = o(N^2)$. As we saw in Sections 4.1 and 4.2, for $\rho < 1$ the queue G/G/1 has a stationary distribution. If interarrival and service time distributions have exponential tails, then it can be shown that $\sup_{0 < t < T} Q(t) = O(\log T)$.

For $\rho \approx 1$, we saw that the diffusion scaling limit is a reflected Brownian motion, with drift $\lambda - \mu$, and netput variance $(\lambda(c_a^2 + c_s^2))^{1/2}$. We will see in Section 7.1.2 that when $\rho < 1$, $\rho \approx 1$, the reflected Brownian motion has a stationary distribution that is exponential, with mean $\frac{\rho}{1-\rho}\frac{c_a^2+c_s^2}{2}$; this approximate mean value is exactly the value derived by Kingman's bound in Section 4.3.

So starting from $Q(0) = N$, with $\rho < 1$, $\rho \approx 1$, we have two approximations: fluid decreasing at rate $\mu - \lambda$ to reach 0 at time $t_0 = \frac{N}{\mu}\frac{1}{1-\rho}$, followed by diffusion approximation of 0, or reflected Brownian motion with stationary distribution which is exponential, with mean $\frac{\rho}{1-\rho}\frac{c_a^2+c_s^2}{2}$. We can formulate a rule of thumb choice between them: For a time horizon $T = O(N^2)$, if $t_0 = O(N)$, use the fluid approximation; if $t_0 = O(N^2)$, use the reflected Brownian motion approximation.

It is instructive to look at the queue as difference between input and output. For simplicity, assume empty queue at time 0. The arrival process $\mathcal{A}(t)$ is a renewal process, and the stationary version of it will have mean λt, and variance $\lambda c_a^2 t$. Under fluid scaling, it will converge to the linear λt, and the deviations from the fluid will under diffusion scaling converge to $(\lambda c_a^2)^{1/2} BM(t)$.

When $\rho < 1$, events of the departure process follow the events of the arrival process after a random but stationary delay. While the departure process is no longer a renewal process, it has a stationary version with mean λt, and variance $\lambda c_a^2 t$, and almost perfect correlation with the arrival process. The delay between the events is always the sojourn time of a customer in the queue, which is of order $O(1)$. Under fluid scaling, the fluid limit of the queue will be 0, and the fluid limits of both arrivals and departures are λt. Under diffusion scaling, the limiting queue length will also be 0, and the limits of the scaled deviations of the arrival and departure processes will be equal, given by $(\lambda c_a^2)^{1/2} BM(t)$.

When $\rho > 1$, the queue will diverge, and from some time onward will be always positive. Therefore, the departure process will now have a stationary version that will be a renewal process, with mean μt and variance $\mu c_s^2 t$, and most important, the departure process and the arrival process are now independent. The fluid and diffusion limits of the arrival and departure processes will be λt, μt and $(\lambda c_a^2)^{1/2} BM_1(t)$, $(\mu c_s^2)^{1/2} BM_2(t)$, where the Brownian motions are independent. The fluid limit of the queue will be $(\lambda - \mu)t$, and its diffusion limit, as the difference between two independent Brownian motions, will be $(\lambda c_a^2 + \mu c_s^2)^{1/2} BM_3(t)$.

When $\rho = 1$, or when $\lambda - \mu$ is of order $1/\sqrt{N}$, the fluid limits will be λt for both arrival and departure, and the fluid limit of the queue length will be 0. Now the delay between the arrival and departure is of order $\sqrt{N}$, and the diffusion limit of the queue length is a reflected Brownian motion. We can write the departure in two ways: $\mathcal{D}(t) = \mathcal{A}(t) - Q(t)$, or as $\mathcal{D}(t) = S(t) - \mathcal{Y}(t)$, but of course neither $\mathcal{A}(t)$ nor $S(t)$ are independent of $Q(t)$. For the special case of $\theta = 0$, it can be shown that for the diffusion limits, $\mathbb{V}\text{ar}(\hat{D}(t)) = \mathbb{V}\text{ar}(\hat{Q}(t)) = \lambda(c_a^2 + c_s^2)(1 - \frac{2}{\pi})t$.

6.8 Another Approach: Strong Approximation

The assumption that service times are i.i.d. and therefore $S(t)$ is a renewal process is quite reasonable. The assumption that the arrival process $\mathcal{A}(t)$ is a renewal process makes sense if the process is a Poisson process, since arrivals are often the result of many possible customers deciding independent of each other when to arrive (a superposition of many independent stationary point processes, when scaled, converges to a Poisson process). However, it is not a reasonable model for generally distributed interarrival times. Nevertheless, the scaling limits of the queue length process may still be reasonable approximations under weaker assumptions on the arrival and service processes. Assume stationary arrival and service processes, with interarrival times and service times that possess finite moments of order $r > 2$. We also make the plausible assumption that the two processes are independent, though that is not essential for the results. Instead of making the renewal assumption (recall Theorems 5.15), assume now that the two processes satisfy a *strong approximation assumption*: We can produce copies $\mathcal{A}'(t)$ of $\mathcal{A}(t)$ together with a Brownian motion $BM(t)$ such that:

$$\sup_{0 < t < T} \left(\mathcal{A}'(t) - \lambda t - (\lambda c_a^2)^{1/2} BM(t) \right) = o(T^{1/r}) \quad \text{as } T \to \infty,$$

with a similar assumption on $S(t)$. That is, the arrival stream and the service process both can be approximated on the fluid scale by λt and μt (a.s. u.o.c.) and deviation from the fluid can be approximated by a scaled Brownian motion (in distribution), and the rate of the approximation is $T^{1/r}$.

We consider again:

$$X(t) = Q(0) + (\lambda - \mu)t + (\mathcal{A}(t) - \lambda t) + (S(\mathcal{T}(t)) - \mathcal{T}(t)).$$

We then have for a copy of $X(t)$ the approximation

$$\sup_{0 < t < T} \left(X'(t) - (Q(0) + \theta t + (\lambda c_a^2 + \mu \min(1, \rho)c_s^2)^{1/2} BM(t)) \right)$$

$$= o(T^{1/r}) \quad \text{as } T \to \infty. \tag{6.17}$$

The queue length $Q(t) = \psi(X(t))$ is approximated by the Skorohod reflection of the Brownian motion, and $\mathcal{Y}(t) = \varphi(X(t))$ is approximated by the

regulator of the Brownian motion. Furthermore, from (6.17) we obtain the
rate of the approximation.

6.9 Diffusion Limits for G/G/∞ in Heavy Traffic

In this section, we discuss approximation to the G/G/∞ system. We will
return to the G/G/∞ system for further results in Chapter 18. We note that
in G/G/∞ there is no waiting, and $Q(t)$ equals the number of customers
in service, which is also the number of busy servers. When the arrival
rate is high, $Q(t)$ is large, and we discuss its approximation. The scaling
now is different from that used for G/G/1: we only scale the number in the
system, and time remains in the original scale. As a result, the approximation
depends on the service time distribution. We will consider again a sequence
of systems, indexed by n, with arrival rates λn so that interarrival times will
be $u_j^n = u_j/n$ arising from scalings of a single process of u_j, $j = 1, 2, \ldots$,
i.i.d. with mean $1/\lambda$ and coefficient of variation c_a. So the arrival process
of the nth system is a renewal process $\mathcal{A}^n(t)$ with arrival rate $n\lambda$, and as
$n \to \infty$, we have $\mathcal{A}^n(t)/n \to \lambda t$ u.o.c. a.s., and $n^{1/2}(\mathcal{A}^n(t)/n - \lambda t) \to_w$
$\lambda^{1/2} c_a BM(t)$. This type of scaling, many arrivals during a single unscaled
service period (and many servers, in this case infinite servers), is called
many-server scaling.

We consider first the queue G/D/∞ with fixed processing time x. We
assume that the process starts with 0 customers in the system at time 0.
Since every customer departs a time x after its arrival,

$$Q^n(t) = \mathcal{A}^n(t) - \mathcal{D}^n(t) = \mathcal{A}^n(t) - \mathcal{A}^n((t-x) \wedge 0)).$$

We therefore have a fluid limit: $Q^n(t)/n \to_{u.o.c.}^{a.s.} \lambda(x \wedge t)$. We then obtain
that for the diffusion scaled process,

$$\hat{Q}^n(t) = n^{1/2}(Q^n(t)/n - \lambda(x \wedge t)) \to_w \lambda^{1/2} c_a (BM(t) - BM((t-x) \wedge 0)).$$

This limit is a Gaussian process that is stationary for all $t > x$. It has mean
0, variance $\lambda c_a^2 x$, and autocovariance function (see Exercise 6.14):

$$\begin{aligned}
&\mathbb{Cov}(\hat{Q}(t+s), \hat{Q}(t)) \\
&= \lambda c_a^2 \mathrm{Cov}(BM(t+s) - BM(t+s-x), BM(t) - BM(t-x)) \\
&= \lambda c_a^2 (x-s)^+, \qquad t > x.
\end{aligned} \tag{6.18}$$

Note, this is not a Markovian process; the future from time t depends on
the history $Q(s)$, $t - x < s < t$.

Consider next the G/G/∞ with processing time X, which is a discrete random variable taking values $x_1 < \cdots < x_m$, with probabilities $p_1, \ldots, p_m$. This queue is the sum of m G/D/∞ queues with dependent arrival streams. Each of them, under the same scaling, converges to a Gaussian process, and so their sum also converges to a Gaussian process, stationary for $t > x_m$. It is then a straightforward calculation to obtain the autocovariance function for this process (see Exercise 6.15). This can be generalized to any processing time distribution G with mean $1/\mu$ and finite variance. One then obtains the limits for the scaled process as $n \to \infty$, for specified t. When we also let $t \to \infty$ we get a stationary version, for which (see Exercise 6.16):

$n^{1/2}[Q^n(t)/n - \lambda/\mu] \to_w$ Gaussian process $\hat{Q}(t)$, mean zero,

$\mathbb{E}(Q^n(t)) = n\lambda/\mu,$

$$\mathbb{V}\mathrm{ar}(\hat{Q}(t)) = \lambda/\mu + \lambda(c_a^2 - 1) \int_0^\infty (1 - G(x))^2 dx, \qquad (6.19)$$

$$\mathbb{C}\mathrm{ov}(\hat{Q}(t), \hat{Q}(t+s)) = \lambda \Big[\int_0^\infty (1 - G(x+s)) dx$$

$$+ (c_a^2 - 1) \int_0^\infty (1 - G(x))(1 - G(x+s)) dx \Big].$$

We consider now three special cases:

Example 6.8 (The M/M/∞ queue.) We have seen in Section 1.3 that the stationary distribution of $Q(t)$ is Poisson. We are now able also to approximate not just the distribution at a single time point, but also the whole process, with a diffusion process. With Poisson arrivals of rate $n\lambda$ and exponential service with rate μ, we have $c_a = 1$, $c_s = 1$ and the equations (6.19) simplify:

$Q^n(t)$ is Poisson$(n\dfrac{\lambda}{\mu})$, with mean and variance $n\dfrac{\lambda}{\mu}$,

$\hat{Q}(t) = \sqrt{n} \, [\dfrac{Q^n(t)}{n} - \dfrac{\lambda}{\mu}] \to_w$ Gaussian process, mean zero,

$\mathbb{C}\mathrm{ov}(\hat{Q}(t), \hat{Q}(t+s)) = \dfrac{\lambda}{\mu} e^{-\mu s}.$

The limiting process is the unique continuous time Gaussian and Markovian process, known as the *Ornstein–Uhlenbeck process*. See Example 7.3 for another derivation.

Example 6.9 (The G/M/∞ queue.) We will see that here also, the stationary process $Q^n(t)$ is approximated by an Ornstein–Uhlenbeck process. We now have renewal arrivals of rate $n\lambda$, with interarrival distribution that has coefficient of variation c_a, and service times are exponential with rate μ. Substituting into equation (6.19) we have

$$
\begin{aligned}
\mathbb{C}\mathrm{ov}(\hat{Q}(t+s), \hat{Q}(t)) \\
&= \lambda \Big[\int_0^\infty e^{-\mu(x+s)} dx + (c_a^2 - 1) \int_0^\infty e^{-\mu x} e^{-\mu(x+s)} dx \Big] \\
&= \lambda \Big[\frac{1}{\mu} e^{-\mu s} + (c_a^2 - 1) \frac{1}{2\mu} e^{-\mu s} \Big] \\
&= \frac{\lambda}{\mu} \frac{c_a^2 + 1}{2} e^{-\mu s}.
\end{aligned}
$$

It can be shown that the only service distributions for which the queue length of G/G/∞ can be approximated by a Markov process, in fact by the Ornstein–Uhlenbeck process, is service time distribution $G(t) = 1 - \theta e^{-\mu t}$, i.e. probability $1 - \theta$ to be 0, and exponential service time otherwise.

Example 6.10 (The M/G/∞ queue)   For the queueing system M/G/∞, we have $c_a^2 = 1$, and so the terms $c_a^2 - 1$ fall away. Furthermore, the stationary distribution of $Q(t)$ is, by insensitivity (see Section 8.8), equal to that of $M/M/\infty$. We get, for the stationary process,

$$
Q^n(t) \text{ is Poisson}(n\lambda/\mu), \text{ with mean and variance } n\frac{\lambda}{\mu},
$$

$$
\mathbb{C}\mathrm{ov}(\hat{Q}(t), \hat{Q}(t+s)) = \lambda \int_s^\infty (1 - G(x)) dx = \frac{\lambda}{\mu}(1 - G_{eq}(s)).
$$

It is interesting to see that the autocorrelation decays like the tail of the equilibrium distribution of the service times.

6.10 Sources

Fluid and diffusion limit theorems for queues were pioneered independently by Iglehart (1965, 1973) and by Borovkov (1964). The Skorohod reflection mapping originated in Skorokhod (1961). Comprehensive treatment of asymptotic results for queues is found in the texts *Asymptotic Methods in Queuing Theory*, Borovkov (1984) and in *Stochastic Process Limits*, Whitt (2002). Parts of this chapter again follow the survey paper of Glynn (1990). The surprising result on the asymptotic variance of departures in critically loaded queues is derived in Al Hanbali et al. (2011). The use of strong approximation is advocated by Chen and Mandelbaum (1994b). The diffusion approximation for G/G/∞ is due to Borovkov (1967); see also Whitt (1982) and Glynn and Whitt (1991).

Exercises

6.1 Prove by induction that the implicit conditions of the dynamics, the non-negativity, and work conservation, uniquely determine the queue length process.

6.2 Show that $y(t) = -\inf\{0, x(s) : 0 \le s \le t\}$ satisfies conditions (i)–(iii) of the Skorohod reflection mapping.

6.3 Show that $y(t) = -\inf\{0, x(s) : 0 \le s \le t\}$ is the minimal function that satisfies conditions (i) and (ii) of the Skorohod reflection mapping

6.4 Show that $y(t) = -\inf\{0, x(s) : 0 \le s \le t\}$ is the unique function that satisfies conditions (i)–(iii) of the Skorohod reflection mapping.

6.5 Show that conditions (i), (ii), and (iii) of the Skorohod reflection mapping are equivalent to conditions (i), (ii), and (iii').

6.6 Show that for the single-server queue, under work-conserving policy, the busy time $\mathcal{T}(t) \to \infty$ as $t \to \infty$.

6.7 Derive the fluid limit for the workload directly from the fluid limit of the queue length.

6.8 Derive the fluid limit for the workload by scaling and using Skorohod reflection on (6.3)–(6.4).

6.9 Show that if $Q^n(0)/n \to 0$, $\lambda^n \to \lambda$, and $\lambda^n - \mu^n \to 0$, then $\bar{Q}(t) = 0$ and $\bar{T}(t) = t$.

6.10 Show that when $Q^n(0)/n \to 0$, $\lambda^n \to \lambda$ and $\lambda^n - \mu^n \to 0$, then $\frac{1}{\sqrt{n}}(\mathcal{A}^n(nt) - \lambda nt) \to_w (\lambda c_a^2)^{1/2} BM(t)$ as well as $\frac{1}{\sqrt{n}}(\mathcal{S}^n(\mathcal{T}(nt)) - \lambda nt) \to_w (\lambda c_s^2)^{1/2} BM(t)$, and the joint distribution of the diffusion scaled arrival and service processes converges weakly jointly to the joint distribution of two independent Brownian motions.

6.11 Obtain the fluid and diffusion scaling limits of $\mathcal{S}_{\mathcal{A}(t)}$.

6.12 Show that for all three cases, when $\rho > 1$, when $\rho < 1$, and when ρ^n satisfies the conditions of heavy traffic (6.12), we have $\mu \hat{W}^n(t)/\hat{Q}^n(t) \to_w 1$ (where we let $0/0 = 1$).

6.13 (*) Obtain fluid and diffusion scaling and limits for G/G/s, fixed s, $\rho \nearrow 1$ [Iglehart and Whitt (1970a); Borovkov (1965)].

6.14 Prove equation (6.18) for the autocovariance of the diffusion limit of $Q(t)$ for G/D/∞.

6.15 Calculate the autocovariance function for the stationary diffusion limit of the queue length process $Q(t)$ for G/Discrete/∞ system.

6.16 Prove equation (6.19) for the autocovariance function of the stationary diffusion limit of G/G/∞.

7

Diffusions and Brownian Processes

In this chapter we describe in more detail Brownian motion that we used to approximate the G/G/1 queue, and some related diffusion and Brownian processes that will be useful in the approximation of queueing networks. We will also show their use to control a manufacturing process. We will mainly quote the relevant results on properties of Brownian processes, and give some more details in the supplement Section 7.7.

7.1 Diffusion Processes, Brownian Motion, and Reflected Brownian Motion

A diffusion process is a continuous time Markov process on the real line, with continuous sample paths. Special cases include Brownian motion, reflected Brownian motion, and the Ornstein–Uhlenbeck process. A general representation using Ito calculus is (see Section 7.7.4)

$$dX(t) = m(X(t))dt + \sigma(X(t))dBM(t), \qquad (7.1)$$

where $m(\cdot)$ represents a state-dependent drift, and $\sigma(\cdot)^2$ a state-dependent diffusion coefficient. As before, we use BM to denote standard Brownian motion. Diffusion processes can also be represented by the Kolmogorov backward equation, by the Fokker–Planck equation, and by the Feynman–Kac formula.

The basic diffusion process is the standard Brownian motion process $BM(t)$, defined by the following properties:

  (i) BM(t) has continuous sample paths, starting at $BM(0) = 0$.
 (ii) BM(t) has stationary independent increments, that is for any $s_1 < t_1 < \cdots < s_n < t_n$, $BM(t_1) - BM(s_1), \ldots, BM(t_n) - BM(s_n)$ are independent random variables, and the distribution of $B(t) - B(s)$ depends on $t - s$ alone.
(iii) BM(t) has a Gaussian, $N(0, t)$, distribution.

Wiener's theorem states that such a process exists and is unique, and it is in fact fully characterized by (i) and (ii) alone.

7.1.1 Properties of Brownian motion

Among the elementary properties of standard Brownian motion are its symmetry: $BM(t) =_w -BM(t)$ and its square root scaling: $aBM(t) =_w BM(a^2t)$. $BM(t)$ is strongly Markovian, that is, for any stopping time T, the process $BM^*(t) = BM(T + t) - BM(T)$ is itself a standard Brownian motion.

The variation of a function x over $[0, t]$ is defined by $v_t(x) = \sup\{\sum_{i=1}^n | x(t_i) - x(t_{i-1})|\}$ where the supremum is over all finite partitions $0 = t_0 < t_1 < \cdots < t_n = t$. For a monotone function, it is simple $|x(t) - x(0)|$. A function is of bounded variation if its variation is finite, and every function of bounded variation can be represented as the difference of two finite monotone functions. While the paths of the standard BM are continuous, they almost surely have infinite variation, that is, $v_t(BM(\cdot, \omega)) = \sup\{\sum_{i=1}^n |BM(t_i, \omega) - BM(t_{i-1}, \omega)|\}$, is infinite for almost all ω.

The quadratic variation of a function x over $[0, t]$ is defined by $q_t(x) = \sup\{\sum_{i=1}^n (x(t_i) - x(t_{i-1}))^2\}$ where the supremum as before is over all finite partitions. For any function of bounded variation, it is equal to 0. An important property of the continuous but extremely variable paths of Brownian motion is that their quadratic variation is given precisely (see Exercise 7.1) by

$$q_t(BM(\cdot, \omega)) = \sup \left\{ \sum_{i=1}^n (BM(t_i, \omega) - BM(t_{i-1}, \omega))^2 \right\} = t. \quad (7.2)$$

This property is the basis for Ito calculus (Section 7.7.4).

The general Brownian motion with initial state $x \geq 0$, drift m, and diffusion coefficient σ^2 is defined as

$$BM_x(t; m, \sigma^2) = x + mt + \sigma BM(t). \quad (7.3)$$

Clearly, $BM_x(t; m, \sigma^2) \sim N(x + mt, \sigma^2 t)$.

A celebrated calculation (see Section 7.7.1), using the reflection principle, shows that

$$\mathbb{P}(\sup_{0<s<t} BM(s) \leq y) = \mathbb{P}(|BM(t)| \leq y) = \Phi(yt^{-1/2}) - \Phi(-yt^{-1/2}), \quad (7.4)$$

where $\Phi(\cdot)$ is the cumulative Gaussian distribution. Another celebrated calculation (see Section 7.7.2), using a change of measure argument, shows that for a general Brownian motion,

$$\mathbb{P}(\sup_{0<s<t} BM_x(s; m, \sigma^2) \leq y) = \Phi\left(\frac{y - x - mt}{\sigma t^{1/2}}\right) - e^{2m(y-x)/\sigma^2}\Phi\left(\frac{-y + x - mt}{\sigma t^{1/2}}\right).$$

(7.5)

7.1.2 Reflected Brownian Motion

Next we define reflected Brownian motion. Consider a general Brownian motion, $BM_x(\cdot; m, \sigma^2)$, and consider its Skorohod reflection mapping as defined in Section 6.3: $Y(\cdot, \omega) = \varphi(BM_x(\cdot, \omega; m, \sigma^2))$ and $Z(\cdot, \omega) = \psi(BM_x(\cdot, \omega; m, \sigma^2))$. Recall that $Y(t, \omega) = -\inf_{0<s<t}\{0, BM_x(s, \omega; m, \sigma^2)\}$, so $Y(t, \omega)$ is non-negative with monotone non-decreasing paths. In the special case of $BM_0(\cdot, m, \sigma^2)$, when $Y(t, \omega) = -\inf_{0<s<t}\{BM_x(s, \omega; m, \sigma^2)\}$ the marginal distribution of $Y(t, \omega)$ is

$$\mathbb{P}(Y(t) \leq y) = \Phi\left(\frac{y + mt}{\sigma t^{1/2}}\right) - e^{-2my/\sigma^2}\Phi\left(\frac{-y + mt}{\sigma t^{1/2}}\right).$$

(7.6)

The process $\psi(BM_x(\cdot, \omega; m, \sigma^2))$ is a *Reflected Brownian Motion*, and we will denote it by $RBM_x(t; m, \sigma^2)$. We have:

$$RBM_x(t, \omega; m, \sigma^2) = BM_x(t, \omega; m, \sigma^2) + Y(t, \omega).$$

(7.7)

This process is non-negative, and at times at which it is > 0 its sample paths behave like sample paths of the Brownian motion that generated it. However, when it hits the value 0, it is "reflected", or "pushed" away toward the positive values. The process $Y(t, \omega)$ is the cumulative amount of pushing, and although it is continuous and monotone increasing, it is not absolutely continuous; in other words, it is not an integral of some pushing rate. Descriptively, whenever $RBM_x(t, \omega; m, \sigma^2)$ hits 0, $Y(t, \omega)$ will have an uncountable number of points of increase in the neighborhood, but the Lebesgue measure of all the points of increase of $Y(t, \omega)$ is 0; on the other hand, it will accumulate to a positive quantity. For that reason, $Y(t)$ is referred to as a *singular control*.

The marginal distribution of $RBM_0(t; m, \sigma^2)$ is calculated from that of Y by a time-reversal argument (see Section 7.7.3). It is

$$\mathbb{P}(RBM_0(t; m, \sigma^2) \leq y) = \Phi\left(\frac{y - mt}{\sigma t^{1/2}}\right) - e^{2my/\sigma^2}\Phi\left(\frac{-y - mt}{\sigma t^{1/2}}\right).$$

(7.8)

When $m < 0$, the distribution of RBM converges as $t \to \infty$ to a stationary distribution for an RBM with negative drift:

$$\mathbb{P}(RBM_x(t; m, \sigma^2) \leq y) \to 1 - e^{2my/\sigma^2} \quad \text{as } t \to \infty, \qquad (7.9)$$

i.e. an exponential distribution with parameter $-2m/\sigma^2$ and mean $\sigma^2/(-2m)$. For the distribution of $RBM_x(t; m, \sigma^2)$ and of its regulator, see Exercise 7.8.

7.1.3 Some Additional Related Processes

We have already encountered the *Ornstein–Uhlenbeck process* in Section 6.9, as the diffusion limit for the G/M/∞ system. It is the process given by:

$$dX(t) = -b[X(t) - m] + \sigma BM(t), \qquad (7.10)$$

with parameters $b > 0, m, \sigma > 0$. It is the unique process that is a diffusion process (i.e. it is Markovian with continuous sample paths), is stationary, and has marginal Gaussian distributions, $X(t) \sim N(m, \frac{\sigma^2}{2b})$, with autocovariance function $\mathbb{C}\text{ov}(X(t + s), X(t)) = \frac{\sigma^2}{2b} e^{-bs}$. Paths of the Ornstein–Uhlenbeck process are similar to paths of Brownian motion, with drift toward the mean value m that is proportional to the distance from m. The discrete time process $X_t = X(t), t = 0, 1, \ldots$, is also a Markov process, given by

$$X_t - m = e^{-b}(X_{t-1} - m) + a_t, \quad a_t \text{ are i.i.d.} \sim N\left(0, (1 - e^{-2b})\frac{\sigma^2}{2b}\right). \quad (7.11)$$

It is known as the first-order autoregressive process.

Another family of processes related to Brownian motion are stationary Gaussian processes. These have continuous paths, Gaussian marginal distributions, and are characterized by the mean and variance of the marginal distribution, and by the autocovariance function. As we saw, the stationary diffusion scale limit to G/G/∞ is such a Gaussian stationary process, which is characterized by its autocovariance function, given by (6.19).

In addition, we shall discuss multivariate Brownian motion and its Skorohod reflection in Section 7.6. We will encounter the Brownian bridge and the Kiefer process in Section 18.1.

7.2 Diffusions and Birth and Death Processes

It is instructive to see that quite general diffusion processes can be obtained as limits of centered and scaled Markovian discrete state birth and death processes. We slightly extend the definition of birth and death processes:

Definition 7.1 A generalized birth and death process on the real line, $X(t)$, is a continuous time discrete state Markov process, with real valued states $\alpha_0 < \alpha_1 < \cdots$, and transitions only to neighboring states, with birth rates λ_i to go from state α_i to α_{i+1} and death rates μ_i to go from α_i to α_{i-1}. The infinitesimal mean and variance of $X(t)$ are defined for $-\infty < x < \infty$ as

$$
\begin{aligned}
m(x) &= \lim_{h \to 0} \tfrac{1}{h} \mathbb{E}\left(X(t+h) - X(t) \,|\, X(t) = \alpha_i\right) \\
&= \lambda_i(\alpha_{i+1} - \alpha_i) - \mu_i(\alpha_i - \alpha_{i-1}), \\
\sigma^2(x) &= \lim_{h \to 0} \tfrac{1}{h} \mathrm{Var}\left(X(t+h) - X(t) \,|\, X(t) = \alpha_i\right) \\
&= \lim_{h \to 0} \tfrac{1}{h} \mathbb{E}\left((X(t+h) - X(t))^2 \,|\, X(t) = \alpha_i\right) \\
&= \lambda_i(\alpha_{i+1} - \alpha_i)^2 + \mu_i(\alpha_i - \alpha_{i-1})^2, \\
&\quad\quad \alpha_i \le x < \alpha_{i+1}.
\end{aligned}
\tag{7.12}
$$

Consider now a sequence of birth and death processes $X^n(t)$, with states α_i^n and rates λ_i^n, μ_i^n, and infinitesimal means and variances $m_{i,n}(x), \sigma_{i,n}^2(x)$. It is then possible to show, similar to Donsker's theorem,

Theorem 7.2 (Stone (1963)) *Let $X^n(t)$ be a sequence of birth and death processes as above, and let $X(t)$ be a diffusion process with drift $m(x)$ and diffusion coefficient $\sigma(x)$. $X^n(t) \to_w X(t)$ as $n \to \infty$ if and only if:*
  (i) $X^n(0) \to_w X(0)$ as $n \to \infty$,
  (ii) the states α_i^n become dense in $(-\infty, \infty)$ as $n \to \infty$,
  (iii) $m_n(x) \to m(x)$ and $\sigma_n^2(x) \to \sigma^2(x)$ u.o.c. as $n \to \infty$.

Example 7.3 (The M/M/∞ queue) Consider a sequence of M/M/∞ birth and death queues, with arrival rates λn and service rate μ. Let $Q^n(t)$ be its queue length, and let $\hat{Q}^n(t) = \frac{1}{\sqrt{n}}\left(Q^n(t) - \frac{\lambda}{\mu}n\right)$.

For $\hat{Q}^n(t) = x$, we have $Q^n(t) = \sqrt{n}x + \frac{\lambda}{\mu}n$. The process $\hat{Q}^n(t)$ is a birth and death process with jumps of $\pm\frac{1}{\sqrt{n}}$, which from state x have rates λn up and $(\sqrt{n}x + \frac{\lambda}{\mu}n)\mu$ down. We then have

$$
\begin{aligned}
m_n(x) &= (\lambda n)\frac{1}{\sqrt{n}} - (\sqrt{n}x\mu + \lambda n)\frac{1}{\sqrt{n}} \\
&= -\mu x,
\end{aligned}
$$

$$
\begin{aligned}
\sigma_n^2(x) &= (\lambda n)\frac{1}{n} + (\sqrt{n}x\mu + \lambda n)\frac{1}{n} \\
&= 2\lambda + x\mu\frac{1}{\sqrt{n}} \\
&\to 2\lambda, \quad \text{as } n \to \infty.
\end{aligned}
$$

We obtain that $\hat{Q}^n(t)$ converges to the Ornstein–Uhlenbeck process:

$$d\hat{Q}(t) = -\mu\hat{Q}(t) + \sqrt{2\lambda}BM(t),$$

as we already saw in Example 6.8.

7.3 Approximation of the G/G/1 Queue

We now put together results of previous sections to discuss approximations of the G/G/1 queue. We also state results for the workload process, $\mathcal{W}(t)$.

We consider first the fluid approximation. Assuming that interarrival times and processing times have finite variance, we have by the law of the iterated logarithm that for large values of T

$$\sup_{0\leq t\leq T} |Q(t) - (Q(0) + (\lambda - \mu)t)^+| = O(\sqrt{T\log\log T}) \quad \text{a.s.,}$$

$$\sup_{0\leq t\leq T} |\mathcal{W}(t) - (\mathcal{W}(0) + (\rho - 1)t)^+| = O(\sqrt{T\log\log T}) \quad \text{a.s.,} \tag{7.13}$$

and starting from $Q(0) = 0$,

$$\sup_{0\leq t\leq T} |\mathcal{T}(t) - (\rho \wedge 1)t| = O(\sqrt{T\log\log T}) \quad \text{a.s.} \tag{7.14}$$

Next, we assume that interarrival times and processing times have finite moments of order $r > 2$ and use strong approximations, and we consider the case when λ and μ are close. Then we can have versions of the queueing process and a Brownian process for which, for large T

$$\sup_{0\leq t\leq T} |Q(t) - RBM_{Q(0)}(t; \lambda-\mu, \lambda c_a^2+\mu(1\wedge\rho)c_s^2)| = o(T^{1/r'}) \quad \text{a.s.,} \tag{7.15}$$

where $r' = \min(r, 4 - \delta)$ and $\delta > 0$ is arbitrary. This means that the queue length process is close in distribution to a reflected Brownian motion, starting from $Q(0)$, with drift $\lambda-\mu$ and diffusion coefficient $\lambda c_a^2+\mu(1-\rho)c_s^2$. When $\lambda < \mu$ (so the queue is stable), the stationary distribution of the queue length can be approximated by that of the RBM, which is exponential with mean $\frac{\lambda(c_a^2+c_s^2)}{2(\mu-\lambda)}$. Note the similarity to Kingman's bound.

7.4 Two Sided Regulation of Brownian Motion

Consider an RCLL process $x(t)$ with $0 \leq x(0) \leq b$, which we wish to keep in the range $(0, b)$. To do so, we push up at 0 and down at b. What we get are the processes Z, L, U such that

(i) $0 \leq Z(t) = x(t) + L(t) - U(t) \leq b$,

(ii) $L(t), U(t)$ non-decreasing, start at 0,

(iii) $Z(t)\, dL(t) = 0, \quad (b - Z(t))\, dU(t) = 0,$

In words, the process is driven by x, and the processes L and U keep Z in $[0, b]$. They start at 0, are non-decreasing, and increase only when $Z(t) = 0$ or $Z(t) = b$, respectively.

There is an explicit formula to calculate U, L directly from x, but it is not very simple. However, it is immediate to see that, similar to one-sided reflection,

$$L(t) = \sup_{0<s\leq t} (x(s) - U(s))^-, \qquad U(t) = \sup_{0<s\leq t} (b - x(s) - L(s))^-. \quad (7.16)$$

Theorem 7.4 *For any x RCLL with $0 \leq x(0) \leq b$, there exist processes L, U as required, and they are unique. The processes L, U are non-anticipating, and are pathwise minimal. The mapping of x to corresponding Z, L, U is continuous in $\mathbb{D}$.*

Exercise 7.5 discusses existence, uniqueness, and minimality of L, U.

Consider now a Brownian motion, $X(t) = BM_x(t; \theta, \sigma^2)$, with $0 \leq X(0) = x \leq b$. Applying the two-sided regulation to $X(t)$, we obtain $\mathcal{Z}, \mathit{Ł}, \mathcal{U}$, the reflected Brownian motion, and its lower and upper barrier regulators,

$$\mathcal{Z}(t) = BM_x(t; m, \sigma^2) + \mathit{Ł}(t) - \mathcal{U}(t) = X(t) + \mathit{Ł}(t) - \mathcal{U}(t). \quad (7.17)$$

$\mathcal{Z}(t)$ behaves like Brownian motion when in $(0, b)$ and is reflected at the boundaries, with minimal cumulative reflections $\mathit{Ł}(t), \mathcal{U}(t)$.

Recall that for one-sided reflection of Brownian motion we could calculate the transient distribution of $\mathit{Ł}(t)$ and $\mathcal{Z}(t)$. We are not able to obtain similar expressions for the two sided regulated Brownian motion. However, it turns out that we can obtain expressions for useful quantities when we consider $t \to \infty$. It is possible to calculate discounted costs of the form:

$$E_x\left\{ \int_0^\infty e^{-\gamma t} [u(\mathcal{Z}(t))dt - rd\mathit{Ł}(t) + cd\mathcal{U}(t)] \right\},$$

for discount factor γ, constants c, r, continuous function u, and starting point x, see Exercise 7.11.

We can also calculate the long-term average of $\mathit{Ł}$ and $\mathcal{U}$ and the stationary distribution of $\mathcal{Z}$. We outline this calculation. Define the following sequence of times: T_1 is the first time that $\mathcal{Z}(t)$ hits zero, and T_{n+1} is the first time after T_n such that $\mathcal{Z}(t)$ hits 0 after first hitting b. Then T_n are regeneration times of the process $\mathcal{Z}$. Furthermore, it can be shown that $\mathbb{E}(T_n) < \infty$ for all n. We then have by the renewal reward theorem that

$$\lim_{t \to \infty} Ł(t)/t = \frac{\mathbb{E}(Ł(T))}{\mathbb{E}(T)}, \quad \lim_{t \to \infty} \mathcal{U}(t)/t = \frac{\mathbb{E}(\mathcal{U}(T))}{\mathbb{E}(T)},$$

$$\lim_{t \to \infty} \mathbb{P}(Z(t) \le x) = \frac{\mathbb{E}[\int_0^T \mathbb{1}(Z(t) \le x)dt]}{\mathbb{E}(T)}, \tag{7.18}$$

where $T = T_{n+1} - T_n$.

The expected quantities can be calculated using Ito calculus. We perform this calculation in Section 7.7.5. The results are:

$$\lim_{t \to \infty} Ł(t)/t = \alpha = \begin{cases} \sigma^2/2b, \\ \dfrac{m}{e^{2mb/\sigma^2} - 1}, \end{cases} \quad \lim_{t \to \infty} \mathcal{U}(t)/t = \beta = \begin{cases} \sigma^2/2b, \\ \dfrac{m}{1 - e^{-2mb/\sigma^2}}, \end{cases}$$

$$\lim_{t \to \infty} \mathbb{P}(Z(t) \le y) = \begin{cases} y/b, & m = 0 \\ \dfrac{e^{2my/\sigma^2} - 1}{e^{2mb/\sigma^2} - 1}, & m \ne 0, \end{cases}$$

$$\tag{7.19}$$

This is easily interpreted: If $m = 0$, then $Z(t) \sim U(0, b)$, i.e. it is uniformly distributed across the range. Otherwise, if $m < 0$ it is distributed like a truncated exponential random variable, and if $m > 0$ then $b - Z(t)$ is distributed like a truncated exponential random variable. The rate of this exponential distribution is $2|m|/\sigma^2$, which is the rate of the stationary exponential distribution of the one-sided reflected Brownian motion, with drift $-|m|$.

As for the amount of regulation, for $m = 0$ it is symmetric on both sides, and totals σ^2/b. For $m \ne 0$ the total rate of push in the positive direction, which equals $Ł - \mathcal{U}$ totals exactly $-m$, so the rate of regulation is equal and opposite the drift, which is as expected, since the regulation is necessary to eliminate the drift out of range. The ratio of the two sides is $1 : e^{-2mb/\sigma^2}$, which is equal to the ratio of the probability densities of $Z(t)$ at the two boundaries.

7.5 Optimal Control of a Manufacturing System

We now formulate a production planning problem for a manufacturing plant, and we use the results of Section 7.4 to optimize the operation of the plant.

We consider a manufacturing system that produces a single item. Cumulative demand for this item is given by $X(t, \omega)$, and demand is supplied immediately out of inventory, and is lost otherwise. We assume a constant work force with production capacity k, and a finite bound on the inventory b. The values k and b are the most important overall decision variables.

We assume that the demand process is stationary and has independent increments. This means that the state of the system at time t is completely determined by the inventory level $Z(t)$. We argue that if $Z(t) > 0$, we should supply the total demand rate $dX(t)$, since otherwise demand will be lost, and we will pay extra inventory costs. Also, we argue that we should always produce at full capacity up to some desired level of inventory: If we determine a maximal inventory level b, and use less than full production capacity to replenish inventory when at level $z < b$, then we can achieve higher profit by lowering b and using full capacity (see Exercise 7.9).

Once k and b are determined, the policy is as follows: Supply at full rate $dX(t)$ and produce at full capacity while $0 < Z(t) < b$, and only reduce supply when $Z(t) = 0$ and reduce production when $Z(t) = b$.

We now introduce the parameters of the system:
- Cumulative demand is $X(t, \omega)$ with mean rate of a items per unit time.
- Production capacity, i.e. workforce, is k items per unit time.
- Maximal inventory level, i.e. storage capacity, is b items.
- Product sells at a price of r per item.
- Product requires material and direct work costs of c per item.
- Work force fixed costs are w per unit capacity per unit time.
- Inventory holding costs are h per item per unit time.
- Inventory drift is $m = k - a$ per unit time.

Denote now in addition to the processes $X(t, \omega)$, $Z(t, \omega)$,
- Cumulative unused production capacity $U(t, \omega)$.
- Cumulative lost sales $L(t, \omega)$.

We then have

$$\text{Production:} \quad kt - U(t), \qquad \text{Sales:} \quad X(t) - L(t),$$

$$\text{Inventory:} \quad Z(t) = kt - X(t) + L(t) - U(t), \qquad (7.20)$$

and the long-range profit per unit time is

$$\lim_{T \to \infty} \frac{1}{T} \left(r(X(T) - L(T)) - wkT - c(kT - U(t)) - h \int_0^T Z(t)dt \right). \quad (7.21)$$

Assume now that the process $kt - X(t, \omega)$ is a Brownian motion with drift $m = k - a$ and variance (diffusion coefficient) σ^2. Then, using the policy of full supply and full production, the inventory level $Z(t)$ behaves as a regulated Brownian motion reflected between 0 and b. We can now determine the long-term profit rate for the system, given by equation (7.21), for any pair of values k, b, by using the expressions obtained for $\lim_{t\to\infty} L(t)/t$, $\lim_{t\to\infty} U(t)/t$, and $\mathbb{P}(Z(t) \le y)$ from equations (7.19). Once we can calculate the profit rates as a function of (k, b), we can optimize long-term performance numerically. Exercise 7.10 presents an example.

Remark $kt - X(t)$ is a difference of two monotone functions and as such is of bounded variation, so it cannot be a Brownian motion. Nevertheless, Brownian approximation does make sense for representing a stationary system with independent increment demands, if the system operates on a large scale.

7.6 Oblique Reflection and Multivariate RBM

To discuss queueing networks we will need a more general, d-dimensional version of Skorohod's problem, and of the reflected Brownian motion.

We first define Skorohod reflection in the positive d-dimensional orthant of $\mathbb{R}^d$; the positive orthant consists of the points in $\mathbb{R}^d$ with non-negative coordinates. Consider a function x in $\mathbb{D}^d$, the space of RCLL functions from $(0, \infty)$ to $\mathbb{R}^d$, with $x(0) \ge 0$. The Skorohod problem in $\mathbb{R}^d$ is to find, for a fixed matrix R, a mapping of $x(t)$ to two vector functions, y, z, so that

$$z(t) = x(t) + Ry(t) \ge 0,$$
$$y(0) = 0, \quad y \text{ non-decreasing}, \qquad (7.22)$$
$$\int_0^t z_j(s)dy_j(s) = 0, \quad j = 1, \ldots, d,$$

where R is a $d \times d$ matrix, called the reflection matrix. This is similar to the one-dimensional Skorohod problem. We can think of the path of x in $\mathbb{R}^d$, which we try to keep within the non-negative orthant. Whenever x reaches one of the boundary hyper-planes of the orthant, it is pushed back to stay in the orthant by the appropriate component y_j of y, and the direction of the reflection from each face of the orthant is determined by the jth column of the reflection matrix R.

Theorem 7.5 (Harrison and Reiman (1981)) *If $R = (I - P)D$ where D is diagonal and P is non-negative with spectral radius less than 1, then Skorohod's problem defines a mapping $\tilde{y} = Ry = \varphi(x)$, $z = \psi(x)$, such that $\tilde{y}, z$ are well defined and unique, $\tilde{y}(t), z(t)$ depend only on $x(s)$, $0 \le s \le t$, and the mapping is continuous: if $x_n \to x$, then $\varphi(x_n), \psi(x_n) \to \varphi(x), \psi(x)$.*

Next we define multivariate Brownian motion in $\mathbb{R}^d$:

Definition 7.6   A standard Brownian motion in $\mathbb{R}^d$ is a stochastic process $BM^d(t) \in \mathbb{R}^d$, $t \geq 0$, where each component $BM_j^d(t)$ is a standard Brownian motion, and the d components are independent.

A general Brownian motion in $\mathbb{R}^d$ is obtained by $X(t) = x_0 + \theta t + A\,BM^d(t)$, where A is a non-singular square matrix, and x_0, $\theta \in \mathbb{R}^d$ are initial value and drift vectors. $X(t)$ has mean $x_0 + mt$, and a variance covariance matrix Γt where $\Gamma = A^\mathsf{T} A$. It is denoted by $X(t) = BM_{x_0}(t; \theta, \Gamma)$.

The reflected Brownian motion in $\mathbb{R}^d$ is defined by

Definition 7.7   Let $X(t) = BM_{x_0}(t; \theta, \Gamma)$ be a d-dimensional Brownian motion, starting at $x_0 \geq 0$, with d-dimensional drift vector θ and diffusion variance covariance matrix Γ, and let R be a given reflection matrix. Then the process with paths $\mathcal{Z}(\cdot) = \psi(X(\cdot))$ is called a reflected Brownian motion, denoted $\mathcal{Z}(t) = RBM_{x_0}(t; \theta, \Gamma, R)$, and $\mathcal{Y}(\cdot) = \varphi(X(\cdot))$ is the regulator of X.

Theorem 7.8 $RBM_{x_0}(t; \theta, \Gamma, R)$ *has a stationary distribution if and only if* $R^{-1}\theta < 0$.

In spite of the existence of a stationary distribution, calculation of moments and probabilities from this distribution are quite intractable and are currently subject to intensive research.

7.7 Supplement: Calculations for Brownian Motion and Derived Diffusions

7.7.1 A Joint Distribution, Reflection Principle

Consider a standard Brownian motion $BM(t)$. Let $M(t)$ be its supremum in $[0,t]$, $M(t) = \sup\{BM(s), 0 \leq s \leq t\}$. We calculate: $G_t(x, y) = \mathbb{P}(BM(t) \leq x, M(t) \leq y)$.

$$G_t(x, y) = \mathbb{P}(BM(t) \leq x) - \mathbb{P}(BM(t) \leq x, M(t) > y)$$
$$= \Phi(xt^{-1/2}) - \mathbb{P}(BM(t) \leq x, M(t) > y),$$

where Φ is the standard Gaussian cumulative distribution function. The second term is the probability of paths that cross y and return to $x \leq y$ at time t. We now use a reflection principle argument to say that this is equal to the probability of paths that cross y and eventually at time t reach a height that is equal to or exceeds $y + (y - x) = 2y - x$ (see Figure 7.1). For the exact argument, one needs to consider the stopping time $T = \min\{t : BM(t) \geq y\}$,

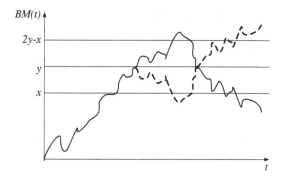

Figure 7.1 The reflection principle.

and use the strong Markov property and the symmetry of Brownian motion, as follows:

$$\mathbb{P}(BM(t) \le x, M(t) > y) = \mathbb{P}(T < t, \; BM(t) \le x)$$
$$= \mathbb{P}(T < t, \; BM^*(t - T) \le x - y)$$
$$= \mathbb{P}(T < t, \; BM^*(t - T) \ge y - x)$$
$$= \mathbb{P}(BM(t) \ge 2y - x)$$
$$= \Phi((x - 2y)t^{-1/2}),$$

where BM^* is the process starting at T, which by the strong Markov property is an independent standard BM.

Hence,

$$G_t(x, y) = \mathbb{P}(BM(t) \le x, M(t) \le y) = \Phi(xt^{-1/2}) - \Phi((x - 2y)t^{-1/2}),$$

with density $g_t(x, y)dx = \mathbb{P}(BM(t) \in dx, M(t) \le y)$:

$$g_t(x, y) = [\phi(xt^{-1/2}) - \phi((x - 2y)t^{-1/2})]t^{-1/2},$$

where $\phi(x) = \frac{1}{\sqrt{2\pi}}e^{-\frac{x^2}{2}}$ is the standard normal density.

From this we obtain that

$$\mathbb{P}(M(t) \le y) = \mathbb{P}(M(t) \le y, BM(t) \le y) = \Phi(yt^{-1/2}) - \Phi(-yt^{-1/2}),$$

with density

$$\mathbb{P}(M(t) \in dy) = 2t^{-1/2}\phi(yt^{-1/2}).$$

Note that $\Phi(yt^{-1/2}) - \Phi(-yt^{-1/2})$ also equals $\mathbb{P}(|BM(t)| \le y)$, that is, the maximum of standard Brownian motion in the interval $[0, t]$ has the same distribution as the absolute value at time t.

7.7.2 *Change of Drift via Change of Measure*

We calculated $g_t(x, y)dx = \mathbb{P}(BM(t) \in dx, M(t) \leq y)$ where $BM(t)$ was a standard Brownian motion. We now wish to calculate $f_t(x, y)dx = \mathbb{P}(BM(t; m, 1) \in dx, M(t) \leq y)$, where $BM(t; m, 1)$ is a Brownian motion starting at 0, with drift m and diffusion coefficient 1. Writing the likelihood ratio,

$$\frac{f_{BM(t;m,1)}(z)}{f_{BM(t)}(z)} = \frac{\frac{1}{\sqrt{2\pi t}}e^{-\frac{(z-mt)^2}{2t}}}{\frac{1}{\sqrt{2\pi t}}e^{-\frac{z^2}{2t}}} = e^{zm-m^2t/2},$$

we use this as the Radon–Nikodym derivative, the random variable $\xi = e^{X(t)m-m^2t/2}$ where $X(t)$ is the random value of $BM(t)$. Hence

$$\mathbb{P}(BM(t; m, 1) \leq x, M(t; m, 1) \leq y)$$
$$= \mathbb{E}\left[\mathbb{1}\{BM(t; m, 1) \leq x, M(t; m, 1) \leq y\}\right]$$
$$= \mathbb{E}\left[\xi\,\mathbb{1}\{BM(t) \leq x, M(t) \leq y\}\right]$$
$$= \mathbb{E}\left[e^{X(t)m-m^2t/2}\mathbb{1}\{X(t) \leq x, M(t) \leq y\}\right]$$
$$= \int_{-\infty}^{x} e^{zm-m^2t/2}\mathbb{P}(X(t) \in dz, M(t) \leq y)$$
$$= \int_{-\infty}^{x} e^{zm-m^2t/2}g_t(z, y)dz.$$

It follows that

$$f_t(x, y) = \mathbb{P}(BM(t; m, 1) \in dx, M(t) \leq y) = e^{xm-m^2t/2}g_t(x, y)$$
$$= t^{-1/2}e^{xm-m^2t/2}\left[\phi(xt^{-1/2}) - \phi((x - 2y)t^{-1/2})\right]$$
$$= (2\pi t)^{-1/2}\left[e^{-(x-mt)^2/2t} - e^{2my}e^{-(x-2y-mt)^2/2t}\right]$$
$$= t^{-1/2}\phi((x - mt)t^{-1/2}) - t^{-1/2}e^{2my}\phi((x - 2y - mt)t^{-1/2}).$$

We also get by scaling, for general $BM(t; m, \sigma^2)$, that

$$F_t(x, y) = \mathbb{P}(BM(t; m, \sigma^2) \leq x, M(t) \leq y)$$
$$= \Phi\left(\frac{x - mt}{\sigma t^{1/2}}\right) - e^{2my/\sigma^2}\Phi\left(\frac{x - 2y - mt}{\sigma t^{1/2}}\right).$$

We now get a first passage time and the distribution of the supremum, by taking $x = y$. Let $T(y)$ be the first t for which $BM(t; m, \sigma^2) = y$, then

$$\mathbb{P}(T(y) > t) = \mathbb{P}(M(t) < y) = F_t(y, y)$$
$$= \Phi\left(\frac{y - mt}{\sigma t^{1/2}}\right) - e^{2my/\sigma^2}\Phi\left(\frac{-y - mt}{\sigma t^{1/2}}\right).$$

The distribution of $T(y)$ is called the inverse Gaussian distribution.

The density of the hitting time:

$$f_{T(y)}(t)dt = \mathbb{P}[T(y) \in (t, t+dt)] = \frac{y}{\sigma\sqrt{2\pi t^3}} \exp\left(-\frac{(y - mt)^2}{2\sigma^2 t}\right) dt. \quad (7.23)$$

7.7.3 Reflected Brownian Motion

Consider now $X(t) = BM(t; m, \sigma^2)$ starting at 0, with drift m and diffusion coefficient σ^2. Consider the Skorohod reflection with regulator and reflected Brownian motion (RBM):

$$Ł(t) = -\inf_{0 \le s \le t} X(t), \qquad Z(t) = X(t) + Ł(t).$$

We look first at $Ł(t)$. We have $Ł(t) = -\inf_{0 \le s \le t} X(s) = \sup_{0 \le s \le t} X(s)^- =_D \sup_{0 \le s \le t} BM(s; -m, \sigma^2)$. Hence, the distribution of $Ł(t)$ is given by:

$$\mathbb{P}(Ł(t) \le y) = \Phi\left(\frac{y + mt}{\sigma t^{1/2}}\right) - e^{-2my/\sigma^2}\Phi\left(\frac{-y + mt}{\sigma t^{1/2}}\right). \quad (7.24)$$

Next we consider $Z(t)$; recall that we assume $Z(0) = 0$. We observe that $Z(t) = X(t) + Ł(t) = \sup_{0 \le s \le t}(X(t) - X(s)) = \sup_{0 \le s \le t}(X(t) - X(t - s))$. We now look at $X^*(s) = X(t) - X(t - s)$, for $0 \le s \le t$. Then X^* has independent increments and $X^*(s) \sim N(ms, \sigma^2 s)$ so X^* itself is again a $BM(t; m, \sigma^2)$.

$$\begin{aligned} Z(t) &= \sup_{0 \le s \le t}(X(t) - X(s)) = \sup_{0 \le s \le t}(X(t) - X(t - s)) \\ &= \sup_{0 \le s \le t} X^*(s) =_D \sup_{0 \le s \le t} X(s) = M(t). \end{aligned}$$

Hence

$$\mathbb{P}(Z(t) \le z) = \Phi\left(\frac{z - mt}{\sigma t^{1/2}}\right) - e^{2mz/\sigma^2}\Phi\left(\frac{-z - mt}{\sigma t^{1/2}}\right), \quad t \ge 0. \quad (7.25)$$

When $t \to \infty$ we get a stationary distribution:

$$\mathbb{P}(Z(t) \le z) \to \begin{cases} 1 - e^{2mz/\sigma^2} & \text{if } m < 0, \\ 0 & \text{if } m \ge 0. \end{cases} \quad (7.26)$$

In the special case that $m = 0$, the distribution of $Z(t)$ is the distribution of the absolute value of a mean 0 Gaussian variable. It has mean and variance (see Exercise 7.2) given by $\sigma\sqrt{\frac{2}{\pi}t}$ and variance $\sigma^2\left(1 - \frac{2}{\pi}\right)t$.

7.7.4 A Glimpse into Ito Calculus

We consider a Brownian motion $BM(t, \omega)$, and let $\mathcal{F}_t$, $t > 0$ be the σ-field generated by $BM(s)$, $s < t$. This roughly means that $\mathcal{F}_t$ is the history of $BM(\cdot)$ up to time t. Let $f(t, \omega)$ be a stochastic process that is adapted to $\mathcal{F}_t$. This means that $f(t, \omega)$ for every t is a random variable that is $\mathcal{F}_t$ measurable, and in particular it means that it is determined by $BM(s, \omega)$, $s < t$ and is independent of $BM(t + s) - BM(t)$, $s > 0$. Assume also that $\mathbb{E}[\int_0^T f(t, \omega)^2 dt] < \infty$. For such f, one can define the *Ito integral*

$$\int_S^T f(t, \omega) dBM(t, \omega) = \lim \sum_j f(t_j, \omega)(BM(t_{j+1}, \omega) - BM(t_j, \omega)),$$

where $S = t_0 < t_1 < \cdots < t_n = T$ and the limit is defined as a mean square limit, for partitions with $\max\{t_{j+1} - t_j\} \to 0$.

We calculate one example:

$$\int_0^T BM(t, \omega) dBM(t, \omega) = \lim \sum_j BM(t_j, \omega)(BM(t_{j+1}\omega) - BM(t_j, \omega))$$

$$= \frac{1}{2} BM(T, \omega)^2 - \frac{1}{2}T, \tag{7.27}$$

which is seen as follows: Denoting $B_j = BM(t_j, \omega)$, we get

$$B_{j+1}^2 - B_j^2 = (B_{j+1} - B_j)^2 + 2B_j(B_{j+1} - B_j) = (\Delta B_j)^2 + 2B_j \Delta B_j,$$

so

$$\sum_j B_j \Delta B_j = \frac{1}{2}B_T^2 - \frac{1}{2}\sum_j (\Delta B_j)^2,$$

and the second summation converges to the quadratic variation of $BM(t, \omega)$, which equals T.

We note the following properties of an Ito integral:

$$\int_0^t f(s) dBM(s) \text{ is measurable } \mathcal{F}_t, \qquad \mathbb{E}\left[\int_0^t f(s) dBM(s)\right] = 0,$$

and in fact $\mathcal{M}(t) = \int_0^t f(s) dBM(s)$ is a martingale.

We now define an *Ito process* (or a *stochastic integral*):

$$X(t, \omega) = X(0, \omega) + \int_0^t u(s, \omega) ds + \int_0^t v(s, \omega) dB(s, \omega),$$

where the second term is an Ito integral and the first term is an ordinary integral, with $\mathbb{E}[\int_0^T |u(t,\omega)|dt] < \infty$. This is written in shorthand as:

$$dX(t) = udt + vdB(t).$$

Ito processes require a new type of calculus, expressed by the *Ito formula*: Let $g(t,x)$ be twice continuously differentiable, and $X(t)$ an Ito process. The Ito derivative of $Y(t) = g(t, X(t))$ is given by

$$dY(t) = \frac{\partial}{\partial t}g(t, X(t))dt + \frac{\partial}{\partial x}g(t, X(t))dX(t) + \frac{1}{2}\frac{\partial^2}{\partial x^2}g(t, X(t))(dX(t))^2,$$

where $(dX(t))^2 = (dX(t)) \cdot (dX(t))$ is computed using the following rules

$$dt \cdot dt = dt \cdot dBM(t) = dBM(t) \cdot dt = 0, \quad dBM(t) \cdot dBM(t) = dt.$$

This makes sense, because the quadratic variation of $BM(t)$ is t and so in a Taylor expansion of $g(t, X(t))$ the second-order term $(dBM(t))^2$ is of the same order of magnitude as the first-order terms.

Example 7.9   Take $g(t, x) = x^2$, and $X(t) = BM(t)$. Then $\frac{\partial g}{\partial t} = 0$, $\frac{\partial g}{\partial x} = 2x$, $\frac{\partial^2 g}{\partial x^2} = 2$

$$d(BM(t)^2) = 2BM(t)dBM(t) + 2(dBM(t))^2 = 2BM(t)dBM(t) + dt,$$

which agrees with (7.27).

By its definition, an Ito process consists of the sum of an Ito integral (which is a martingale, and has unbounded variation and well-defined quadratic variation), and a stochastic process whose paths are of bounded variation. Any process that has two such components is an Ito process.

7.7.5 Calculations for Two Sided Regulated Brownian Motion

We now use Ito calculus to prove the formulas given in equation (7.19).

We consider a Brownian motion $X(t) = BM_0(t; m, \sigma^2)$ and the two-sided regulated Brownian motion $\mathcal{Z}(t) = X(t) + \mathcal{L}(t) - \mathcal{U}(t)$.

We note that $\mathcal{Z}(t)$ is an Ito process, with martingale component $\sigma BM(t)$ and bounded variation component $mt + \mathcal{L}(t) - \mathcal{U}(t)$. Let $f(x)$ be twice differentiable. Then, using the Ito formula for $f(\mathcal{Z}(t))$,

$$df(\mathcal{Z}) = f'(\mathcal{Z})d\mathcal{Z} + \frac{1}{2}f''(\mathcal{Z})(d\mathcal{Z})^2$$

$$= f'(\mathcal{Z})(\sigma dBM + mdt + d\mathcal{L} - d\mathcal{U}) + \frac{1}{2}\sigma^2 f''(\mathcal{Z})dt$$

$$= \sigma f'(\mathcal{Z})dBM + \Gamma f(\mathcal{Z})dt + f'(\mathcal{Z})d\mathcal{L} - f'(\mathcal{Z})d\mathcal{U}$$

where we use the notation Γ for the operator $\Gamma f = (mf' + \frac{1}{2}\sigma^2 f'')$. This says that:

$$f(\mathcal{Z}(t)) = f(\mathcal{Z}(0)) + \sigma \int_0^t f'(\mathcal{Z})dBM + \int_0^t \Gamma f(\mathcal{Z})ds$$
$$+ \int_0^t f'(\mathcal{Z})d\mathcal{L} - \int_0^t f'(\mathcal{Z})d\mathcal{U}.$$

Recall now that $\mathcal{L}$ increases only when $\mathcal{Z}(t) = 0$ and $\mathcal{U}$ increases only when $\mathcal{Z}(t) = b$. Hence,

$$\int_0^t f'(\mathcal{Z})d\mathcal{L} = f'(0) \int_0^t d\mathcal{L} = f'(0)\mathcal{L}(t),$$
$$\int_0^t f'(Z)d\mathcal{U} = f'(b) \int_0^t d\mathcal{U} = f'(b)\mathcal{U}(t),$$

and we have

$$f(\mathcal{Z}(t)) = f(\mathcal{Z}(0)) + \sigma \int_0^t f'(\mathcal{Z})dBM + \int_0^t \Gamma f(\mathcal{Z})ds$$
$$+ f'(0)\mathcal{L}(t) - f'(b)\mathcal{U}(t).$$

Recall the definition of the stopping time T in Section 7.4, as the time for $\mathcal{Z}(\cdot)$ to go from 0 to b and return to 0. Starting from $\mathcal{Z}(0) = 0$, we will calculate the expected value of $f(\mathcal{Z}(T))$:

$$\mathbb{E}_0(f(\mathcal{Z}(T))) = \mathbb{E}_0\left[f(\mathcal{Z}(0)) + \sigma \int_0^T f'(\mathcal{Z})dBM\right.$$
$$\left. + \int_0^T \Gamma f(\mathcal{Z})ds + f'(0)\mathcal{L}(T) - f'(b)\mathcal{U}(T)\right].$$

By definition of T, $\mathcal{Z}(T) = 0$ so $\mathbb{E}_0(f(\mathcal{Z}(T)))$ and $\mathbb{E}_0(f(\mathcal{Z}(0)))$ cancel. The expected value of an Ito integral is 0, so $\mathbb{E}_0[\int_0^T f'(\mathcal{Z})dBM] = 0$. We obtain the relation:

$$0 = \mathbb{E}_0\left[\int_0^T \Gamma f(\mathcal{Z})ds + f'(0)\mathcal{L}(T) - f'(b)\mathcal{U}(T)\right].$$

Denote by $\pi_{\mathcal{Z}}(dz)$ the stationary distribution of $\mathcal{Z}(t)$. Using the renewal reward theorem, we have that

$$\int_0^b \Gamma f(z)\pi_{\mathcal{Z}}(dz) = \frac{\mathbb{E}_0[\int_0^T \Gamma f(\mathcal{Z})ds]}{\mathbb{E}_0(T)}.$$

Recall again by the renewal reward theorem, and using the notation of

Section 7.4, that for the stationary version of $\mathcal{Z}(t)$,

$$\alpha = d\mathcal{L}(t) = \frac{\mathbb{E}_0[\mathcal{L}(T)]}{\mathbb{E}_0(T)}, \qquad \beta = d\mathcal{U}(t) = \frac{\mathbb{E}_0[\mathcal{U}(T)]}{\mathbb{E}_0(T)},$$

to get, for any twice differentiable f, the equation

$$0 = \int_0^b \Gamma f(z)\pi_Z(dz) + f'(0)\alpha - f'(b)\beta.$$

This is now solved for α, β, π_Z by substituting appropriate f.

Proposition 7.10 *The quantities $\pi_Z(dz)$, α, β are given by (7.19).*

Proof For $m = 0$, use $f(z) = z$, $f(z) = z^2$, and $f(z) = e^{\zeta z}$ to solve for α, β and the Laplace transform of π_Z.

For $m \neq 0$, use $f(z) = z$ and $f(z) = e^{-\theta z}$ with $\theta = 2m/\sigma^2$ to solve for α, β, and $f(z) = e^{\zeta z}$ to obtain the Laplace transform of π_Z.

Here are the details:

The case $m = 0$:

(i) $f(z) = z$, $f'(z) = 1$, $f''(z) = 0$, $\Gamma f(z) = 0$, equation is $0 = 0 + \alpha - \beta$.

(ii) $f(z) = z^2$, $f'(z) = 2z$, $f''(z) = 2$, $\Gamma f(z) = \sigma^2$, equation is $0 = \sigma^2 + 0 - 2b\beta$.

(iii) $f(z) = e^{\zeta z}$, $f'(z) = \zeta e^{\zeta z}$, $f''(z) = \zeta^2 e^{\zeta z}$, $\Gamma f(z) = \frac{1}{2}\sigma^2\zeta^2 e^{\zeta z}$, equation is $0 = \frac{1}{2}\zeta^2\sigma^2\mathbb{E}[e^{\zeta Z}] + \zeta\alpha - \zeta e^{\zeta b}\beta$.

From which, $\alpha = \beta = \frac{\sigma^2}{2b}$, and the Laplace transform of the stationary $\mathcal{Z}(t)$ is $\mathbb{E}[e^{\zeta Z}] = \frac{1}{\zeta b}\left(e^{\zeta b} - 1\right)$, from which we get that $Z \sim \text{Uniform}(0, b)$.

The case $m \neq 0$, with $\theta = \frac{2m}{\sigma^2}$:

(i) $f(z) = z$, $f'(z) = 1$, $f''(z) = 0$, $\Gamma f(z) = m$, equation is $0 = m + \alpha - \beta$.

(ii) $f(z) = e^{-\theta z}$, $f'(z) = -\theta e^{-\theta z}$, $f''(z) = \theta^2 e^{-\theta z}$, $\Gamma f(z) = -m\theta e^{-\theta z} + \frac{1}{2}\sigma^2\theta^2 e^{-\theta z} = 0$, equation is $0 = 0 - \theta\alpha + \theta e^{-\theta b}\beta$.

From which we obtain $\alpha = \frac{m}{e^{\theta b} - 1}$, $\beta = \frac{m}{1 - e^{-\theta b}}$.

(iii) $f(z) = e^{\zeta z}$, $f'(z) = \zeta e^{\zeta z}$, $f''(z) = \zeta^2 e^{\zeta z}$, $\Gamma f(z) = m\zeta e^{\zeta z} + \frac{1}{2}\sigma^2\zeta^2 e^{\zeta z}$, equation is $0 = \left(m\zeta + \frac{1}{2}\zeta^2\sigma^2\right)\mathbb{E}[e^{\zeta Z}] + \zeta\alpha - \zeta e^{\zeta b}\beta = \frac{1}{2}\zeta\sigma^2(\theta + \zeta)\mathbb{E}[e^{\zeta Z}] + \zeta\alpha - \zeta e^{\zeta b}\beta$.

From which we obtain $\mathbb{E}[e^{\zeta Z}] = \frac{\theta}{\theta + \zeta}\frac{e^{(\theta+\zeta)b} - 1}{e^{\theta b} - 1}$, which is the transform of the density $\mathbb{P}(\mathcal{Z}(t) \in dz) = \frac{\theta e^{\theta z}}{e^{\theta b} - 1}$, $0 < z < b$.   □

7.8 Sources

Our presentation in this chapter follows material in J. Michael Harrison's book *Brownian Motion and Stochastic Flow Systems*, Harrison (1985, 2013). The limiting diffusions for scaled birth and death processes come from Stone (1963); see also Iglehart (1965). Explicit form of the two-sided regulated Brownian motion is obtained by Kruk et al. (2007), and the reflected Brownian motion in the orthant originated in Harrison and Reiman (1981). The optimal policy for the manufacturing system of Section 7.5 is derived in Harrison and Taylor (1978) and Harrison and Taksar (1983). Approximation schemes to calculate moments and probabilities for multivariate reflected Brownian motion are derived by Dai and Harrison (1991, 1992), see also Dai and Miyazawa (2011). An extended version of Skorohod reflection is discussed by Ramanan (2006). Ito calculus is best studied in the book *Stochastic Differential Equations*, Oksendal (2013).

Exercises

7.1 Let $X(t)$ be a standard Brownian motion. Let $t = t_0 < t_1 < \cdots < t_n = t + \tau$ and let $\epsilon = \min_{k=1,\ldots,n} (t_k - t_{k-1})$. Show that for all t and τ:

$$\lim_{\epsilon \to 0, \, n \to \infty} \mathbb{E}\left(\sum_{k=1}^{n} \left(X(t_k) - X(t_{k-1}) \right)^2 - \tau \right)^2 = 0.$$

In words, the quadratic variation of $BM(t)$ converges in mean square to t [Breiman (1992), section 12.8].

7.2 Let $Z \sim N(0, \sigma^2)$ be a mean 0 normal random variable. Calculate the mean and variance of $|Z|$.

7.3 Consider a sequence of M/M/1 birth and death queues, with arrival rates λ_n and service rate μ_n, where $\lambda_n \to \lambda$ and $\sqrt{n}(\mu_n - \lambda_n) \to \theta$. Let $Q^n(t)$ be its queue length, and let $\hat{Q}^n(t) = \frac{Q^n(nt)}{\sqrt{n}}$. Write the decomposition into netput and regulator, and consider the netput process as a birth and death process. Then use Stone's theorem to show that $\hat{Q}_n(t)$ converges to a reflected Brownian motion.

7.4 Obtain the reflection mapping (solution of the Skorohod reflection problem) for the following two functions (you can give a formula for $\varphi(x), \psi(x)$ or make a drawing):
(i) $x(t) = -0.5 + \cos(t)$, $0 \le t \le 4\pi$.
(ii) $x(t) = t \sin(t)$, $0 \le t \le 4\pi$.

7.5 Show existence, uniqueness, and minimality of the two sided regulators, and verify the recursive equation (7.17).

7.6 Find the long-term average $\lim_{T \to \infty} \frac{1}{T} \int_0^T L(t) dt$ for the one-sided regulated Brownian motion (reflected Brownian motion).

7.7 Calculate the expectation of the stationary two-sided regulated Brownian motion $\mathcal{Z}(t)$.

7.8 The equations (7.24), (7.25) give the distributions of the regulator $Ł(t)$ and the reflected Brownian motion $\mathcal{Z}(t)$, with drift m and diffusion coefficient σ^2, starting at $X(t) = \mathcal{Z}(t) = 0$. Obtain the distributions of $Ł(t)$ and $\mathcal{Z}(t)$ for $\mathcal{Z}(0) = x_0 > 0$.

7.9 Provide a mathematical proof that the optimal control of stationary manufacturing with stationary independent increments demand is to use an upper inventory bound, and produce at maximal rate anywhere below that bound.

7.10 Here is some data for a manufacturing system: average demand rate is $a = 100$, with standard deviation $\sigma = 15$, sales price $r = 14$, material cost $c = 5$, workforce cost $w = 6$, inventory holding cost $h = 1$. Determine the optimal workforce k, and the optimal upper inventory level bound b, and calculate the long-term average profit V.

How would the solution change if you vary any one of r, c, w, h (say in what direction would V, k and b move, and if you can obtain the rates).

7.11 (∗) The calculations in Section 7.7.5 were for optimal control of the stationary manufacturing system. However, often one wants to take into account the current initial state of the system. In that case, it is more reasonable to optimize the discounted profit, with some discount rate γ. For the same policy of upper bound inventory and full production, calculate the discounted infinite horizon profit for initial state z_0, and given k, b, γ [Harrison (1985), chapter 5, or Harrison (2013), chapter 6].

7.12 The following is the Skorohod embedding problem that was discussed in Section 5.2.2: Let X be a random variable with $\mathbb{E}(X) = 0$, $\mathbb{V}\mathrm{ar}(X) < \infty$. Let $BM(t)$ be a standard Brownian motion. Find a stopping time T such that $BM(T) =_D X$ (equal in distribution) and $\mathbb{E}(T) = \mathbb{V}\mathrm{ar}(X)$.

The following exercises 7.13–7.19 lead to an answer to this problem. This answer was found by Dubins (1968). There are many other answers, including the original one by Skorohod; a survey of results related to this problem is Obłój (2004).

7.13 Quote a theorem that shows: $\sup_{0 \leq s \leq t} BM(s) \to \infty$ and $\inf_{0 \leq s \leq t} BM(s) \to -\infty$ as $t \to \infty$, almost surely. This means that almost every path of a Brownian motion visits all of the values on the real line.

7.14 Show that $BM(t)$ and $BM(t)^2 - t$ are martingales.

7.15 Let Y be a random variable with distribution concentrated on two points, $a < 0 < b$ and mean zero. Find the distribution of Y and its variance.

7.16 Let $T_x = \inf\{t : BM(t) = x\}$. Let $T = \min(T_a, T_b)$. Then T solves the Skorohod embedding problem for the two-point random variable Y of the previous exercise. Use the martingale $BM(t)$ to prove that $BM(T) =_D Y$, and the martingale $BM(t)^2 - t$ to calculate $\mathbb{E}(T)$.

7.17 Let X have zero mean and finite variance. Let $m^p = \mathbb{E}(X|X > 0)$, $m^n =$

$\mathbb{E}(X|X \leq 0)$. Define Y as the two point distribution on m^p and m^n, with mean zero. Show that $\mathbb{P}(Y = m^p) = \mathbb{P}(X > 0)$, and $\mathbb{P}(Y = m^n) = \mathbb{P}(X \leq 0)$.

7.18   Define a sequence of stopping times $T^{(k)}$ as follows: Define $m_{0,1} = \mathbb{E}(X) = 0$. Start with $m_{1,1} = m^p$, $m_{1,2} = m^n$, and let $T^{(1)}$ the stopping time on $BM(t)$ that stops at $\min(T_{m_{1,1}}, T_{m_{1,2}})$. Then $\{m_{0,1}, m_{1,1}, m_{1,2}\}$ divide the real line into $4 = 2^2$ intervals, $I_{2,j}, j = 1, \ldots, 2^2$. Let $m_{2,j} = \mathbb{E}(X|X \in I_{2,j})$. Define the stopping time $T^{(2)} = \min\{t : t > T^{(1)}, t = m_{2,j} \text{ for some } j\}$. Next, proceed inductively: given $T^{(k)}$ and the set of values $\{m_{i,j}: i = 0, \ldots, k, j = 1, \ldots, 2^i\}$, these values divide the real line into 2^{k+1} intervals, $I_{k+1,1}, \ldots, I_{k+1,2^{k+1}}$. Let $m_{k+1,j} = \mathbb{E}(X|X \in I_{k+1,j}), j = 1, \ldots, 2^{k+1}$, and define $T^{(k+1)} = \min\{t : t > T^{(k)}, t = m_{k+1,j} \text{ for some } j\}$.

Prove that

(i) $\mathbb{P}(BM(T^k) = m_{k,j}) = \mathbb{P}(X \in I_{k,j})$.

(ii) Let Y^k take the values $m_{k,j}$ with probability $\mathbb{P}(X \in I_{k,j})$. Then $\mathbb{E}(T^k) = \mathrm{Var}(Y^k)$.

Hint: Use the strong Markov property of $BM(\cdot)$, and rules for calculating means and variances from conditional means and variances.

7.19   Show that $T = \lim_{k \to \infty} T^{(k)}$ is a stopping time. Show that it solves the Skorohod embedding problem.

Part III

Queueing Networks

So far we have considered only a single queue, but in reality systems may contain many queues, forming a queueing network. In a queueing network each customer waits in queue and receives service at a succession of service stations. This is the case in a manufacturing process, in communication networks, in traffic systems, and in computing. In Part III we describe networks in which each service station has a single queue of customers, and each customer follows his own route through the service stations. In Chapter 8 we discuss the case where arrivals, service, and routing are all memoryless, which is the classic Jackson network, and some related systems. For all of these, the stationary distribution is obtainable and is of product form. In Chapter 9 we discuss the same network with general i.i.d. interarrivals and service times, the generalized Jackson network. Like the G/G/1 system, the generalized Jackson network cannot be analyzed in detail, and we discuss fluid and diffusion approximations to the network processes.

8

Product-Form Queueing Networks

A queueing network consists of several service stations (nodes of the network), and each customer may visit a sequence of stations (routes through the network), where he waits in queue and receives service at each one of them. When each service station has a single queue, we refer to it as a single class network. Jackson networks are single class networks in which arrivals, service times and routing are memoryless. We discuss classic Jackson networks and related models in this chapter. The distinguishing feature of these networks is that we can write the stationary distribution of the queue lengths explicitly, in product form, in fact as a product of geometric terms. We introduce the concepts of time reversibility and of partial balance, which allow us to derive these stationary distributions.

8.1 The Classic Jackson Network

Definition 8.1   The (single servers) Jackson network consists of processing nodes $i = 1, \ldots, I$. External customers arrive to node i in a Poisson process of rate α_i, and service at node i is exponential with rate μ_i. Service at each node is work conserving and FCFS. Upon completion of service at node i, the customer moves to node j with probability $p_{i,j}$, or leaves the system with probability $1 - \sum_j p_{i,j}$ (see Figure 8.1). Interarrivals, service times, and routing decisions are all independent.

Strictly speaking, such networks, in which each node acts as a single-server queue, are referred to in the literature as *migration processes*. The extension to what is usually termed Jackson networks, with more general nodes, is delayed to Section 8.3, Definition 8.7.

We consider first the routing probabilities, which summarize how customers circulate in the network. We denote by P the matrix of the routing probabilities $p_{i,j}$. The following theorem states the major property of positive matrices:

125

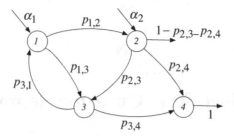

Figure 8.1 A Jackson network.

Theorem 8.2 (Perron–Frobenius)   *Let A be a positive square matrix (with all elements positive). Let $\eta_j, j = 1, 2, \ldots$ be its eigenvalues. Then the following holds:*

  (i) $r = \max\{|\eta_j|\}$ *is a real valued eigenvalue of A. It is called* the spectral radius *of A.*

 (ii) *r is isolated, i.e. $r > |\eta_j|$ for all $\eta_j \neq r$.*

(iii) *r has a single eigenvector (up to a constant multiplier) that is all positive.*

(iv) *Let u, w be the positive length 1 left and right eigenvectors of r (u a row vector, w a column vector). Then $\frac{1}{r^n} A^n$ converges to the rank one matrix wu.*

Several additional related properties of positive matrices also hold.

The conclusions of the Perron–Frobenius theorem continue to hold for non-negative A, if there exists m such that A^m is positive (see Exercise 8.1). The Perron–Frobenius theorem can also be generalized to square matrices with infinite number of rows. We note that the routing matrix P is non-negative, with sum of rows ≤ 1. Its spectral radius is then ≤ 1. Similarly, any transition matrix of a finite state Markov chain is non-negative, and has rows that add up to 1, and its spectral radius is $r = 1$.

We assume that the routing matrix P has spectral radius < 1. This implies (by the Perron–Frobenius theorem) that $\sum_{k=0}^{\infty} P^k$ converges (at a geometric rate equal to the spectral radius), and is equal to $(I - P)^{-1}$.

Recall the definition of *rate stability*, (2.4).

Proposition 8.3 (Rate stability and traffic equations)   *Assume that all the queues in a queueing network are rate stable. In that case, the rates of inflow and outflow from the various nodes, $\lambda_i, i = 1, \ldots, I$, satisfy the following* *traffic equations:*

$$\lambda_i = \alpha_i + \sum_{j=1}^{I} p_{ji} \lambda_j, \qquad i = 1, \dots, I. \tag{8.1}$$

Proof If the queue at node i, $Q_i(t)$, is rate stable, then the long-term rates of inflow and outflow must be equal.

$$\lim_{T \to \infty} \frac{\text{Inflow } (0, T]}{T} - \lim_{T \to \infty} \frac{\text{Outflow } (0, T]}{T} = \lim_{T \to \infty} \frac{Q(T)}{T} = 0.$$

Putting the outflow on the left of (8.1), it is equated with the sum of the external arrival rate and the rates at which customers are routed to node i from the other nodes. □

In matrix form, the traffic equations are written as

$$\lambda = \alpha + P^\top \lambda,$$

and they are solved by

$$\lambda = (I - P^\top)^{-1} \alpha = (I + P^\top + P^{\top 2} + \cdots) \alpha.$$

Here α are rates of customers moving into the system, $P^{\top k} \alpha$ are the rates of customers moving after the kth processing step, $k = 1, 2, \dots$. The i, j element of $(I - P^\top)^{-1} = I + P^\top + P^{\top 2} + \cdots$ expresses, for a job currently in the queue of node i, the expected number of current and future visits to node j before leaving the system. The convergence of the series and invertibility of $I - P^\top$ are assured by spectral radius < 1. We will return to the traffic equations, in Section 9.2.

8.2 Reversibility and Detailed Balance Equations

We define the vector queue length process $Q(t) = (n_1, \dots, n_I)$, where n_i is the number of customers at node i at time t. Because arrivals are Poisson, services exponential, and routing random, and all are independent, $Q(t)$ is a continuous-time discrete-state Markov chain. Note that the assumption that processing is FCFS is irrelevant to the state of the process, as any work-conserving non-preemptive or even preemptive policy will have customers leaving at rate μ_i when node i is non-empty. FCFS only helps to keep track of additional information on the identity of the customers in the node. $Q(t)$ is in fact a multivariate birth and death process, with state change of ± 1 at customer arrivals, departures, or moves between nodes. We next look for its stationary distribution.

In general, the stationary distribution π of a continuous time discrete state Markov chain is determined by its global balance equations:

$$\sum_y \pi(x)q(x,y) = \sum_y \pi(y)q(y,x), \quad \text{for all } x \in S, \qquad (8.2)$$

where S is the discrete state space and $q(x,y)$ is the transition rate from state x to state y. There is one equation for each state; it says that the total flow (sometimes called flux, in analogy with electrical networks) out of each individual state x equals the total flow into x. For an irreducible chain, these equations always have a unique, up to a constant, non-negative, $\neq 0$, solution. If it is convergent, then the Markov chain is ergodic, and the normalized solution is the stationary distribution of the chain. Global balance equations are hard to solve in general, especially when the state space is countably infinite. Often the solution is intractable, and there is no way to express it efficiently.

Definition 8.4 A stationary stochastic process $X(t)$ is time reversible if the joint distributions of $X(t_1), \ldots, X(t_n)$ and $X(T-t_1), \ldots, X(T-t_n)$ are the same for every $T, t_1, \ldots, t_n$.

For time-reversible Markov chains, the stationary distribution can be obtained by solving the detailed balance equations (see Exercise 8.3),

$$\pi(x)q(x,y) = \pi(y)q(y,x), \quad \text{for all pairs } x \neq y, \ x, y \in S, \qquad (8.3)$$

in which the flow between any pair of states is balanced. There are many more equations in (8.3) than in (8.2), one for each possible transition, but their solution is immediate.

The M/M/1 queue is time reversible, as are all one-dimensional birth and death processes (see Exercise 8.5). In particular, this implies

Theorem 8.5 (Burke's theorem) *In a single queue with Poisson arrivals and exponential service, with one, or with s, or with an infinite number of servers, the departure process $\mathcal{D}(t)$ is a Poisson process. Furthermore, for a stationary M/M/· queue, the queue length $Q(t)$ is independent of previous departures $\mathcal{D}(s)$, $s < t$.*

Proof All M/M/· systems are time reversible (when stationary). The departures from the stationary M/M/· system are the arrivals of the time-reversed process; hence, they are Poisson. Furthermore, in the M/M/· system $Q(t)$ is independent of the arrival process $\mathcal{A}(s)$, $s > t$ (recall all the processes are RCLL). So the same holds for the reversed process, in reversed time, but the arrivals of the reversed process for reversed time $s > t$ are the departure times of the original M/M/· system, $\mathcal{D}(s)$, $s < t$. □

In a feed-forward Jackson network, in which a node is visited by each customer at most once, each node (service station) behaves like an M/M/1 queue, and all traffic consists of Poisson streams. This holds as a special case for tandem queues, also known as flow-shops, in which each customer goes through the nodes in order from 1 to I. The stationary distribution of the state of such networks is then a product of the M/M/1 stationary distributions of the nodes (for a more general result, see Exercise 8.6).

8.3 Partial Balance and Stationary Distribution of Jackson Networks

Jackson networks are in general not time reversible. However, they do have tractable, product-form, stationary distributions. The reason is that they satisfy partial balance equations: For each state x, the flow (flux) of customers that move out of node i (to all the other nodes or outside) is balanced with the flow (flux) into node i (from any other node or from outside). We now set up and present the solution of these partial balance equations.

To set up the partial balance equations, denote by $T_{i,j}$ a transition in which a single customer moves from node i to node j. We add node 0 to represent the outside and let $T_{0,j}$ be an exogenous arrival to node j, and $T_{i,0}$ a departure from the system out of node i. We then write transitions from state x to state y, via the movement of a customer from node i to node j as $y = T_{i,j}(x)$. Partial balance equations for a Jackson network are:

$$\sum_{j \neq i} \pi(x) q(x, T_{i,j}(x)) = \sum_{j \neq i} \pi(T_{i,j}(x)) q(T_{i,j}(x), x), \text{ for } x \in S, 0 \leq i \leq I.$$

(8.4)

There are several equations for each state $x \in S$, one for each node $i = 0, 1, \ldots, I$. In these equations, on the left we have a transition out of x by a customer moving out of node i, and on the right a transition into state x by a customer moving into node i, where the summation over j includes node 0. The rates for the various transitions are:

$$
\begin{aligned}
q(x, T_{i,j}(x)) &= \alpha_j, & i &= 0, \\
q(x, T_{i,j}(x)) &= \mu_i p_{i,j}, & i &\neq 0, \ x_i > 0, \ j \neq 0, \\
q(x, T_{i,j}(x)) &= \mu_i \left(1 - \sum_{k \neq i,0} p_{i,k} \right), & i &\neq 0, \ x_i > 0, \ j = 0, \\
q(x, T_{i,j}(x)) &= 0, & i &\neq 0, \ x_i = 0.
\end{aligned}
$$

(8.5)

Theorem 8.6 *The solution to the partial balance equations for the Jackson network is given by*

$$\pi(n_1, \ldots, n_I) = B \prod_{i=1}^{I} \left(\frac{\lambda_i}{\mu_i} \right)^{n_i}, \tag{8.6}$$

where λ_i are the rates obtained from the traffic equations. This converges if and only if for every node i, $\rho_i = \frac{\lambda_i}{\mu_i} < 1$. The stationary distribution is then:

$$\pi(n_1, \ldots, n_I) = \prod_{i=1}^{I} (1 - \rho_i) \rho_i^{n_i}. \tag{8.7}$$

Proof Clearly, if $\pi(\cdot)$ satisfies the detailed or the partial balance equations, then it satisfies the global balance equations. So all that is needed is to verify that (8.6) satisfies (8.4) when substituting (8.5).

We now perform the check: For $i \neq 0$ with $n_i > 0$, the partial balance for an item leaving node i and an item returning to node i, we obtain:

$$\pi(x)\mu_i = \pi(x)\left[\frac{\mu_i}{\lambda_i}\alpha_i + \sum_{j \neq i, 0} \frac{\mu_i}{\lambda_i}\frac{\lambda_j}{\mu_j}\mu_j p_{j,i} \right],$$

which, after canceling and multiplying by λ_i yields the traffic equation for node i:

$$\lambda_i = \alpha_i + \sum_{j \neq i, 0} \lambda_j p_{j,i}.$$

For node 0, arrivals to the system out of node 0 need to balance with departures, and on substituting

$$\pi(x) \sum_{j \neq 0} \alpha_j = \pi(x) \left[\sum_{j \neq 0} \frac{\lambda_j}{\mu_j}\mu_j \left(1 - \sum_{k \neq 0, j} p_{j,k} \right) \right],$$

which is simply

$$\mathbf{1}^\top (I - P^\top)\lambda = \mathbf{1}^\top \alpha,$$

i.e. summation of the traffic equations.

We verified that (8.6) solves the balance equations. Clearly, adding these up will converge if and only if $\lambda_i < \mu_i$, $i = 1, \ldots, I$. When the sum of (8.6) converges, (8.7) is obtained by normalizing the sum to 1. $\square$

We see in (8.7) that the number of customers in each node has the stationary distribution of an M/M/1 queue, and furthermore, at time t the numbers of customers in the various nodes are independent. Note however that this is only true for the stationary system, observed at time t. Clearly n_i at time t and n_j at time s where $s \neq t$ are not independent.

At this point, we extend the definition of Jackson networks to allow more general assumptions on processing times.

Definition 8.7 In the (classic) Jackson network in addition to Poisson arrivals and Markovian routing as in Definition 8.1, we assume that processing requirements are exponential, and processing at node i is at rate $\mu_i(n_i)$, i.e. the processing rate depends on the number n_i of customers present at the node

It follows, using again partial balance, (see Exercise 8.7):

Theorem 8.8 *The stationary distribution of the queue lengths of a Jackson network with processing rates $\mu_i(n_i)$ is given by:*

$$\pi(n_1, \ldots, n_I) = B \prod_{i=1}^{I} \frac{\lambda_i^{n_i}}{\prod_{k=1}^{n_i} \mu_i(k)}, \tag{8.8}$$

where $B > 0$ is the normalizing constant.

Proposition 8.9 *A stationary Jackson network is ergodic.*

Proof Starting from any state, if there are no arrivals for a long time, the state of the system will be 0, and hence a Jackson network is irreducible. Hence, $\rho_i < 1$ for all i is necessary and sufficient for ergodicity. □

8.4 Time Reversal and the Arrival Theorem

A Markov chain is defined by the property that past and future states are independent, given the present state. In this definition, past and future play symmetric roles. It follows that

Theorem 8.10 *If $X(t)$ is a stationary Markov chain, then the time-reversed process, $X(T - t)$, is also a stationary Markov chain.*

This is true also for Markov chains that are not time reversible. For a stationary continuous time discrete state Markov chain $X(t)$ with transition rates $q(x, y)$, let $X^*(t) = X(T - t)$ be the time-reversed stationary Markov chain. Denote by $q^*(x, y)$ its transition rates. Then we have

Theorem 8.11 (Kelly's lemma) *For a Markov chain with transition rates*

$q(x, y)$, *if we can find a collection of positive numbers* $\pi(x)$, *and non-negative* $q^*(x, y)$, *that satisfy for every* $x, y \in S$,

$$\pi(x)q(x, y) = \pi(y)q^*(y, x),$$
$$\sum_{y \in S} q(x, y) = \sum_{y \in S} q^*(x, y),$$
(8.9)

then $q^*(x, y)$ *are the transition rates of the time-reversed process, and* $\pi(x)$ *is the stationary distribution of both* $X(t)$ *and* $X^*(t)$.

The proof is left as Exercise 8.8. This lemma allows us to verify easily a guess of the stationary distribution for a given Markov chain, which is often much easier than verifying balance equations by substitution. Alternatively, we may be able to guess the transition rates of the time-reversed chain, and use them to immediately compute the stationary probabilities. We will make use of this in Chapter 22.

For a stationary Jackson network, we can calculate the transition rates of the reversed process. We note that it also has transitions that move a single item between two nodes, or has a single item enter (departure in the original network) or leave (arrival in the original network). Using the $T_{i,j}$ notation, by Exercise 8.9:

$$
\begin{array}{ll}
q^*(x, T_{i,j}(x)) = \lambda_j \left(1 - \sum_{k \neq j, 0} p_{j,k}\right), & i = 0, \\
q^*(x, T_{i,j}(x)) = \mu_i(n_i) \frac{\lambda_j}{\lambda_i} p_{j,i}, & i \neq 0, \ x_i > 0, \ j \neq 0, \\
q^*(x, T_{i,j}(x)) = \mu_i(n_i) \frac{\alpha_i}{\lambda_i}, & i \neq 0, \ x_i > 0, \ j = 0, \\
q^*(x, T_{i,j}(x)) = 0, & i \neq 0, \ x_i = 0.
\end{array}
$$
(8.10)

From this, we immediately have

Theorem 8.12 *The time-reversed Jackson network is in itself a Jackson network, with arrival rates* $\alpha_i^* = \lambda_i \left(1 - \sum_{j \neq i, 0} p_{i,j}\right)$, *routing probabilities* $p_{i,j}^* = \frac{\lambda_j}{\lambda_i} p_{j,i}$, *and processing rates* $\mu_i(n_i)$.

This is verified in Exercise 8.9. The surprising part of the theorem is that the processing rate for the reversed network is the same as for the original one. An immediate corollary to the theorem is that all departure streams out of the stationary reversed network (which are arrival streams for the original network) are Poisson streams. This proves that in a Jackson network, output streams of customers leaving the system out of the various nodes, are independent Poisson processes. Note, however, that streams of customers moving between nodes, e.g. customers moving from node i to node j, are not Poisson (see Exercise 8.10).

Even though arrivals into a node from outside and from other nodes are not Poisson, the following important property of *Jackson arrivals see time average* is similar to PASTA (the proof is left as Exercise 8.13):

Theorem 8.13 (Arrival theorem)   *In a stationary Jackson network, a customer in transit at time t sees the system without him in state x with probability $\pi(x)$, i.e. he sees the steady state of the system with the exclusion of himself.*

8.5 Sojourn Times in Jackson Networks

The most important information for a customer in a queueing system is his sojourn time. Unfortunately the calculation of sojourn times, even for the memoryless Jackson network, is far from simple.

By the arrival theorem 8.13 in a stationary Jackson network, a customer that arrives at node i will see it in its stationary state. For single-server Jackson networks, this will be a statioanary M/M/1 node with geometric distribution of the number of customers. In this case, his sojourn time at the node is exponential, with rate $\mu_i - \lambda_i$. For more general processing rates, an arriving customer will see the node as a stationary birth and death process, and the sojourn time of the customer at the node may then be computed using the parameters λ_i and $\mu_i(n_i)$; e.g. for M/M/s nodes, recall Exercise 1.3.

It is then straightforward to calculate the expected travel times. Denote by W_i the sojourn time of a customer from entering node i until his departure, and let $\overline{W}_i = E(W_i)$. The expected sum of sojourn times over the random route can be calculated, using Wald's equation, from:

$$\overline{W}_i = \theta_i + \sum_j p_{i,j} \overline{W}_j,$$

where θ_i is the sojourn time of a single visit to node i. In particular for the single-server case, we have $\theta_i = \frac{1}{\mu_i - \lambda_i}$. The equation is solved by

$$\overline{W} = (I - P)^{-1}\theta.$$

Note the similarity to the traffic equations.

If a customer is moving along a path where there is no possibility that he will be overtaken by a customer that has entered the path later, we say that there is no overtaking along the path. On a path with no overtaking, sojourn times at the nodes along the path are independent. As a result, travel along a path with no overtaking is simply the sum of independent sojourn times at the nodes (exponential for single servers). However, if overtaking is possible, then sojourn times at successive nodes along the path are not independent.

The simplest example of a system with overtaking is a single M/M/1 queue with feedback, where each customer after completion of service will leave the system with probability $1 - p$, and will require additional service, where he will rejoin the end of the queue, with probability p. Clearly, a customer that is fed back to the end of the queue will be overtaken by all the customers that arrived during his service time. In that case, the combined arrival stream into the queue, of exogenous and feedback customers, is not Poisson, and sojourn times in successive visits to the queue are not independent; see Exercises 8.10 and 8.11).

A recursive representation of the moments of the sojourn times W_i is derived by Lemoine (1987) and Daduna (1991). Several methods to approximate the distribution and the moments of travel times exist, and are quite accurate, and it seems that the assumption of independence of sojourn times does not introduce a large error. For a survey, see Boxma and Daduna (1990).

8.6 Closed Jackson Networks

When the network contains a fixed total of N customers in its nodes, with no arrivals from outside or departures to outside, it is called a closed Jackson network. The queue length process $Q(t) = (n_1, \ldots, n_I)$ lists the number of customers in each node at time t, with $\sum_i n_i = N$. It is again a continuous time Markov chain, with single customers moving between nodes. A customer moves from node i to node j with rate $\mu_i(n_i)p_{i,j}$, where the processing rates $\mu_i(n_i)$ depend on the number of customers at the node, and $p_{i,j}$ are the routing probabilities.

The traffic equations are now given by:

$$\lambda_i = \sum_{j=1}^{I} p_{ji} \lambda_j, \quad i = 1, \ldots, I,$$
$$\sum_{i=1}^{I} \lambda_i = 1, \tag{8.11}$$

which are solved by the stationary distribution of the Markov chain of a single customer performing a random walk between the nodes. The values λ_i again appear in the stationary distribution.

Theorem 8.14 *The stationary distribution of $Q(t)$ for a closed Jackson network is given by*

$$\pi_N(n_1, \ldots, n_I) = B(N) \prod_{i=1}^{I} \frac{\lambda_i^{n_i}}{\prod_{k=1}^{n_i} \mu_i(k)}, \tag{8.12}$$

where the states satisfy $\sum_i n_i = N$, and $B(N)$ is a normalizing constant.

The proof is again by showing that these probabilities satisfy partial balance, where we now do not have transitions to the outside-of-system state 0. We leave the details to Exercise 8.12.

We note that the calculation of $B(N)$, the normalizing constant, is often laborious, since it involves summation over a large number of states, all the partitions of N customers into I nodes, i.e. $\binom{N+I-1}{I-1}$.

An interesting property, analogous to PASTA and to the arrival theorem for open Jackson networks, is the following theorem (see Exercise 8.13):

Theorem 8.15 (Arrival theorem) *In a stationary closed Jackson network with N customers, a customer in transit at time t sees the system in state x with probability $\pi_{N-1}(x)$, i.e. he sees the steady state of the system with the exclusion of himself.*

8.7 Kelly Networks

Kelly networks generalize Jackson networks by allowing several types of customers. Without loss of generality, one can assume that each type has a deterministic route through the nodes. Customers of type c arrive in the system at rate α_c and move through nodes $r(c,1), r(c,2), \ldots, r(c, S_c)$ and leave the system after the S_c (last) step. Figure 8.2 illustrates a Kelly network, with four nodes, and with three types of customers, each with its own deterministic route. Note that probabilistic Markovian routing can then be modeled by defining a type for each possible route that the random routing might induce. This requires a countably infinite number of types, but all the results remain valid.

The system is again described by a continuous time Markov chain, and it obeys partial balance if the following condition holds: The processing at each of the nodes does not distinguish between customers of different types. This means that from the moment that a customer of type c enters node i, on the $r(c, s)$ step of its route, until it leaves node i, he is treated exactly the same way as customers of all other types that are in that node. In other words, each node maintains a single queue. The type of the customer is only used at the end of its service, to determine the rest of its route.

The rules for processing of customers by node i are as follows: The total amount of processing of every customer is $\sim exp(1)$, and all processing steps of all customers are independent. The processing is worked off at a

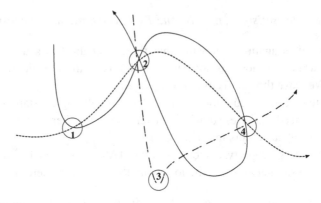

Figure 8.2 Kelly network with four nodes and three routes.

variable rate which depends on the node, the number of customers in the node, and the position of the customer in the queue at the node, as follows:

  (i) The total service rate at node i is $\mu_i(n_i) > 0$ when there are $n_i > 0$ customers at the node.

 (ii) A fraction $\gamma_{l,n}$ of the total service capacity is given to the customer in position l in the queue.

(iii) When the customer in position l leaves, customers in positions $l + 1, \ldots, n$ move to positions $l, \ldots, n - 1$.

(iv) On arrival, a new customer is inserted in position l in the queue with probability $\delta_{l,n+1}, l = 1, \ldots, n + 1$.

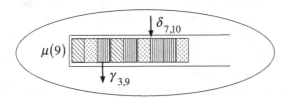

Figure 8.3 The queue in a Kelly network node.

Figure 8.3 illustrates the state and transition rates at a node of a Kelly network. In it the node has an ordered queue of nine customers of three types, the total rate at which customers are processed is $\mu(9)$, specifically the customer in position 3 will then depart at rate $\mu(9)\gamma_{3,9}$, and any new customer that should arrive will enter in position 7 with probability $\delta_{7,10}$.

Let $t(l, i)$ be the type of the customer in position l at node i. Let $s(l, i)$ be the step along its route. Let $c(l, i) = (t(l, i), s(l, i))$; we call this the class of the customer in position l in node i. The vector $C(i) = (c(1, i), \ldots, c(n_i, i))$ describes the state of node i. Specifically, if we look at the node in Figure 8.3, its state will inform us that the first customer is of type diagonal striped, and also will inform us at which step along the route of diagonal-striped customers this customer is, with similar information on the customers in positions $2, \ldots, 9$. Finally, $Q(t) = (C(1), \ldots, C(I))$ is the state of the whole system at time t.

Next we obtain the flow rates through the nodes. For step s on the route of customers of type c, we have at node i a flow:

$$\lambda_i(c, s) = \begin{cases} \alpha_c & \text{if } r(c, s) = i, \\ 0 & \text{otherwise} . \end{cases}$$

This is the arrival rate of customers of type c, on step s of their route, if that step is through node i. We then have the total flow rate in and out of node i:

$$\lambda_i = \sum_c \sum_{s=1}^{S_c} \lambda_i(c, s).$$

Note that it includes summation only over customer types that go through node i, and counts the rate multiplied by the number of times that the route of a customer visits the node. Let

$$b_i^{-1} = \sum_{n=0}^{\infty} \frac{\lambda_i^n}{\prod_{l=1}^{n} \mu_i(l)}. \tag{8.13}$$

A birth and death process with birth rate λ_i and death rate $\mu_i(n)$ is stable if and only if $b_i > 0$, i.e. if the above series converges.

Theorem 8.16 *If $b_1, \ldots, b_I > 0$, then the stationary probabilities for the state of node i are:*

$$\pi_i(C(i)) = b_i \prod_{l=1}^{n_i} \frac{\lambda_i(t(l, i), s(l, i))}{\mu_i(l)}, \tag{8.14}$$

and the stationary distribution of $Q(t)$ is

$$\pi(C) = \prod_{i=1}^{I} \pi_i(C(i)). \tag{8.15}$$

The proof of the theorem is by partial balance of the nodes, and is left as Exercise 8.14.

Remark Note that the stationary distribution does not distinguish between policies, beyond using $\mu_i(n_i)$, the control values $\gamma_{l,n}$, $\delta_{l,n}$ have no effect on the expressions (8.13)–(8.15). By the argument of Theorems 2.23 and 3.7, the average sojourn times at each node, and hence for the whole route, are independent of the policy, although the distribution of sojourn times will depend on the policy.

Special cases of the policy of processing at a node i with n customers are

PS processor sharing, when $\gamma_{l,n} = \frac{1}{n}$.

FCFS single server first in first out, $\mu_i(n) = \mu_i$, $\gamma_{1,n} = 1$, $\delta_{n+1,n+1} = 1$.

LCFS single server last in first out, $\mu_i(n) = \mu_i$, $\gamma_{n,n} = 1$, $\delta_{n+1,n+1} = 1$.

k servers $\mu_i(n) = (n \wedge k)\mu_i$, $\gamma_{l,n} = 1/(n \wedge k)$, $l = 1, \ldots, (n \wedge k)$, $\delta_{n+1,n+1} = 1$.

Random serve customers in a random order, $\mu_i(n) = \mu_i$, $\gamma_{1,n} = 1$, $\delta_{l,n+1} = \frac{1}{n+1}$, $l = 1, \ldots, n$.

Infinite servers unlimited number of servers, $\mu_i(n) = \mu_i n$, $\gamma_{l,n} = \frac{1}{n}$.

Loss system finite number of servers, and no waiting room, $\mu_i(n) = n$, $\gamma_{l,n} = \frac{1}{n}$ for $n \leq K$, and $\mu_i(K+1) = \infty$, $\gamma_{n,n} = 1$, for $n > K$ (the rate of ∞ means that an arrival leaves immediately when all K servers are busy).

8.8 Symmetric Queues and Insensitivity

Consider the model of Kelly networks, with the following modifications for some of its nodes, which are called symmetric service nodes:
- The service discipline is symmetric in the sense that $\delta_{l,n+1} = \gamma_{l,n+1}$.
- The service times of customers of type c on step s of their route are distributed with a general distribution $G_{c,s}$, with finite mean $1/\mu_{c,s}$.

Examples of symmetric queues are processor sharing, LCFS-preemptive, also known as the stack, infinite server queues, and loss systems. Figure 8.4 represents the service policy of the two types of symmetric nodes. In an LCFS node, arrivals and departures are always at the last customer in the queue, while for PS, ∞-server, and K-loss nodes, all the customers are served at equal rates (and order in the queue is irrelevant). For another example of a symmetric node, see Exercise 8.15.

The remarkable thing about these queues is that they are insensitive: The stationary distribution of the state of the system is the same as one gets for exponential processing times and is of product form. We summarize this in the following theorem, which we present without proof. Here we refer to

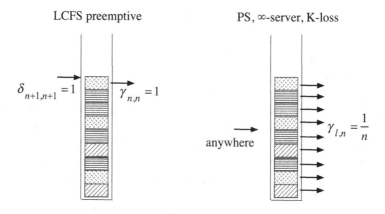

Figure 8.4 Examples of symmetric queues.

customer classes, where a class consists of customers of some type, on a specific step on their path,

Theorem 8.17 (Insensitivity)   *Let i be a symmetric node in a Kelly network, with customer classes $c \in C(i)$. Assume customers of class c have arrival rate α_c and average service time m_c. Let $\rho_i = \sum_{c \in C(i)} \alpha_c m_c$, and let $\mu_i(n)$ be the service rate of node i when there are n customers in the node. Then:*

(i) The stationary distribution of the number of customers at node i is:

$$\mathbb{P}(Q_i(t) = n) = b_i \frac{\rho_i^n}{\prod_{l=1}^{n} \mu_i(l)}$$

  (here $\mu_i(l)$ expresses the speed of processing, and in particular, if it is a constant $\mu_i(l) = 1$, then we get $(1 - \rho_i)\rho_i^n$).

(ii) $Q_i(t)$ is independent of the state of the other nodes at time t.

(iii) Given that there are n customers at node i, their classes are independent and the probability that a customer in a given position is of type c is $\alpha_c m_c / \rho_i$.

(iv) Given the number of customers at node i and the class of each customer in the different positions in the queue, the remaining service times of the customers are independent, and the remaining service times of customers of type c are given by the equilibrium (excess) time distribution for class c, with density $g_c^{eq}(x) = (1 - G_c(x))/m_c$.

8.9 Multi-Class Kelly-Type Networks

A generalization of Kelly networks is to allow customers of type c to follow Markovian stochastic routes. Instead of a deterministic route listing a deterministic sequence of nodes $r(c, 1), \ldots, r(c, S_c)$, type c when at node i will proceed from there to node j with probability $p^c_{i,j}$. In that case, instead of class $k = (c, s)$, which is at node $i = r(c, s)$, being followed deterministically by class $l = (c, s + 1)$ at node $j = r(c, s + 1)$, we will now have that customer of class k at node i will move to class l at node j with probability $p_{k,l}$. We now have classes $k = 1, \ldots, K$, located at nodes $i = 1, \ldots, I$, with $i = s(k)$ to denote that class k is served at node i, and we have a $K \times K$ routing matrix P with elements $p_{k,l}$. Arrivals are now at rates α_k, $k = 1, \ldots, K$, and the traffic equations $\lambda = (I - P^{\mathsf{T}})^{-1} \alpha$ give us the flows λ_k for class k at node $s(k)$.

In a Kelly-type network of this form, service and service policy are as defined in Section 8.7 and are determined as follows: Processing requirements of jobs are exponential rate 1. Node i with n customers has processing capacity $\mu_i(n)$. It maintains a single queue of customers of types $\{k : s(k) = i\}$, and customer in position l receives a fraction $\gamma_{l,n}$ of the processing capacity. New customers join into position $\delta_{l,n+1}$. We now write the stationary probabilities for a Kelly-type network: If the queue of node i consists of n_i customers of classes $C(i) = (k(1), \ldots, k(n_i))$ then, similar to (8.14), (8.15):

$$\pi_i(k(1), \ldots, k(n_i)) = b_i \prod_{l=1}^{n_i} \frac{\lambda_{k(l)}}{\mu_i(l)}, \qquad (8.16)$$

with b_i a normalizing constant, and

$$\pi(C(1), \ldots, C(I)) = \prod_{i=1}^{I} \pi_i(C(i)). \qquad (8.17)$$

Parts IV, V, and VI of the book will mainly be concerned with multi-class queueing networks. Kelly-type networks are the only examples of multi-class queueing networks with tractable stationary distributions.

8.10 Sources

The material in this section is covered by the first four chapters of Frank Kelly's incredible book (published when he was in his 20s) *Reversibility and Stochastic Networks*, Kelly (1979). That book revolutionized the whole direction of queueing theory research. Open Jackson networks were pioneered by Jackson (1963) and closed Jackson networks by Gordon and Newell (1967). Kelly networks are often also-called BCMP (Baskett–Chandy–Muntz–Palacios) networks after the original paper, Baskett et al. (1975). A pioneering beautiful proof of insensitivity appears in Sevast'Yanov (1957). It was a hot topic of research in the 1980s, see Schassberger (1986) and Henderson (1983), and is also a major topic in the book *Systems in Stochastic Equilibrium*, Whittle (1986). Properties of positive and non-negative matrices, with finite or countable dimension, is the subject of Seneta (1981) and Berman and Plemmons (1994). Much more material on product-form networks is contained in the books *An Introduction to Queueing Networks*, Walrand (1988); *Queueing Networks and Product Forms*, by Van Dijk (1993); *Queueing Networks: Customers, Signals and Product Form Solutions*, by Chao et al. (1999); *Introduction to Stochastic Networks*, by Serfozo (1999).

Exercises

8.1 Show that the conclusions of the Perron–Frobenius theorem 8.2 continue to hold for A that is non-negative if there exists A such that A^m is positive.

8.2 Show that if the transition matrix P of a discrete time, discrete state Markov chain satisfies that P^m is positive, then the chain is irreducible and a-periodic.

8.3 Show that a Markov chain is time reversible if and only if it satisfies the detailed balance equations.

8.4 Prove the Kolmogorov criterion: A stationary Markov chain is time reversible if and only if for any finite sequence of states $x_1, x_2, \ldots, x_k$, the transition rates satisfy:

$$p(x_1, x_2)p(x_2, x_3) \cdots p(x_k, x_1) = p(x_1, x_k)p(x_k, x_{k-1}), \cdots p(x_2, x_1).$$

8.5 Show that every stationary birth and death process is time reversible.

8.6 Write down the formula for the steady-state distribution of the queue length vector for Jackson tandem queueing systems and feed-forward systems (including solution of the traffic equations). Show that if node i precedes node j, then $Q_i(t)$ is independent of $Q_j(s)$ for all $s < t$.

8.7 Verify the stationary distribution for the Jackson network with processing rates $\mu_i(n_i)$ as given by Theorem 8.8.

8.8 Prove Kelly's lemma 8.11.

8.9 Prove that a Jackson network considered in reversed time is again a Jackson network, and calculate its parameters.

8.10 The M/M/1 queue with feedback is a single queue with rate λ external Poisson arrivals and rate μ exponential service times, where upon completion of service a customer rejoins the queue with probability θ, and leaves the system with probability $1 - \theta$. Calculate the stationary distribution of the queue $Q(t)$ and show it is the same as M/M/1 with $\rho = \frac{\lambda}{\mu(1-\theta)}$, and that customers leave the system as a Poisson process of rate λ. However, show that the stream of customers entering service (new arrivals and returns) is not Poisson.

8.11 Consider the M/M/1 queue with feedback as a Kelly-type multi-class network, where customers on their kth visit are class k customers. Obtain the stationary distribution of this system, and the sojourn time distributions for each type [Van den Berg and Boxma (1989, 1993)].

8.12 Prove Theorem 8.14, on the stationary distribution of closed Jackson networks, by verifying partial balance.

8.13 Prove the arrival theorem, customers in transit see a stationary state, for both open and closed Jackson networks [Kelly (1979); Sevcik and Mitrani (1981)].

8.14 Verify the expressions for the stationary distribution of Kelly networks and Kelly-type multi-class networks as given by (8.14), (8.15) and (8.16), (8.17).

8.15 (∗) Consider an M/M/K/K Erlang loss system, where service rates of the servers differ, say server k has service rate μ_k. Use the policy of assign to longest idle server (ALIS), so arriving customers go to the server that has been idle for the longest time. Show that this system is insensitive [Adan and Weiss (2012b); Adan et al. (2010)].

8.16 (∗) The following is a model for a Jackson network with positive as well as negative customers: Nodes are $i = 1, \ldots, I$, service rates μ_i. Positive customers arrive at rate α_i^+ and are added to the queue at node i. Negative customers arrive at rate α_i^-, and on arrival they eliminate a customer from node i if not empty. On completion of service, positive customers move as positive customers with routing probabilities $p_{i,j}^+$, and move as negative customers with probabilities $p_{i,j}^-$, where on arrival at node j they eliminate a customer if not empty, and depart at rate $d_i = 1 - \sum_j (p_{i,j}^+ + p_{i,j}^-)$. Arrivals, processings, and routings are independent and memoryless. The state of the system is given by the queue lengths of positive customers, $Q(t) = (n_1, \ldots, n_I)$. Prove that $Q(t)$ has a product-form stationary distribution $\prod_i \rho_i (1 - \rho_i)^{n_i}$, where $\rho_i = \frac{\lambda_i^+}{\mu_i + \lambda_i^-}$, and λ_i^+, λ_i^-, $i = 1, \ldots, I$ solve the non-linear equations:

$$\lambda_i^+ = \alpha_i^+ + \sum_j \mu_j \rho_j p_{j,i}^+, \qquad \lambda_i^- = \alpha_i^- + \sum_j \mu_j \rho_j p_{j,i}^-,$$

whenever these equations have a unique solution. Such systems are motivated as modeling networks of neurons: A positive signal arriving at a neuron increases its total signal count or potential by one; a negative signal reduces it by one if the potential is positive. When its potential is positive, a neuron "fires", sending positive or negative signals at random intervals to other neurons or to the outside. Positive signals represent excitatory signals and negative signals represent inhibition [Gelenbe (1991)].

9

Generalized Jackson Networks

We now consider networks with the same structure as Jackson networks, but in which interarrival and service times have general distributions. Such queueing networks can no longer be described by continuous time discrete state Markov chains, and with the exception of networks with symmetric nodes, stationary distributions for such networks are intractable. In this chapter we study the dynamics of these more general networks, and their fluid and diffusion approximations. The main idea is to represent the queue as a reflection of a netput process, where the netput process is a linear combination of the arrival processes, the service processes, and the routing processes. The steps are similar to the analysis of the G/G/1 process, in Chapter 6.

9.1 Primitives and Dynamics

The primitives of the generalized Jackson network are input, service, and routing processes. With each of the nodes $i = 1, \ldots, I$, we associate an arrival counting process, $\mathcal{A}_i(t)$, which counts the number of exogenous arrivals over $(0, t]$, and a service process, $\mathcal{S}_i(t)$, which counts the number of job completions at node i for a total processing time of t. Jobs are routed from node i to other nodes, and we let $\xi_i(r)$ be a unit vector that indicates the destination of the rth job completed at node i, so $\xi_{i,j}(r) = 1$ if the job goes to node j and $\xi_{i,j}(r) = 0$ otherwise. For the first m jobs processed at node i, the vector $\mathcal{R}_i(m) = \sum_{r=1}^{m} \xi_i(r)$ counts the cumulative number of jobs routed to each node and $m - \sum_{j=1}^{I} \mathcal{R}_{i,j}(m)$ is the number of jobs that leave the system.

Let $\mathcal{T}_i(t)$ be the cumulative time that node i is processing jobs over $(0, t]$. Then, starting from the vector of initial queues $Q(0)$, the queue lengths (including jobs in service) $Q(t)$ at time t are

$$Q_i(t) = Q_i(0) + \mathcal{A}_i(t) - \mathcal{S}_i(\mathcal{T}_i(t)) + \sum_{j=1}^{I} \mathcal{R}_{j,i}(\mathcal{S}_j(\mathcal{T}_j(t))), \qquad (9.1)$$

consisting of initial jobs plus exogenous arrivals minus job completions plus jobs routed from other nodes over $(0, t]$. In this equation, we make the assumption that processing is non-preemptive head of the line (HOL), so that $\mathcal{S}_i(\mathcal{T}_i(t))$ indeed counts the number of departures from node i. We make the additional assumption that processing is work conserving (i.e. non-idling), so:

$$Q_i(t) \geq 0, \qquad \int_0^t Q_i(t)d\mathcal{I}_i(t) = 0, \quad i = 1, \ldots, I, \qquad (9.2)$$

where $\mathcal{I}_i(t)$ is the total idling time of node i over $(0, t]$.

Given the primitives, under work-conserving non-preemptive policy, the implicit equations (9.1), (9.2) uniquely determine the evolution of the system (see Exercise 9.1).

Our assumptions for the stochastic behavior of the network are the following: $\mathcal{A}_i(t)$ and $\mathcal{S}_i(t)$ are renewal processes, $\xi_{i,j}$ are multi-Bernoulli variables, and all are independent. The arrivals are at rates α_i, processing is at rates μ_i, and routing probabilities are given by the matrix P where $p_{i,j}$ is the probability that a job completed by node i is routed to node j. We assume that P has spectral radius < 1 so that every customer leaves the system after a finite number of steps a.s. We assume existence of second moments for interarrivals and processing times, which are quantified by coefficients of variation $c_{a,i}, c_{s,i}$ respectively. For strong approximations, we assume existence of moments of order $r > 2$.

9.2 Traffic Equations and Stability

In Proposition 8.3 we derived the traffic equations

$$\lambda = \alpha + P^{\mathsf{T}}\lambda, \qquad (9.3)$$

which are solved by $\lambda = (I - P^{\mathsf{T}})^{-1}\alpha = (I + P^{\mathsf{T}} + P^{\mathsf{T}2} + \cdots)\alpha$. These will be the flow rates in and out of the nodes in the system if the system is rate stable. The condition for stability for product-form Jackson networks was $\mu_i > \lambda_i$. Clearly, the flow out of node i cannot exceed μ_i. We now derive more general traffic equations, that do not assume rate stability of the network, and describe their solution.

We need to revise our definitions: We define λ_i as the rate of inflow into node i, and we define v_i as the rate of outflow from node i, and we no longer assume that they are equal. With this definition, the correct traffic equations are

$$\lambda = \alpha + P^{\mathsf{T}} v, \qquad (9.4)$$

where the right-hand side decomposes the input rate to each node, as the sum of exogenous input and feedback input. The output, which is the rate of job completions, needs to satisfy $v_i \leq \lambda_i \wedge \mu_i$ for each node. Under a work-conserving policy, it should actually achieve $v_i = \lambda_i \wedge \mu_i$, in vector form: $v = \lambda \wedge \mu$. Hence, we define the general traffic equations as:

$$\lambda = \alpha + P^{\mathsf{T}} (\lambda \wedge \mu). \qquad (9.5)$$

To solve the general traffic equations (9.5), let: $z = (\lambda - \mu)^+$ and $w = (\lambda - \mu)^-$, so that $\lambda = v + z$, $\mu = v + w$ and therefore $\lambda = \mu + z - w$, $v = \mu - w$. The unknown z and w need to satisfy the following conditions:

(i) $z = [\alpha - (I - P^{\mathsf{T}})\mu] + (I - P^{\mathsf{T}})w$,   (ii) $w, z \geq 0$,   (iii) $w^{\mathsf{T}} z = 0$.   (9.6)

The three conditions, linear relation, non-negativity, and complementary slackness, define a *linear complementarity problem* (LCP).

The general LCP is to find vectors z, w that satisfy $z = b + Bw$, $z, w \geq 0$, $w^{\mathsf{T}} z = 0$, for given constant vector b and matrix B. In general, the solution of LCP is NP-hard, i.e. we have no way to solve it efficiently. However, in our case the matrix $B = I - P^{\mathsf{T}}$ where P is non-negative with spectral radius < 1. In this special case, LCP has a unique solution for any b, and the solution can be computed efficiently. Exercises 9.3, 9.4 outline an algorithm to solve (9.6) efficiently.

The vector $[\alpha - (I - P^{\mathsf{T}})\mu]$ in (9.6) is called the *netput rate* of the network. It is the rate of inventory accumulation or depletion if processing is at full rate in all the nodes, it will be the rate if all the nodes are non-empty and processing is work conserving. We can see that with input rates λ and output rates v, z_i expresses the rate at which work accumulates in node i, and w_i expresses excess processing rate at node i. If there is excess processing rate in a node, then the node will be idle a fraction w_i/μ_i of the time. If $z_i > 0$, node i will accumulate jobs linearly, and by work conservation it will never idle. Hence, work conservation is equivalent to complementary slackness.

The solution of the LCP classifies the nodes into three groups:

A – Stable nodes: with $\mu_i > \lambda_i$.
B – Critical nodes: with $\mu_i = \lambda_i$.
C – Overloaded nodes: with $\mu_i < \lambda_i$.

In the case of memoryless Jackson networks, it can be shown that nodes of type A on their own form a stable queueing network that is fed by the exogenous input and by constant flows of rate μ_j from the nodes B and C, and this stable sub-network has a product-form stationary distribution. To be more precise, since the network is transient it does not possess a stationary distribution; however, as $t \to \infty$, the distribution of the queues at the nodes in A will converge to the stationary distribution of the corresponding stable Jackson network. At the same time, the nodes in group B have null recurrent queues, and nodes in group C are transient; see Exercises 9.5 – 9.7.

9.3 Centering and Skorohod Oblique Reflection

We now rewrite equations (9.1), and decompose and center some of the terms, to obtain for the dynamics of the queues:

$$Q(t) = X(t) + (I - P^\mathsf{T})\mathrm{diag}(\mu)I(t) = X(t) + \mathcal{Y}(t), \qquad (9.7)$$

where, in more detail,

$$X_i(t) = Q_i(0) + \left(\alpha_i - \mu_i + \sum_{j=1}^{I} p_{j,i}\mu_j\right)t + [\mathcal{A}_i(t) - \alpha_i t]$$

$$-[S_i(\mathcal{T}_i(t)) - \mu_i \mathcal{T}_i(t)] + \sum_{j=1}^{I} p_{j,i}[S_j(\mathcal{T}_j(t)) - \mu_j \mathcal{T}_j(t)] \quad (9.8)$$

$$+ \sum_{j=1}^{I} [\mathcal{R}_{j,i}(S_j(\mathcal{T}_j(t))) - p_{j,i}S_j(\mathcal{T}_j(t))].$$

$$\mathcal{Y}_i(t) = \mu_i(t - \mathcal{T}_i(t)) - \sum_{j=1}^{I} p_{j,i}\mu_j(t - \mathcal{T}_j(t)), \qquad (9.9)$$

and work conservation is expressed by

$$I_i(t) = t - \mathcal{T}_i(t) = \int_0^t \mathbb{1}\{Q_i(s) = 0\}ds, \qquad (9.10)$$

where $\mathbb{1}\{\cdot\}$ denotes characteristic function, and $\mathrm{diag}(\mu)$ the diagonal matrix of the processing rates μ_i.

The matrix $R = (I - P^\mathsf{T})\mathrm{diag}(\mu)$ is called *the input-output matrix* of the network, with elements $R_{i,j}$ that give the rate of depletion in the queue of node i, measured in expected number of customers per unit time, due to processing at node j for one time unit (here negative depletion is actually an increase). The inverse matrix R^{-1} has elements $m_{i,j}$, where $m_{i,j}$ is the expected amount of work to be done by node i, per item that is present in the queue of node j. In particular, the vector of expected work in the system, $\mathbb{E}(\mathcal{W}(t)|Q(t)) = R^{-1}Q(t)$.

In this decomposition, we refer to $\mathcal{X}(t)$ as the *netput process*. $\mathcal{X}(t) - Q(0)$ is the amount of increase in the queues resulting from arrivals minus service completions if all the nodes are processing continuously from 0 to t (it may well be negative). The other part of the decomposition is given by $\mathcal{Y}(t) = R\mathcal{I}(t)$, where $\mathcal{I}(t)$ is the vector of cumulative idle times, and it consists of the expected lost number of service completions because of the actual idling time.

It is immediately seen that $\mathcal{X}$, Q, and $\mathcal{Y} = R\mathcal{I}$ satisfy the conditions of Skorohod's oblique reflection, with reflection matrix R as given by (7.22) in Section 7.6. Therefore, $\mathcal{X}$, Q, $\mathcal{Y}$ satisfy $\mathcal{Y}(\cdot) = \varphi(\mathcal{X}(\cdot))$ and $Q(\cdot) = \psi(\mathcal{X}(\cdot))$.

9.4 Fluid Limits

Recall, as in Section 6.2, that fluid scaling is obtained by scaling time and space by a factor of n, so time is measured in units of n, and space in units of n. To avoid starting with 0 at time 0, we use, similar to Section 6.4, a sequence of systems, sharing the same arrival times, processing times, and routings, but with different initial conditions. Thus, we will have for system n, the vector and matrix quantities:

$$\bar{Q}^n(0) = \frac{1}{n}Q^n(0), \quad \bar{\mathcal{A}}^n(t) = \frac{1}{n}\mathcal{A}(nt), \quad \bar{S}^n(t) = \frac{1}{n}S(nt),$$

$$\bar{\mathcal{T}}^n(t) = \frac{1}{n}\mathcal{T}^n(nt), \quad \bar{\mathcal{R}}^n(u) = \frac{1}{n}\mathcal{R}(\lfloor nu \rfloor).$$

Note that while initial state and busy times depend on n, $\mathcal{A}$, S, $\mathcal{R}$ do not.

Theorem 9.1 *Assume that $\bar{Q}^n(0) \to \bar{Q}(0)$. Then, as $n \to \infty$, the fluid scaled queue length and busy time processes converge almost surely to*

$$\bar{Q}(t) = \psi\left(\bar{Q}(0) + (\alpha - (I - P^\mathsf{T})\mu)t\right), \tag{9.11}$$

$$\bar{\mathcal{T}}(t) = t - R^{-1}\varphi\left(\bar{Q}(0) + (\alpha - (I - P^\mathsf{T})\mu)t\right).$$

Proof We use the centering decomposition of the previous section, equations (9.7), (9.8) and perform fluid scaling to get

$$\bar{X}^n_i(t) = \bar{Q}^n_i(0) + (\alpha_i - \mu_i + \sum_{j=1}^{I} p_{j,i}\mu_j)t + [\bar{\mathcal{A}}^n_i(t) - \alpha_i t]$$

$$- [\bar{S}^n_i(\bar{\mathcal{T}}^n_i(t)) - \mu_i\bar{\mathcal{T}}^n_i(t)] + \sum_{j=1}^{I} p_{j,i}[\bar{S}^n_j(\bar{\mathcal{T}}^n_j(t)) - \mu_j\bar{\mathcal{T}}^n_j(t)]$$

$$+ \sum_{j=1}^{I} [\bar{\mathcal{R}}^{nj}_i(\bar{S}^n_j(\bar{\mathcal{T}}^n_j(t))) - p_{j,i}\bar{S}^n_j(\bar{\mathcal{T}}^n_j(t))]. \tag{9.12}$$

Note that we used $S_i(\mathcal{T}_i^n(nt)) = S_i(n\bar{\mathcal{T}}^n_i(t))$ and $\mathcal{R}_{j,i}(S_j(\mathcal{T}_j^n(nt))) = \mathcal{R}_{j,i}(S_j(n\bar{\mathcal{T}}^n_j(t))) = \mathcal{R}_{j,i}(n\bar{S}_j(\bar{\mathcal{T}}^n_j(t)))$. The other fluid scaled components of the decomposition are:

$$\bar{I}^n_i(t) = t - \bar{\mathcal{T}}^n_i(t) = \int_0^t \mathbb{1}\{\bar{Q}^n_i(s) = 0\}ds,$$

$$\bar{\mathcal{Y}}^n(t) = R\bar{I}^n(t) = (I - P^\top)\mathrm{diag}(\mu)\bar{I}^n(t), \tag{9.13}$$

and in summary we have:

$$\bar{Q}^n(t) = \bar{X}^n(t) + R\bar{I}^n(t), \tag{9.14}$$

and as in the previous section, $\bar{Q}^n(t)$ is the oblique reflection of $\bar{X}^n(t)$ with regulator $\bar{\mathcal{Y}}^n(t)$.

We now let $n \to \infty$ to obtain the fluid limit of $\bar{X}^n(t)$. We assume that $Q^n(0)/n \to \bar{Q}(0)$. By the FSLLN we have $\bar{\mathcal{A}}^n_i(t) - \alpha_i t \to 0, \bar{S}^n_i(t) - \mu_i t \to 0, \bar{\mathcal{R}}^{ni}_j(u) - p_{i,j}u \to 0$, all of them u.o.c. a.s. We now use the same argument as in Section 6.4: Since $\bar{\mathcal{T}}^n_j(t) \le t$, it follows from the u.o.c. convergence that also $\bar{S}^n_i(\bar{\mathcal{T}}^n_i(t)) - \mu_i\bar{\mathcal{T}}^n_i(t) \to 0$ u.o.c. a.s. Furthermore, a.s. $\bar{S}^n_i(\bar{\mathcal{T}}^n_i(t)) < \mu_i t + \epsilon$ for n large, and so by the same argument, $\bar{\mathcal{R}}^{nj}_i(\bar{S}^n_j(\bar{\mathcal{T}}^n_j(t))) - p_{j,i}\bar{S}^n_j(\bar{\mathcal{T}}^n_j(t)) \to 0$ u.o.c. a.s. Therefore,

$$\bar{X}^n(t) \to^{a.s.}_{u.o.c.} \bar{X}(t) = \bar{Q}(0) + (\alpha - (I - P^\top)\mu)t.$$

By the continuity of the oblique reflection: $\bar{Q}(\cdot) = \psi(\bar{X}(\cdot)), \bar{Y}(\cdot) = \varphi(\bar{X}(\cdot))$. □

Starting from any initial fluid levels $\bar{Q}(0)$, the paths of the fluid limit are continuous piecewise linear functions of t in $\mathbb{R}^I$. It behaves as follows: The fluid limit for the underloaded nodes in $i \in A$, $\bar{Q}_i(t)$ decreases along a piecewise linear path to 0; for critical nodes in $i \in B$, the fluid level reaches a constant level and remains constant; and for overloaded nodes in $i \in C$, $\bar{Q}_i(t)$

eventually increases linearly. Once all the nodes in A reach 0, the nodes in B will remain constant, and the nodes in C will have fluid accumulating at rate $\lambda_i - \mu_i$, where λ_i is obtained from the solution of the generalized traffic equations (9.5) (see Exercise 9.2).

It follows that:

Proposition 9.2 *There exists a t_0 such that for $t > t_0$:*

$$\bar{Q}i(t) = \begin{cases} 0 & i \in A, \\ \bar{Q}_i(t_0) & i \in B, \\ \bar{Q}_i(t_0) + (\lambda_i - \mu_i)(t - t_0) & i \in C, \end{cases} \tag{9.15}$$

$$\bar{\mathcal{T}}_i(t) = \bar{\mathcal{T}}_i(t_0) + \min(\rho_i, 1)(t - t_0),$$

$$\bar{\mathcal{W}}_i(t) = \mu_i \bar{Q}i(t),$$

where λ is the solution of the traffic equations (9.5), and $\rho_i = \frac{\lambda_i}{\mu_i}$.

9.5 Diffusion Limits

The fluid limits describe the behavior of the system on a fluid scale, i.e. with time and space (customers or work) in units of n. We will now discuss diffusion limits, which measure the deviations from the fluid limits, in time still scaled by n but space scaled by $\sqrt{n}$, i.e. deviations of queue length and workload from the fluid limit measured in units of $\sqrt{n}$.

In particular, for nodes that have fluid > 0, the fluid limits describe the fluid dynamics of the queues and workloads as piecewise linear functions (see Exercise 9.2). In particular, for overloaded nodes, the fluid will be positive always. When the fluid is positive, the deviations measured in units of $\sqrt{n}$ will converge as $n \to \infty$ to a Brownian motion. The proof of that is similar to the proof for the case of a single G/G/1 queue with $\rho > 1$, Theorem 6.2. However, these deviations will be negligible compared to n, and the fluid limit will give the correct approximation to the queue length and the workload.

In contrast, for nodes of type A (stable nodes, underloaded nodes), the fluid limit will be 0 from some time point onward. For these nodes, once their fluid is 0, the actual deviations from 0, measured in units of $\sqrt{n}$, will as $n \to \infty$ converge in probability to 0. The proof of that is the same as the proof for the case of a single G/G/1 queue with $\rho < 1$, Theorem 6.3.

The analysis for critical nodes is more delicate, and this is where we will actually be looking at nodes that are in *balanced heavy traffic*, i.e. the traffic intensity at these nodes, as obtained from the solution of the traffic equations (9.5), is $\rho_i \approx 1$. We will in this section focus on the case that all nodes in the network have $\rho_i \approx 1$, and develop the diffusion limit for those.

To obtain the diffusion limits, we again consider a sequence of systems indexed by n, as indicated in Section 9.1, where exogenous arrivals are at rates α^n and services are at rates μ^n, and the solution of the traffic equations (9.5) is the vector λ^n.

We assume the following as $n \to \infty$:

$$\alpha^n \to \alpha, \quad \mu^n \to \mu,$$
$$\sqrt{n}(\alpha^n - (I - P^\mathsf{T})\mu^n) \to \theta, \tag{9.16}$$
$$Q^n(0)/\sqrt{n} \to \hat{Q}(0).$$

Proposition 9.3　*The conditions (9.16) imply*

(a) $\lambda^n - \mu^n \to 0$, *as a result $\rho_i^n \to 1$, so all the nodes are close to critical, and $\lambda^n \to \lambda = \mu$.*

(b) *Because $\rho_i^n \to 1$, $\bar{\mathcal{T}}^n(t) \to t$ u.o.c. a.s.*

(c) *Because $Q^n(0)/\sqrt{n} \to \hat{Q}(0)$, $Q^n(0)/n \to 0$.*

(d) *As a result, $\bar{Q}(t) = 0$ for all $t > 0$.*

Proof　To show (a) we write the traffic equations (9.4): $\lambda^n = \alpha^n + P^\mathsf{T}(\mu^n \wedge \lambda^n)$ in the form of the LCP (9.5):

$$\text{(i)} \quad (\lambda^n - \mu^n)^+ = [\alpha^n - (I - P^\mathsf{T})\mu^n] + (I - P^\mathsf{T})(\lambda^n - \mu^n)^-,$$

$$\text{(ii)} \quad (\lambda^n - \mu^n)^+, \ (\lambda^n - \mu^n)^- \geq 0,$$

$$\text{(iii)} \quad (\lambda^n - \mu^n)^{+\mathsf{T}}(\lambda^n - \mu^n)^- = 0.$$

We note that this type of LCP has a unique solution and the solution is continuous in its parameters. By (9.16), $\alpha^n - (I - P^\mathsf{T})\mu^n \to 0$, so the unique limiting solutions as $n \to \infty$ will be

$$(\lambda - \mu)^+ = 0 + (I - P^\mathsf{T})(\lambda - \mu)^- \implies \lambda = \mu.$$

Next, (c) follows immediately from (9.4), and (b) and (d) follow from Theorem 9.1 by substituting in (9.11). □

We refer to these conditions as *balanced heavy traffic* conditions. We now have that λ_i is the limiting flow rate in and out of node i, and θ_i is the rate of deviation from these flows, in the diffusion scale. We refer to θ as the drift. Because $\bar{Q}(t) = 0$, we define the diffusion scaled queue length vector as

$$\hat{Q}^n(t) = \frac{1}{\sqrt{n}} Q^n(nt) = \sqrt{n}\bar{Q}^n(t).$$

Theorem 9.4 *Under balanced heavy traffic conditions, as $n \to \infty$ the diffusion scaled queue length process converges to a multivariate reflected Brownian motion:*

$$\hat{Q}^n(t) \to_w RBM_{\hat{Q}(0)}(t; \theta, \Gamma, R), \text{ where:} \qquad (9.17)$$

$$\Gamma_{i,i} = \alpha_i c_{a,i}^2 + \mu_i c_{s,i}^2 (1 - 2p_{i,i}) + \sum_{k=1}^{I} \mu_k p_{k,i} (1 - p_{k,i} + c_{s,k}^2 p_{k,i}),$$

$$\Gamma_{i,j} = -\left[\mu_i c_{s,i}^2 p_{i,j} + \mu_j c_{s,j}^2 p_{j,i} + \sum_{k=1}^{I} \mu_k (1 - c_{s,k}^2) p_{k,i} p_{k,j} \right].$$

Proof Similar to (9.7), (9.8), and (9.12)–(9.14), and using Proposition 9.2, we have, after multiplying by $\sqrt{n}$,

$$\hat{Q}_i^n(t) = \sqrt{n}\bar{Q}_i^n(t) = \sqrt{n}\bar{X}_i^n(t) + \sqrt{n}\bar{Y}_i^n(t) \qquad (9.18)$$

$$= \sqrt{n}\Big[\bar{Q}_i^n(0) + (\alpha_i^n - \mu_i^n + \sum_{j=1}^{I} p_{j,i}\mu_j^n)t + [\bar{\mathcal{A}}_i^n(t) - \alpha_i^n t]$$

$$-[\bar{S}_i^n(\bar{\mathcal{T}}_i^n(t)) - \mu_i^n \bar{\mathcal{T}}_i^n(t)] + \sum_{j=1}^{I} p_{j,i}[\bar{S}_j^n(\bar{\mathcal{T}}_j^n(t)) - \mu_j^n \bar{\mathcal{T}}_j^n(t)]$$

$$+ \sum_{j=1}^{I} [\bar{\mathcal{R}}_i^{nj}(\bar{S}_j^n(\bar{\mathcal{T}}_j^n(t))) - p_{j,i}\bar{S}_j^n(\bar{\mathcal{T}}_j^n(t))]\Big]$$

$$+ \sqrt{n}\Big[\mu_i^n(t - \bar{\mathcal{T}}_i^n(t)) - \sum_{j=1}^{I} p_{j,i}\mu_j^n(t - \bar{\mathcal{T}}_i^n(t)) \Big].$$

We now consider $\sqrt{n}\bar{X}_i^n(t)$, and let $n \to \infty$. We then have:
- By assumption, $\sqrt{n}\bar{Q}_i^n(0) \to \hat{Q}_i(0)$.
- By the heavy traffic conditions, $\sqrt{n}(\alpha_i^n - \mu_i^n + \sum_{j=1}^{I} p_{j,i}\mu_j^n)t \to \theta t$.
- By the FCLT for renewal processes and continuous mapping, $[\bar{\mathcal{A}}_i^n(t) - \alpha_i^n t] \to_w (\alpha_i c_{a,i}^2) BM(t)$.
- By $\bar{\mathcal{T}}_i^n(t) \to t$ and non-decreasing, using FCLT and time change, the remaining components also converge to Brownian motion.

It follows that $\hat{X}^n(t) \to_w BM_{\hat{Q}(0)}(t; \theta, \Gamma)$. Since $\hat{Q}^n(t)$ is the oblique reflection of $\hat{X}^n(t)$ with reflection matrix R, by the continuity of the oblique reflection, $\hat{Q}^n(t) \to_w RBM_{\hat{Q}(0)}(t; \theta, \Gamma, R)$.

It remains to calculate Γ. We look at the simpler process $\tilde{X}(t)$, given by:

$$\tilde{X}_i(t) = Q_i(0) + \mathcal{A}_i(t) - S_i(t) + \sum_{k=1}^{I} \mathcal{R}_{k,i}(S_k(t)),$$

with renewal $\mathcal{A}$, $\mathcal{S}$ and multinomial $\mathcal{R}$ and calculate the variance covariance matrix of $\tilde{X}(t)$.

The first term is constant and contributes nothing. The second term has variance $\alpha_i c_{a,i}^2$ and $\mathrm{Cov}(\mathcal{A}_i(t), \mathcal{A}_j(t)) = 0$, $i \neq j$, and it is independent of the remaining terms. To calculate the variance of the remaining terms, we first condition on the values of the random vector $\mathcal{S}(t)$, to obtain

$$\mathbb{E}\Big[\mathcal{S}_i(t) - \sum_{k=1}^{I} \mathcal{R}_{k,i}(\mathcal{S}_k(t)) \Big| \mathcal{S}(t)\Big] = \mathcal{S}_i(t) - \sum_{k=1}^{I} p_{k,i} \mathcal{S}_k(t),$$

$$\mathrm{Var}\Big[\mathcal{S}_i(t) - \sum_{k=1}^{I} \mathcal{R}_{k,i}(\mathcal{S}_k(t)) \Big| \mathcal{S}(t)\Big] = \sum_{k=1}^{I} \mathcal{S}_k(t) p_{k,i}(1 - p_{k,i}),$$

$$\mathrm{Cov}\Big[\mathcal{S}_i(t) - \sum_{k=1}^{I} \mathcal{R}_{k,i}(\mathcal{S}_k(t)), \mathcal{S}_j(t) - \sum_{k=1}^{I} \mathcal{R}_{k,j}(\mathcal{S}_k(t)) \Big| \mathcal{S}(t)\Big]$$

$$= -\sum_{k=1}^{I} \mathcal{S}_k(t) p_{k,i} p_{k,j},$$

and from this we obtain (using $\mathrm{Var}(B) = \mathrm{Var}(\mathbb{E}(B|A)) + \mathbb{E}(\mathrm{Var}(B|A))$):

$$\mathrm{Var}\Big[\mathcal{S}_i(t) - \sum_{k=1}^{I} \mathcal{R}_{i,k}(\mathcal{S}_k(t))\Big]$$

$$= \Big[\mu_i c_{s,i}^2 (1 - p_{i,i})^2 + \sum_{k \neq i} \mu_k c_{s,k}^2 p_{k,i}^2 + \sum_{k=1}^{I} \mu_k p_{k,i}(1 - p_{k,i})\Big] t.$$

To get the covariance, we first calculate

$$\mathrm{Cov}\Big[\mathcal{S}_i(t) - \sum_{k=1}^{I} p_{k,i} \mathcal{S}_k(t), \mathcal{S}_j(t) - \sum_{k=1}^{I} p_{k,j} \mathcal{S}_k(t)\Big]$$

$$= \Big[-\mu_i c_{s,i}^2 p_{i,j} - \mu_j c_{s,j}^2 p_{j,i} + \sum_{k=1}^{I} \mu_k c_{s,k}^2 p_{k,i} p_{k,j}\Big] t,$$

and using $\mathrm{Cov}(B, C) = \mathrm{Cov}(\mathbb{E}(B|A), \mathbb{E}(C|A)) + \mathbb{E}(\mathrm{Cov}(B, C|A))$, we then obtain:

$$\mathrm{Cov}\Big[\mathcal{S}_i(t) - \sum_{k=1}^{I} \mathcal{R}_{i,k}(\mathcal{S}_k(t)), \mathcal{S}_j(t) - \sum_{k=1}^{I} \mathcal{R}_{j,k}(\mathcal{S}_k(t))\Big]$$

$$= -\Big[\mu_i c_{s,i}^2 p_{i,j} + \mu_j c_{s,j}^2 p_{j,i} + \sum_{k=1}^{I} (\mu_k - \mu_k c_{s,k}^2) p_{k,i} p_{k,j}\Big] t.$$

Clearly, these calculations are valid for the renewal sequences of interarrivals, services, and routings indexed by n, and for their limits. Scaling time by n and space by $\sqrt{n}$ will leave the variance at the same value. Finally, we note that by the time change lemma (Theorem 5.6), $S_i^n(t) - \sum_{k=1}^{I} \mathcal{R}_{i,k}^n(t)$ and $S_i^n(\mathcal{T}^n(t)) - \sum_{k=1}^{I} \mathcal{R}_{i,k}^n(\mathcal{T}^n(t))$, under diffusion scaling, converge to the same limit. $\qquad\square$

By Theorem 7.8, the multivariate reflected Brownian motion that approximates the queue lengths has a stationary distribution if and only if $R^{-1}\theta < 0$. The approximation will give us a general idea about the behavior of the network. As noted before, calculating moments or probabilities for this stationary distribution requires sophisticated numerical methods.

Recall the alternative approach of strong approximation assumptions discussed in Section 6.8. This can also be used for generalized queueing networks.

9.6 Stability of Generalized Jackson Networks

We have seen that if $\lambda_i < \mu_i$ for all nodes, then the fluid limit will decrease to 0, and the diffusion limit will also be 0. What can be said about the process $Q(t)$ itself? Is it stable? Stability can be described in many different ways, and to prove stability of any type, one needs to make the appropriate assumptions. Recall that for the G/G/1 queue, the Loynes construction required only stationarity of the arrival and service stream, and ergodicity for their rates, in order to show that waiting times possess a stationary distribution if $\rho < 1$. Things are much harder for Jackson-type networks, and there is no Loynes construction (there is a Loynes construction for feedforward networks). One difficulty is that unlike the single queue case, even if the system is stable, it may never be empty of customers; see, for example, Exercises 9.10 and 9.11. Stability of generalized Jackson networks was considered a hard problem in the 1980s, and several ways to prove different types of stability under varying assumptions emerged in the late 1980s.

The generalized Jackson network with renewal arrivals, i.i.d. services, Markovian routing, and work-conserving non-preemptive HOL policy, can be described by a Markov process in continuous time, with a general state space as follows: At time t we need to know for each node the queue length, and in addition, the residual time until the next external arrival and the residual time until the next service completion (defined as 0 if the server at node i is idle and no customers are waiting). This is summarized by $\mathcal{Z}(t) = (Q_i(t), U_i(t), V_i(t), i = 1, \ldots, I)$. The paths of this process are again

RCLL in $\mathbb{Z}^J \times \mathbb{R}^{2J}$, and it is a *piecewise deterministic Markov process*, where $Q_i(\cdot)$ remain constant, and $U_i(\cdot), V_i(\cdot)$ decrease at rate 1 between events, and the appropriate U_i or V_i jump up at the instant of an arrival or job completion, with corresponding changes in Q. We state here:

Theorem 9.5 *For a generalized Jackson queueing network, if $\rho_i < 1$ at every node, then the queue length process is stable in the sense of Definition 2.6, and in fact, $\mathcal{Z}(t) = (Q_i(t), U_i(t), V_i(t), i = 1, \ldots, I)$ possess a stationary distribution.*

We give a modern proof for this theorem in Chapter 11, Section 11.3.

9.7 Sources

A textbook on the linear complementarity problem (LCP) is Cottle et al. (2009). Goodman and Massey (1984) derive an algorithm to solve LCP for Jackson networks, and show that nodes of type A have a limiting stationary distribution. Local stability, where a transient Markov chain has a subset of states that have a limiting stationary distribution, is discussed in Adan et al. (2020). The fluid and diffusion approximations of generalized Jackson networks are developed in the pioneering paper by Reiman (1984a); see also Harrison and Williams (1987a,b), which prove conditions for existence of stationary diffusion approximations and derive conditions for product-form distribution. Further details are given by Chen and Mandelbaum (1994a,b), where they use a strong approximation approach. Stability of generalized Jackson networks was proved by Sigman (1990), Baccelli and Foss (1994), and Meyn and Down (1994). The classic book on stability of general state-space Markov chains is *Markov Chains and Stochastic Stability*, Meyn and Tweedie (1993).

Exercises

9.1   Prove by induction that the implicit conditions of the dynamics, the non-negativity, and work conservation (9.1), (9.2), uniquely determine the queue length process.

9.2   Consider the Jackson network with the following data:

$$\alpha = \begin{bmatrix} 60 \\ 12.5 \\ 0 \\ 0 \\ 0 \end{bmatrix}, \quad P = \begin{bmatrix} 0 & 0.4 & 0.2 & 0.4 & 0 \\ 0 & 0 & 0.5 & 0 & 0.5 \\ 0 & 0 & 0 & 0 & 0.5 \\ 0 & 0.2 & 0 & 0 & 0.8 \\ 0.2 & 0 & 0 & 0 & 0 \end{bmatrix}, \quad \mu = \begin{bmatrix} 62.5 \\ 30 \\ 50 \\ 30 \\ 75 \end{bmatrix}.$$

(i) Draw the network.

(ii) Classify the states into stable, overloaded, and balanced and describe the long-term behavior.

(iii) Find the steady-state limiting distribution of the stable part of the network.

(iv) Assume that initial limiting fluid, under fluid scaling is

$$\bar{Q}(0) = \begin{bmatrix} 62.5 \\ 30 \\ 50 \\ 30 \\ 75 \end{bmatrix},$$

and calculate the fluid paths of the network.

9.3 Suppose that in the solution of the LCP (9.5), (9.6) the identities of the nodes of type A (underloaded) are known, i.e. $U = \{i : w_i > 0\}$ is given. Derive the solution of the LCP problem and the rate of inflow λ and outflows ν for the network.

9.4 Prove that if you define $\tilde{\lambda} = \alpha + P^{\mathsf{T}}\mu$, then $\tilde{\lambda}_i > \lambda_i$, i.e. $\tilde{\lambda}$ provides an upper bound for the vector of inflows. Generalize this statement for subsets of nodes, and use it to derive an algorithm to solve the LCP (9.5), (9.6) in at most I iterations, each involving one matrix inversion.

9.5 Consider a memoryless Jackson network $Q(t)$ and let $U = \{i : \rho_i = \lambda_i/\mu_i < 1\}$ be its type A nodes. Define the network $Q^+(t)$ by:

$$\alpha_i^+ = \begin{cases} \alpha_i + \sum_{j \notin U} \mu_j p_{j,i}, & i \in U, \\ \alpha_i, & i \notin U, \end{cases}$$

$$p_{i,j}^+ = \begin{cases} 0, & i \notin U \text{ and } j \in U, \\ p_{i,j}, & \text{otherwise.} \end{cases}$$

(i) Show that $Q_i^+(t) \geq_{ST} Q_i(t)$.

(ii) Find the stationary distribution of the process $Q_U^+(t)$, that includes only the queues at the nodes in i.

9.6 Consider a memoryless Jackson network $Q(t)$, and let $U = \{i : \rho_i = \lambda_i/\mu_i < 1\}$ be its type A nodes. Define the network $Q^{-,\epsilon}(t)$ by:

$$\alpha_i^{-,\epsilon} = \alpha_i,$$
$$\mu_i^{-,\epsilon} p_{i,j}^{-,\epsilon} = \mu_i p_{i,j},$$
$$\mu_i^{-,\epsilon} q_i^{-,\epsilon} = \begin{cases} \mu_i q_i, & i \in U, \\ \mu_i q_i + \lambda_i - \mu_i + \epsilon, & i \notin U, \end{cases}$$

where λ_i is the inflow rate obtained from the traffic equations for Q, and $q_i, q_i^{-,\epsilon}$ are the fraction of completed jobs at node i that leave the system.

(i) Show that $Q_i^{-,\epsilon}(t) \leq_{ST} Q_i(t)$.

(ii) Show that $Q^{-,\epsilon}(t)$ is stable, and find its stationary distribution.

9.7   Use the results of the previous exercises to prove:

Theorem 9.6 (Goodman and Massey (1984))   *In a Jackson network, for the set of nodes $U = \{i: \rho_i = \lambda_i/\mu_i < 1\}$ (nodes of type A, stable nodes),*

$$\lim_{t \to \infty} \mathbb{P}(Q_i(t) = n_i : i \in U) = \prod_{i \in U}(1 - \rho_i)\rho_i^{n_i},$$

by showing that $Q^{-,\epsilon}(t)$, $Q(t)$, $Q^+(t)$ can be coupled so that: $Q^{-,\epsilon}(t) \leq Q(t) \leq Q^+(t)$.

9.8   We have derived the diffusion limits for a stable G/G/1 queueing network, where the queue length converges to $RBM(t; \theta, \lambda(c_a^2 + c_s^2))$, where $\theta = \lim \sqrt{n}(\lambda^n - \mu^n)$. For a queue with input consisting of the superposition of several renewal processes, and with s servers, a similar result was derived by Iglehart and Whitt (1970a,b). The proof requires quite a bit of technique, but can you just write down what you think is the result (recall Exercise 6.13)?

9.9   In Section 9.5 we derived the components of the variance covariance matrix Γ when all the stations are heavily loaded. Generalize this for the case when the nodes of the network are of types A, B, and C [Chen and Mandelbaum (1991); Chen and Mandelbaum (1994b)].

9.10 Consider a tandem queueing network with two nodes. Construct an example with deterministic arrivals and deterministic service times, for which $\rho_i < 1$ but the system never empties.

9.11 Consider a parallel service system with two servers, in which arrivals are assigned randomly to one of the servers, and leave at service completion. Construct an example with deterministic arrivals and deterministic service times, for which $\rho_i < 1$ but the system never empties.

Part IV

Fluid Models of Multi-Class Queueing Networks

In this part of the book, we introduce multi-class queueing networks (MCQN) and study their fluid approximations. In MCQN, customers in each node are differentiated into classes, and each class has its own queue at the node. Each entering customer then moves through its own sequence of classes, from node to node. We can now control the scheduling of customers at each node, the routing of customers between classes, and the admission of customers into the system. These controls influence stability and performance. Fluid approximations track the transient behavior of the networks, and are used to obtain stability and to optimize performance of the network.

In Chapter 10 we present surprising examples of MCQN with $\rho < 1$ that are unstable under some policies, and prepare the background for rigorous treatment of stability. In Chapter 11 we define fluid limits, and show that their stability implies stability of the stochastic system. This enables us to study stability of MCQN under various policies. Chapter 12 introduces more general processing networks, and the maximum pressure policy, which uses local information for decentralized control of the network, and can guarantee stability whenever $\rho < 1$.

The remaining two chapters discuss control of transient MCQN. In Chapter 13 we introduce queues with infinite supply of work, which can help a system achieve stability with $\rho = 1$. In Chapter 14 we formulate a fluid optimization problem that we can solve using a separated continuous linear programming (SCLP) algorithm. We use it to find the optimal fluid solution over a finite time horizon and we track this optimal fluid solution, using infinite virtual queues and a maximum pressure policy, to obtain asymptotically optimal controls. We illustrate the application of this method to large-volume production in semi-conductor wafer fabrication plants.

10

Multi-Class Queueing Networks, Instability, and Markov Representations

10.1 Multi-Class Queueing Networks

The queueing networks we have considered so far, Jackson networks and Kelly-type networks, had a single queue at every node, and all customers at each node received the same treatment, with no consideration of how much work they require and where they came from or are going next. We can think of them as single class queueing networks. We now consider multi-class queueing networks (MCQN), where each node has several queues, one for each class at the node, differentiated by their processing requirements and their routes; we will refer to nodes from now on as workstations or simply stations.

We can think of MCQN as a generalization of Kelly networks with nodes $i = 1, \ldots, I$, in which each type of customers, say c, on step r of its route, at a given node, say i, will wait in a special queue, or buffer, specific to type and step (c, r). We refer to (c, r) as a class, and we renumber all classes as $k = 1, \ldots, K$, where k that corresponds to (c, r), is served by station $s(k)$. We define the constituency matrix C of dimension $I \times K$ with 0 or 1 elements, where $C_{i,k} = 1$ indicates that class k is processed at station i, and we also denote by $C_i = \{k : s(k) = i\}$ the constituency of stations i. MCQN are more general than Kelly networks, in that we can now have exogenous inputs to each class and different processing time distributions for each class, and we no longer assume fixed routes between classes.

The dynamics of MCQN are again governed by an equation similar to the single class dynamics equation (9.1):

$$Q_k(t) = Q_k(0) + \mathcal{A}_k(t) - \mathcal{S}_k(\mathcal{T}_k(t)) + \sum_{l=1}^{K} \mathcal{R}_{l,k}(\mathcal{S}_l(\mathcal{T}_l(t))), \qquad (10.1)$$

where as before, $\mathcal{A}_k(t)$ counts arrivals into class k, $\mathcal{S}_k(t)$ is the total number of class k service completions for total service duration t, $\mathcal{R}_{k,l}(m)$ is the number of class k routed to class l out of the first m job completions, and

159

$\mathcal{T}_k(t)$ is the total time devoted by station $s(k)$ to the processing of class k. In addition, we have the constraints

$$\sum_{k \in C_i} (\mathcal{T}_k(t) - \mathcal{T}_k(s)) \le t - s, \ s < t, \quad \sum_l (\mathcal{R}_{k,l}(n) - \mathcal{R}_{k,l}(m)) \le n - m, \ m < n,$$

$$(10.2)$$

where the constraints on $\mathcal{T}_k$ for each station i now replace the single class constraint $\mathcal{T}_i(t) - \mathcal{T}_i(s) \le t - s$.

In equation (10.1) we assume HOL (head of the line) processing, which means that departures from class k are $\mathcal{D}_k(t) = \mathcal{S}_k(\mathcal{T}_k(t))$. However, we may occasionally allow processing at a station to be split between the classes served at that station, while keeping HOL within each class. We will often assume work-conserving policies; idle time and work conservation are expressed by

$$\mathcal{I}_i(t) = t - \sum_{k \in C_i} \mathcal{T}_k(t), \qquad \int_0^t \sum_{k \in C_i} Q_k(s) \, d\mathcal{I}_i(s) = 0. \qquad (10.3)$$

Compared to single class networks, we have many more options for control of MCQN. In particular, we have three modes of control: scheduling control, by deciding in what sequence jobs at station i are processed, and this is done by controlling $\mathcal{T}_k(t)$; routing control, where should a customer leaving class k go, and this is done by controlling $\mathcal{R}_{k,l}(n)$; and admissions control, when do we allow customers of type k to enter the system, which is done by controlling $\mathcal{A}_k(t)$.

In the rest of this chapter and in the following Chapter 11, we will focus on sequencing control, and keep the long-term parameters of exogenous input rates α_k, routing fractions $p_{k,l}$ and processing requirements $m_k = 1/\mu_k$ fixed, so that our control consists of determining $\mathcal{T}_k(t)$. We will consider more general controls in Chapters 12–14 and in Part V.

Remark   In Kelly networks, we allowed processing rates in station i to depend on the number of customers at the station, $\mu_i(n)$. We now assume constant processing capacity of each station.

Given the vectors α, μ and the matrix P with spectral radius < 1, we can calculate the offered load by calculating λ_k, the long-term flow rate of class k customers by solving the traffic equations (8.1), and then the offered load for station i is $\rho_i = \sum_{k \in C_i} \frac{\lambda_k}{\mu_k}$. In matrix form:

$$\rho = CR^{-1}\alpha, \qquad (10.4)$$

where $R = (I - P^{\mathsf{T}})\text{diag}(\mu)$ is the input-output matrix defined in Section 9.3, $M = R^{-1}$ is the mean work requirement matrix, with $m_{k,l}$ the work required at class k for the processing of a customer currently in class l until it leaves the system. Here, $R^{-1}\alpha$ is the vector of offered loads per class, and multiplying by C sums up the offered load for each station.

We will assume as before that exogenous arrivals to class k are independent renewal processes, that processing times of class k are i.i.d., and routing is Markovian.

As expected, the policy chosen to control $\mathcal{T}_k(t)$ will influence the performance of the MCQN. In particular, the policy may influence whether the network is stable or not. It was a shock to the queueing community to discover that, most surprisingly, $\rho_i < 1$ for all stations no longer guarantees stability of all work-conserving policies. In Section 10.2 we present examples of MCQN with $\rho < 1$ that are unstable under some sequencing policies. In Section 10.3 we present Markovian models that describe MCQN under specific sequencing policies; these will be used to study stability. In Section 10.4 we introduce the general Markov process theory that is necessary to discuss stability. Stability will then be discussed in the following Chapters 11 and 12.

10.2 Some Unstable MCQN

We now consider three examples of MCQN that have $\rho < 1$, but are unstable under some given sequencing policy. In the first two examples all the customers in the networks follow a single route that may visit the same node (station) several times. Such networks are typical of some manufacturing lines, e.g. aluminum sheet or steel sheet production (with three or more passages of heating and pressing) and on a much more elaborate scale computer semi-conductor wafer fabrication, which we discuss in Section 14.1. They are called *re-entrant lines*, a term coined by P.R. Kumar.

10.2.1 The Lu–Kumar Network

The Lu–Kumar network is described in Figure 10.1. It is a re-entrant line consisting of two machines (nodes, stations), $i = 1, 2$, and four classes, $k = 1, 2, 3, 4$. All parts (customers, items), follow the same route $1 \rightarrow 2 \rightarrow 3 \rightarrow 4$ and exit. Classes 1,4 are served at node 1, classes 2,3 at node 2, i.e. $s(1) = s(4) = 1$, $s(2) = s(3) = 2$ and $C_1 = \{1, 4\}$, $C_2 = \{2, 3\}$. Arrivals are at rate α, and mean service times are $m_2 = m_4 = 2/3$, $m_1 = m_3 = \epsilon$. We consider the following buffer priority policy: At station 1 priority is given to class 4 (the exit class), and at station 2 priority is given to class 2.

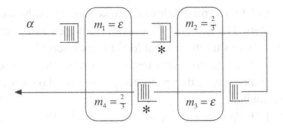

Figure 10.1 The Lu–Kumar re-entrant line.

Assume interarrivals as well as service times are deterministic, and consider the case that $\alpha = 1$. Then the offered loads at the two nodes are:

$$\rho_1 = \alpha(m_1 + m_4) = 2/3 + \epsilon, \qquad \rho_2 = \alpha(m_2 + m_3) = 2/3 + \epsilon.$$

We now show that under this policy the total number of parts in the system diverges. For simplicity, we take $\epsilon = 0$.

Consider the following deterministic sample path: Start at time 0^- with x parts in buffer 1. These parts will move immediately to buffer 2. Thereafter a single part will arrive at times $0, 1, 2, \ldots$, to buffer 1, and move immediately on to buffer 2. At time $t_1 = 2x^-$ buffer 2 will be empty. Here, t_1 solves the equation $t_1 = \frac{2}{3}x + \frac{2}{3}t_1$. Because buffer 2 has priority over buffer 3, during all this time station 2 will not process parts in buffer 3. Therefore, the initial parts and all arrivals up to time t_1, a total of $3x$ parts, will at that time be in buffer 3, and will then move immediately to buffer 4. Buffer 4 will complete their processing at time $t_2 = (t_1 + \frac{2}{3}3x)^- = 4x^-$, and at that time buffer 1 will contain $2x$ parts, which have arrived during the interval (t_1, t_2), and which have not been processed by station 1, because buffer 4 has priority over buffer 1. At time $t_2 = 4x^-$, all the other buffers will be empty. So at time $4x^-$ we have double the load we had at time 0^-. The system will cycle from that time onward, with the inventory blowing up.

It is easy to see that under any other buffer priority policy (of which thee are an additional three), this system will not diverge. In particular, if priority is given to buffers 1 and 3, the system will reach buffer 2 empty at time $2x^-$ with a single job in buffer 4, and thereafter there will never be more than two parts in the system (see Exercise 10.1). We will return to the Lu–Kumar network in Section 11.4.

10.2.2 A FCFS Re-Entrant Line

Examples of systems misbehaving under buffer priority policies raised the question of whether the same can happen under first come first served (FCFS, also know as FIFO, first in first out) policy. Bramson has shown that indeed a re-entrant line with just two machines and $\rho_i < 1$ can be unstable even under FCFS.

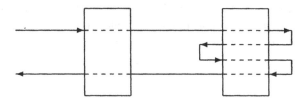

Figure 10.2 A FCFS unstable re-entrant line.

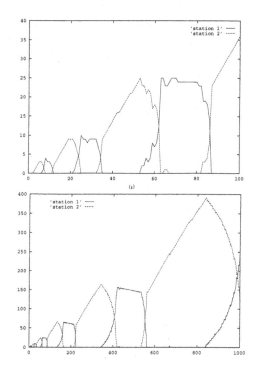

Figure 10.3 Simulation of the FCFS re-entrant line, reproduced with permission from Dai (1995).

Dai considered the following example of the Bramson FCFS network: The network is a six-step re-entrant line, with $C_1 = \{1, 6\}$, $C_2 = \{2, 3, 4, 5\}$, with Poisson input of rate 1, and exponential processing times with means $m_2 = 0.897$, $m_6 = 0.899$, $m_1 = m_3 = m_4 = m_5 = 0.01$, so that $\rho_1 = \rho_2 = 0.9$ (see Figure 10.2).

Figure 10.3, taken from Dai (1995), shows the results of a simulation of this network, operated under FCFS policy at both stations. The figure shows the total inventory at station 1 (solid line) and at station 2 (dashed line). The simulation shows how workload fluctuates moving from one station to the other with ever-growing inventory. While every station returns to 0 infinitely often, the total work in the system increases linearly.

10.2.3 The Kumar–Seidman Rybko–Stolyar Network

This third example, introduced by Kumar and Seidman, was the first example that showed that MCQN may be unstable under some given work-conserving policy, even when the system has enough processing capacity. Kumar and Seidman considered the network with deterministic interarrival and service times. The same network was considered later by Rybko and Stolyar, under Poisson arrivals and exponential service times, and their rigorous analysis showed that the total number of items in the system diverges almost surely.

The Kumar–Seidman Rybko–Stolyar (KSRS) network has two stations, four classes, and two routes. Customers of type 1 enter machine 1, pass to machine 2, and exit; they form classes 1 and 2. Customers of type 2 enter machine 2, pass to machine 1, and exit; they form classes 3 and 4 (see Figure 10.4). Arrival rates are α_1, α_3. Mean service times are $m_k, k = 1, 2, 3, 4$. With these parameters, a necessary condition for stability (we will show that $\rho_i < 1$ is necessary in Theorem 11.12) is:

$$\rho_1 = \alpha_1 m_1 + \alpha_3 m_4 < 1, \qquad \rho_2 = \alpha_1 m_2 + \alpha_3 m_3 < 1. \qquad (10.5)$$

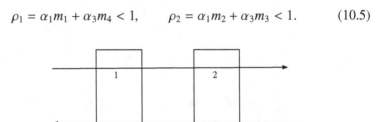

Figure 10.4 The Kumar–Seidman Rybko–Stolyar MCQN.

A reasonable policy for this system is to give priority to classes 2 and 4, which are the exit buffers, since every service completion in these buffers reduces the inventory, while service completions in buffers 1 and 3 do not reduce the inventory, since they only move parts inside the system. As it happens, this policy may lead to diverging inventory even with $\rho_1, \rho_2 < 1$.

We now explain why that may happen. We note that from the first moment that one of the exit buffers is empty, we will always work only on one of these buffers at any time. This is because while an exit buffer is not empty, the station will process it and input to the other exit buffer is blocked, so if it is empty, it will remain empty at least until the other exit buffer is empty. Hence, under the policy of priority to the exit buffers, the following condition is also necessary for stability:

$$\alpha_1 m_2 + \alpha_3 m_4 < 1. \tag{10.6}$$

Under the policy of priority to the exit buffers, the two exit buffers form what is called *a virtual machine*; as we just explained, from some time onward, the virtual machine can only process one job at a time. We will see (in Section 11.4, Theorem 11.20 and Exercise 11.22) that if the three conditions in (10.5), (10.6), hold the system is globally stable: It is positive Harris recurrent (we define this in Section 10.4) for every stationary work-conserving policy.

10.3 Policy Driven Markovian Description of MCQN

Stability of a MCQN when the necessary conditions of $\rho_i < 1$ hold depends, as we have seen, on the policy used. To be able to analyze this, we need a Markov description of the MCQN that takes into account the policy that is used.

As described in Section 9.6, the state of a network so far included the vector of queue lengths $Q(t)$, the vector of remaining times until next arrival, $U(t)$, and the remaining processing times of the customers in service, $V(t)$. In generalized Jackson networks and the related types of single class networks described in Chapters 8 and 9, $(Q(t), U(t), V(t))$, which is moving in the uncountable state space $\mathbb{Z}^K \times \mathbb{R}^K \times \mathbb{R}^K$, is sufficient to describe the state and evolution of the queue length process under the essentially unique work-conserving HOL policy.

To describe the evolution of a MCQN by a Markov process, the values of $Q(t), U(t), V(t)$ are not sufficient, and we need to add more elements, that depend on the policy. We still focus on policies that are work conserving and HOL, but we also allow preemption and splitting of service effort between classes at a node. We now define for each policy an augmented state space, and consider the process $\mathbf{X}(t) = (\mathbf{Y}(t), a(t))$, with the following elements: $\mathbf{Y}(t)$ includes history of the system, given by $\mathbf{Y}(t) = (Q(t), U(t), V(t), Y(t))$, where $Y(t)$ is any additional information from the past that is needed to continue the processing after time t according to the policy, and $a(t)$ is the information about the current action. Usually $Y(t)$ will include information about each customer in the system, which will include his class and his "age". Age may be the time since he joined the class, or the time since he arrived in the system, or the time since his processing started, or his remaining processing time. $a(t)$ will be the information about the current processing activity, with $a_k(t)$ the fraction of the processing capacity of station $i = s(k)$ allocated to customers of class k; this information may be necessary when preemptions and processor splitting are used, and also when preemptions are not allowed.

The state space is contained in the space (listed in order of components)

$$\mathbb{Z}^K \times \mathbb{R}^{2K} \times (\mathbb{Z}^K \times \mathbb{R}^K)^\infty \times \mathbb{R}^K \qquad (10.7)$$

where $(\mathbb{Z}^K \times \mathbb{R}^K)^\infty$ is the space of finitely terminating sequences (finite sequences of any length). For HOL policies, $a(t)$ consists of K real numbers; for policies that are not head of the line, but allow partial processing within each class, $a_k(t)$ needs to include effort for each customer in the system, and this information also moves in the space of finitely terminating sequences of real numbers.

The space of finitely terminating sequences of real numbers is metrizable. It is therefore possible to define a norm for the state space of $\mathbf{X}$, under which the state space is separable, locally compact, and complete. We will not make use of this theoretical possibility, since we do not need such a norm. We will instead use a *surrogate-norm* for $\mathbf{X}(t) = x$, given by

$$|x| = \sum_k (Q_k(t) + U_k(t) + V_k(t)). \qquad (10.8)$$

It is adequate for our purpose to serve as a Lyapunov function, which is always non-negative, and is 0 only when the system is empty.

The elements $Y(t)$ and $a(t)$ are specific to each policy, and we now list some examples.

Allow processor splitting Components of $a(t)$ are given by $a_k(t)$ where $a_k(t)$ is the current rate of processing of class k; $a_k(t) \geq 0$, $\sum_{k \in C_i} a_k(t) \leq$

1, and if $Q_k(t) = 0$ then $a_k(t) = 0$. For work-conserving policies, $\sum_{k \in C_i} a_k(t) = 1$ unless station i is empty.

non-preemptive policies When processor splitting is not allowed, $a_k(t) > 0$ is only changed when the processing of a class k job is complete.

Priority policies With or without preemption, no additional information is needed on the history of the process beyond $(Q(t), U(t), V(t))$, and at each non-empty station i, $a(t)$ will consist of $a_k(t) = 1$ when k is the non-empty class of highest priority among classes of i.

FCFS $Y(t)$ will be a finitely terminating sequence, where we list for each customer his class and the last time that he arrived in that class (see Exercise 10.6 for a more compact description). Under FCFS, $a_k(t) = 1$ if class k contains the longest waiting (the earliest arriving) customer in station $s(k)$.

Time stamped priority This is similar to FCFS, but the age of each job is now recorded as the time that it arrived originally in the system (rather than arrival in the class). Similar to this is EDD (earliest due date) where each job arrives with its own due date.

HLPPS Head of the line proportional processor sharing: This allows processing at a rate proportional to the number of customers of each class at the station, but all the processing goes to the HOL customer. Here $a_k(t) = Q_k(t) / \sum_{l:s(l)=s(k)} Q_l(t)$.

PS Processor sharing: Similar to HLPPS but all the jobs in each station are processed at the same rate. We now need to record for each customer in the system his elapsed processing time, as additional information in $Y(t)$.

LCFS-preemptive Last come first served preemptive: This requires both the time since arrival and the elapsed processing time of each job in the system.

SPT and SRPT Shortest processing time first (non-preemptive) and shortest remaining processing time first (preemptive): These policies assume that processing time is revealed upon arrival, and $Y(t)$ then includes the remaining processing time of each customer in the system. See also Exercise 10.7.

Policies PS, LCFS-preemptive, SPT, and SRPT are not head of the line policies, so we no longer have $\mathcal{D}(t) = \mathcal{S}(\mathcal{T}(t))$, and the dynamics under these policies are more complicated. We will exclude them from our main discussion (see Exercises 10.7 and 10.8 for dynamics under such policies).

10.4 Stability and Ergodicity of Uncountable Markov Processes

In this section we give some definitions and summarize some properties of Markov processes with uncountable state space. These will be needed to formulate and prove Theorem 11.8, which will serve as the main tool in verifying stability of queueing systems.

We now consider a general Markov process $X(t)$ in continuous time with uncountable state space S, defined on a probability space (Ω, Σ, P). We assume that S has a norm, denoted by $|x|$ for $x \in S$, and let $\mathcal{B}$ be the Borel sets of S. We denote:

- $\mathbb{P}_x^t(B)$ is the transition probability from state x to a state in the set $B \in \mathcal{B}$ in a time interval t, i.e. $\mathbb{P}_x^t(B) = \mathbb{P}(X(s+t) \in B | X(s) = x)$.
- $\tau_A = \inf\{t \geq 0 : X(t) \in A\}$ and, for $\delta > 0$, $\tau_{\delta,A} = \inf\{t > \delta : X_t \in A\}$, are the first time to reach the set A, and to reach (or return to) set A after δ. We let τ_A, $\tau_{\delta,A}$ be infinite if A is never reached.

Remarks

- τ_A, $\tau_{\delta,A}$ are stopping times under some mild conditions, which hold for piecewise deterministic Markov chains.
- Furthermore, piecewise deterministic Markov chains have the strong Markov property: The past and the future of the process given its value at a stopping time τ are independent.
- All the processes of interest to MCQN are piecewise deterministic Markov chains: They have piecewise constant or continuous piecewise linear paths, with breakpoints at times of arrivals or service completions (this is true of all age related data).

Stability of Markov processes with general state space is much more complicated than stability of Markov chains that move on a countable state space. Recall the main properties of continuous time discrete (countable) state space Markov chains:

- The chain is irreducible if all the states communicate, i.e. can be reached with positive probability in a finite number of steps.
- A state is transient if it is visited only a finite number of times.
- A state is recurrent if it is visited infinitely often.
- A state is positive recurrent if the expected time between visits is finite.
- If a set of communicating states is positive recurrent, then a stationary distribution exists, with $\pi(i)$ proportional to (expected time to re-visit i)$^{-1}$.

For uncountable state space, an uncountable number of states are almost surely never revisited, and transitions between any two specific states will have zero probability. Hence, the above definitions of irreducibility, of transience, and of recurrence become meaningless. It is still the case that some general Markov processes possess a stationary distribution (and may or may not be ergodic), in which case we will also say that queueing systems that are described by such Markov processes are stable (and may or may not then be ergodic). However, we will need a different approach before we can talk about stationarity, stability, and ergodicity.

The main idea in going from countable to uncountable state space, is to locate a set of states A that plays the role of a single state (an atom) in a countable chain. To act like a single state, A must have the property that from every point in A all other states are in a sense "equally accessible". We make the following definition of such sets

Definition 10.1   We single out a set $A \in \mathcal{B}$ for which there exists T and there exists a non-zero non-negative measure ξ such that

$$\mathbb{P}_x^T(B) \geq \xi(B) \text{ for every } x \in A \text{ and all } B \in \mathcal{B}.$$

The time T may depend on A but not on x, and:
- If T is a random time, then A is said to be *a petite set.*
- If T is fixed, A is said to be *a small set.*
- If T can be chosen as any fixed value in an interval $[a, b]$, then A is said to be *a uniformly small set.*

The measure ξ is a defective probability measure; it gives a probability for each event, but it sums up to $0 < \epsilon < 1$. We can then decompose transitions out of A as a mixture of two transition probabilities:

$$\mathbb{P}_x^T(B) = \epsilon \frac{\xi(B)}{\epsilon} + (1 - \epsilon) \frac{[\mathbb{P}_x^T(B) - \xi(B)]}{1 - \epsilon}, \qquad x \in A, \ B \in \mathcal{B},$$

where the first transition probability does not depend on x, so transitions out of A, after a time interval T, will with probability ϵ behave like transitions out of a single state (independent of which $x \in A$ was visited). Such a set A will then behave somewhat like a state in a countable Markov chain (an atom).

Definition 10.2   Assume that a Markov process has such a *closed set A which is petite, or small, or uniformly small.*
- It is a *Harris Markov process* if A is accessible from everywhere, i.e. $\mathbb{P}_x(\tau_A < \infty) > 0$ for all initial states x.
- It is *Harris recurrent* if it is reached in finite time from everywhere, i.e. $\mathbb{P}_x(\tau_A < \infty) = 1$ for all initial states x.
- It is *positive Harris recurrent* if it is Harris recurrent and the time to return to A has finite expectation, i.e. $\sup_{x \in A} \mathbb{E}_x(\tau_{\delta,A}) < \infty$.

The following theorem, presented without proof, summarizes the consequences of these concepts:

Theorem 10.3 *Harris recurrence implies the existence of a unique (up to a constant multiplier) invariant measure π, that is to say*

$$\pi(B) = \int \mathbb{P}^t_x(B)\pi(dx), \quad \text{for every } t > 0 \text{ and } B \in \mathcal{B},$$

and $\pi(A) > 0$. If this measure is finite, we normalize it to a probability measure and it is a stationary distribution of the process. This is the case if and only if the process is positive Harris recurrent.

If the petite set is actually uniformly small, and the process is positive Harris recurrent, then the process is ergodic, as defined by:

$$\lim_{t \to \infty} ||\mathbb{P}^t_x(\cdot) - \pi(\cdot)|| = 0, \quad \text{for all } x,$$

where $|| \, ||$ is the total variation norm of probability measures, defined as $||P - Q|| = \sup_{B \in \mathcal{B}} |P(B) - Q(B)|$.

To verify the existence of a stationary measure and ergodicity, we therefore need to do two things: Identify a petite, small, or uniformly small set A, and prove that τ_A is finite, and possibly also that $\tau_{\delta,A}$ has finite expectation.

To prove that a set is petite, small, or uniformly small (we refer to such a set as atom-like), is in general extremely technical. The following theorem gives sufficient conditions that any bounded neighborhood of the empty state of a MCQN is such an atom-like set. We need the following definitions:

Definition 10.4 Let F be a probability distribution.
(i) We say that F has infinite support if $F(x) < 1$ for all $x > 0$.
(ii) We say that F is spread out if for some ℓ, and some $q \geq 0$ such that $\int_0^\infty q(x)dx > 0$, the ℓ-fold convolution of F satisfies

$$F^{\star \ell}(d) - F^{\star \ell}(c) \geq \int_c^d q(x)dx, \quad \text{for all } d > c.$$

(This means that the sum of ℓ independent arrival times has an absolutely continuous component, i.e. a component that has a probability density).

Theorem 10.5 *Assume that the inter-arrival time distributions F_k for all classes k with $\alpha_k > 0$: (i) have infinite support and (ii) are spread out. Assume also a work-conserving HOL policy, so that the MCQN is modeled by a Markov process $\mathbf{X}(t)$ with norm $|x|$. Then for any $\kappa > 0$, the set of states $B_\kappa = \{x : |x| \leq \kappa\}$ is uniformly small.*

Remarks:

- Conditions (i), (ii) are satisfied if interarrival times are unbounded continuous random variables, in particular they are satisfied for Poisson arrivals.
- Unfortunately, condition (i) excludes bounded interarrival times, and condition (ii) excludes lattice interarrival times, i.e. $0, a, 2a, \ldots$.
- Condition (i) may not be relaxed completely, because systems with finite interarrival times may never be empty, even when $\rho_i < 1$; see Exercises 9.10 and 9.11.
- Condition (ii) may not be relaxed completely, because systems with more than one input stream and integer interarrival times cannot be ergodic; see Exercise 10.11
- In systems with a single input class, if condition (ii) is relaxed, then B_κ is still petite for all $\kappa > 0$.
- On the other hand, conditions (i) and (ii) are by no means necessary. It is customary to assume that sets of the form B_κ are uniformly small. This assumption is perhaps no less reasonable than assuming that interarrivals form a renewal process.

Before we discuss the behavior of $\tau_{\delta, A}$ for general Markov processes, we recall the situation for discrete-time discrete-state Markov chains. We can prove positive recurrence by solving the global balance equations, and normalizing to find the stationary distribution. However, in reality solving the global balance equations, if the state space is countable but not finite, is only possible in very special cases, e.g. birth and death chains, time-reversible Markov chains, or product-form networks. Fortunately, solving for the stationary distribution is not necessary to show positive recurrence. The main tool to prove stability (positive recurrence) is to find a Lyapunov function:

A Lyapunov function is non-negative, and is 0 only at the origin, i.e. at the point where the norm is $|x| = 0$ (usually the state when the queue is empty), we denote this state as state 0. As a special case, one can use the norm itself as a Lyapunov function. The state 0 is a positive recurrent state of the Markov chain $X(t)$ if:

$$\mathbb{E}_0(|X(1)|) < \infty, \qquad \mathbb{E}_x(|X(1)|) - |x| \le -\epsilon < 0, \quad x \neq 0.$$

In words, the first condition says that starting from 0, one does not go too far in one time unit, while the second condition says that starting from any x away from 0, one expects to get closer to 0 by at least ϵ in each time unit. If the chain is irreducible, then this implies all the states are positive recurrent.

For general Markov processes, the procedure is analogous. We state without proof:

Theorem 10.6 (Multiplicative Foster criterion) *Let $X(\cdot)$ be a continuous time general state space Markov process, such that: for some $c > 0$, $\epsilon > 0$, $\kappa > 0$,*

$$\mathbb{E}_x \left| X(c(|x| \vee \kappa)) \right| \leq (1 - \epsilon)(|x| \vee \kappa), \quad \text{for all } x.$$

If $B_\kappa = \{x: |x| \leq \kappa\}$ is petite then the process is positive Harris recurrent. If B_κ is uniformly small then the process is ergodic.

We note that this means that for $|x| \leq \kappa$, starting from x, in a time proportional to κ we do not expect to go too far:

$$\mathbb{E}_x |X(c\kappa)| < \kappa, \quad |x| \leq \kappa,$$

and starting from any $|x| > \kappa$, in a time proportional to the current norm, we expect to get closer to 0 by a factor of $(1 - \epsilon)$:

$$\mathbb{E}_x |X(c(|x|))| \leq (1 - \epsilon)|x|, \quad |x| > \kappa,$$

so instead of 0 being positive recurrent in the discrete state case, it is the set B_κ that is returned to in a finite time.

A direct consequence of this theorem is:

Theorem 10.7 *Assume that B_κ is petite (uniformly small) for all $\kappa > 0$. If there exists a $\delta > 0$ such that*

$$\lim_{|x| \to \infty} \frac{1}{|x|} \mathbb{E}_x \left| \mathbf{X}(|x|\delta) \right| = 0, \tag{10.9}$$

then $\mathbf{X}(\cdot)$ is positive Harris recurrent (ergodic).

This theorem states that once we know that bounded sets act like atoms, stability is determined by expected behavior under fluid scaling. We elaborate on that in the next chapter, culminating with Theorem 11.8.

Lyapunov functions that are more general than the norm can be used to verify the conditions for Theorems 10.6, 10.7. We shall demonstrate the use of Lyapunov functions in Sections 11.3, 11.4, 12.3.

At this point, we reiterate Definitions 2.2, 2.6:

Definition 10.8 A queueing system is stable under a given policy, if $Q(t)$ possesses a stationary distribution. It is ergodic if starting from any initial state, $Q(t)$ converges to the stationary distribution.

It is rate stable if $\lim_{t \to \infty} Q(t)/t = 0$ a.s.

Thus, if we can describe a queueing system by a Markov process $\mathbf{X}(t)$, and if we can show that it is positive Harris recurrent, then we deduce that $Q(t)$ is stable. If in addition there exists a uniformly small atom for $\mathbf{X}(t)$, we deduce that $Q(t)$ is ergodic.

Rate stability of $Q(t)$ requires less restrictive conditions, and can be verified from SLLN assumptions, without a Markovian description of the system.

Verifying stability, ergodicity, or rate stability is the subject of the next two chapters.

Remark For countable state Markov chains, Harris recurrence and positive Harris recurrence are somewhat different from recurrence and positive recurrence. If a countable state Markov chain is irreducible, i.e. all the states communicate, then the concepts are equivalent. Note that it is exactly the fact that it is unclear how to define communicating states for uncountable state space that necessitates the definition of Harris recurrence.

10.5 Sources

The discovery that MCQN with $\rho < 1$ can be unstable under some policies is due to Kumar and Seidman (1990) who used the Kumar–Seidman Rybko–Stolyar network example, with deterministic processing times. An important paper that discusses the same example in a stochastic setting with Poisson arrivals and exponential service times is Rybko and Stolyar (1992). Kumar (1993) is a comprehensive introduction to re-entrant lines. Lu and Kumar (1991) presented the example discussed in the Section 10.2.1. Examples of instability under FIFO are due to Bramson (1994). The simulation in Section 10.2.2 is due to Dai (1995). In that paper, Dai also formulates the Markovian framework for analyzing stability of MCQN under given policies, which we discuss in Section 10.3. Policies that are not head of the line are analyzed, using measure-valued processes, in Gromoll et al. (2002), Gromoll (2004), and Down et al. (2009). The discussion of general Markov processes in Section 10.4 follows the monograph *Stability of Queueing Networks*, Bramson (2008). The proof of Theorem 10.5, of conditions that closed bounded neighborhoods of the empty state are petite or uniformly small was initially derived in Meyn and Down (1994), and is masterfully described in Bramson (2008). Foster's criterion is due to Foster (1953), and its generalization, the multiplicative Foster criterion of Theorem 10.6, is proved in Bramson (2008). The authoritative book on stability of Markov

processes is Meyn and Tweedie (1993). A more recent overview of stability
of stochastic systems is given by Foss and Konstantopoulos (2004).

Exercises

10.1 Prove that the Lu–Kumar system, with $\alpha = 1$ and deterministic arrivals and
      service times, starting with x parts at time 0^-, does not diverge under any
      buffer priority policy except priority to buffers 2 and 4. Analyze the case that
      priority is given to buffers 1 and 3 in detail.

10.2 The Lu–Kumar network has immediate feedback from class 2 to class 3
      at the same node. Show that a similar example can be constructed without
      immediate feedback.

10.3 Explain that the FIFO network of Bramson operates similarly to the Lu–
      Kumar network under class 2,4 priorities.

10.4 Consider the Lu–Kumar network with Poisson input of rate 1, exponential
      services with rates μ in buffers 2 and 4, and negligible processing time at
      buffers 1 and 3. Argue that under priority to buffers 2 and 4, it is stable for
      $\mu < 2$ and blows up for $\mu > 2$.

10.5 Analyze the KSRS network with deterministic interarrivals and services,
      under priority to buffers 2 and 4, when it satisfies all three conditions, and
      when it violates the virtual machine condition.

10.6 Propose a Markov process to describe MCQN under FCFS that does not
      require the ages of all the customers in the system.

10.7 (∗) To describe the dynamics of a single queue under SPT or SRPT, one
      requires keeping track of the remaining processing times of all jobs. This
      requires that the queue be described by a measure-valued process. Describe
      the dynamics of the single queue under SPT or SRPT [Down et al. (2009)].

10.8 (∗) Consider a single-server queue. Denote by $\xi(x) = 1/x$, $x > 0$, $\xi(0) = 0$.
      Define the attained service process $\eta(t) = \int_0^t \xi(Q(s))ds$. It is the cumulative
      amount of service per customer delivered by the server under PS policy.
      Use it to express the measure-valued process and the dynamics of the queue
      [Gromoll et al. (2002); Gromoll (2004)].

10.9 Consider the M/G/1 queue, with state described by (n, x) where n is the
      number of customers in the system, and x the age of the customer in service
      (0 if empty). Show that $\{(0, x) : 0 \le x < \kappa\}$ is a uniformly small set.

10.10 Consider the G/G/1 queue, with state described by (n, x, y) where n is the
      number of customers in the system, and x the age of the customer in service (0
      if empty), and y the time since the last arrival. Assume that interarrival times
      have infinite support. Show that $\{(0, x, y) : 0 \le x + y < \kappa\}$ is a uniformly
      small set.

10.11 Show that if there is more than one input to a MCQN (or even a generalized
      Jackson network), and interarrival times are integer, then the Markov process
      describing the network cannot be ergodic.

11

Stability of MCQN via Fluid Limits

In this chapter we find out how to verify stability of MCQN by examining fluid limits. Fluid limits for generalized Jackson networks were derived in Section 9.4: Under fluid scaling of time and space by n, we used FSLLN of the arrival and service processes, to show that the queue converged to a unique fluid limit. For MCQN this is no longer the case, since fluid limits may now depend on policy. Our first task is to show, in Section 11.1, that under any work-conserving HOL policy, fluid limits for the queue lengths in a MCQN always exist (they may often not be unique). Under a given policy, we can write fluid model equations, which need to be satisfied almost surely by every fluid limit. Deterministic paths that satisfy these fluid model equations will be called fluid solutions. In Section 11.2 we prove the main theorem: That if all fluid solutions are stable, in the sense that starting from non-zero values they reach 0 in a bounded finite time, then the MCQN is stable under that policy. This is followed in Section 11.3 by verification of stability for several types of MCQN, under various policies, when $\rho < 1$. These include the proof that generalized Jackson networks are stable, feed-forward MCQN networks are stable under any work-conserving HOL policy, and re-entrant lines are stable under the first buffer first served (FBFS) policy. Further examples are suggested in exercises. Finally, in Section 11.4 we discuss global stability regions, defined for a general MCQN as a set of conditions (more stringent than $\rho < 1$) on the system parameters, under which the system is stable for every work-conserving HOL policy. Global stability regions may be specified by piecewise linear Lyapunov functions.

11.1 System Equations, Fluid Limit Model, Fluid Equations and Fluid Solutions

Consider a MCQN with primitives given by i.i.d. sequences of interarrivals, processing times, and Markovian routings, as defined in Section 10.1. As-

175

sume the policy is work-conserving HOL, and the system is described by a Markov process $\mathbf{X}(t) = ((Q(t), U(t), V(t), Y(t)); a(t))$ as described in Section 10.3. Recall that a state $\mathbf{X}(t) = x$ can be normed, and we use the surrogate-norm $|x| = \sum_{k=1}^{K} (Q_k(t) + U_k(t) + V_k(t))$. The queue lengths and the busy times will then satisfy the following stochastic queueing system equations, which are valid for any work-conserving HOL policy:

$$Q_k(t) = Q_k(0) + \mathcal{A}_k(t) - \mathcal{S}_k(\mathcal{T}_k(t)) + \sum_{l=1}^{K} \mathcal{R}_{l,k}(\mathcal{S}_l(\mathcal{T}_l(t))), \quad k = 1, \ldots, K,$$

$$\sum_{k \in C_i} \mathcal{T}_k(t) + \mathcal{I}_i(t) = t, \quad i = 1, \ldots, I, \qquad (11.1)$$

$$Q_k(t) \geq 0, \quad \mathcal{T}_k(0), \mathcal{I}_i(0) = 0, \quad \mathcal{T}_k, \mathcal{I}_i \text{ non-decreasing},$$

$$\mathcal{I}_i(t) \text{ increases only when } \sum_{k \in C_i} Q_k(t) = 0.$$

Remark Here $\mathcal{A}_k(\cdot), \mathcal{S}_k(\cdot)$ are delayed renewal processes, starting with the initial residual interarrival and service times $U_k(0), V_k(0)$.

We consider a sequence of systems, all sharing the same interarrival, service, and routing processes, but with different initial conditions, given by $Q^n(0), U^n(0), V^n(0)$. We assume that the initial states satisfy

$$U^n(0)/n \to 0, \quad V^n(0)/n \to 0, \quad Q^n(0)/n \to \bar{Q}(0) < \infty, \quad \text{as } n \to \infty,$$
$$(11.2)$$

so at time 0 buffers contain initial scaled customers (we refer to those as fluid), but the impact of the initial residual interarrival and processing times disappears as $n \to \infty$. The assumption that the scaled initial residual times converge to 0 is made for simplicity and we will relax it later. By the FSLLN and (11.2),

$$\mathcal{A}(nt)/n \to \alpha t, \quad \mathcal{S}(nt)/n \to \mu t, \quad \mathcal{R}(n)/n \to P, \quad \text{as } n \to \infty \text{ u.o.c. a.s..}$$
$$(11.3)$$

Equation (11.3) holds almost surely. We will denote by $\mathfrak{G}$ the set of ω of probability 1 for which it holds. For the rest of this chapter and throughout the following chapters, we will only consider $\omega \in \mathfrak{G}$. All the results obtained for $\{\omega : \omega \in \mathfrak{G}\}$ are then valid almost surely.

We now look at the fluid scaling of the nth system, scaling time and space by n, for each sample path. Then the scaled quantities $(\bar{Q}^n, \bar{\mathcal{T}}^n, \bar{\mathcal{I}}^n)$ will satisfy a scaled version of (11.1) for each n.

Definition 11.1 If for some sample path ω, and some diverging sub-sequence of values of n, which we denote by r, $(\bar{Q}^r(t, \omega), \bar{\mathcal{T}}^r(t, \omega), , \bar{\mathcal{I}}^r(t, \omega))$

converges u.o.c. to functions $(\bar{Q}(t), \bar{\mathcal{T}}(t), \bar{\mathcal{I}}(t))$ as $r \to \infty$, then we call $(\bar{Q}(t), \bar{\mathcal{T}}(t), \bar{\mathcal{I}}(t))$ a fluid limit.

Note that there may be different fluid limits for different ω, and for each ω there may be different limits for different subsequences. Each of those limits will be called a fluid limit.

Theorem 11.2 *Fluid limits of $Q, \mathcal{T}, \mathcal{I}$ exist and are Lipschitz continuous, and for work-conserving HOL policies, every fluid limit for $\omega \in \mathfrak{G}$ satisfies the fluid model equations for $k = 1, \ldots, K$:*

$$\bar{Q}_k(t) = \bar{Q}_k(0) + \alpha_k t - \mu_k \bar{\mathcal{T}}_k(t) + \sum_{j=1}^{K} p_{j,k} \mu_j \bar{\mathcal{T}}_j(t),$$

$$\sum_{k \in C_i} \bar{\mathcal{T}}_k(t) + \bar{\mathcal{I}}_i(t) = t, \quad i = 1, \ldots, I, \tag{11.4}$$

$$\bar{Q}_k(t) \geq 0, \quad \bar{\mathcal{T}}_i(0), \bar{\mathcal{I}}_i(0) = 0, \quad \bar{\mathcal{T}}_i, \bar{\mathcal{I}}_i \text{ non-decreasing,}$$

$$\bar{\mathcal{I}}_i(t) \text{ increases only when } \sum_{k \in C_i} \bar{Q}_k(t) = 0.$$

Proof The main step is to show that fluid limits of $\mathcal{T}_k(t, \omega)$ exist for every fixed ω. We know that $\mathcal{T}_k(t, \omega)$ is non-decreasing, and for all $t > s > 0$, $\mathcal{T}_k(t, \omega) - \mathcal{T}_k(s, \omega) \leq t - s$, so it is Lipschitz continuous. The same holds for the sequence of scaled functions $\mathcal{T}_k^n(nt, \omega)/n$; so for every ω this is a sequence of equicontinuous functions. Hence, by the Arzela–Ascoli theorem, it has a sub-sequence $\mathcal{T}_k^r(rt, \omega)/r$ that converges u.o.c. as $r \to \infty$. We can then, for every fixed ω, find a further sub-sequence such that $\mathcal{T}^r(rt, \omega)/r$ converges on all coordinates. Clearly, the limit $\bar{\mathcal{T}}(t)$ is also non-decreasing and Lipschitz continuous with Lipschitz coefficient 1.

Now consider only ω for which $\mathcal{A}, \mathcal{S}, \mathcal{R}$ satisfy FSLLN convergence, i.e. $\omega \in \mathfrak{G}$, which excludes only a set of probability 0. For all of these, by (11.2), (11.3), we then have:

$$\mathcal{A}_k(rt, \omega)/r \to \alpha_k t, \quad \mathcal{S}_k(\mathcal{T}_k^r(rt, \omega), \omega)/r \to \mu_k \bar{\mathcal{T}}_k(t),$$

$$\mathcal{R}_{j,k}(\mathcal{S}_j(\mathcal{T}_j^r(rt, \omega), \omega), \omega)/r \to p_{j,k} \mu_j \bar{\mathcal{T}}_j(t), \quad \text{as } r \to \infty \text{ u.o.c.}$$

Substituting these limits into the scaled versions of (11.1), we get the first two equations of (11.4), and it follows that limits of $\bar{Q}^r, \bar{\mathcal{I}}^r$ exist, and are Lipschitz continuous. Also, clearly $\bar{\mathcal{I}}$ is non-decreasing.

Finally, we need to show that $\bar{\mathcal{I}}_i$ only increases when $\sum_{k \in C_i} \bar{Q}_k(t) = 0$. Assume that $\sum_{k \in C_i} \bar{Q}_k(t) > 0$ for some t. By continuity of $\bar{Q}$ it is > 0 over some $[t_1, t_2]$ where $t_1 < t < t_2$. But then for some n_0, $\sum_{k \in C_i} Q_k^r(s, \omega) > 0$

for $rt_1 \leq s \leq rt_2$, for every $r > n_0$, which implies that $I_i^r(s, \omega)$ is not increasing in the interval $rt_1 < s < rt_2$ for all $r > n_0$, and hence $\bar{I}_i(t)$ is constant for $t_1 < t < t_2$, and is not increasing at t. □

Remark The argument that we used to show that $\bar{I}_i(t)$ increases only when station i is empty, which used the fact that (t_1, t_2) in the fluid scale corresponds to the much wider (nt_1, nt_2) in the original system, will be used again to obtain (11.6), and in the proof of Proposition 12.7 in Chapter 12.

The next corollary states an important consequence of Theorem 11.2, and some extensions of the conditions under which fluid limits exist.

Corollary 11.3 *(i) For almost all $\omega \in \Omega$, there is a sub-sequence of the sequence $\frac{1}{n}Q^n(nt, \omega), \frac{1}{n}\mathcal{T}^n(nt, \omega), \frac{1}{n}I^n(nt, \omega)$, that converges u.o.c. to a fluid limit that satisfies (11.4).*

(ii) Theorem 11.2 and (i) continue to hold when instead of scaling by n scaling is done by any sequence a_n such that $a_n \to \infty$ when $n \to \infty$, where we now denote the fluid scaling of a function $z(t)$ by $\bar{z}^{a_n}(t) = \frac{1}{a_n}z(a_n t)$.

(iii) Theorem 11.2 and (i) continue to hold when instead of $\frac{1}{a_n}U^{a_n}(0) \to$ 0, $\frac{1}{a_n}V^{a_n}(0) \to 0$, we assume $\frac{1}{a_n}U^{a_n}(0) \to \bar{U}(0)$, $\frac{1}{a_n}V^{a_n}(0) \to \bar{V}(0)$.

Proof (i) We assume (11.2) holds, and we assume $\omega \in \mathfrak{G}$, i.e. we exclude the set of ω for which (11.3) fails, which is a set of probability 0. The corollary follows from the proof of Theorem 11.2.

(ii) The existence of a subsequence $r \to \infty$ such that $\frac{1}{a_r}\mathcal{T}^{a_r}(a_r t) \to \bar{\mathcal{T}}(t)$ follows by the same argument as for Theorem 11.2. The remainder of the proof needs no change.

(iii) The only difference is that instead of $\frac{1}{a_n}\mathcal{A}_k^n(a_n t) \to \alpha_k t$ and $\frac{1}{a_n}S_k^n(a_n t) \to \mu_k t$, we now have $\frac{1}{a_n}\mathcal{A}_k^n(a_n t) \to \alpha_k(t - \bar{U}_k(0))$ and $\frac{1}{a_n}S_k^n(a_n t) \to \mu_k(t - \bar{V}(0))$ u.o.c. a.s. and equations (11.4) need to be modified accordingly (see Exercise 11.1). □

Definition 11.4 (i) The collection of all fluid limits for paths $\omega \in \mathfrak{G}$ of a MCQN, under a given policy, is called *the fluid limit model*.

(ii) Equations that need to be satisfied by every fluid limit with $\omega \in \mathfrak{G}$ are called *fluid model equations*.

(iii) Solutions of the fluid model equations are called *fluid model solutions*.

The equations (11.4) are referred to as the *standard fluid model equations*. They correspond to the stochastic model equations (11.1), and apply to any work-conserving HOL policy. These standard equations do not include any information about the policy (apart from work-conserving HOL), and so they certainly do not determine all fluid limits of the system.

For any specific policy for the MCQN, we need to supplement the stochastic model equations (11.1) by additional equations, which will determine the sample paths of the stochastic $Q(t), \mathcal{T}(t)$ generated by the policy from the primitives. We may then be able also to add some fluid model equations to (11.4), and verify that all fluid limits satisfy them.

We give two examples:

Static priority policies: Within each service station, classes are ordered by priorities, and the customers of the highest priority class at each station are served. Without loss of generality, we renumber all classes so that if $k_1, k_2 \in C_i$ and $k_1 < k_2$, then class k_1 has higher priority than class k_2. Denote $Q_k^+(t) = \sum_{l \leq k, s(l)=s(k)} Q_l(t)$, and $\mathcal{T}_k^+(t) = \sum_{l \leq k, s(l)=s(k)} \mathcal{T}_l(t)$, so that $Q_k^+(t)$ includes all customers at the station serving class k that are of the same or higher priority, and $\mathcal{T}_k^+(t)$ is the service time devoted to these customers by time t. Then we add the supplementary stochastic model equations:

$$\int_0^t Q_k^+(s)d(s - \mathcal{T}_k^+(s)) = 0, \quad k = 1, \ldots, K, \qquad (11.5)$$

which say that while there are any customers at station i with priority k or higher, there will be no allocation of processing time to any class except those of priority k or higher.

The analog supplementary fluid model equations are

$$\int_0^t \bar{Q}_k^+(s)d(s - \bar{\mathcal{T}}_k^+(s)) = 0, \quad k = 1, \ldots, K. \qquad (11.6)$$

The proof that (11.6) holds for every fluid limit (with $\omega \in \mathfrak{G}$) follows the same argument used to show that under work conservation, $\bar{I}_i(t)$ can increase only when station i is empty (see Exercise 11.2).

FCFS policy: Let $\mathcal{W}_i(t)$ be the immediate workload at station i, i.e. the amount of time needed to empty station i if no work comes into station i after t. Let $\mathcal{D}_k(t)$ be the service completions out of class k over $(0, t]$. Then we add for each station i the equations:

$$\mathcal{D}_k(t + \mathcal{W}_i(t)) = Q_k(0) + \mathcal{A}_k(t) + \sum_{l=1}^{K} \mathcal{R}_{l,k}(\mathcal{D}_l(t)), \quad k \in C_i, \qquad (11.7)$$

which say that all the input to station i prior to time t, must leave by time $t + \mathcal{W}_i(t)$, so that all of $\mathcal{W}_i(t)$ is processed before any input later than t.

The fluid analog is

$$\bar{\mathcal{D}}_k(t + \bar{\mathcal{W}}_i(t)) = \bar{Q}_k(0) + \alpha_k t + \sum_{l=1}^{K} p_{l,k} \bar{\mathcal{D}}_l(t), \quad k \in C_i, \qquad (11.8)$$

where

$$\bar{\mathcal{W}}_i(t) = \sum_{k \in C_i} \frac{1}{\mu_k} \bar{Q}_k(t). \qquad (11.9)$$

For details of this case, see Exercises 11.3 and 11.4.

Equations (11.1) and (11.5) uniquely determine the sample path of $Q(t, \omega), \mathcal{T}(t, \omega), \mathcal{I}(t, \omega)$ under priority policy (they are implicit, but can be shown to do so recursively, see Exercise 11.2). One might think that (11.4) and (11.6) will determine the deterministic paths of the fluid solutions, however, this is not the case, as we shall demonstrate by example in Section 11.4.

Similarly, equations (11.1) and (11.7) determine the sample paths of the stochastic system, but (11.4) and (11.8) do not determine the fluid solution, since in the input of mixed fluids it is not possible to determine what is first come.

It is important to note that:
- Once a policy is well defined, it uniquely determines the stochastic sample paths of $Q(t), \mathcal{T}(t), \mathcal{I}(t)$, from the primitives $\mathcal{A}(t), \mathcal{S}(t), \mathcal{R}(n)$. One can then often set up a set of stochastic model equations that uniquely determine these paths.
- Under fluid scaling, the MCQN may have many distinct fluid limits even when we consider only a single path $\omega \in \mathfrak{G}$.
- We may be able to set up fluid model equations analog to the stochastic model equations. All the fluid limits from all $\omega \in \mathfrak{G}$ must satisfy the fluid model equations, i.e. the fluid limit model is contained in the collection of all fluid model solutions. However, it may well be that the collection of all fluid solutions is larger than the fluid limit model.
- By discovering more fluid model equations for a given policy, we get a better approximation of the fluid limit model.
- We have no general techniques to determine whether our collection of fluid model solutions equals the fluid limit model.
- The ideal case is when the fluid model equations have a unique solution, which implies this solution is the unique fluid limit for all $\omega \in \mathfrak{G}$.

11.2 Stability via Fluid Models

In this section we derive a fluid criterion for stability of MCQN.

We will need the following well-known general result (see Exercise 11.5):

Lemma 11.5 (Bramson (2008) lemma 4.13) *Let $X_i \geq 0$, $i = 1, 2, \ldots$ be i.i.d. random variables with finite mean. Then*

$$\lim_{n \to \infty} \frac{1}{n} \max\{X_1, \ldots, X_n\} = 0 \ a.s.,$$

$$(11.10)$$

$$\lim_{n \to \infty} \frac{1}{n} \mathbb{E}\left(\max\{X_1, \ldots, X_n\}\right) = 0.$$

We will also need the following proposition:

Proposition 11.6 *Assume that the scaled initial queue lengths are bounded, i.e. $\frac{1}{a_n} Q_k^{a_n}(0) \leq \kappa$. Then the sequence of scaled queue lengths $\frac{1}{a_n} Q_k^{a_n}(a_n t)$ is uniformly integrable.*

Proof We have the inequality:

$$\frac{1}{a_n} Q_k^{a_n}(a_n t) \leq \kappa + \sum_{l=1}^{K} \frac{1}{a_n} \mathcal{A}_l^{a_n}(a_n t).$$

By the elementary renewal Theorem 2.8, $\frac{1}{a_n} \mathbb{E}(\mathcal{A}_l^{a_n}(a_n t)) \to \alpha_l t$, which implies that $\frac{1}{a_n} \mathcal{A}_l^{a_n}(a_n t)$ are uniformly integrable. The proposition follows.

□

We are now ready for the main theorem. We consider a MCQN under some work-conserving HOL policy, which is described by a Markov process $\mathbf{X}(\cdot)$. Recall the surrogate-norm for a state $\mathbf{X}(t) = x$ given by $|x| = \sum_{k=1}^{K} (Q_x(t) + U_k(t) + V_k(t))$, and let $|Q(t)| = \sum_{k=1}^{K} Q_x(t)$. We define:

Definition 11.7 The fluid limit model is stable if there exists a δ such that for every fluid limit, $\bar{Q}(t) = 0$ for all $t > \delta|\bar{Q}(0)|$. This will hold in particular if it holds for every fluid solution.

Recall also the definition of closed bounded sets around 0, $B_\kappa = \{x : |x| \leq \kappa\}$.

Theorem 11.8 *Consider a MCQN under some fixed HOL policy, which is described by the Markov process $\mathbf{X}(\cdot)$. Assume that bounded sets B_κ are petite (uniformly small). If the fluid limit model under this policy is stable, then $\mathbf{X}(\cdot)$ is positive Harris recurrent (ergodic). In particular, this will be the case if every fluid solution is stable.*

Proof To prove the theorem we need, by Theorem 10.7, to show that $\lim_{|x|\to\infty}\frac{1}{|x|}\mathbb{E}_x\big|\mathbf{X}(|x|\delta)\big| = 0$. In this limit, we consider the process with initial state $\mathbf{X}(0) = x$, and let x vary in such a way that $|x| \to \infty$, and scale time and space by the very same $|x|$. We now take any sequence of initial states x_n such that $\lim_{n\to\infty}|x_n| = \infty$, and look at the sequence of queueing systems, sharing the same $\mathcal{A}(\cdot), \mathcal{S}(\cdot), \mathcal{R}(\cdot)$, and with the sequence of initial values x_n. Under the given policy, we have a sequence of processes $Q^{x_n}(\cdot), \mathcal{T}^{x_n}(\cdot)$, where the superscript denotes for each element in the sequence the initial state as well as the scaling of the system.

We now look at this sequence of systems, and we scale time and space by the norm of the initial states. The crucial step in the proof is to consider the scaled sequences:

$$\frac{1}{|x_n|}Q^{x_n}(|x_n|t),\quad \frac{1}{|x_n|}U^{x_n}(|x_n|t),\quad \frac{1}{|x_n|}V^{x_n}(|x_n|t),\quad n = 1, 2, \ldots.$$

We need to show that the expected value of each of these three non-negative sequences of vector processes converges to 0 as $n \to \infty$ for all $t > \delta$.

We note first that by definition of the surrogate-norm $|\cdot|$,

$$\frac{1}{|x_n|}Q_k^{x_n}(0) \le 1,\ \frac{1}{|x_n|}U_k^{x_n}(0) \le 1,\ \frac{1}{|x_n|}V_k^{x_n}(0) \le 1,$$

and since the sequences of scaled initial values are bounded, we can find a subsequence x_r of the x_n for which:

$$\frac{1}{|x_r|}Q^{x_r}(0) \to \bar{Q}(0),\ \frac{1}{|x_r|}U^{x_r}(0) \to \bar{U}(0),\ \frac{1}{|x_r|}V^{x_r}(0) \to \bar{V}(0).$$

Consider now $U_k^{x_r}(|x_r|t)$, for classes with $\alpha_k > 0$. For s such that $\bar{U}_k(0) < s < t$, we have:

$$\frac{1}{|x_r|}U_k^{x_r}(|x_r|s) \le \frac{1}{|x_r|}\max\{u_{k,\ell} : 1 \le \ell \le \mathcal{A}_k(|x_r|t)\}$$

$$= \frac{\mathcal{A}_k(|x_r|t)}{|x_r|}\frac{1}{\mathcal{A}_k(|x_r|t)}\max\{u_{k,\ell} : 1 \le \ell \le \mathcal{A}_k(|x_r|t)\}.$$

and by Lemma 11.5 this implies that

$$\frac{1}{|x_r|}U_k^{x_r}(|x_r|t) \to 0 \quad \text{u.o.c., a.s. for } t > \bar{U}_k(0),$$

$$\frac{1}{|x_r|}\mathbb{E}[U_k^{x_r}(|x_r|t)] \to 0 \quad \text{u.o.c. for } t > \bar{U}_k(0).$$

Taking into account $\bar{U}(0)$ we showed:

$$\frac{1}{|x_r|} U_k^{x_r}(|x_r|t) \to (\bar{U}_k(0) - t)^+ \quad \text{u.o.c., a.s.,}$$

$$\frac{1}{|x_r|} \mathbb{E}[U_k^{x_r}(|x_r|t)] \to (\bar{U}_k(0) - t)^+ \quad \text{u.o.c.} \tag{11.11}$$

Similarly

$$\frac{1}{|x_r|} V_k^{x_r}(|x_r|t) \to (\bar{V}_k(0) - t)^+ \quad \text{u.o.c., a.s.,}$$

$$\frac{1}{|x_r|} \mathbb{E}[V_k^{x_r}(|x_r|t)] \to (\bar{V}_k(0) - t)^+ \quad \text{u.o.c.} \tag{11.12}$$

It is left to look at the queue lengths. We fix a sample path $\omega \in \mathfrak{G}$, and consider the sequence $\frac{1}{|x_r|} Q^{x_r}(|x_r|t, \omega)$. By Theorem 11.2 and Corollary 11.3, for this ω we can find a subsequence of x_r, say x_s with $|x_s| \to \infty$, such that $\frac{1}{|x_s|} Q^{x_s}(|x_s|t, \omega) \to \bar{Q}(t)$. By our assumption that the fluid limit model is stable, this fluid limit satisfies: $\bar{Q}(t) = 0$ for $t > \epsilon > 0$. Furthermore, by Proposition 11.6 $\frac{1}{|x_s|} Q^{x_s}(|x_s|t)$ is uniformly integrable, so the same convergence holds for expected values.

We have now shown that any sequence x_n has a subsequence x_r for which $\lim_{|x_r| \to \infty} \frac{1}{|x_r|} \mathbb{E}_{x_r} |\mathbf{X}(|x_r|t)| = 0$ for all $t > t_0 = \epsilon + \bar{U}(0) + \bar{V}(0)$. It follows that there can be no sequence x_n for which $\lim_{|x_n| \to \infty} \frac{1}{|x_n|} \mathbb{E}_{x_n} |\mathbf{X}(|x_n|t)| > 0$ for any $t > t_0$. Choosing $\delta = t_0$, it then follows that $\lim_{|x| \to \infty} \frac{1}{|x|} \mathbb{E}_x |\mathbf{X}(|x|\delta)| = 0$.

Adding the assumption that B_κ are petite (uniformly small), the theorem now follows from Theorem 10.7. □

As stated in Definition 10.8, if $\mathbf{X}(t)$ is positive Harris recurrent (ergodic), we say the MCQN is stable (ergodic). In particular, $Q(t)$ is stable in the sense that it has a stationary distribution (and converges to this distribution as $t \to \infty$).

Remark We note that Theorems 11.2 on existence of fluid limits and Theorem 11.8 on existence of stationary distribution and ergodicity do not require that interarrival and service times possess finite variances.

Remark The assumption that the policy is work conserving is not necessary for Theorem 11.2 to hold. However, it is needed for the corresponding fluid model equations. It is not necessary for the proof of Theorem 11.8. However, if the policy is not work conserving, we have no criteria to verify that B_k are petite. In particular, maximum pressure policy discussed in the next chapter is not work conserving. The assumption that the policy is HOL is used in deriving the stochastic system equations. It is necessary in the proof of Theorem 11.2, since it assures that we can use $|\sum_k (Q_k(t) + U_k(t) + V_k(t))|$ as a surrogate norm. For this reason, Theorem 11.8 cannot be applied directly to PS or to preemptive LCFS, which are not HOL.

Remark   The existence of fluid limits that satisfy fluid model equations as shown in Theorem 11.2, requires only that SLLN limits exist, i.e. it is enough to require that interarrivals, service times, and routings, will form jointly distributed stationary and ergodic sequences, such that $\mathcal{A}_k(t)/t \to \alpha_k$, $S_k(t)/t \to \mu_k$, $\mathcal{R}_{k,j}(\ell)/\ell \to p_{k,j}$ u.o.c. a.s. The following two theorems hold under these weaker conditions.

Definition 11.9   We say that the fluid limit model is weakly stable if for every fluid limit we have that if $\bar{Q}(t_0) = 0$, then $\bar{Q}(t) = 0$, $t > t_0$, i.e. once a fluid limit is 0 at some time, it will remain at 0 thereafter.

Recall Definition 2.2 of rate stability.

Theorem 11.10   *If the fluid limit model of a MCQN under some fixed policy is weakly stable, then the MCQN is rate stable, in the sense that starting with any fixed $Q(0)$, one has $\lim_{t\to\infty} Q(t)/t = 0$ for all $\omega \in \mathfrak{G}$.*

Proof   Assume by contradiction that for some $\omega \in \mathfrak{G}$, and $a_n \to \infty$, $Q_k(a_n, \omega)/a_n > c > 0$. By Theorem 11.2, for a subsequence a_r we have $\lim_{r\to\infty} Q_k(a_r t, \omega)/a_r = \bar{Q}_k(t)$ u.o.c., where $\bar{Q}_k(t)$ is a fluid limit. Since we are looking at a single process $Q(t)$, $Q(0)$ is fixed, and therefore $\bar{Q}(0) = 0$. On the other hand $\bar{Q}_k(1) > c$, which contradicts the assumption that the fluid limit model is weakly stable.   □

Definition 11.11   We say that the fluid limit model is unstable if for every fluid limit that has $\bar{Q}(0) = 0$ there exists δ such that $\bar{Q}(\delta) > 0$.

Theorem 11.12   *(a) If the fluid limit model of a MCQN under some fixed policy is unstable, then for any initial fixed state $Q(0) = x$ and every $\omega \in \mathfrak{G}$, $\liminf_{t\to\infty} Q(t)/t > 0$ for all $\omega \in \mathfrak{G}$.*

(b) If for some station i, $\rho_i > 1$, then every fluid limit is unstable, and the MCQN is unstable in the stronger sense that there exists positive c such that $\liminf_{t\to\infty} |Q(t)|/t \geq c$ on for all $\omega \in \mathfrak{G}$.

The proof is similar to that of Theorem 11.10, and is left as Exercise 11.6.

We have shown that the question of stability of a MCQN under a given policy can be studied by analyzing its fluid limit model. If we can show that the fluid limit model is stable, this will imply stability of the stochastic system. In the following sections we will use these results to investigate stability of several types of MCQN, under various policies.

Remark Stability of the fluid limit model is not a necessary condition; there are examples of stochastic Markovian queueing systems that are stable, but their fluid limit models are not stable.

11.3 Some Fluid Stability Proofs

In this section we put to use the criteria developed in Theorems 11.8, 11.10 and 11.12 to find out when various MCQN are stable under given policies. By Theorem 11.8, to prove stability and ergodicity of the network the main task is to show stability of the fluid limit model. The main tool to do that is to use an appropriate Lyapunov function. We assume renewal arrival and service processes, and consider only work-conserving HOL policies. The various results span several special classes of MCQN, and several policies. For each of those we will consider the fluid model equations, and devise an appropriate Lyapunov function that is positive for all fluid states except at $\bar{Q}(t) = 0$, where it is zero, and show that it reaches 0 in a finite time, from any starting point. Once we show that, fluid stability follows, and if B_κ are petite (uniformly small), stability (ergodicity) of the MCQN is proved.

We start with a useful lemma. Every Lipschitz continuous function is absolutely continuous (in other words, it is the integral of a density), and therefore it has a derivative at all t except at a set of t of Lebesgue measure 0. Call t a regular point of f if $\dot{f}(t)$ exists.

Lemma 11.13 *Let f be a non-negative absolutely continuous function, with derivative $\dot{f}(t)$ at regular t. (a) If $f(t) = 0$ at regular t, then also $\dot{f}(t) = 0$. (b) Assume that whenever $f(t) > 0$ at a regular point then its derivative $\dot{f}(t) < -\kappa < 0$. Then for all $t > f(0)/\kappa$, $f(t) = 0$. (c) Assume instead $\dot{f}(t) \leq 0$. If $f(0) = 0$, then $f(t) = 0$ for all $t > 0$.*

Proof See Exercise 11.7. □

11.3.1 Stability of Generalized Jackson Networks

We studied generalized Jackson networks in Chapter 9. We left the question of stability open, and we will now prove Theorem 9.5. As noted, Jackson networks are MCQN with a single class in each station. We reformulate the theorem here (recall Definition 10.4 of unbounded support and spread out):

Theorem 11.14 *If bounded sets are petite (uniformly small), in particular, if interarrival time distributions have infinite support and are spread out, and if $\rho_i < 1$ for all nodes, then a generalized Jackson network is stable (ergodic) under any work-conserving HOL policy.*

Proof Let α, μ, P be the parameters of the generalized Jackson network. Denote $\rho = (\rho_1, \ldots, \rho_I)$. Let $\theta = \alpha - (I - P^\mathsf{T})\mu$, and $R = (I - P^\mathsf{T})\mathrm{diag}(\mu)$; recall that these are the vector of netput rates and the input-output matrix, respectively. The theorem assumption is that $\rho = R^{-1}\alpha < 1$ component-wise, and so $\theta = R(\rho - 1) < 0$ componentwise. As we saw, fluid solutions for the fluid model are given by

$$\bar{Q}(t) = \bar{Q}(0) + \theta t + R\bar{I}(t).$$

We claim first that at all regular points $\dot{\bar{I}}_i(t) \le 1 - \rho_i$. We write: $\bar{Q}(s+t) = \bar{Q}(s) + \theta t + R(\bar{I}(s+t) - \bar{I}(s))$. Let $\tilde{x}(t) = \bar{Q}(s) + \theta t$, then $\bar{Q}(s + \cdot) = \psi(\tilde{x}(\cdot), R)$, $\bar{I}(s + \cdot) - \bar{I}(s) = \varphi(\tilde{x}(\cdot), R)$, where $\psi(\cdot, R), \varphi(\cdot, R)$ are the Skorohod oblique reflection and regulator with the reflection matrix R (as defined in Section 7.6). To evaluate $\tilde{x}(t)$, we consider $x^0(t) = \theta t$. The oblique reflection of $x^0(t)$ is $\psi(\theta t, R) = 0$ and $\varphi(\theta t, R) = (1 - \rho)t$, so $x^0(t) + R(1 - \rho)t = 0$. But $\tilde{x}(t) \ge x^0(t)$, and so

$$\tilde{x}(t) + R(1 - \rho)t \ge x^0(t) + R(1 - \rho)t = 0,$$

from which it follows, by the minimality of the regulators in oblique reflection, that:

$$\bar{I}(s + t) - \bar{I}(s) = \varphi(\tilde{x}(t), R) \le (1 - \rho)t,$$

and we showed $\dot{\bar{I}}_i(t) \le 1 - \rho_i$. We now have also shown, since $\bar{I}_i(0) = 0$ and $\dot{\bar{I}}_i(t) \le 1 - \rho_i$, that $\bar{I}(t) \le (1 - \rho)t$ componentwise.

Define now as our Lyapunov function:

$$f(t) = 1^\mathsf{T} R^{-1} \bar{Q}(t) = 1^\mathsf{T} \left[R^{-1}\bar{Q}(0) + (\rho - 1)t + \bar{I}(t) \right].$$

Note that $f(t) \ge 0$, since R^{-1} is non-negative, and $f(t) = 0$ only if $\bar{Q}(t) = 0$. In fact, $f(t)$ is the total expected processing time of all items in the system at t.

Assume $f(t) > 0$ at a regular point t of $\bar{Q}, \bar{I}$. Then for some i, $\bar{Q}_i(t) > 0$ and hence $\dot{\bar{I}}_i(t) = 0$, while for all the other components $\dot{\bar{I}}_j(t) \le 1 - \rho_j$.

Then:

$$\dot{f}(t) = \mathbf{1}^{\mathsf{T}}\left[\rho - \mathbf{1} + \dot{\bar{I}}(t)\right] = \rho_i - 1 + \sum_{j \neq i}\left(\rho_j - 1 + \dot{\bar{I}}_j(t)\right) \leq \rho_i - 1 < 0.$$

Letting $\kappa = \min_{j=1,\dots,I} 1 - \rho_j$, we showed that $\dot{f}(t) \leq -\kappa < 0$. Let $d = \mathbf{1}^{\mathsf{T}}R^{-1}\mathbf{1}$. Then for $|\bar{Q}(0)| = 1$, $f(0) \leq d$. Hence, by Lemma 11.13 $f(t) = 0$ as well as $\bar{Q}(t) = 0$ for $t > t_0 = d/\kappa$. The theorem follows. □

Remark $\bar{Q}(t)$ need not be a decreasing function in all coordinates while $t < t_0$. This is the reason why the proof is not straightforward (see Exercise 11.8). This is related to questions of emptying a network in minimal time.

11.3.2 Multi-Class Single-Station Networks, and Feed-Forward Networks

A multi-class single-station network has several classes, with arrivals and feedbacks (see Figure 11.1).

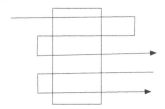

Figure 11.1 A multi-class single-station network with two routes.

Theorem 11.15 *If bounded sets are petite (uniformly small) and $\rho < 1$, then a multi-class single-station network is stable (ergodic) under any work-conserving HOL policy.*

Proof We now have that ρ and $I(t)$ are scalars. Consider

$$f(t) = \mathbf{1}R^{-1}\bar{Q}(t) = f(0) + (\rho - 1)t + \bar{I}(t).$$

It is ≥ 0 (again by R^{-1} non-negative), and it is $= 0$ only if $\bar{Q}(t) = 0$. Consider a regular t for which $f(t) > 0$. Then $Q_k(t) > 0$ for some k. But then, by work conservation, $\frac{d}{dt}\bar{I}(t) = 0$, and so $\dot{f}(t) = \rho - 1 < 1$. Hence, f and $\bar{Q}(t)$ are 0 for all $t > f(0)/(1 - \rho)$. □

A feed-forward MCQN is a network in which we can number the stations in such a way that if $k \in C_i$ and $l \in C_j$, then $i < j$ implies $p_{l,k} = 0$, i.e. once customers leave station i and go to another station, say j, they can never return to station i.

Corollary 11.16   *If bounded sets are petite (uniformly small) and $\rho < 1$, then a feed-forward MCQN is stable (ergodic) under any work conserving head of the line policy.*

For a proof, see Exercise 11.9.

11.3.3 Stability of Re-Entrant Lines under FBFS and LBFS Policies

A re-entrant line is a MCQN in which all customers follow a single route, i.e. there is a single arrival stream to class $k = 1$, and at completion of service at class k customers move to class $k + 1$ for $k = 1, \ldots, K - 1$, and leave the system from class K (see Figure 11.2). We scale the system so $\alpha = 1$, and use the convention that $\mathcal{T}_0(t) = t$ and $\mu_0 = \alpha = 1$. The standard dynamics (11.4) for the fluid model now have the form:

$$\bar{Q}_k(t) = \bar{Q}_k(0) + \mu_{k-1}\bar{\mathcal{T}}_{k-1}(t) - \mu_k\bar{\mathcal{T}}_k(t).$$

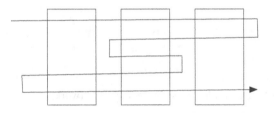

Figure 11.2 A re-entrant line.

A static buffer priority policy for a re-entrant line assigns priority $r(k)$ to class k, and servers at each station process the non-empty queue of the class with the smallest $r(k)$. We assume that the policy is applied in a preemptive resume fashion, i.e. when a high-priority job arrives in a station it preempts current lower priority jobs whose remaining processing time is then frozen. We introduce the following additional notations:

$$H_k = \{l : s(l) = s(k), r(l) \leq r(k)\},$$

$$\mathcal{T}_k^+(t) = \sum_{l \in H_k} \mathcal{T}_l(t), \quad I_k^+(t) = t - \mathcal{T}_k^+(t), \quad \mathcal{W}_k^+(t) = \sum_{l \in H_k} m_l Q_l(t).$$

The policy requires that all the service capacity be devoted to high-priority customers, and so $I_k^+(t)$ can only increase when $\mathcal{W}_k^+(t) = 0$. The same constraint holds for the fluid limits, and so we add to the standard fluid model equations the constraint:

$$\int_0^\infty \bar{W}_k^+ \, d\bar{I}_k^+(t) = 0, \quad k = 1, \ldots, K.$$

We denote: $a_k(t) = \mu_{k-1}\dot{\bar{T}}_{k-1}(t)$, $d_k(t) = \mu_k\dot{\bar{T}}_k(t)$, so $\dot{\bar{Q}}_k(t) = a_k(t) - d_k(t)$.

Lemma 11.17   *For a preemptive resume priority policy at a re-entrant line the following properties of the fluid model hold at every regular point t: (a) $a_k(t) = d_{k-1}(t)$ where $d_0(t) = 1$. (b) If $\bar{Q}_k(t) = 0$, $d_k(t) = a_k(t)$. (c) If k is the highest priority class with non-empty queue in station $i = s(k)$ then $\sum_{l \in H_k} m_l d_l = 1$. (d) Under conditions of (c), for all buffers at station i with $r(l) > r(k)$, $d_l = 0$.*

Proof   (a) follows from the definition. If $\bar{Q}_k(t) = 0$ by Lemma 11.13 $\dot{\bar{Q}}_k(t) = 0$, which implies (b). Since all processing at station i goes to classes in H_k, $\dot{\bar{T}}_k^+(t) = 1$, and from $\dot{\bar{T}}_k^+(t) = \sum_{l \in H_k} m_l(\mu_l \dot{\bar{T}}_l(t))$, we get (c). Recall $\dot{\bar{T}}_l \geq 0$ and their sum ≤ 1. So for $\{l : s(l) = s(k), r(l) > r(k)\}$, we have $1 \geq \dot{\bar{T}}_l^+ \geq \dot{\bar{T}}_k^+ = 1$ and (d) follows.   □

Under FBFS policy, at each station, priority is given to the non-empty queue of the lowest number class, i.e. $r(k) = k$. This is also called a push policy, as it tends to push work into the stations rather than empty the stations.

Theorem 11.18   *A re-entrant line with $\rho_i < 1$ is stable under FBFS policy.*

Proof   The theorem's assumption is that $\rho_i = \sum_{k \in C_i} m_k < 1$. We will show that there exist $t_0 = 0 \leq t_1 \leq \cdots \leq t_K$ so that for all times $t \geq t_k$ all buffers $l \leq k$ are empty. The proof is by induction on the hypothesis that the fluid buffers $1, \ldots, k-1$ are empty for all $t \geq t_{k-1}$.

With the convention of $\bar{Q}_0(t) = 0$, $t_0 = 0$, we have that initially all processing capacity of station $s(1)$ is devoted to class 1, so the buffer of class 1 will be empty for $t \geq t_1 = 1/(\mu_1 - 1)$.

If at t_{k-1}, $\bar{Q}_k(t_{k-1}) > 0$, then k is the first non-empty buffer so by parts (a),(b) of Lemma 11.17, $a_k(t) = d_{k-1}(t) = a_{k-1}(t) = \cdots = a_1(t) = d_0(t) = 1$, and so at time t_{k-1} and at $t \geq t_{k-1}$, k is the highest priority non-empty buffer. Combining this with part (c) of Lemma 11.17, we obtain for as long as $\bar{Q}_k(t) > 0$

$$m_k d_k(t) = \left(1 - \sum_{l \in H_k^+ \backslash k} m_l d_l(t)\right)$$
$$= \left(1 - \sum_{l \in H_k^+ \backslash k} m_l\right)$$
$$> m_k \implies d_k(t) > 1,$$

where the inequality follows from the theorem assumption that $1 > \rho_{s(k)} = \sum_{l:s(l)=s(k)} m_l$. Therefore, for $t \geq t_{k-1}$, while $\bar{Q}_k(t) > 0$, we have $\dot{\bar{Q}}_k(t) = a_k - d_k < 0$, and so $\bar{Q}_k(t)$ will be empty after $t_k = t_{k-1} + \bar{Q}_k(t_{k-1})/(d_k - 1)$. □

Under LBFS policy, at each station priority is given to the non-empty queue closest to the exit, i.e. $r(k) = K + 1 - k$. This is also called a pull policy, as it tends to pull work out of the system.

Theorem 11.19 *A re-entrant line with $\rho_i < 1$ is stable under LBFS policy.*

See Exercise 11.10 for the proof, and also Exercise 11.11.

11.3.4 Additional Stable Policies

Stability of the following policies has also been proved. The proofs use similar methods but are substantially harder, a type of entropy is used as a Lyapunov function. They will not be presented here.

The policy of HLPPS, head of the line proportional processor sharing, in which processing at each station is split between classes, at a rate proportional to the number of customers in the class, and the processing is given to the head of the line customer in the class, is stable for general MCQN with $\rho < 1$; see Exercise 11.24.

The policy of EDD, earliest due date, where each customer is given a due date upon arrival, and this date is his priority throughout his time in the system, is stable for general MCQN with $\rho < 1$. This policy is sometimes referred to as time stamp policy. When the arrival time is taken as the due date, it is referred to as global-FCFS; see Exercise 11.25.

Recall the definition of Kelly-type networks, in Section 8.7. Kelly-type networks, in which processing times at station i have general processing time distributions with rate μ_i, are stable under FCFS policy if $\rho < 1$; see Exercise 11.26.

11.4 Piecewise Linear Lyapunov Functions and Global Stability

We return to the Lu–Kumar network of Section 10.2.1. We saw that if $m_2 + m_4 > 1$, then the discrete stochastic queueing network can be unstable even though $\rho_i < 1$. We now consider the fluid model for this system. Figure 11.3 describes the behavior of a fluid solution. In this figure, the 45° angle line is the fluid arrival process $\bar{\mathcal{A}}(t) = t$, and the lower envelope is the fluid departure process $\bar{\mathcal{D}}(t)$, and the triangular/trapezoid shapes represent the fluid contents in the four buffers (class 1 - clear, class 2 - dark, class 3 - shaded, class 4 - dark).

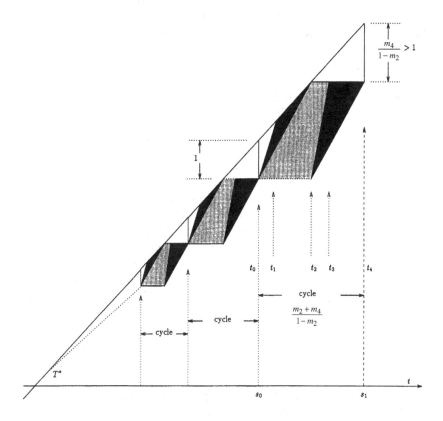

Figure 11.3 Fluid solution of the Lu–Kumar network, reproduced with permission from Dai and Weiss (1996).

We describe a cycle of the fluid, starting with initial fluid $\bar{Q}(s_0)$, and evolving over $s_0 < t < s_1$: Starting at s_0 with fluid $\bar{Q}(s_0) = (1, 0, 0, 0)$, it blows up, so that at time $s_1 = s_0 + \frac{m_2+m_4}{1-m_2}$ the state is $\bar{Q}(s_1) = (\frac{m_4}{1-m_2}, 0, 0, 0)$.

In between we see buffer 1 emptying into 2 ($s_0 < t < t_1$), buffer 2 filling ($s_0 < t < t_1$), and emptying ($t_1 < t < t_2$), buffer 3 filling ($s_0 < t < t_2$) and emptying into 4 ($t_2 < t < t_3$), and while 4 is first filling up and then emptying ($t_2 < t < s_1$), buffer 1 is again filling up. Note that in the period $s_0 < t < t_2$ buffer 4 is empty, because station 2 works on buffer 2 and output from buffer 3 is blocked, and in the period $t_2 < t < s_1$ buffer 2 is empty, because station 1 works on buffer 4 and output from buffer 1 is blocked (see Exercises 11.19 and 11.20).

More interesting, if we extrapolate backwards we get that at time $T^* = s_0 - \frac{m_2+m_4}{m_2+m_4-1}$ the state was $(0,0,0,0)$. Starting at state $\bar{Q}(0) = 0$, $\bar{Q}(t) = 0$, $t > 0$ is a stable fluid solution. However, at any moment T^* it may flare up to a divergent fluid path. Note that starting from $\bar{Q}(0) \neq 0$, the fluid solution is unique and divergent, as described in Figure 11.3.

To complete our study of the Lu–Kumar queueing network, we will prove

Theorem 11.20 *The Lu–Kumar queueing network will be stable under every work-conserving HOL policy if the following conditions are satisfied:*

$$
\begin{aligned}
m_1 + m_4 &< 1/\alpha, \\
m_2 + m_3 &< 1/\alpha, \\
m_2 + m_4 &< 1/\alpha.
\end{aligned}
\tag{11.13}
$$

We note: If $\rho_i > 1$, then the queue lengths in the Lu–Kumar network will diverge for any policy, by Theorem 11.12, part (b), and if $m_2 + m_4 > 1/\alpha$, the queue lengths will diverge under the policy of priority to buffers 2 and 4, at least for some examples of interarrival and processing time distributions. We say in this case that the conditions of Theorem 11.20 define the global stability zone for the Lu–Kumar queueing network. More generally, if we consider a MCQN with parameters (α, μ, P), we define

Definition 11.21 A closed region $\mathbb{K}$ of parameter values is the global stability region of a MCQN, if for every set of parameters (α, μ, P) in the interior of $\mathbb{K}$ the MCQN is stable under every work-conserving HOL policy, regardless of interarrival and processing time distributions, while outside $\mathbb{K}$, there exist interarrival and processing time distributions and work-conserving HOL policies under which the MCQN will diverge.

In general, it is difficult to locate the global stability regions of a MCQN. To prove Theorem 11.20, we will use *piecewise linear Lyapunov functions*. We show first:

Lemma 11.22 *For $i = 1, \ldots, I$, let $\bar{\mathcal{W}}_i(t) = \sum_{k \in C_i} m_k \bar{Q}_k(t)$ be the expected immediate workload at station i, and let $G_i(t)$ be a non-negative*

linear combination of $(\bar{Q}_1(t), \ldots, \bar{Q}_K(t)) \geq 0$, *such that:*
(a) For each i, if $\bar{W}_i(t) > 0$, *then* $\dot{G}_i(t) < -\epsilon < 0$.
(b) If $\bar{W}_i(t) = 0$, *then either* $G_i(t) < \max_j G_j(t)$, *or* $\min_j G_j(t) = \max_j G_j(t)$. *Let* $G(t) = \max(G_1(t), \ldots, G_I(t))$. *Then G is an absolutely continuous non-negative function, and* $\dot{G}(t) < -\epsilon$ *at all regular points where* $G(t) > 0$, *and hence* $G(t) = 0$ *for* $t > G(0)/\epsilon$.

Proof First note that $G(t)$ is absolutely continuous, with derivatives almost everywhere. If t is a regular point of G and $G(t) = G_j(t)$, then $\dot{G}(t) = \dot{G}_j(t)$ (property of the max of linear functions). Assume then that at a regular point, $G(t) > 0$. Then $\bar{W}_i(t) > 0$ for some i. By property (b), we can find j such that $G_j(t) = G(t)$ and $\bar{W}_j(t) > 0$. But then $\dot{G}(t) = \dot{G}_j(t) < -\epsilon$ (using (a)). By Lemma 11.13, $G(t) = 0$ for $t > G(0)/\epsilon$. $\quad\square$

Proof of Theorem 11.20 W.l.g. we take $\alpha = 1$. We already showed that if $m_2 + m_4 > 1$, then the fluid model is unstable for some policies. Assume now $m_2 + m_4 < 1$. Denote $\bar{Q}_k^+ = \sum_{l=1}^k \bar{Q}_l$ (this is the total amount of fluid in the system that still needs to go through class k). Consider the following linear combinations for some $0 < \theta_1, \theta_2 < 1$:

$$G_1(t) = \theta_1 \bar{Q}_1^+(t) + (1 - \theta_1)\bar{Q}_4^+(t), \qquad G_2(t) = \theta_2 \bar{Q}_2^+(t) + (1 - \theta_2)\bar{Q}_3^+(t).$$

We wish to find θ_1, θ_2, and ϵ so that (a) and (b) of Lemma 11.22 hold.
    To obtain condition (b):
If $\bar{W}_1(t) = 0$, then $\bar{Q}_1(t) = 0$, $\bar{Q}_4(t) = 0$, and so we need to have:

$$G_1(t) = (1 - \theta_1)(\bar{Q}_2(t) + \bar{Q}_3(t)) \leq G_2(t) = \bar{Q}_2(t) + (1 - \theta_2)\bar{Q}_3(t).$$

If $\bar{W}_2(t) = 0$, then $\bar{Q}_2(t) = 0$, $\bar{Q}_3(t) = 0$, and so we need to have:

$$G_1(t) = \bar{Q}_1(t) + (1 - \theta_1)\bar{Q}_4(t) \geq G_2(t) = \bar{Q}_1(t).$$

To obtain (a), we need to consider the derivatives of G_1, G_2. Using the notation of Lemma 11.17:

$$\dot{G}_1(t) = 1 - \theta_1 d_1(t) - (1 - \theta_1)d_4(t), \qquad \dot{G}_2(t) = 1 - \theta_2 d_2(t) - (1 - \theta_2)d_3(t).$$

It can be checked that (a) (b) hold for $\theta_1, \theta_2, \epsilon$ given by the solution of the LP:

$$\begin{aligned}
\max \ & \epsilon \\
\text{s.t. } & \epsilon \leq \mu_1\theta_1 - 1, \quad \epsilon \leq \mu_4(1 - \theta_1) - 1, \\
& \epsilon \leq \mu_2\theta_2 - 1, \quad \epsilon \leq \mu_3(1 - \theta_2) - 1, \qquad (11.14) \\
& 0 \leq \theta_2 \leq \theta_1 \leq 1, \quad \epsilon \geq 0.
\end{aligned}$$

This LP is feasible for some $\epsilon > 0$ if and only if $m_1 + m_4 < 1$, $m_2 + m_3 < 1$, $m_2 + m_4 < 1$, see Exercise 11.21 to complete the proof.   □

11.5 Sources

The use of a multiplicative Foster-Lyapunov criterion for stability of MCQN with countable state Markov description is due to Rybko and Stolyar (1992), who have used it to show that the KSRS network is unstable under priority to buffers 2 and 4, but is stable under FCFS. This led to the proof that fluid stability implies stochastic stability for general uncountable state space MCQN in the pioneering paper of Dai (1995). Further results on fluid criteria are in Dai and Meyn (1995) and Dai (1996). The monograph *Stability of Queueing Networks*, Bramson (2008) is a comprehensive treatment of the subject.

Stability of re-entrant lines under LBFS is first discussed in Kumar (1993). Proofs of stability based on fluid stability in this chapter are mostly from Dai and Weiss (1996). Further papers that deal with similar MCQN models and stability proofs are Chen (1995) and Chen and Zhang (1997, 2000). The stability of the HLPPS policy is proved in Bramson (1996), the stability of EDD policy is proved in Bramson (2001) and Yang (2009), and the stability of Kelly-type networks under FCFS is proved in Bramson (1996).

An example of a stable stochastic queueing system with unstable fluid limit was found by Bramson (1999). Further discussion of the relation between the fluid and the stochastic system are in Dai et al. (2004) and in Gamarnik and Hasenbein (2005). The use of piecewise linear Lyapunov functions was pioneered by Botvich and Zamyatin (1992). A full discussion of stability of two-station networks and derivation of their global stability region is Dai and Vande Vate (2000). Linear programming formulations to verify stability and locate global stability regions are developed in Kumar and Meyn (1996) and Bertsimas et al. (1996, 2001). Dai et al. (1999) give an example of a non-monotone global stability region.

Exercises

11.1 Derive the standard fluid model equations under the assumption that $\frac{1}{n}U_k^n(0) \to \bar{U}_k(0) > 0$, and $\frac{1}{n}V_k^n(0) \to \bar{V}_k(0) > 0$, for $k = 1, \ldots, K$, as $n \to \infty$.

11.2 Explain the stochastic and the fluid model equations (11.5), (11.6) that are added for static priority policies. Show that they determine the stochastic queueing process, and verify that their fluid versions are satisfied by every fluid limit.

11.3 Write down the stochastic system equations for class and station immediate workload, and the resulting standard fluid model equations for them.

11.4 Explain the stochastic and the fluid model equations (11.7) – (11.9) that are added for FCFS policies. Show that they determine the stochastic workload and queueing processes, and verify that their fluid versions are satisfied by every FCFS fluid limit.

11.5 Prove Lemma 11.5.

11.6 Prove Theorem 11.12.

11.7 Prove Lemma 11.13.

11.8 Show that in a stable Jackson network, $\bar{Q}(t)$ need not be a decreasing function in all coordinates while $t < t_0$.

11.9 Show that a feed-forward MCQN is stable under any work-conserving HOL policy.

11.10 Prove that a re-entrant line with $\rho < 1$ is stable under LBFS.

11.11 Consider a fluid re-entrant line with input rate 1, and initial buffer contents $Q_1(0) = 1$, and all the other buffers empty. Show that this fluid re-entrant line under FBFS policy will be empty by time

$$ t_K = \sum_{k=1}^{K} m_k \frac{\prod_{l=1}^{k-1} \left(1 - \sum_{j \in H_l \setminus l} m_j \right)}{\prod_{l=1}^{k} \left(1 - \sum_{j \in H_l} m_j \right)}. $$

11.12 Find a lower bound to the time needed to empty a fluid re-entrant line and suggest a policy that will achieve this time [Weiss (1995)].

11.13 For the following two examples of fluid re-entrant lines, draw the fluid levels for FBFS, LBFS, and minimum time to empty, and compare time to empty, and inventory:

(i) $C_1 = \{1, 2\}$, $C_2 = \{3\}$, $\alpha = 0$, $m = (1, 0.5, 1)$, $Q(0) = (1, 1, 0)$.

(ii) $C_1 = \{1, 3\}$, $C_2 = \{2\}$, $\alpha = 0$, $m = (1, 2, 0.5)$, $Q(0) = (1, 0, 1)$.

11.14 Consider a fluid MCQN with $\rho < 1$. Calculate a bound on the minimum time to empty the network, and devise a policy that will achieve that lower bound (use processor splitting, predictive policy). Use this to suggest a policy that will be stable for any MCQN with $\rho < 1$.

11.15 Consider a single-station fluid re-entrant line with $\rho < 1$. Show that the total amount of fluid in the network is minimized pathwise under LBFS policy, and is maximized pathwise under FBFS policy.

11.16 For a MCQN with several types of customers following deterministic routes, devise static priority policies analog to FBFS and to LBFS, and prove their stability. This could be termed a *path priority policy*.

11.17 Consider a two-station, three classes fluid re-entrant line, where $C_1 = \{1, 3\}$, $C_2 = \{2\}$, and assume that $\alpha = 1$, $m_1 + m_3 < 1$, $m_2 < 1$. Show it is stable under all work-conserving HOL policies [Dai and Weiss (1996)].

11.18 Write down the fluid model equations for the Lu–Kumar network, under the static priority policy of priority to buffers 2 and 4. Identify the properties of all the fluid solutions of these equations.

11.19 For the Lu–Kumar network with input rate α and initial fluid x in buffer 1, show that when $\rho_i < 1$ and $m_1 + m_2 > 1/\alpha$ then $m_2 > m_1$, $m_4 > m_3$, so that Figure 11.3 is correct. With $s_0 = 0$, calculate the values of t_1, t_2, t_3, s_1 and $\bar{Q}(t_1), \bar{Q}(t_2), \bar{Q}(t_3), \bar{Q}(s_1)$.

11.20 Plot a fluid solution for the Lu–Kumar network when $\rho_i < 1$ and $m_1 + m_2 < 1/\alpha$, for some initial fluid in buffer 1, and priority to buffers 2 and 4.

11.21 For the Lu–Kumar network, verify that the function G satisfies properties (i) and (ii) necessary for the piecewise linear Lyapunov function to prove stability, and that the linear program (11.14) is feasible with $\epsilon > 0$, if $m_1 + m_4 < 1$, $m_2 + m_3 < 1$, $m_2 + m_4 < 1$ [Dai and Weiss (1996)].

11.22 Consider a KSRS network with input rates α_1, α_2 and mean service times m_1, m_2, m_3, m_4. Show that the conditions:
$$\alpha_1 m_1 + \alpha_2 m_4 < 1, \quad \alpha_1 m_2 + \alpha_2 m_3 < 1, \quad \alpha_1 m_4 + \alpha_2 m_2 < 1,$$
define a global stability region for the network [Botvich and Zamyatin (1992)].

11.23 Consider the example of the re-entrant line that is unstable under FCFS, of Section 10.2. Plot an unstable fluid solution for it.

11.24 (∗) *HLPPS policy:* Under head of the line proportional processor sharing policy, each station is splitting its service capacity between classes in proportion to the number of customers present, and is then serving the head of the line of each class. Show that a fluid MCQN with $\rho < 1$ is stable under HLPPS [Bramson (1996, 2008)].

11.25 (∗) *Early due date policy:* Consider the policy in which each customer receives a due date upon arrival, and the policy is to give priority to the customer with the earliest due date (EDD). Show that MCQN under EDD policy is stable [Bramson (2001, 2008)].

11.26 (∗) *Kelly-type networks under FCFS:* Consider a Kelly-type network as in Section 8.9, with general renewal arrivals and general processing times of rate μ_i at node i. Show that with $\rho_i < 1$ it is stable under FCFS [Bramson (1996, 2008)].

12

Processing Networks and Maximum Pressure Policies

In this chapter (Section 12.1) we introduce processing networks that are more general than MCQN. These processing networks can have activities that use more than one resource (machine, station), and process items from more than one buffer (queue) simultaneously. While we have so far only examined control of MCQN by scheduling of classes at each station, processing networks also allow us to control input into the system, and routing between buffers.

In the previous two chapters we saw that $\rho < 1$ is not enough to determine stability of MCQN, and the same network may be stable under some policies and unstable under others. We then saw that stability of a MCQN under a given policy may often be determined by examining its fluid model. We would now like to find a good policy that will be stable for any queueing network with $\rho < 1$. We have in fact already suggested several such policies: Two of those are the policy of emptying the fluid network in minimum time (Exercise 11.14), and the policy of head of the line proportional processor sharing (HLPPS) (Exercise 11.24). Both these policies require processor sharing; the former one also requires knowing all processing times in advance, and both policies are inefficient in that they do not attempt to reduce waiting times of customers. Two other policies that are more sensible and stable for $\rho < 1$ are global first come first served (time stamp policy, and earliest due date (Exercise 11.25). However, these policies are not designed for more general processing networks.

The maximum pressure policy, which we introduce next in Section 12.2, is designed for general processing networks. We show in Section 12.3 that the maximum pressure policy is rate stable for most processing networks with $\rho \leq 1$, and in particular for MCQN. Policies that are rate stable at $\rho = 1$ are called maximum throughput, and they achieve full utilization of at least some of the resources. To find the maximal utilization of resources, we introduce a linear program, called the static planning program, which finds the maximum workload ρ that the network can carry, for given input rates. Additional advantages of maximum pressure policy are that it can be implemented in a decentralized way, it does not require knowledge of input rates, and it has good performance in terms of waiting times. Some extensions of maximum pressure properties are discussed in Section 12.4.

Finally, in Section 12.5 we discuss applications. These include the implementation of maximum pressure to MCQN and to MCQN with discretionary routing, and an important application to input queued communication switches.

12.1 A More General Processing System

So far, our MCQN have consisted of service stations $i = 1, \ldots, I$, and classes (or buffers) $k = 1, \ldots, K$, where each class k had a single station $s(k)$ which served it, summarized by the $I \times K$ constituency matrix C. Routing between classes was uncontrolled, Markovian, and summarized by the $K \times K$ routing probabilities matrix P, or by fixed deterministic routes as a special case. Input was also uncontrolled, given by an independent input stream of customers arriving at each one of a subset of the classes.

We now define a more general system, by adding another layer that is intermediate between stations and classes, given by activities, $j = 1, \ldots, J$, where each activity is associated with a subset of the stations, and a subset of the classes. We will now refer to the stations as the processors, which are our limited resources, and to the classes as buffers, which contain items that are processed according to HOL. To perform activity j, the associated subset of the processors is used simultaneously for a period of time, and a single item out of each of the associated subset of the buffers is processed, and at the end of this time period, the processed items move to other buffers or leave the system.

This model will be referred to as a processing network, and it can model a wide range of systems. It can be a manufacturing system, in which $i = 1, \ldots, I$ are tools, $k = 1, \ldots, K$ are materials/parts, and activity j uses a subset of tools, and a set of materials/parts, as a step in the production of finished goods. Examples are production lines, assembly systems, chemical processes, sorting machines, etc. It can also model control of health systems such as emergency rooms in a hospital, parallel computing systems, or networks of input queued communication switches.

We now specify the system. The processing system consists of activities $j \in \mathcal{J} = \{1, \ldots, J\}$, buffers $k \in \mathcal{K} = \{0, 1, \ldots, K\}$, and processors $i \in \mathcal{I} = \{1, \ldots, I\}$. Buffer 0 is the outside world with infinite supply of items. Buffers $k \neq 0$ contain queues of items. Processors associated with activities are described by the $I \times J$ resource consumption matrix A, where $A_{i,j}$ is the indicator that activity j uses processor i. Buffers associated with activities are described by the $J \times (K + 1)$ constituency matrix C, where $C_{j,k}$ is the indicator that activity j is processing buffer k, and we let C_j denote the set of buffers processed by activity j, the constituency of activity j. We separate the activities into input activities and service activities. Input activities only process buffer 0, and service activities only process buffers $k \neq 0$. Processors are also divided into input processors that are used only by input activities, and service processors that are used only by service activities (see Figure 12.1). These general processing networks allow control of input and routing in MCQN, as well as modeling of systems far more general than MCQN.

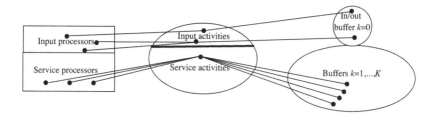

Figure 12.1 A more general system.

The rth activation of activity j requires a processing amount $\upsilon_j(r)$, at the completion of which one item leaves each of the buffers C_j in the constituency of the activity and is then routed to another buffer. We let $\xi^j_{k,l}(r)$ be the indicator that on the rth activation of activity j, an item from buffer k is moved to buffer l; these would be arrivals if activity j is an input activity, in which case $k = 0$; they would be a departure from the queue if $l = 0$. We let $\mathcal{S}_j(t)$ be the number of processing completions in the first t time units that activity j is applied; for input activities it counts arrivals, for service activities it counts the number of service completions. We let $\mathcal{R}^j_{k,l}(r)$ be the count of the number of items moved from buffer k to buffer l by the first r activations of activity j; a move from 0 is an *arrival*, a move to 0 is a *departure*, and we take $\mathcal{R}^j_{0,0}(r) = 0$.

We let $Q_k(t)$, $k = 1, \ldots, K$ be the number of items in buffer k at time t, and we let $\mathcal{T}_j(t)$ be the total activation time of activity j, over $(0, t)$. The dynamics of the stochastic model, for any HOL policy, are now

$$Q_k(t) = Q_k(0) - \sum_{j \in \mathcal{J}} C_{j,k} \mathcal{S}_j(\mathcal{T}_j(t)) + \sum_{j \in \mathcal{J}} \sum_{l \neq k} C_{j,l} \mathcal{R}^j_{l,k}(\mathcal{S}_j(\mathcal{T}_j(t))). \quad (12.1)$$

We allow processor splitting, and let $0 \le a_j \le 1$ be the allocated processing for activity j. Each of the processors $\{i: A_{i,j} = 1\}$ that activity j is using will then devote a fraction a_j of its capacity to activity j, and a_j is the rate at which time will be allocated to activity j, so the processing requirement v_j will be processed at rate a_j. The allowable allocations satisfy the following set of linear constraints::

$$\sum_{j \in \mathcal{J}} A_{i,j} a_j = 1 \text{ for input processors } i,$$

$$\sum_{j \in \mathcal{J}} A_{i,j} a_j \le 1 \text{ for service processors } i, \tag{12.2}$$

$$0 \le a_j \le 1, \quad j \in \mathcal{J}.$$

Note that we allow idling of service processors, but require full utilization of the input processors. We denote the set of allowable allocations by $\mathcal{A}$; it is a convex polytope with a finite number of extreme points. We denote the set of extreme points (vertices) by $\mathcal{E}$; these are extreme allocations.

Service policies are defined as follows: Each job (consisting of applying an activity to a set of items from the constituency of the activity) is processed by a single activity, and if the activity is preempted, then the job is frozen until the activity is resumed; on the resumption of the activity, the preempted job is taken up again. A new activity can only start if each of the buffers in its constituency contains at least one item that is not in service (by another activity) or frozen. At time t we let $\mathcal{A}(t)$ be the available feasible allocations and $\mathcal{E}(t)$ the set of available feasible extreme allocations (see Exercise 12.2).

Recall that $\mathcal{T}_j(t)$ are Lipschitz continuous, and therefore are integrals of their derivatives that exist almost everywhere. Under any policy we then have the following constraints on $\mathcal{T}_j(t)$, expressed in terms of their derivatives:

$$\sum_{j \in \mathcal{J}} A_{i,j} \dot{\mathcal{T}}_j(t) = 1, \quad \text{input processors } i.$$
$$\sum_{j \in \mathcal{J}} A_{i,j} \dot{\mathcal{T}}_j(t) \le 1, \quad \text{service processors } i. \tag{12.3}$$
$$0 \le \dot{\mathcal{T}}_j(t) \le 1, \quad j \in \mathcal{J}.$$

Note that all this is saying is that $(\dot{\mathcal{T}}_j(t))_{j \in \mathcal{J}} \in \mathcal{A}$.

The primitives of the processing network are sequences of service requirements $v_j(r)$, $r = 1, 2, \ldots$, and sequences of routing decisions $\xi_{k,l}^j(r)$, $r = 1, 2, \ldots$. In contrast to previous chapters, we make only minimal assumptions: we assume that $v_j(r)$, $\xi_{k,l}^j(r)$, $r = 1, 2, \ldots$ form jointly a stationary ergodic sequence, and they satisfy a SLLN:

$$\lim_{n \to \infty} \sum_{r=1}^{n} v_j(r)/n = m_j, \quad \text{almost surely,}$$
$$\lim_{n \to \infty} \sum_{r=1}^{n} \xi_{k,l}^j(r)/n = p_{k,l}^j, \quad \text{almost surely,} \tag{12.4}$$

where m_j are average processing times and P^j is a $K + 1 \times K + 1$ matrix of routing probabilities, associated with activity j (this is a stochastic matrix, including routing into buffer 0, which accounts for departures). We let $\mu_j = 1/m_j$ be the processing rate of activity j.

12.2 Maximum Pressure Policies

We consider policies that depend only on the vector of buffer contents, $Q = (Q_1, \ldots, Q_K)$. Define the $K \times J$ input-output matrix R as:

$$R_{k,j} = \mu_j \left(C_{j,k} - \sum_{l=0}^{K} C_{j,l} \, p_{l,k}^j \right), \tag{12.5}$$

so that as before, $R_{k,j}$ is the expected rate at which contents of buffer k are depleted by activity $j \in \mathcal{J}$, the first term on the right-hand side are departures, and the second term includes exogenous arrivals and feedback arrivals from other buffers. In particular, if j is an input activity, then $-R_{k,j} = \mu_j p_{0,k}^j$ is the rate of arrivals into k by input activity j.

The pressure of allocation $a \in \mathcal{A}$ in state $z \in \mathbb{R}^K$ is defined as:

$$p(a, z) = z \cdot Ra, \tag{12.6}$$

where $\cdot$ denotes the scalar product of the vectors z and Ra. We wish to employ allocations that achieve maximum pressure, so we have candidate allocations $a^* \in \arg\max_{a \in \mathcal{A}} p(a, Q(t))$. Because the pressure is a linear function, there exist maxima that are extreme points. We will then look for $a^* \in \arg\max_{a \in \mathcal{E}} p(a, Q(t))$. However, the solution to this may not be feasible. Since we are looking for a feasible allocation, we will restrict the search to available feasible extreme allocations, $\mathcal{E}(t)$. Note that $\mathcal{A}(t)$ is also a convex polyhedron, and the extreme allocation of $a_j = 0$ for all service activities is always available, so $\mathcal{E}(t)$ is a bounded finite non-empty set, and feasible extreme maximum pressure allocations are $a^*(t) \in \arg\max_{a \in \mathcal{E}(t)} p(a, Q(t)) \subseteq \arg\max_{a \in \mathcal{A}(t)} p(a, Q(t))$.

Definition 12.1 A service policy is a maximum pressure policy if at any time t it uses an allocation $a^* \in \arg\max_{a \in \mathcal{E}(t)} p(a, Q(t))$. Ties can be broken in an arbitrary way, e.g. by listing all the points of $\mathcal{E}$ in a priority list.

To be able to analyze this we need to understand the maximization over points of $\mathscr{E}(t)$, which may be hard. It would be easier to just solve $\max_{a \in \mathscr{E}} p(a, Q(t))$, as a linear program. We would like therefore to require that for our processing network $\max_{a \in \mathscr{E}} p(a, Q(t)) = \max_{a \in \mathscr{E}(t)} p(a, Q(t))$ should hold. However, we can never guarantee that, since $Q_k(t) > 0$ may consist entirely of jobs that are in service or frozen by preemption. Instead, we impose a weaker condition that implies $\max_{a \in \mathscr{E}} p(a, Q(t)) = \max_{a \in \mathscr{E}(t)} p(a, Q(t))$ if all non-empty buffers contain many jobs.

Assumption 12.2 (EAA assumption) *Extreme allocation available* (EAA) requires that for any state Q there exists an extreme pressure allocation $a^* \in \arg\max_{a \in \mathscr{E}} p(a, Q)$ such that if $a_j^* > 0$ then also $Q_k > 0$ for all k in the constituency of j.

EAA assumption is a property of the network. It holds for some types of networks but does not hold for all. Maximum pressure policy is effective under EAA. In particular, EAA holds for *strict Leontief networks*, which we define now.

Definition 12.3   A processing network is a strict Leontief network if each activity serves only a single buffer, i.e. each row of the matrix C has only a single 1 entry.

Strict Leontief networks include MCQN and *input queued switches*, which we discuss later, in Section 12.5.2.

Proposition 12.4   *EAA assumption is satisfied by strict Leontief networks.*

Proof   For each service activity j, let $k(j)$ be the unique buffer that is served by j. Then for all $k \neq k(j)$, $C_{j,k} = 0$ and by (12.5), $R_{k,j} \leq 0$. Consider a state vector z, and let $\mathcal{J}_0$ be the set of all the service activities j such that $z_{k(j)} = 0$. Then by the above, $(Rz)_j \leq 0$ for all $j \in \mathcal{J}_0$. To show EAA we need to show that there exists an allocation in $\arg\max_{a \in \mathscr{E}} p(a, z)$ for which $a_j = 0$ for all $j \in \mathcal{J}_0$. Let $\hat{a} \in \arg\max_{a \in \mathscr{E}} p(a, z)$. We define a modified allocation,

$$\tilde{a}_j = \begin{cases} \hat{a}_j & j \notin \mathcal{J}_0 \\ 0 & j \in \mathcal{J}_0 \end{cases}.$$

Then it is immediate to see that $\tilde{a} \in \mathscr{A}$, and by the above $p(\tilde{a}, z) \geq p(\hat{a}, z)$, so $\tilde{a} \in \arg\max_{a \in \mathscr{A}} p(a, z)$.

If $\tilde{a}$ is an extreme point of $\mathscr{A}$, the proof is complete. Otherwise, because it maximizes the pressure it must be a convex combination of some extreme points. But then each one of these, say a^* will be a maximal extreme allocation for z, with $a_j^* = 0$, $j \in \mathcal{J}_0$. $\qquad\qquad\qquad\qquad\square$

Define the *static planning problem* (as suggested by Harrison (2000)):

$$\min \rho$$

$$\text{s.t. } Rx = 0,$$

$$\sum_{j \in \mathcal{J}} A_{i,j} x = 1, \quad \text{for input processors } i, \qquad (12.7)$$

$$\sum_{j \in \mathcal{J}} A_{i,j} x \le \rho, \quad \text{for service processors } i,$$

$$x \ge 0,$$

where we look for x which is an allocation. Here the first constraint requires input rate equal to output rate throughout the system, the second constraint requires that input processors never idle, and the third constraint requires that the utilization (fraction of capacity used) of each service processor does not exceed ρ. In an optimal solution, we will have that processors i for which $\rho_i = \rho$ are critically loaded. The unknown vector x represents a *nominal allocation* $x = \dot{\mathcal{T}}(t)$. If $\dot{\mathcal{T}}(t)$ is maintained at this allocation then the system will maintain a balance of input and output flows: in fact, if $\lim_{t \to \infty} \mathcal{T}(t)/t = x$, then the processing network is rate stable, i.e. $\lim_{t \to \infty} Q(t)/t = 0$. We will say more about nominal allocations in Chapters 16 and 17. The unknown ρ measures traffic intensity, and the static planning problem finds the right combination of input activities and routing decisions so that the input processors will work all the time, the system will have balanced rates, and the traffic intensity ρ will be minimal.

Similar to Theorem 11.12 part (b), if $\rho > 1$, then, under any policy, for some buffer k $\liminf_{t \to \infty} Q_k(t)/t \ge c > 0$ a.s., i.e. the processing network diverges (see Exercise 12.3). In other words, $\rho \le 1$ is a necessary condition for existence of a rate-stable policy. The main result on maximum pressure policies is:

Theorem 12.5 *Assume that a processing network has EAA property and that the solution of the static planning problem (12.7) has optimal $\rho \le 1$. Then under a maximum pressure policy that allows processor splitting and preemption, the processing network is rate stable, i.e. $\lim_{t \to \infty} Q(t)/t = 0$ almost surely.*

Remark EAA assumption cannot be relaxed, as some examples show. See Exercise 12.4 for such an example.

We shall prove Theorem 12.5 in the next section.

12.3 Rate Stability Proof via the Fluid Model

To prove Theorem 12.5 we will again use the fluid limit model. In fact, Theorem 12.5 will follow directly, by using the results of Section 11.2, and in particular, Theorem 11.10, from the following:

Theorem 12.6 *Assume that a processing network has EAA property and that $\rho \leq 1$. Then under a maximum pressure policy that allows processor splitting and preemption, the fluid model of the network is weakly stable, i.e. starting from 0, it stays at 0.*

Following the steps in the proof of Theorem 11.2, we can show that fluid limits for $Q(t), \mathcal{T}(t)$ exist and are Lipschitz continuous, and for all $\omega \in \mathfrak{G}$ (i.e. the set of probability 1 for which SLLN equations (12.4) hold), we have that every fluid limit satisfies

$$\bar{Q}_k(t) = \bar{Q}_k(0) - \sum_{j \in \mathcal{J}} C_{j,k} \mu_j \bar{T}_j(t) + \sum_{l \neq k} \sum_{j \in \mathcal{J}} C_{j,l} \, p^j_{l,k} \mu_j \bar{T}_j(t), \quad k = 1, \ldots, K,$$

and we can write this more succinctly as

$$\bar{Q}(t) = \bar{Q}(0) - R\bar{T}(t). \tag{12.8}$$

In addition, we also have the constraints on the allocations:

$$\dot{\bar{T}}(t) \in \mathscr{A}. \tag{12.9}$$

Maximum pressure policies are not necessarily work conserving, so we do not have an equation that states that idling of a processor happens only if all the buffers that it can serve are empty.

From (12.9) we see that for any HOL policy, for all $t > 0$:

$$R\dot{\bar{T}}(t) \cdot \bar{Q}(t) \leq \max_{a \in \mathscr{A}} Ra \cdot \bar{Q}(t).$$

The following crucial lemma, which we prove later, imposes a stronger constraint on the fluid limit model under maximum pressure policy. This is the special fluid model equation for the maximum pressure policy:

Lemma 12.7 *Under a maximum pressure policy that allows processor splitting and preemption, if EAA holds for the network, then every fluid limit (of $\omega \in \mathfrak{G}$) must satisfy the stronger equation:*

$$R\dot{\bar{T}}(t) \cdot \bar{Q}(t) = \max_{a \in \mathscr{A}} Ra \cdot \bar{Q}(t). \tag{12.10}$$

Using this lemma, we can now prove weak fluid stability under maximum pressure.

Proof of Theorem 12.6 Let $\bar{Q}(t), \bar{\mathcal{T}}(t)$ be a fluid solution. We consider the quadratic Lyapunov function:

$$f(t) = \sum_{k=1}^{K} (\bar{Q}_k(t))^2 = \bar{Q}(t) \cdot \bar{Q}(t),$$

and we calculate its derivative:

$$\begin{aligned}
\dot{f}(t) &= 2\dot{\bar{Q}}(t) \cdot \bar{Q}(t) \\
&= -2R\dot{\bar{\mathcal{T}}}(t) \cdot \bar{Q}(t) \qquad \text{by equation (12.8),} \\
&= -2 \max_{a \in \mathscr{A}} Ra \cdot \bar{Q}(t) \qquad \text{by Lemma 12.7.}
\end{aligned}$$

Let x^* be an optimal solution of the static planning problem (12.7). Then $Rx^* = 0$ and because $\rho \leq 1$, we have $x^* \in \mathscr{A}$. We then have:

$$\begin{aligned}
\dot{f}(t) &= -2 \max_{a \in \mathscr{A}} Ra \cdot \bar{Q}(t) \\
&\leq -2Rx^* \cdot \bar{Q}(t) \qquad \text{since } x^* \in \mathscr{A} \\
&= 0 \qquad \text{since } Rx^* = 0.
\end{aligned}$$

The Lyapunov function f then has the properties: $f(t) \geq 0$ and is 0 only when $\bar{Q}_k(t) = 0$, $k = 1, \ldots, K$, and it satisfies $\dot{f}(t) \leq 0$. It follows by Lemma 11.13 that if $f(0) = 0$, then $f(t) = 0$ for all $t > 0$. Hence, $\bar{Q}_k(0) = 0$, $k = 1, \ldots, K$ implies $\bar{Q}_k(t) = 0$ for all $t > 0$, so the fluid limit model is weakly stable. □

It is now clear how maximum pressure works: The proof uses as Lyapunov function the sum of squares of the buffer contents, and it is then seen, by Lemma 12.7 that maximum pressure maximizes the rate at which this sum of squares decreases, where the maximization is over all possible allocations. So maximum pressure does two things: To decrease the sum of squares, it tries to equalize the buffer contents in the various buffers, and at the same time, if $\rho < 1$, the total fluid buffer content will decrease to 0 and stay at 0. Furthermore, starting from 0, the fluid will stay at 0 also for $\rho = 1$.

Proof of Lemma 12.7 Define $\mathcal{T}^a(t)$ as the cumulative time that allocation $a \in \mathscr{E}$ was used. Under maximum pressure, only extreme allocations are used (i.e. $a \in \mathscr{E}(t) \subseteq \mathscr{E}$), so that $\mathcal{T}_j(t) = \sum_{a \in \mathscr{E}} a_j \mathcal{T}^a(t)$. Similar to Theorem 11.2 we can show that fluid limits of $\mathcal{T}^a$ exist and that $\bar{\mathcal{T}}^a(t)$ is Lipschitz continuous so it has derivatives almost everywhere, and is the integral of its derivative. Consider a regular time t and assume that $p(a, \bar{Q}(t)) < \max_{a' \in \mathscr{E}} p(a', \bar{Q}(t))$. We claim that in that case $\dot{\bar{\mathcal{T}}}^a(t) = 0$.

Let a^* be an allocation such that $p(a^*, \bar{Q}(t)) = \max_{a' \in \mathscr{E}} p(a', \bar{Q}(t))$ and such that for all j for which $a_j^* > 0$ also $\bar{Q}_k(t) > 0$ for all $k \in C_j$. Such a^* exists by EAA. Consider now only j for which $a_j^* > 0$.

We now use an argument similar to the one used in the proof of Theorem 11.2 to show that fluid limits of work-conserving policies are work-conserving. By continuity of $\bar{Q}(t)$, for some $\epsilon, \delta > 0$ and for all $s \in (t - \epsilon, t + \epsilon)$, we have $p(a, \bar{Q}(s)) < \max_{a' \in \mathscr{E}} p(a', \bar{Q}(s)) - \delta$, and $\bar{Q}_k(s) > \delta$ for all $k \in C_j$. But then for large enough n_0, and for all $s \in (nt - n\epsilon, nt + n\epsilon)$, $p(a, Q(s)) < \max_{a' \in \mathscr{E}} p(a', Q(s)) - n\delta/2$ and $Q_k(s) > n\delta/2$ for all $k \in C_j$, holds for all $n > n_0$, and in particular we can take n_0 such that $n_0 \delta/2 > J$. Note that J is the maximal number of items in any queue that are served not by a^* or are frozen. Hence $a^* \in \mathscr{E}(s)$ for all $s \in (nt - n\epsilon, nt + n\epsilon)$, and hence $p(a, \bar{Q}(s)) < p(a^*, \bar{Q}(s))$, for $s \in (nt - n\epsilon, nt + n\epsilon)$. This implies that allocation a is not maximum pressure in $s \in (nt - n\epsilon, nt + n\epsilon)$, and so $\mathcal{T}^a(nt + n\epsilon) - \mathcal{T}^a(nt - n\epsilon) = 0$, and hence $\dot{\bar{\mathcal{T}}}^a(t) = 0$.

From $\sum_{a \in \mathscr{E}} \dot{\bar{\mathcal{T}}}^a(t) = 1$, and from $\dot{\bar{\mathcal{T}}}^a(t) = 0$ when $p(a, \bar{Q}(t))$ is not maximal, it follows that:

$$\sum_{a \in \mathscr{E}} \dot{\bar{\mathcal{T}}}^a(t) p(a, \bar{Q}(t)) = \max_{a \in \mathscr{A}} Ra \cdot \bar{Q}(t).$$

But clearly,

$$R\dot{\bar{\mathcal{T}}}(t) \cdot \bar{Q}(t) = \sum_{k=1}^{K} \bar{Q}_k(t) \sum_{j \in \mathcal{J}} R_{k,j} \dot{\bar{\mathcal{T}}}_j(t)$$

$$= \sum_{k=1}^{K} \bar{Q}_k(t) \sum_{j \in \mathcal{J}} R_{k,j} \sum_{a \in \mathscr{E}} a_j \dot{\bar{\mathcal{T}}}^a(t)$$

$$= \sum_{a \in \mathscr{E}} \dot{\bar{\mathcal{T}}}^a(t) p(a, \bar{Q}(t)),$$

and the proof is complete.   □

12.4 Further Stability Results under Maximum Pressure Policy

12.4.1 *Parametrization of Maximum Pressure Policies*

We defined pressure by $p(a, Q(t))$. However, all the results proved so far in this chapter continue to hold if we use a parametrized form of pressure, with pressure defined as $p(a, \tilde{Q}(t))$, where $\tilde{Q}_k(t) = \gamma_k Q_k(t) + \theta_k$, where $\gamma_k > 0$, and θ_k are real. The proofs are similar.

12.4.2 Fluid Stability When $\rho < 1$.

We extend the result of Theorem 12.6 to show that if $\rho < 1$, then not just weak fluid stability holds, but also full fluid stability.

Theorem 12.8 *Assume that a processing network has EAA property and that $\rho < 1$. Assume in addition that there exists $x \geq 0$ for which $Rx > 0$. Then under a maximum pressure policy that allows processor splitting and preemption, the fluid model of the network is stable, i.e. there exists δ such that starting from any fluid $\bar{Q}(0)$ we will have $\bar{Q}(t) = 0$, $t > |Q(0)|/\delta$.*

Remark The assumption that there exists $x \geq 0$ for which $Rx > 0$ that we need here, comes instead of the assumption that P has spectral radius < 1, which we made for MCQN (see Exercise 12.5).

Proof Let $(\tilde{x}, \rho)$ be a solution of the static planning problem (12.7) with $\rho < 1$. Let $\hat{x}$ satisfy $R\hat{x} > 0$. For any input activity j, $C_{j,k} = 0$, $k = 1, \ldots, K$, and therefore $R_{k,j} \leq 0$, so we can take $\hat{x}_j = 0$ and still have $R\hat{x} > 0$. We can also scale all of $\hat{x}$ so that for all service processors i, $\sum_{j \in \mathcal{J}} \mathcal{A}_{i,j} \hat{x}_j \leq 1 - \rho$. Define $x^* = \tilde{x} + \hat{x}$. Then, by our choice of $\hat{x}$, we have $x^* \in \mathcal{A}$. Furthermore, $Rx^* = R\tilde{x} + R\hat{x} = R\hat{x} \geq \delta \mathbf{1}$, where $\delta = \min_k (R\hat{x})_k > 0$. We now have

$$R\dot{\bar{T}}(t) \cdot \bar{Q}(t) = \max_{a \in \mathcal{A}} Ra \cdot \bar{Q}(t) \quad \text{by Lemma 12.7,}$$

$$\geq Rx^* \cdot \bar{Q}(t) \quad \text{by } x^* \in \mathcal{A}$$

$$\geq \delta \mathbf{1} \cdot \bar{Q}(t) \quad \text{by } Rx^* \geq \delta \mathbf{1}$$

$$\geq \delta \|\bar{Q}(t)\| \quad \text{where } \|\bar{Q}(t)\| \text{ is the Euclidean norm.}$$

With $f(t) = \|\bar{Q}(t)\|^2$, we now have

$$\dot{f}(t) = 2\dot{\bar{Q}}(t) \cdot \bar{Q}(t) = -2R\dot{\bar{T}}(t) \cdot \bar{Q}(t) \leq -2\delta \|\bar{Q}(t)\| = -2\delta \sqrt{f(t)},$$

i.e. $\frac{d}{dt} \sqrt{f(t)} \leq -\delta$. It follows that $Q(t) = 0$ for $t > \|\bar{Q}(0)\|/\delta$, and therefore for $t > |\bar{Q}(0)|/\delta$. $\qquad\qquad\square$

12.4.3 Rate Stability of Processing Networks under Non-Splitting Maximum Pressure Policies

Let $\mathcal{N} \subseteq \mathcal{A}$ denote all the allocation vectors that allow only $a_j \in \{0, 1\}$, i.e. allocations that do not use processor splitting; note that $\mathcal{N} \subseteq \mathcal{E}$ (see Exercise 12.6). Let $\mathcal{N}(t)$ be the non-splitting allocations available in state $Q(t)$. Maximum pressure policies that do not allow processor splitting are defined as using an allocation that is in $\arg\max_{a \in \mathcal{N}(t)} p(a, Q(t))$. It can be seen that the standard EAA assumption, and the necessary condition of $\rho \leq 1$ are not sufficient for rate stability. See Exercise 12.7 for an example of a network that is stable when processor splitting is allowed, but is unstable if splitting is not allowed. It is possible to modify the EAA assumption, and reformulate the static planning problem for non-splitting policies, so that rate stability can again be obtained under the stricter modified EAA and $\rho \leq 1$ assumptions.

A special case for which no modification of Theorem 12.5 is needed, is when $\mathcal{N} = \mathcal{E}$. This will be the case in particular if the network is a *reverse Leontief network*:

Definition 12.9   A processing network is a reverse Leontief network if each activity is using only a single processor, i.e. each column of the matrix A has only a single 1 entry.

Proposition 12.10   *In a reverse Leontief network, all extreme allocations satisfy $a_j \in \{0, 1\}$, i.e. $\mathcal{N} = \mathcal{E}$.*

For a proof, see Exercise 12.8. The following corollary is immediate

Corollary 12.11   *For a reverse Leontief network, assume that the network has EAA property and that $\rho \leq 1$. Then under a maximum pressure policy that allows preemption but does not allow processor splitting, the processing network is rate stable, i.e. $\lim_{t \to \infty} Q(t)/t = 0$ almost surely.*

12.4.4 Rate Stability of Processing Networks under Non-Preemptive Maximum Pressure Policies

If splitting processors is not allowed and preemption is not allowed, then there may be states in which no extreme allocation exists (see Exercise 12.7), and in that case maximum pressure is not defined. Even if maximum pressure policies exist, there are examples where EAA holds and $\rho < 1$, yet maximum pressure that allows processor splitting but does not allow preemptions diverges; see Exercise 12.9 for an example. It is however possible to prove (we do not give the proof here):

Theorem 12.12   *Assume that a processing network is reverse Leontief and has EAA property and that $\rho \leq 1$. Then under a maximum pressure policy that does not allow processor splitting and does not allow preemptions, the processing network is rate stable, i.e. $\lim_{t \to \infty} Q(t)/t = 0$ almost surely.*

12.5 Applications

12.5.1 Multi-Class Networks and Unitary Networks

MCQN are a special case of processing networks, in which each buffer has a single activity and a single processor. It is therefore strictly Leontief as well as reverse Leontief. More generally, we define

Definition 12.13 A processing network is unitary if it is both strictly Leontief and reverse Leontief, i.e. each activity j has a single processor $s(j)$ that it is using, and a single buffer $k(j)$ that it is processing. All columns of A and all rows of C are unit vectors.

In unitary networks, the input buffer 0 may have several input activities, each of them using a single resource, but each input resource may be used by several activities. Each service buffer may be served by several different activities, which are using different resources. So input rates are controlled by the choice of input activities, and routing is controlled by service activities, and these choices determine resource usage. The static planning problem then finds the best level of utilization, so as to minimize the maximal workload on the resources. We would expect that optimal solutions of the static planning problem will tend to equalize the workloads of the various processors. We would also expect for given parameters μ_j, $p^j_{k,l}$ that if ρ is minimized, the actual waiting times will be improved (according to the heuristic value of $1/(1 - \rho)$). Alternatively, we could increase input rates, μ_j for input activities j to get higher ρ and higher utilization. Again, these networks are both strictly Leontief and reverse Leontief. By Theorems 11.8, 12.8, and 12.12 we can therefore state:

Proposition 12.14 *Assume that the solution of the static planning problem (12.7) for a unitary processing network has $\rho < 1$. Assume that every bounded set of states B_κ is petite (uniformly small). Then under maximum pressure policy, the network is stable (ergodic).*

We now investigate how maximum pressure is implemented in unitary networks, and in particular in MCQN.

We note that the constraints of $a \in \mathscr{A}$ are separate for each resource: They require $a_j \geq 0$ and $\sum_{j \in \mathcal{J}} A_{i,j} a_j \leq 1$ for each service processor, and they require $a_j \geq 0$ and $\sum_{j \in \mathcal{J}} A_{i,j} a_j = 1$ for each input processor.

For input processor i, the choice is:

$$\arg\max_{j : A_{i,j}=1} \mu_j \left(-\sum_{l=1}^{K} p^j_{0,l} Q_l(t) \right) \tag{12.11}$$

where the activity with the least negative value is chosen.

For service processor i, the choice is:

$$\arg\max_{j:A_{i,j}=1} \left\{0, \mu_j\left(Q_{k(j)}(t) - \sum_{l=1}^{K} p^j_{k(j),l}Q_l(t)\right)\right\}. \tag{12.12}$$

Pressure is a function of the queue lengths, multiplied by appropriate values of rates. Maximum pressure makes choices as follows: For input activities, each input resource will direct input to the shorter queues. For service activities, each resource will choose an activity where the difference between the queue length at the buffer processed by the activity and at the buffers directly downstream from it is small. Priority is given to buffers with long queues, or to buffers where downstream buffers may face starvation. On the other hand, processing is withheld from buffers that are close to empty, or from buffers where the downstream buffers are congested. Note that input resources never idle, while service resources will idle if pressure for each available activity is negative.

A big advantage of this policy is that it is almost local: At each station, we need to examine only the states of the queues at the local buffers and at the buffers that are directly downstream. This allows decentralized control at each resource (machine). Furthermore, the policy does not require knowledge about the input rates.

Exercise 12.10 presents a system where under HOL policy of join shortest queue, stability depends on the distribution of the processing time, and a round-robin policy that is stable needs to be adjusted according to the arrival rates, while maximum pressure is stable whenever $\rho < 1$.

12.5.2 Input Queued Switches under Maximum Weight Policy

Communication networks are controlled by switches that direct the movement of data packets between terminals. An input queued switch has N input ports and N output ports, and operates in discrete time slots, where time slot r lasts from $r - 1$ to r. Packets arrive at the input ports at the beginning of a time slot, and are moved to output ports by the end of the time slot, from where they continue on their routes. In our model, input ports have infinite space, and store items in virtual output queues, with $Q_{i,j}(r)$ the number of items in input port i destined for output port j, at the start of time slot r. The cumulative number of arrivals by the beginning of time slot r, at input port i destined for outport port j, is $\mathcal{A}_{i,j}(r)$. We assume that arrivals at input port i destined for output port j are jointly stationary ergodic sequences $\mathcal{A}_{i,j}(r)$ that satisfy SLLN:

$$\lim_{r\to\infty}\frac{1}{r}\mathcal{A}_{i,j}(r) = \lambda_{i,j} \quad \text{a.s.} \tag{12.13}$$

In each time slot there is a matching of input to output ports, given by a permutation matrix π (with a single 1 in each row and in each column, and 0's elsewhere), and a single item is then moved from input port i to output port j, if $\pi_{i,j}Q_{i,j}(r) > 0$, simultaneously from all input ports. We denote by Π the set of all $N!$ permutation matrices.

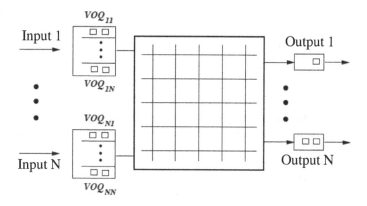

Figure 12.2 Input queued switch, reproduced with permission from Dai and Prabhakar (2000).

A matching algorithm m will determine a sequence $\pi^m(r)$ for time slots $r = 1, 2 \ldots$. Let $\mathcal{D}_{i,j}(r)$ be the cumulative number of packets that moves to output port j from input port i by the end of time slot r. Under matching algorithm m, the switch is rate stable (see Definition 2.2) if $\lim_{r\to\infty} Q_{i,j}(r)/r = 0$ a.s., or equivalently, if $\lim_{r\to\infty} \mathcal{D}_{i,j}(r)/r = \lambda_{i,j}$ a.s.

We say that a *matching algorithm is efficient* if the switch is rate stable under any set of arrival rates that satisfy:

$$\sum_i \lambda_{i,j} \le 1, \quad \sum_j \lambda_{i,j} \le 1. \tag{12.14}$$

Clearly, these conditions are necessary for stability.

Note We require that the algorithm will be independent of the input rates, and will be rate stable for all rates that satisfy (12.14).

We define weight (pressure) by

$$f_\pi(r) = \langle \pi, Q(r) \rangle = \sum_{i,j} \pi_{i,j} Q_{i,j}(r).$$ (12.15)

Definition 12.15   Maximum weight policy is using, at state $Q(r)$, a matching $\pi \in \arg\max_{\pi \in \Pi} f_\pi(r)$.

Proposition 12.16   *Maximum weight (maximum pressure) policy is efficient.*

Proof   It can be seen that the input queued switch can be modeled as a processing network, and that the maximum weight policy is in fact a maximum pressure policy for this model; see Exercise 12.11. One can then verify that conditions (12.14) are exactly the conditions that the static planning problem (12.7) has solution $\rho \le 1$; see Exercise 12.12. Furthermore, the network is strictly Leontief, so EAA assumption holds. It is then seen that for this model, $\mathcal{E} = \mathcal{N}$; see Exercise 12.13. Therefore the policy does not require processor splitting. Furthermore, since the model is in discrete time, it does not involve preemptions. It follows that maximum weight policy is indeed efficient.   □

To implement maximum weight policy for the switch, one needs to solve at every time slot the problem $\max_{\pi \in \Pi} f_\pi(r)$. Because there are $N!$ permutations, this looks like a hard problem, but in fact it is *an assignment problem* that can be solved efficiently. One algorithm to solve it is the so-called Hungarian method, which is polynomial. It can also be solved by the simplex method for transportation problems. Since the assignment problem needs to be solved in each time slot, the optimal solution at the previous time slot should be used to solve for the next time slot more efficiently; see Exercise 12.14.

12.5.3 Networks of Data Switches

We now consider a network of input queued switches. Such networks of switches form the backbone of communication systems, where flows of data, consisting of data packets, follow a route through a sequence of switches. Figure 12.3, from Dai and Lin (2005), shows a network of three switches, each with two input and two output ports, and three flows through the network. We use c to denote a specific flow, l_1 to denote input port, and l_2 to denote output port. Each flow goes through a sequence of switches, using a particular pair of input and output ports (l_1, l_2) at each switch, and

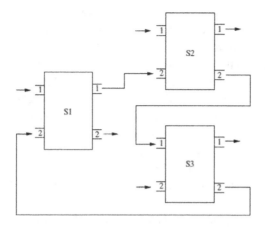

Figure 12.3 A network of input queued switches, reproduced with permission from Dai and Lin (2005).

then continuing to another switch along the route; we assume that each route passes a switch at most one time (this is a realistic assumption in a communication network, and is important in the following proof). We then have queues at the input ports, and we label by (c, l_1, l_2) the queue of packets of flow c queued at input port l_1 and destined for outport port l_2 of the switch. Let

$$H(c, l_1, l_2) = \begin{cases} 1 & \text{if flow } c \text{ uses pair } (l_1, l_2) \text{ of one of the switches,} \\ 0 & \text{otherwise.} \end{cases}$$

(12.16)

At every time slot in each of the switches a matching is established. Then for each matched pair of input and output ports a flow that is using that pair is chosen, and a single packet from that flow is transmitted (if the buffer is not empty). The packet is then queued up at the next input port on the route of that flow.

We assume that flow c has a stationary ergodic arrival stream with rate of α_c packets per time slot. We then have the following capacity constraints of the switches:

$$\sum_{c, l_2} \alpha_c H(c, l_1, l_2) \leq 1 \quad \text{for every input port } l_1,$$

$$\sum_{c, l_1} \alpha_c H(c, l_1, l_2) \leq 1 \quad \text{for every output port } l_2. \quad (12.17)$$

We can model this network of switches as a processing network, as follows: There is a single input buffer, and one internal buffer for each $k = (c, l_1, l_2)$, for a total of L internal buffers. There is one input processor and one input activity per flow c, for a total of C input flows. Each input activity is active all the time, and provides a stationary and ergodic sequence of packets into the system at rate α_c for every time slot. Service processors include the input ports and the output ports of all the switches. There is one service activity per internal buffer, and the service activity $j = k$ for buffer $k = (c, l_1, l_2)$ is using two service processors, input processor l_1 and output processor l_2. Each application of a service activity takes one time slot, and will move one packet from buffer k (if it is not empty) to a buffer k', if buffer k' is the next switch of the route of flow c, or out, if buffer k is the last switch on flow c.

The input-output matrix (12.5) for this network is an $L \times (C + L)$ matrix, with C the number of flows, and L the number of internal buffers, and

$$R_{k,j} = \begin{cases} -\alpha_c & \text{if } j \text{ is the input activity and } k \text{ is the first buffer of flow } c, \\ 1 & \text{if } j \text{ is service activity for buffer } k, \text{ i.e. } j = k = (c, l_1, l_2), \\ -1 & \text{if } j = k', \text{ and buffer } k \text{ follows buffer } k' \text{ on flow } c. \end{cases}$$

(12.18)

Similar to the analysis of the single switch, it can be shown (see Exercise 12.15) that the conditions (12.17) are exactly the conditions that the static planning problem 12.7 has a solution $\rho \le 1$. We have:

Proposition 12.17   *If (12.17) hold, then maximum pressure is rate stable for the network of switches; i.e. maximum pressure is efficient.*

The proof is left as Exercise 12.16.
We now give an exact description of the maximum pressure policy.

Proposition 12.18   *Maximum pressure policy for a network of input queued switches is solved by using maximum weight separately for each switch. The weights for permutation π are calculated as $\sum_{i,j} \pi_{i,j} Z_{i,j}$ where $Z_{i,j}$ is calculated as follows: Given input-output pair (i, j) at the switch, consider the queue (c, i, j), $c \in C$ of various flows that use (i, j). Let $Q_{(c,i,j)}(r)$ be the queue at that buffer. Let (c, k, l) be the next step of flow c after (c, i, j), and let $Q_{(c,k,l)}(r)$ be the queue at that buffer. If (c, i, j) is last on this flow, use 0 for the queue length of the next step. Then:*

$$Z_{i,j}(r) = \max_c \left(Q_{(c,i,j)}(r) - Q_{(c,k,l)}(r) \right).$$

The proof is left as Exercise 12.17.

12.6 Sources

A formulation of stochastic processing networks, involving activities that need several processors, and that serve several classes simultaneously, with comprehensive discussion is given in the papers of Harrison and Van Mieghem (1997) and Harrison (2000, 2002, 2003). Maximum pressure policies originated under the name of back pressure policies in an important paper by Tassiulas and Ephremides (1992), followed by Tassiulas (1995). Properties of maximum pressure policies are investigated in the pioneering paper of Stolyar (2004), where he defined max-weight scheduling of a generalized switch and showed its optimality under complete state space collapse. Our presentation here is based on Dai and Lin (2005), and several counter examples mentioned in the text and described in the exercises are also from that paper. The use of maximum pressure for input queued data switches comes from Dai and Prabhakar (2000). Some further research on input queued switches operating under maximum pressure (max-weight) policy, including fluid and diffusion approximations, state space collapse, and calculation of expected queue lengths, is derived by Shah and Wischik (2012), Kang and Williams (2012), and Maguluri and Srikant (2016). The extension to networks of switches is from Dai and Lin (2005). Asymptotic optimality of maximum pressure policies for networks with complete state apace collapse is derived in Dai and Lin (2008).

Exercises

12.1 Show that there is always a feasible extreme allocation, i.e. if $\mathscr{A} \neq \emptyset$, then $\mathscr{E}(t)$ is not empty at any t.

12.2 Show that $\mathscr{E}(t)$ defined as the extreme points of the available allocations $\mathscr{A}(t)$, are also extreme points of all allocations, $\mathscr{A}$, i.e. $\mathscr{E}(t) \subseteq \mathscr{E}$.

12.3 Formulate and verify the analog of Theorems 11.10, 11.12 for processing networks.

12.4 Show that the following processing network (taken from Dai and Lin (2005)) with the following data has $\rho < 1$, and is stable under some allocation policy, but it is unstable under maximum pressure policy, because it does not satisfy EAA assumptions.

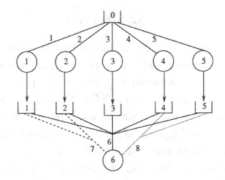

In this network, circles are processors, open boxes are buffers, and lines represent activities. Activities 1–5 are input activities, each with its own input processor. Processor 6 is a service processor with activities 6,7,8. Activity 6 processes an item from each buffer, activity 7 (activity 8) processes an item from buffers 1,2 (buffers 4,5). Each processed item leaves the system. Activity durations are deterministic:

$$v_1(\ell), v_2(\ell), v_3(\ell) = 2, \ \ell \geq 1, \quad v_4(1), v_5(1) = 1, \ v_4(\ell), v_5(\ell) = 2, \ \ell \geq 2,$$

$$v_6(\ell), v_7(\ell), v_8(\ell) = 1, \ \ell \geq 1, \quad \mu_i = \begin{cases} 0.5, & i = 1 \ldots, 5, \\ 1, & i = 6, 7, 8. \end{cases}$$

12.5 Show that in a MCQN, the following conditions are equivalent:
- There exists $x \geq 0$ such that $Rx > 0$ in all components.
- The routing matrix P has spectral radius < 1.

12.6 Show that the set of $\{0, 1\}$ allocations is composed only of extreme points of $\mathscr{A}$, i.e. $\mathscr{N} \subseteq \mathscr{E}$.

12.7 Show that the following processing network (taken from Dai and Lin (2005)) with the following data has $\rho < 1$ and is stable under maximum pressure policy that allows processor splitting and preemptions, but is not stable under non-splitting allocations.

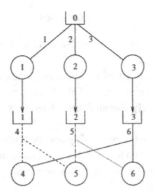

In this network, circles are processors, open boxes are buffers, and lines represent activities. Activities 1–3 are input activities each with its own input processor. Activities 4, 5, 6 are service activities, each uses two processors, in order to process one item out of buffers 1, 2, 3. The input activities have processing rates $\mu_1 = \mu_2 = \mu_3 = 0.4$; the service activities have processing rates $\mu_4 = \mu_5 = \mu_6 = 1$.

12.8 A processing network is reverse Leontief if every activity is using only a single processor. Show that for reverse Leontief networks, all extreme allocations are integer, i.e. processors are not split, with each allocation $a_j \in \{0, 1\}$, $j = 1, \ldots, J$.

12.9 Consider the following processing network (taken from Dai and Lin (2005)) with the following data:

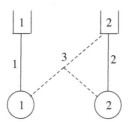

In this network input processors are not included, circles are service processors, open boxes are buffers, and lines represent activities. Arrivals to buffer 1 occur at times $0.5 + 2n$, $n = 1, 2, \ldots$, Arrivals to buffer 2 occur at times $1 + 1.5n$, $n = 1, 2, \ldots$. Activity 1 uses processor 1 and serves buffer 1, with $m_1 = 1$. Activity 2 uses processor 2 and serves buffer 2, with $m_2 = 2$. Activity 3 uses both processors, and serves buffer 2, with $m_3 = 1$. All processing times are deterministic, and the system is empty at time 0.

Solve the static planning problem for this network to calculate $\rho < 1$, explain why EAA holds. Show that using processor splitting non-preemptive maximum pressure policy the system diverges. Describe how the system operates under preemptive processor splitting policy.

12.10 Consider the following network (taken from Dai and Lin (2005)). It is modeled as a processing network with three service buffers, each with its own processor and service activity, and with four input processors, one of which has two input activities, which provide discretionary routing to buffers 1 or 2. All input processors have exponential service times (Poisson inputs) with the rates indicated in the figure. Processing for the service activities are rate 1, the processing times of servers 2, 3 are exponential, while processing time of buffer 1 is generally distributed. Explain or show the following properties:

(i) Under HOL policy, and join the shortest queue for the discretionary routing, the system is unstable if buffer 1 processing times are exponential, but it is stable if it is hyper-exponential with large enough variability.

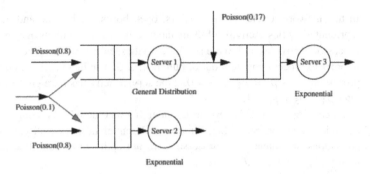

(ii) It is stable under round-robin policy that directs a fraction of 90% of the discretionary input to buffer 1, but if the input rates are changed slightly, we will need to change the fraction routed to buffer 1.

(iii) Formulate the static planning problem and show that $\rho < 1$. Find how maximum pressure is implemented here and check that it is stable whenever $\rho < 1$.

12.11 Model the input queued switch as a processing network, and show that the maximum weight policy is in fact maximum pressure for this processing network.

12.12 Show that the conditions (12.14) are exactly the conditions that the static planning problem (12.7) has a solution $\rho \le 1$.

12.13 Show that for the processing network modeling the input queued switch, $\mathscr{E} = \mathscr{N}$.

12.14 (*) Discuss solution of the pressure maximization as an assignment problem, using the Hungarian method, and using the transportation simplex algorithm, and examine their online implementation with efficient use of the previous slot solution for the next slot solution.

12.15 Show that for a network of switches, the conditions (12.17) are exactly the conditions under which the static planning problem (12.7) is feasible with $\rho \le 1$.

12.16 Prove Proposition 12.17.

12.17 Prove Proposition 12.18.

13

Processing Networks with Infinite Virtual Queues

13.1 Introduction, Motivation, and Dynamics

Why is a single-server queue unstable with $\rho = 1$? The answer is that unless the arrival and services are deterministic, in any stable system the variability makes the server idle a fraction of the time, and that rules out stability if $\rho = 1$. However, real systems do often manage to achieve full utilization of the server, by the following device: The queue is monitored, and before the queue becomes empty it is replenished. From the point of view of the server, it is serving an infinite queue, even if in reality the queue can be kept finite at all times. We call such a monitored queue *an infinite virtual queue*, IVQ. A single server serving an IVQ can be fully utilized, and work at load $\rho = 1$. Consider now a tandem of two servers. If the initial queue is an IVQ, the first server will be fully utilized. However, the second server, serving the output of the first queue, will need to have $\rho < 1$. We can improve its utilization by adding another IVQ that provides work to the second server. In that case the second server may give priority to the output from the first queue, but when the queue fed by the first server is empty, the second server will serve his own IVQ. A further example is the so-called push-pull network which is the IVQ version of the KSRS network of Section 10.2.3. Such networks with IVQs are illustrated in Figure 13.1.

Infinite virtual queues provide a further generalization of processing networks. In the processing networks of Chapter 12 the network had a single

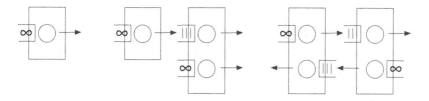

Figure 13.1 Networks with IVQs.

buffer 0 that modeled the outside world and provided virtual infinite input, and we separated input processors and activities from service processors and activities. We now allow several buffers that have infinite virtual supply, and we no longer distinguish between input and service activities. In this new formulation, we no longer have a distinguished buffer 0: instead, the buffers (queues, classes) $\mathcal{K} = \{1, \ldots, K\}$ are partitioned into a subset $\mathcal{K}_\infty$ that includes all the IVQs, and a subset $\mathcal{K}_0$ that includes the standard queues. As before, we have the resource consumption $I \times J$ matrix A, where A_{ij} is the indicator that activity j requires resource i, and the $J \times K$ constituency matrix C where C_{jk} is the indicator that activity j is processing buffer k.

The rth activation of activity j will engage all the resources $\{i : A_{i,j} = 1\}$, and will process one item out of each of the buffers $\{k : C_{jk} = 1\}$, and it will last a time $v_j(r)$, at the end of which each of the processed items will either leave the system, or move into one of the standard buffers, with $\xi^j_{k,l}(r)$ indicating that the item from buffer k moved to buffer l in this rth activation of activity j. As before, we let $\mathcal{S}_j(t)$ count the number of service completions by activity j over service of total duration t, and $\mathcal{R}^j_{k,l}(r)$ the number of items moved from buffer k to buffer l in the first r activations of activity j.

For the standard queues, we then have as before in equation (12.1):

$$Q_k(t) = Q_k(0) - \sum_{j:k \in C_j} \mathcal{S}_j(T_j(t)) + \sum_{l \neq k} \sum_{j:l \in C_j} \mathcal{R}^j_{lk}(\mathcal{S}_j(T_j(t))) \geq 0, \quad k \in \mathcal{K}_0.$$

(13.1)

For the IVQ buffers, there is no true number of items in queue. Instead, we will use a level relative to some (arbitrarily chosen) initial level $Q_k(0)$. We note that we do not count items moving into any of the IVQs, since items that complete processing either move into a standard buffer or leave the system. Instead of feedback from other buffers, we assign a *nominal input rate* α_k into each of the IVQs, $k \in \mathcal{K}_\infty$. Using the initial level and the nominal input rate, the dynamics of the IVQs are then given by:

$$Q_k(t) = Q_k(0) - \sum_{j:k \in C_j} \mathcal{S}_j(T_j(t)) + \alpha_k t, \quad k \in \mathcal{K}_\infty. \qquad (13.2)$$

The notable distinction between standard queues and IVQs is that while the former are constrained to be non-negative, the IVQs are not restricted in sign and can have negative levels. The level $Q_k(t)$ for an IVQ measures the deviation from the nominal input $\alpha_k t$. We refer to it as input, since it is the rate at which the IVQ feeds items into the network of standard queues.

In addition, at any moment in time the allocation of effort to activity j, denoted by a_j, needs to satisfy, for every resource i, similar to (12.2),

$$\sum_{j \in \mathcal{J}} A_{i,j} a_j \le 1, \qquad a_j \ge 0, \tag{13.3}$$

where $0 \le a_j \le 1$ if processor splitting is allowed, and $a_j \in \{0, 1\}$ if processor splitting is not allowed.

Our minimal probabilistic assumptions on processing networks with IVQ are that $\mathcal{S}, \mathcal{R}$ obey SLLN, i.e. almost surely when $n, t \to \infty$

$$\mathcal{S}_j(t)/t \to \mu_j = m_j^{-1}, \quad \mathcal{R}_{k,l}^j(n)/n \to p_{k,l}^j. \tag{13.4}$$

In a special case of processing networks with IVQs each activity uses a single processor, and is processing items from a single buffer. These networks correspond to strict Leontief and reversed Leontief as defined in the last chapter, and we refer to them as unitary. Such unitary networks generalize MCQN, by allowing discretionary routing, and admission control, as in the previous chapter. In addition, we now have that whenever for processor i there exists $k \in \mathcal{K}_\infty$ such that $A_{i,j} C_{j,k} = 1$ for some j, processor i can always be kept busy, and may therefore be fully utilized. We refer to these as MCQN-IVQ networks.

In the next section, we will consider MCQN-IVQ that have Poisson service processes and can be described by countable Markov chains, and derive their properties. Next, we will study fluid scaling of processing networks with IVQ, and formulate a static production planning problem. Finally, we will study another example that illustrates stationary control of systems with full utilization, i.e. working with $\rho = 1$.

13.2 Some Countable Markovian MCQN-IVQ

13.2.1 A two-node Jackson Network with Infinite Supply of Work

Consider the following network, described in Figure 13.2, with two workstations, two IVQs and two standard queues. Processing of an item at station i is exponential with rate μ_i, and customers leaving station i move to the standard queue in station $3 - i$ with probability p_i and leave the system with probability $1 - p_i$. The policy is that station i gives preemptive priority to the standard queue, and otherwise serves a job from the IVQ. Clearly, both servers are fully utilized, and the output from the system consists of two independent Poisson streams of rates $\mu_i(1 - p_i)$, $i = 1, 2$. A condition for stability is that $\rho_i = \frac{\mu_{3-i} p_{3-i}}{\mu_i} < 1$, $i = 1, 2$. The population of customers in the standard queues can be described by a two-dimensional random walk on

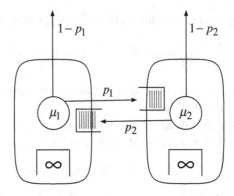

Figure 13.2 A network with two nodes and with IVQs.

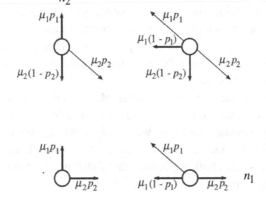

Figure 13.3 Transition rates for the two-node network with IVQs.

the non-negative quadrant, $(Q_1(t), Q_2(t)) = (n_1, n_2)$, with transition rates given in Figure 13.3. The stationary distribution of the queue lengths in the two standard queues can be obtained by the compensation method, which we will describe in more detail in Section 15.1. It is given by an infinite sum of product-form terms. For $(n_1, n_2) \neq (0, 0)$

$$P(n_1, n_2) = \sum_{k=1}^{\infty} (-1)^{k+1} \Big[(1 - \alpha_k)\alpha_k^{n_1}(1 - \beta_{k+1})\beta_{k+1}^{n_2}$$

$$+ (1 - \alpha_{k+1})\alpha_{k+1}^{n_1}(1 - \beta_k)\beta_k^{n_2} \Big],$$

where for $k \geq 1$,

$$\alpha_{k+1}^{-1} = \frac{\mu_1 + \mu_2}{\mu_2 p_2} \beta_k^{-1} - \alpha_{k-1}^{-1} - \frac{1 - p_2}{p_2},$$

$$\beta_{k+1}^{-1} = \frac{\mu_1 + \mu_2}{\mu_1 p_1} \alpha_k^{-1} - \beta_{k-1}^{-1} - \frac{1 - p_1}{p_1},$$

with initially $\alpha_0 = \beta_0 = 1$, $\alpha_1 = p_1$, $\beta_1 = p_2$, and the steady-state probability $P(0, 0)$ is equal to:

$$P(0, 0) = 1 - p_1 - p_2 + \sum_{k=1}^{\infty} (-1)^{k+1} \left(\alpha_k \beta_{k+1} + \alpha_{k+1} \beta_k \right).$$

See Exercises 13.1, 13.2.

This model where the two nodes are giving foreground and background service, can be generalized to general Jackson networks, with some nodes that have IVQs in addition to standard queues, see Exercise 13.3.

13.2.2 A Three-Buffer Re-Entrant Line with Infinite Supply of Work

Consider the three-buffer re-entrant line, with infinite supply of work, where processing times are exponential, and the policy is preemptive priority to buffer 3 over the IVQ buffer 1; see Figure 13.4. Under full utilization of workstation 1 the system is stable if $m_1 + m_3 > m_2$, where $m_i = \mu_i^{-1}$. The system is described by a two-dimensional random walk on the non-negative quadrant, with state $(Q_2(t), Q_3(t)) = (n_2, n_3)$. The transition rates are also shown in Figure 13.4. The stationary distribution for this system is (see

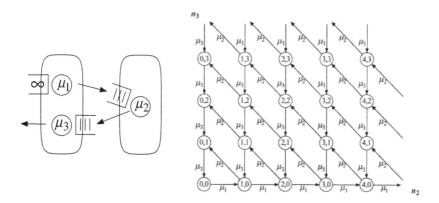

Figure 13.4 A three-buffer re-entrant line with IVQ, and its transition rates.

Exercise 13.4).

$$\pi(n_2, n_3) = \begin{cases} \dfrac{m_1}{m_1 + m_3} (1 - \alpha_2)\alpha_2^{n_2}\alpha_3^{n_3}, & \begin{array}{l} n_2 > 0 \text{ and } n_3 \geq 0 \\ \text{or } (n_2, n_3) = (0, 0), \end{array} \\[3ex] \dfrac{m_3}{m_1 + m_3} (1 - \alpha_2)\alpha_3^{n_3 - 1}, & n_2 = 0 \text{ and } n_3 > 0, \end{cases}$$

$$(13.5)$$

where:

$$\alpha_3 = \frac{\mu_1 + \mu_2 + \mu_3 - \sqrt{(\mu_1 + \mu_2 + \mu_3)^2 - 4\mu_1\mu_3}}{2\mu_3},$$

$$\alpha_2 = \frac{\mu_1}{\mu_2} \frac{-\mu_1 - \mu_2 + \mu_3 + \sqrt{(\mu_1 + \mu_2 + \mu_3)^2 - 4\mu_1\mu_3}}{2\mu_3}.$$

13.2.3 The Countable Markovian Push-Pull Network

Consider the following network, with two workstations and two customer routes, each starting at one of the stations from an IVQ, and moving to the other station, to queue up for a second service, in a standard queue. The network is described in Figure 13.5; it is an IVQ version of the KSRS network of Section 10.2.3. We assume processing times are exponential with rates indicated in the figure. Under full utilization, each station can either provide new work for the other station from the IVQ, which we call a *push activity*, or process its own queue to complete a job, which we call this a *pull activity*. If we assume full utilization and rate stability, then the throughputs of the two streams solve the following traffic equation:

$$\alpha_1 = \theta_1\lambda_1 = (1 - \theta_2)\mu_1, \qquad \alpha_2 = \theta_2\lambda_2 = (1 - \theta_1)\mu_2,$$

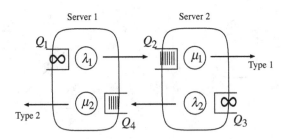

Figure 13.5 A push-pull queueing network.

where θ_i is the fraction of time that station i works on its IVQ. From this, one obtains the throughput rates

$$\alpha_1 = \frac{\mu_1 \lambda_1 (\mu_2 - \lambda_2)}{\mu_1 \mu_2 - \lambda_1 \lambda_2}, \qquad \alpha_2 = \frac{\mu_2 \lambda_2 (\mu_1 - \lambda_1)}{\mu_1 \mu_2 - \lambda_1 \lambda_2},$$

and these are positive if either $\mu_i > \lambda_i, i = 1, 2$, the *inherently stable case*, or $\mu_i < \lambda_i, i = 1, 2$, the *inherently unstable case*. In the remaining combinations of parameter values, the stations are not balanced, and if one works at full utilization, the other must be either under utilized or over utilized.

Consider the policy of preemptive priority to pull activities. Under this policy, states where both standard queues are not empty are transient, and so the recurrent states need to satisfy $Q_2(t) \cdot Q_4(t) = 0$, and at those times that $Q_4(t) = 0$ both stations work on stream 1, and $Q_2(t)$ behaves like an M/M/1 queue, with arrival rate λ_1 and service rate μ_1 with similar behavior for $Q_4(t)$ when $Q_2(t) = 0$. Hence, in the inherently stable case, preemptive priority to pull activities is stable, and the process switches between busy periods of the top stream M/M/1 queue, and of the bottom stream M/M/1 queue (see Figure 13.6 and Exercise 13.5. *This policy is unstable if $\lambda_i > \mu_i, i = 1,2$, i.e. in the inherently unstable case.*

In the inherently unstable case, the following policy is stable: Choose integers s_i such that $\frac{\lambda_{3-i}}{\mu_{3-i}} \left(\frac{\mu_i}{\lambda_i} \right)^{s_i} < 1, i = 1, 2$. Then use the following

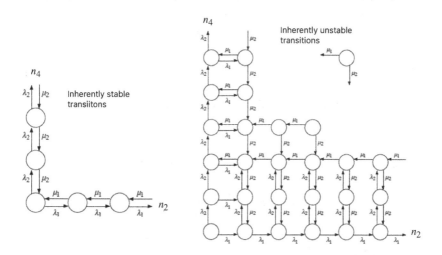

Figure 13.6 Transition rates, for the inherently stable and the inherently unstable case.

actions according to state $(Q_2(t), Q_4(t)) = (n_1, n_2)$ (the transition rates for this policy and for the policy in the inherently stable case are illustrated in Figure 13.6):

(i) In states where $(n_1 < s_1, n_2 < s_2)$. use push activities in both stations. These states will be transient.

(ii) In states where $(n_1 > s_1, n_2 > s_2)$, use pull activities in both stations. These states will also be transient.

(iii) In states where $n_1 = 0$ or when $n_2 = 0$, use push activities in both stations. In states where $n_1 = s_1$ or when $n_2 = s_2$, use pull activities in both stations.

(iv) In states where $0 < n_i < s_i$ and $n_{3-i} \geq s_{3-i}$, work on stream i only, using push and pull activity.

Another way of describing this policy for the inherently unstable system is as follows: Fix thresholds s_1, s_2. At all times use full utilization. Do not serve Q_4 while $n_1 < s_1$, and do not serve Q_2 while $n_2 < s_2$. At all other times, use pull priority (see Exercise 13.6).

The effect of this policy is that the process alternates between "busy periods" in two corridors: When it is in the interior of corridor $0 \leq n_i \leq s_i$ it moves toward the edges, $n_i = 0$ or $n_i = s_i$. At $n_i = s_i$ it may move toward the origin; at $n_i = 0$ it may move away from the origin. Because $\lambda_i > \mu_i$, it will emerge from the interior of the corridor more frequently on the side of $n_i = s_i$ and move back to the origin, which will guarantee stability.

One can write explicit formulas for the stationary distribution of the Markov chain for this policy; see Exercise 13.7. Another policy is suggested in Exercise 13.8, and the two policies are compared in Exercise 13.9.

13.3 Fluid Models of Processing Networks with IVQs

We now study fluid limits of processing networks with IVQs. We will again consider a sequence of systems, differing in their initial values. The stochastic system dynamics for the sequence of systems are then

$$Q_k^n(t) = Q_k^n(0) - \sum_{j: k \in C_j} S_j(T_j^n(t)) + \sum_{l \neq k} \sum_{j: l \in C_j} \mathcal{R}_{lk}^j(S_j(T_j^n(t))) \geq 0, \quad k \in \mathcal{K}_0,$$

$$Q_k^n(t) = Q_k^n(0) - \sum_{j: k \in C_j} S_j(T_j^n(t)) + \alpha_k t, \quad k \in \mathcal{K}_\infty.$$

$$(13.6)$$

To keep a complete record of the system under a given policy, we use as in Section 10.3 the process:

$$\tilde{\mathbf{X}}(t) = (\mathbf{Y}(t), a(t)) = ((\mathbf{Q}(t), V(t), Y(t)), a(t))$$

where $V_j(t)$ are residual processing times of activity j, $Y(t)$ is additional information from the past to determine the policy, and $a_j(t)$ records the allocation to activity j at the time t. Recall that the matrices A and C determine the resource consumption and the buffers processed by each activity. We use as norm for $\mathbf{X}(t) = x$, the quantity $|x| = \sum_{k \in \mathcal{K}} |Q_k(t)| + \sum_{j \in \mathcal{J}} V_j(t)$. Recall that for $k \in \mathcal{K}_\infty$, $Q_k(t)$ may be negative.

We assume as $t, n \to \infty$:

$$S_k(t)/t \to \mu_k = m_k^{-1}, \quad \mathcal{R}^j_{k,l}(n)/n \to p^j_{k,l} \quad k \in \mathcal{K}, \ l \in \mathcal{K}_0, \quad \text{u.o.c. a.s.}$$
$$(13.7)$$

and we denote by $\mathfrak{G}$ the set of measure 1 for which this holds. For simplicity, we assume that $V_j^n(0) = 0$, and we assume

$$\frac{1}{n} Q_k^n(0) \to \bar{Q}_k(0), \ k \in \mathcal{K}, \quad \text{as } n \to \infty. \qquad (13.8)$$

Following the same steps as in Section 11.1, we find that fluid limits exist, and for all $\omega \in \mathfrak{G}$, i.e. almost surely, they are Lipschitz continuous, and they satisfy the corresponding fluid equations (see Exercise 13.10):

$$\bar{Q}_k(t) = \bar{Q}_k(0) - \sum_{j:k \in C_j} \mu_j \bar{\mathcal{T}}_j(t) + \sum_{l \neq k} \sum_{j:l \in C_j} \mu_j p^j_{lk} \bar{\mathcal{T}}_j(t) \geq 0, \quad k \in \mathcal{K}_0,$$

$$\bar{Q}_k(t) = \bar{Q}_k(0) - \sum_{j:k \in C_j} \mu_j \bar{\mathcal{T}}_j(t) + \alpha_k t, \qquad\qquad k \in \mathcal{K}_\infty.$$
$$(13.9)$$

Following the same steps as in Section 11.2, we can prove for processing networks with IVQs (recalling the Definitions 11.7 and 11.9 of fluid stability and weak fluid stability):

Theorem 13.1 *(i) If for some HOL policy, the fluid model for a processing network with IVQs is weakly stable, then the stochastic system is rate stable for that policy, i.e. $\lim_{t \to \infty} Q_k(t)/t = 0$ a.s., for all $k \in \mathcal{K}$.*
(ii) Assume the system under some policy can be described by a Markov process $\mathbf{X}(t)$, and assume that closed bounded sets are petite (uniformly small). If the fluid model of $\mathbf{X}(t)$ is stable, then the stochastic system is positive Harris recurrent, i.e. $Q(t)$ possess a stationary distribution (ergodic, i.e. the distribution of $Q(t)$ converges to the stationary distribution as $t \to \infty$).

Verification of this theorem is left as Exercise 13.11.

Part (i) of Theorem 13.1 is useful. Unfortunately, for general processing networks with IVQs there are no criteria for verifying that bounded sets are petite or uniformly small, so there is no way of verifying whether part (ii), which promises existence of a stationary distribution, is valid. The situation is a little better for unitary networks with IVQs. Using the following definition, one can prove the following theorem. We do not present the proof here.

Definition 13.2   A policy for a unitary network with IVQ is a *weak pull priority policy* if at all times that some standard queues are available for processing, at least one processor is fully allocated to processing standard queues.

Theorem 13.3   *In a unitary network with IVQ, under a weak pull priority policy, if the distributions of the IVQ processing times have infinite support and are spread out, then every bounded set is uniformly small.*

13.4 Static Production Planning Problem and Maximum Pressure Policies

We recall the definition of the input-output matrix R (12.5),

$$R_{k,j} = \mu_j \left(C_{j,k} - \sum_{l=1}^{K} C_{j,l} \, p_{l,k}^j \right), \qquad k \in \mathcal{K}, \, j \in \mathcal{J}. \tag{13.10}$$

As before, $R_{k,j}$ is the rate at which we expect the level of buffer k to decrease by the application of activity j. We note that this is defined both for standard buffers and for IVQs, where for $k \in \mathcal{K}_\infty$ we have that $p_{l,k}^j = 0$, and so rows of R for $k \in \mathcal{K}_\infty$ are all non-negative. We denote by $R_{\mathcal{K}_\infty}$ rows $k \in \mathcal{K}_\infty$ of R, and by $R_{\mathcal{K}_0}$ rows $k \in \mathcal{K}_0$ of R.

We now formulate a linear program to determine the optimal nominal input rates for the IVQs. This is a generalization of the static planning problem (12.7) of Section 12.2. For given rewards, w_k, $k \in \mathcal{K}_\infty$, obtain α_k, $k \in \mathcal{K}_\infty$ that solve:

$$\max_{x,\alpha} \sum_{k \in K_\infty} w_k \alpha_k$$

$$\text{s.t.} \quad R_{\mathcal{K}_\infty} x = \alpha, \tag{13.11}$$

$$R_{\mathcal{K}_0} x = 0,$$

$$A x \leq \mathbf{1},$$

$$x, \alpha \geq 0.$$

Here, instead of determining the workload ρ that is imposed by external input rates α, we impose a constraint of 1 on the offered loads, and determine nominal input rates α that will maximize the *total static expected profit* $w^\mathsf{T}\alpha$ where w_k, $k \in \mathcal{K}_\infty$ are the rewards per customer, from customers introduced by the IVQs. The values of x_j are the static allocations of processors to activity $j \in \mathcal{J}$. Let α^*, x^* denote an optimal solution. From this solution, we obtain the actual workloads of the various processors, as $\rho_i = (A x)_i \leq 1$. Typically, some of the processors will have a workload of $\rho_i = 1$.

Recall that possible allocations are constrained (as in (12.2)), by $\mathcal{A} = \{a : Aa \leq 1, a \geq 0\}$, and we also have extreme allocations $\mathcal{E}$ and available feasible allocations $\mathcal{A}(t)$ and extreme available allocations $\mathcal{E}(t)$. We define the pressure for an allocation a as in (12.6), by

$$p(a, z) = z \cdot Ra, \tag{13.12}$$

where z represent levels of all the buffers, including both $k \in \mathcal{K}_0$ and $k \in \mathcal{K}_\infty$. Similar to Definition 12.1, we define maximum pressure policies for a processing network with IVQs by $a^* \in \arg\max_{a \in \mathcal{E}(t)} p(a, Q(t))$. The EAA assumption needs a slight modification, since IVQs are always available,

Assumption 13.4 (EAA assumption) Extreme allocation available for processing networks with IVQs requires that for any state Q there exists an extreme pressure allocation $a^* \in \arg\max_{a \in \mathcal{E}} p(a, Q)$ such that if $a_j^* > 0$, then also $Q_k > 0$ for all $k \in \mathcal{K}_0$ that belong to the constituency of j.

As before, EAA holds for strictly Leontief networks. Following the steps of the proof of Theorem 12.5, we then have:

Theorem 13.5 *Assume that a processing network with IVQs has the EAA property. Assume also that the nominal input rates of the IVQs, α, satisfy the constraints of (13.11). This will hold in particular if $\alpha \leq \alpha^*$, where α^* is an optimal solution of (13.11). Then under a maximum pressure policy that allows processor splitting and preemption, the processing network is rate stable, i.e. $\lim_{t \to \infty} Q(t)/t = 0$ almost surely.*

Corollary 13.6 *If the network is unitary, i.e. it is a MCQN-IVQ with alternative routing and input control, then maximum pressure policy that does not allow processor splitting and does not allow preemptions, is rate stable for all $\alpha \leq \alpha^*$.*

See Exercises 13.12 and 13.13 to verify these theorems.

As we observed, in the solution of the static production planning problem

(13.11) some of the processors will have a workload of $\rho_i = 1$. If some of this workload consists of processing customers from IVQs, then it may be possible to operate the processing network with $\rho_i = 1$ without congestion. We define:

$$\mathcal{J}_0 = \{j : \sum_{k \in \mathcal{K}_0} C_{j,k} > 0\}, \qquad \tilde{\rho}_i = \sum_{j \in \mathcal{J}_0} A_{ij} x_j^*. \qquad (13.13)$$

Here $\mathcal{J}_0$ lists all the activities that process some standard queues, and $\tilde{\rho}_i$ is the workload of processor i that involves standard buffers, while $\rho_i - \tilde{\rho}_i$ is the workload that is entirely devoted to activities that process only IVQs. We now have:

Key Research Question: *For processing networks with IVQs, for which the solution of the production planning problem (13.11) has $\tilde{\rho}_i < 1$, $i = 1, \ldots, I$, find a policy that works with α^*, and for which the fluid limit model is stable.*

Recall that for processing networks without IVQs, by Theorem 12.8, if $\rho_i < 1$, $i \in \mathcal{I}$, then under maximum pressure policy and appropriate conditions, the fluid limit model is stable. Unfortunately, this is not the case for processing networks with IVQs. In the next section we will show an example where $\tilde{\rho}_i < 1$, $i = 1, \ldots, I$ and under maximum pressure policy the fluid model is weakly stable, but it is not stable. We also find, for that system, a policy that will have a stable fluid model.

13.5 An Illustrative Example: The Push-Pull System

We have already in Section 13.2.3 considered the Markovian push-pull system (Figure 13.5). We now consider it with general processing times. This network is the IVQ version of the Kumar–Seidman Rybko–Stolyar network that was discussed in Section 10.2.3.

The static production planning problem (13.11) for the push-pull network, with unknown flow rates α_1, α_2, and allocations x_j, $j = 1, 2, 3, 4$, is

$$\max_{u,\alpha} \ w_1\alpha_1 + w_2\alpha_2$$

s.t.
$$\begin{bmatrix} \lambda_1 & 0 & 0 & 0 \\ -\lambda_1 & \mu_1 & 0 & 0 \\ 0 & 0 & \lambda_2 & 0 \\ 0 & 0 & -\lambda_2 & \mu_2 \end{bmatrix} \begin{bmatrix} x_1 \\ x_2 \\ x_3 \\ x_4 \end{bmatrix} = \begin{bmatrix} \alpha_1 \\ 0 \\ \alpha_2 \\ 0 \end{bmatrix},$$

$$\begin{bmatrix} 1 & 0 & 0 & 1 \\ 0 & 1 & 1 & 0 \end{bmatrix} \begin{bmatrix} x_1 \\ x_2 \\ x_3 \\ x_4 \end{bmatrix} \leq \begin{bmatrix} 1 \\ 1 \end{bmatrix},$$

$$x, \alpha \geq 0.$$

The feasible region and optimal solution for this LP is seen in Figure 13.7, for the inherently unstable case of $\lambda_i > \mu_i$, $i = 1, 2$. Similar figures are obtained for all other combinations of parameters. According to the values

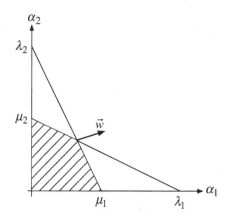

Figure 13.7 Feasible region of the production planning problem.

of the parameters w, λ, μ, the optimal nominal inputs can be one of the following (see Exercise 13.14):

(i)   either $\alpha_1 = \min\{\lambda_1, \mu_1\}$, $\alpha_2 = 0$,
(ii) or $\alpha_1 = 0$, $\alpha_2 = \min\{\lambda_2, \mu_2\}$,
(iii) or $\alpha_1 = \dfrac{\lambda_1\mu_1(\lambda_2 - \mu_2)}{\lambda_1\lambda_2 - \mu_1\mu_2}$, $\alpha_2 = \dfrac{\lambda_2\mu_2(\lambda_1 - \mu_1)}{\lambda_1\lambda_2 - \mu_1\mu_2}$.

The interesting case is (iii). It comprises two cases, the inherently stable case of $\mu_i > \lambda_i$, $i = 1, 2$, and the inherently unstable case of $\lambda_i > \mu_i$, $i = 1, 2$.

We note that in the cases (iii), for both stations $\rho_i = 1$, but $\tilde{\rho}_i = \frac{\alpha_i}{\mu_i} < 1$. We first consider operating the push-pull system with maximum pressure policy.

Proposition 13.7 *The fluid model of the push-pull system in the inherently stable and in the inherently unstable case is weakly stable under maximum pressure policy, but it is not stable.*

Proof The maximum pressure policy for the push-pull system is (see Exercise 13.15):
- For station 1, push if $\lambda_1(Q_1(t) - Q_2(t)) \geq \mu_2 Q_4(t)$, pull if $\lambda_1(Q_1(t) - Q_2(t)) \leq \mu_2 Q_4(t)$. Break ties in some consistent way.
- For station 2, push if $\lambda_2(Q_3(t) - Q_4(t)) \geq \mu_1 Q_2(t)$, pull if $\lambda_2(Q_3(t) - Q_4(t)) \leq \mu_1 Q_2(t)$. Break ties in some consistent way.

It is weakly stable for α_1^*, α_2^* by Theorem 13.5. However, for the fluid limit, when $\lambda_1(\bar{Q}_1(0) - \bar{Q}_2(0)) = \mu_2 \bar{Q}_4(0)$ and $\lambda_2(\bar{Q}_3(0) - \bar{Q}_4(0)) = \mu_1 \bar{Q}_2(0)$, then for all the buffers, $\dot{\bar{Q}}_k(t) = 0$, and $\bar{Q}_k(t) =$ constant$\neq 0$, for $t > 0$ (see Exercise 13.15), so the fluid model is not stable. □

The following proposition can also be proved; we do not give the proof here:

Proposition 13.8 *In the singular cases of $\lambda_1 = \mu_1$ or $\lambda_2 = \mu_2$, the push-pull network cannot be stabilized, that is, it will not be positive Harris recurrent under any Markov policy.*

We can, however, find policies that have stable fluid limits, both in the inherently stable and in the inherently unstable case. For the inherently stable case we have:

Proposition 13.9 *In the inherently stable case, the push-pull system under non-preemptive priority to pull activities has a stable fluid limit. If processing times are i.i.d. with unbounded support and spread out, then the system is ergodic.*

Proof It is easy to see that the fluid limit model is stable (see Exercise 13.16). Also, it is easy to see that this policy is weak pull priority (see Exercise 13.16). Therefore, by Theorems 13.3 and 11.8, it follows that the system is ergodic. □

We now consider the inherently unstable case. We have already seen in Section 13.2 that in the case of exponential service times the policy described in Figure 13.6 as well as the policy described in Exercise 13.8 stabilize the system. The main idea is to set some thresholds for the standard

queues, and push work to these queues when they fall below the threshold, but use pull activities otherwise. We consider the following policy: Choose constants $\beta_i < 0$, $0 < \kappa_i < 1$, $i = 1, 2$, and use linear thresholds, defined by $Q_4 = \beta_1 + \kappa_1 Q_2$, and $Q_2 = \beta_2 + \kappa_2 Q_4$. The linear threshold policy is
- When $Q_2(t) = 0$ or $Q_4(t) = 0$ use push at at both stations,
- for $Q_4(t) < \beta_1 + \kappa_1 Q_2(t)$, use push at station 2, and pull at station 1,
- for $Q_2(t) < \beta_2 + \kappa_2 Q_4(t)$, use push at station 1, and pull at station 2,
- in all the remaining states use pull at both stations.
The policy is illustrated in Figure 13.8.

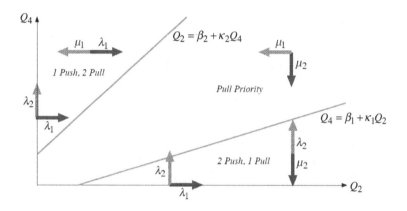

Figure 13.8 Linear threshold policy for inherently unstable case.

Proposition 13.10 *In the inherently unstable case, the push-pull system under the linear threshold policy has a stable fluid limit. If we assume that all bounded sets of states are petite (uniformly small), then this implies that the stochastic system will be positive Harris recurrent (ergodic).*

The proof is left as Exercise 13.17. In Figure 13.9, taken from Nazarathy and Weiss (2010), we show a simulation of the behavior of the inherently unstable push-pull system, under a linear threshold policy; what is plotted is the path of the workloads at the two standard queues.

13.6 Sources

Multi-class queueing networks with infinite virtual queues were suggested in Weiss (2005), and the countable Markovian examples given here were analyzed in Adan and Weiss (2005, 2006) and Kopzon et al. (2009). The

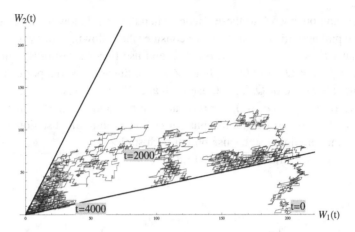

Figure 13.9 A sample path under the linear threshold policy.

analysis of the push-pull system with general processing times is described in Nazarathy and Weiss (2010), where Figure 13.8 appears. The formulation of the static production planning problem and further examples are given in Guo et al. (2014). A proof that the balanced push-pull network cannot have a stable fluid limit is given by Nazarathy et al. (2013).

Exercises

13.1 For the two-node network with IVQs in Section 13.2.1, what are the marginal stationary distributions of each of the standard queues.

13.2 (∗) For the two-node network with IVQs of Section 13.2.1, derive the two-dimensional stationary distribution, using the compensation method of Section 15.1 [Adan and Weiss (2005)].

13.3 The following is a generalization of the two-node IVQ network in Section 13.2.1. Consider a Jackson network, with nodes $i \in \mathcal{I}$, with exogenous input rates α_i, service rates μ_i, and routing probabilities p_{ij}. Assume that the nodes are partitioned into standard nodes $\mathcal{I}_0$, and IVQ nodes $\mathcal{I}_\infty$. Each node $i \in \mathcal{I}_\infty$ has in addition to the queue of items received from outside or from other nodes, also an infinite supply of work. At the IVQ nodes, there is preemptive priority to customers that arrive from outside or from other nodes. However, when the queue of such customers is empty, the IVQ node serves customers from its infinite buffer [Weiss (2005)].

(i) Find the flow rates, and conditions for stability for this system.

(ii) Show that queue lengths at the standard nodes have a product-form joint stationary distribution.

(c) Find the stationary marginal distribution of the queue lengths at the IVQ nodes, and show that they have Poisson input and output.

13.4 For the three-buffer re-entrant line with IVQ in Section 13.2.2, write down the balance equations for the random walk describing the queue length of the standard queues, and derive the stationary distribution of $(Q_2(t), Q_3(t))$ [Adan and Weiss (2006)].

13.5 Obtain the stationary distribution for the push-pull system in Section 13.2.3, with exponential processing times, in the inherently stable case under pull priority.

13.6 Show the equivalence of the two definitions of the policy for the inherently unstable push-pull system in Section 13.2.3.

13.7 Obtain the stationary distribution for the push-pull system with exponential processing times in the inherently unstable case, under the policy described in Section 13.2.3 [Kopzon et al. (2009)].

13.8 Consider the push-pull system in Section 13.2.3, with symmetric rates, i.e. $\lambda_1 = \lambda_2 = \lambda, \mu_1 = \mu_2 = \mu$, with exponential processing times in the inherently unstable case of $\lambda > \mu$, under the following policy: While $|Q_1(t) - Q_2(t)| > 1$, serve the shorter queue, using both push and pull for this queue. When the queues differ by no more than 1, use pull on both queues. Show that this policy is stable, and obtain its stationary distribution. [Kopzon et al. (2009)].

13.9 Compare the performance of the two policies, in exercises 13.7 and 13.8 [Kopzon et al. (2009)].

13.10 Verify that fluid limits for processing networks with IVQs exist and satisfy the standard fluid equations.

13.11 Outline the proof of Theorem 13.1, in analogy with Theorems 11.10 and 11.8.

13.12 Outline the proof of Theorem 13.5, in analogy with Theorem 12.6.

13.13 Outline the proof of Corollary 13.6, in analogy with 12.14.

13.14 Solve the static production planning problem for the push-pull system of Section 13.5, and plot the feasible region for all possible combinations of parameter values.

13.15 Obtain the maximum pressure policy for the push-pull system in Section 13.5, and show that the fluid model is weakly stable but not stable.

13.16 For the inherently stable push-pull system of Section 13.5, under pull priority, show that the fluid model is stable, and show that the policy is a weak pull priority policy.

13.17 For the inherently unstable push-pull system of Section 13.5, show that the linear threshold policy described in Figure 13.8 has a stable fluid limit model [Nazarathy and Weiss (2010)].

13.18 Consider the process $\mathcal{D}(t)$ of departures from the push-pull system. Calculate the correlation between $\mathcal{D}_1(t)$ and $\mathcal{D}_2(t)$. For simplicity, consider the symmetric case $\mu_1 = \mu_2 = \mu, \lambda_1 = \lambda_2 = 1$ [Nazarathy and Weiss (2010)].

14

Optimal Control of Transient Networks

So far in Chapters 10–13 we have used fluid models to establish stability of a queueing network under a given policy. In this chapter we will use fluid models to find policies that optimize its performance. We formulate a problem for the control of a queueing network over a finite time horizon. This problem forms a bridge between deterministic scheduling problems, and long-term average Markov decision problems. Our aim is to minimize inventory costs, equivalently minimize waiting times, and equivalently maximize early departures.

To solve this problem we use a two-stage method: We first approximate the discrete stochastic queueing network by a continuous deterministic fluid model, and solve for the optimal fluid solution. This requires the solution of a continuous linear program, which is a control problem with linear constraints, and it is solved centrally. We then track the optimal fluid solution by using a maximum pressure type policy that is implemented locally. An important feature of the tracking is that while fluid in a buffer is positive, it can be considered as an IVQ. We will show this procedure is asymptotically optimal when the number of jobs in the system is large, and the time horizon is of the same order of magnitude.

To motivate this approach, we describe in Section 14.1 the problem of controlling the flow of wafers in a semi-conductor microchip fabrication plant. This presents a very large queueing network of the re-entrant type that could serve as a perfect candidate for our approach (Section 14.1). We will also use a miniature wafer fab re-entrant line as an illustration of our method throughout the chapter. We will present the stochastic network optimization problem and discuss its significance in Section 14.2, and present its approximating fluid network problem in Section 14.3. The fluid network problem is a separated continuous linear programming problem (SCLP), and in Section 14.4 we digress from queueing models, and discuss SCLP. We describe the structure of their solutions, and a recent simplex-type algorithm to solve them. We then discuss in Section 14.5 the nature of the optimal solution of the fluid network problem: Its main feature is that it requires piecewise constant allocation of resources to activities over well-defined time intervals. The solvability and simple form of the fluid solution is the first major ingredient of our method.

The second major ingredient is that buffers with positive fluid act as IVQs in every interval of the fluid solution, and this is used to track the fluid solution: In Section 14.6 we model the deviations of the actual stochastic network from the approximating optimal fluid solution as a multi-class queueing network with IVQs. We then track the fluid solution by using maximum pressure policy for the deviation process. We show in Section 14.7 that this is asymptotically optimal. We present a simulation of a small system to illustrate the implementation of our method in Section 14.8, and conclude with some remarks on implementation in Section 14.9.

14.1 The Semiconductor Wafer Fabrication Industry

An important motivation for control of MCQN is provided by the semi-conductor wafer fabrication industry (wafer fabs). Wafer fabs are big, so-phisticated and expensive manufacturing plants, that are used to produce miniaturized electronic circuits for computer microchips as well as for DRAMs, flash memories, and LEDs. There are currently some 120 modern 300 mm wafer fabs in the world, with global annual sales of 335×10^9\$.

A wafer fab operates in a clean room environment; for most modern fabs the process is entirely automated, and the plant is kept under a pressur-ized Nitrogen atmosphere. The production starts with a pure single crystal 300mm silicon wafer. Circuits are built in layers, in a process that involves hundreds of steps, revisiting the machines repeatedly, until at the end of the process each wafer carries hundreds of micro-chips. A modern wafer fab costs over 6×10^9\$, it is designed to produce some 10,000 wafers per week and is supposed to return the investment in a few years. The production cycle time is typically 6 weeks, and the sales value of the work in process, consisting of queues of wafers in various degrees of completion in front of the various machines, is of the order of 200×10^6\$. The control of these queues, over a period of 6 weeks, is the type of problem we consider here.

Figure 14.1 is a schematic representation of the process; the horizontal axis presents time, the vertical axis lists the various machines and tools, the zig-zagging line is the production process re-entrant line, and each dot is a production step, where wafers are waiting to be processed through that step. In our terms, the queues, of work in process at each dot, form a re-entrant line MCQN, with $i = 1, \ldots, 30$ workstations, and $k = 1, \ldots, 500$ buffers.

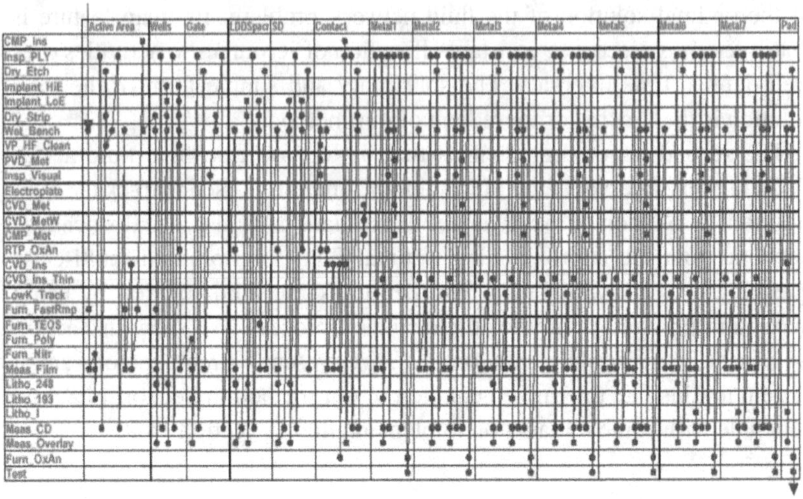

	Active Area	Wells	Gate	LDDSpac	SD	Contact	Metal1	Metal2	Metal3	Metal4	Metal5	Metal6	Metal7	Pad
CMP_Ins														
Insp_PLY														
Dry_Etch														
Implant_HiE														
Implant_LoE														
Dry_Strip														
Wet_Bench														
VP_HF_Clean														
PVD_Met														
Insp_Visual														
Electroplate														
CVD_Met														
CVD_MetW														
CMP_Met														
RTP_OxAn														
CVD_Ins														
CVD_Ins_Thin														
LowK_Track														
Furn_FastRmp														
Furn_TEOS														
Furn_Poly														
Furn_Nitr														
Meas_Film														
Litho_248														
Litho_193														
Litho_I														
Meas_CD														
Meas_Overlay														
Furn_OxAn														
Test														

Figure 14.1 Generic 130-nm semiconductor re-entrant process.

Our aim is to provide a policy that would control the flow of the $200 \times 10^6 \$$ work in process comprising some 60,000 wafers, over a time horizon of a full production cycle of 6 weeks. We will discuss this problem in the next section.

14.2 The Finite Horizon Problem Formulation

In this section we discuss the problem of control of a stochastic processing network over a finite time horizon, as exemplified by the wafer fab.

On the face of this it, it is a very large job-shop scheduling problem, since the data in the fab lists every wafer in its current state, and each operation that is needed for its completion. Job-shop scheduling problems belong to the realm of combinatorial optimization and are known to be NP-hard, and finding an efficient approximate solution is NP-hard as well. So in fact there is no efficient way to obtain the optimal schedule for this job-shop problem.

But even if an optimal schedule could be found, it may not be useful: As it is implemented over time, inaccuracies in the data and unexpected events (many small ones and a few large ones) accumulate and interfere with the optimal schedule, and there is no theory to say how close or far from optimal the result is.

An alternative is to use a stochastic model for the wafer fab, and solve it as a very large Markov decision problem. Again, the problem is too large to be solved optimally. Several approximation methods exist for large Markov decision problems, and it can also be approximated on a diffusion scale by a continuous stochastic Brownian control problem (this is the approach in Part V of this text), but these methods are still limited to problems of moderate size.

However, this stochastic approach focuses on optimizing the steady state of the system, which may not be suitable. Wafer fabs never reach steady state: the mix of products changes with time, technology changes, maintenance influences processing rates, etc. In fact, steady state ignores completely the initial state of the system, whereas for a six-weeks time horizon in a wafer fab the initial state is of utmost importance.

Our approach here is to formulate the problem as a discrete stochastic processing network as defined in Chapter 12, with processors $i \in \mathcal{I}$, buffers $k \in \mathcal{K}$, and activities $j \in \mathcal{J}$, with given initial inventory, which we wish to control over a finite time horizon, i.e. optimize the transient behavior of the system.

The dynamics of the processing network are as in Section 12.1 equation (12.1), with one modification: Because we are only dealing with a finite horizon, we can assume that all the work to be processed over the time horizon is already in the system, where some of the buffers may serve as input buffers that will release material into the system using some input activities. So we need not have an exogenous input process, and need not distinguish input buffers and activities. With consumption matrix A and constituency matrix C, and with $\mathcal{D}(t)$ the departure process we have:

$$
\begin{aligned}
Q_k(t) &= Q_k(0) - \mathcal{D}_k(t) \\
&= Q_k(0) - \sum_{j \in \mathcal{J}} C_{j,k} S_j(\mathcal{T}_j(t)) + \sum_{l \in \mathcal{K}} \sum_{j \in \mathcal{J}} C_{j,l} \mathcal{R}^j_{l,k}(S_j(\mathcal{T}_j(t))) \geq 0.
\end{aligned}
\tag{14.1}
$$

The capacity constraints of the processors are

$$
A\dot{\mathcal{T}}(t) \leq 1, \quad \mathcal{T}(0) = 0, \quad \dot{\mathcal{T}}(t) \geq 0.
\tag{14.2}
$$

The random cost to be minimized over the finite time horizon is

$$
\min \sum_{k=1}^{K} \int_0^T h_k \dot{Q}_k(t) dt.
\tag{14.3}
$$

It can also be presented in a different way:

$$
\min \sum_{k=1}^{K} \int_{0}^{T} h_k Q_k(t)\,dt
$$

$$
= \sum_{k=1}^{K} h_k Q_k(0)T - \max \sum_{k=1}^{K} \int_{0}^{T} h_k \mathcal{D}_k(t)\,dt \tag{14.4}
$$

$$
= \sum_{k=1}^{K} h_k Q_k(0)T - \max \sum_{k=1}^{K} \int_{0}^{T} h_k (T-t)\,d\mathcal{D}_k(t).
$$

In the first form, the objective is to minimize weighted queue lengths, or equivalently minimize weighted waiting times in each buffer. In the second form, the last line, the objective can be viewed as maximizing the added value of each item as it moves through the system, weighted by its useful life from departing the buffer up to the time horizon.

Solving the finite horizon stochastic problem so as to optimize the expected objective is a harder problem than either the job-shop scheduling approach or the long-term average Markov decision approach. We therefore look for approximate solutions, that will be close to optimal for systems with a large volume of items.

Remark   The job-shop scheduling approach can be more realistic if instead of looking for the optimal schedule one looks for a robust schedule. The infinite horizon Markov decision approach can be more realistic if instead of long-term average one optimizes the expected discounted objective over time.

14.3 Formulation of the Fluid Optimization Problem

We now replace the discrete stochastic network by its fluid approximation. This will make sense if the initial number of customers in the system is of the order of N where N is large, and the time horizon is of a similar order of magnitude, i.e. the time horizon is long enough to process all the initial customers, and is not much longer.

We assume that the following FSLLN holds for processing rates and routing probabilities, and let R be the input-output matrix, i.e.:

$$
\lim_{t \to \infty} \mathcal{S}_j(t)/t \to \mu_j, \qquad \lim_{r \to \infty} \mathcal{R}_{k,l}^{j}(r)/r \to p_{k,l}^{j},
$$

$$
R_{k,j} = \mu_j \left(C_{j,k} - \sum_{l=1}^{K} C_{j,l} p_{l,k}^{j} \right). \tag{14.5}
$$

We approximate the discrete stochastic control problem (14.1)–(14.3), by the continuous deterministic fluid optimization problem:

$$\min \int_0^T \sum_{k=1}^K h_k q_k(t)dt \tag{14.6}$$

$$\text{s.t. } q(t) + \int_0^t Ru(s)ds = q(0),$$
$$Au(t) \le 1, \tag{14.7}$$
$$u(t), q(t) \ge 0, \quad 0 \le t \le T.$$

Similar to the derivation of (14.4), integrating by parts the objective (14.6), and using the dynamics constraint of (14.7), we obtain:

$$\int_0^T h^\mathsf{T} q(t)dt = h^\mathsf{T} q(0)T + \int_0^T (T-t)h^\mathsf{T} \dot{q}(t)dt$$
$$= h^\mathsf{T} q(0)T - \int_0^T (T-t)h^\mathsf{T} Ru(t)dt.$$

Figure 14.2 illustrates this integration by parts in terms of areas for the one-dimensional case. This yields an alternative objective

$$\max \int_0^T (T-t)w^\mathsf{T} u(t)dt, \tag{14.8}$$

which is equivalent to (14.6) if $w_j^\mathsf{T} = (h^\mathsf{T} R)_j$, but more generally, w^T need not be equal to $h^\mathsf{T} R$, and in general we can think of $w_j(T-t)$ as the value added of allocating resources to activity j at time t. This form is also the form used in the algorithm described in the next Section 14.4.

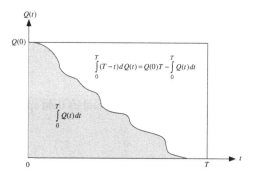

Figure 14.2 Integration by parts of the objective.

The fluid network problem is to determine control flow rates $u_j(t)$ and buffer fluid level $q_k(t)$, so as to optimize the objective of minimum inventory (14.6), or maximum value added over time (14.8), subject to the flow and resource constraints (14.7).

This type of problem is a separated continuous linear programming problem (SCLP). It is called separated because it separates the constraints on instantaneous controls from the constraints on averaged controls. SCLP is a control problem over a finite time horizon, with state variable functions $q_k(t)$ and control variable functions $u_j(t)$, with a linear objective and linear constraints. In Section 14.4 we briefly summarize properties of SCLP, the structure of its solutions, and an algorithm to solve it.

To illustrate our method, we consider a toy problem of optimal control of a re-entrant line with three machines and nine production steps. In Figure 14.3 we show the discrete stochastic system at time 0, with queues of discrete items at the nine buffers, and a supply of additional items in buffer 0. We then also show the continuous deterministic approximating system with fluids in the buffers.

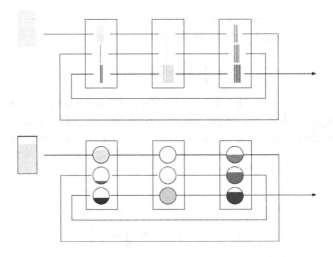

Figure 14.3 A discrete re-entrant line and its fluid approximation.

14.4 Brief Summary of Properties of SCLP and its Solution

In this section we give a brief summary of the properties of SCLP. We will make use of terms and properties from the theory of standard linear programs (LP); readers unfamiliar with those may wish to skip this section on first reading.

Separated continuous linear programming problems (SCLP) are optimal control problems that share all the essential properties of standard linear programming problems. They possess a symmetric dual and satisfy strong duality, their extreme points can be described by sequences of bases, and they can be solved by a simplex-type algorithm. In this section we briefly summarize the current state of SCLP theory.

The following optimization problem is an SCLP:

$$\text{max} \quad \int_0^T ((\gamma^\top + (T-t)c^\top)u(t) + d^\top x(t))\, dt$$

SCLP $\quad$ s.t. $\quad \int_0^t Gu(s)ds + Fx(t) \le \alpha + at,$

$$Hu(t) \quad \le b,$$

$$x(t), u(t) \ge 0, \quad t \in [0, T].$$

Its symmetric dual is:

$$\text{min} \quad \int_0^T ((\alpha^\top + (T-t)a^\top)v(t) + b^\top y(t))dt$$

SCLP* $\quad$ s.t. $\quad \int_0^t G^\top v(s)ds + H^\top y(t) \ge \gamma + ct,$

$$F^\top v(t) \quad \ge d,$$

$$v(t), y(t) \ge 0, \quad t \in [0, T].$$

Here the problem data consists of system matrices G, H, F, $K \times J$, $I \times J$, $K \times L$ dimensional, rate vectors a, b, c, d, initial condition vectors α, γ, and time horizon T. The unknowns are vectors $u(t)$, $v(t)$ of primal and dual control functions, and $x(t)$, $y(t)$ of primal and dual state functions, which include also slacks, so u, y are $J + I$-dimensional, and x, v are $K + L$-dimensional. Here the dual runs in reversed time, so that $v(t)$ is dual to $x(T - t)$, and $y(t)$ is dual to $u(T - t)$. Complementary slackness (see Exercise 14.3) is:

$$\int_0^T x(T-t)^\top v(t)dt = \int_0^T y(T-t)^\top u(t)dt = 0.$$

Under some simple non-degeneracy and feasibility conditions, unique complementary slack strongly dual optimal solutions exist, in which the controls are piecewise constant and the states continuous piecewise linear, with breakpoints $0 = t_0 < t_1 < \cdots < t_{M-1} < t_M = T$; denote the states at the breakpoints as $x^m = x(t_m), y^m = y(T - t_m), m = 0, \ldots, M$. Denote the constant rates in the intervals as $u^m = u(t), \dot{x}^m = \frac{d}{dt}x(t), v^m = v(T - t), \dot{y}^m = \frac{d}{dt}y(T - t)$ for $t_{m-1} < t < t_m, m = 1, \ldots, M$; denote the interval lengths as $\tau_m = t_m - t_{m-1}$. Then we have: The values of $x(0), y(0)$ are obtained from the boundary LPs

$$\text{Boundary-LP} \quad \begin{array}{ll} \max & d^\mathsf{T}x^0 \\ \text{s.t.} & Fx^0 \leq \alpha, \\ & x^0 \geq 0, \end{array} \qquad \text{Boundary-LP*} \quad \begin{array}{ll} \min & b^\mathsf{T}y^M \\ \text{s.t.} & H^\mathsf{T}y^M \geq \gamma, \\ & y^M \geq 0. \end{array}$$

The values of the rates are non-degenerate complementary slack basic solutions of

$$\text{Rates-LP} \quad \begin{array}{l} [G\ 0]u^m + [I\ F]\dot{x}^m = a, \\ [H\ I]u^m = b, \\ u^m \geq 0, \end{array}$$

$$\text{Rates-LP*} \quad \begin{array}{l} [G^\mathsf{T}\ 0]v^m + [-I\ H^\mathsf{T}]\dot{y}^m = c, \\ [F^\mathsf{T}\ -I]v^m = d, \\ v^m \geq 0, \end{array}$$

where the bases $B_m, m = 1, \ldots, M$ are adjacent, with ζ_m leaving the basis and ξ_m entering in the pivot $B_m \to B_{m+1}$. Furthermore, $\dot{x}_k^1 \neq 0$ if $x_k^0 > 0$ and $\dot{y}_j^M \neq 0$ if $y_j^M > 0$. Finally, the values of the interval lengths are solutions of

$$x_k(t_m) = x_k^0 + \sum_{r=1}^{m} \dot{x}_k^r \tau_r = 0, \qquad \text{when } \zeta_m = \dot{x}_k,$$

$$y_j(T - t_m) = y_j^M + \sum_{r=m+1}^{M} \dot{y}_j^r \tau_r = 0, \qquad \text{when } \zeta_m = u_j,$$

for $m = 1, \ldots, M - 1$, and the equation

$$\tau_1 + \cdots + \tau_M = T.$$

In addition, the solution needs to satisfy that

$$x_k(t_m) = x_k^0 + \sum_{r=1}^{m} \dot{x}_k^r \tau_r \geq 0,$$

$$y_j(T - t_m) = y_j^M + \sum_{r=m+1}^{M} \dot{y}_j^r \tau_r \geq 0.$$

The sequence $B_1, \ldots, B_M$ is called the optimal base sequence, and it determines the entire solution uniquely (see Exercise 14.4). In fact, any solution with a base sequence $B_1, \ldots, B_M$ that satisfies all the above conditions is an optimal solution (see Exercise 14.5).

The validity region of a base sequence $B_1, \ldots, B_M$ consists of values (α, γ, T) for which this base sequence is optimal. The validity region is a convex polyhedral cone (it may consist of the origin alone).

The simplex-type algorithm solves SCLP in a parametric sequence of steps, by starting from an optimal solution for some $(\alpha_0, \gamma_0, T_0)$, and moving along a parametric straight line to the problem's (α, γ, T). Along the parametric line, the solution will move through a sequence of validity regions. At the boundary of a validity region one or more intervals may shrink to zero, or one of the > 0 coordinates of x^m, y^m may shrink to zero. This happens at some t_m, and is called a collision. As a result of the collision, it is necessary to update the optimal base sequence, at the time point t_m, in what is called an SCLP pivot. The SCLP pivot at t_m involves eliminating zero, one, or more than one bases from the sequence, and inserting zero, one, or more than one new bases at t_m. The typical update of eliminating one basis and adding one basis to the base sequence is achieved by a single pivot on the Rates-LP. The algorithm is initiated by solving the boundary LPs to obtain x^0, y^M, and a single interval solution with a base sequence of length one. The algorithm terminates at the optimal solution for (α, γ, T) after a finite bounded number of SCLP pivots, and the number of intervals in the optimal solution is bounded.

Recent implementations of this simplex-type algorithm can solve problems of a re-entrant line with dozens of stations and hundreds of buffers in a few seconds, with similar performance for more general processing networks.

14.5 Examination of the Optimal Fluid Solution

The first phase in our method of control is to solve the SCLP fluid approximation problem (14.7)–(14.8), as described in Section 14.4. This is done centrally, for the whole system and the whole time horizon. The data necessary for the solution is quite minimal: the initial buffer levels, the processing rates of the activities, and the expected cost rates.

Figure 14.4 shows a fluid solution of the illustrative example of the three machine nine buffers re-entrant line of Section 14.3 that was described by Figure 11.2. It was solved for some initial fluids $q(0)$, processing rates μ_j, and objective weights w_j. In this figure we plot fluid levels against time. The fluid contents of the nine buffers and the input buffer are plotted, one above the other.

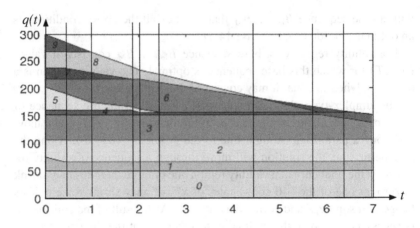

Figure 14.4 Solution of the fluid approximation SCLP problem.

The important feature of the solution, which holds for all solutions of SCLP, is that $q_k(t)$ are all continuous piecewise linear. To illustrate that, we added vertical lines in Figure 14.4 at the breakpoints of the solution. Thus, inside each of the nine intervals plotted in the solution, all the buffers contents change linearly. This is because in each of these intervals the values $u_j(t) = \dot{\mathcal{T}}_j(t)$ are constant.

The fluid solution therefore divides the time horizon into intervals, with pre-calculated end points $0 = t_0 < t_1 < \cdots < t_M = T$. In interval (t_{m-1}, t_m), the optimal fluid solution uses a constant allocation of the processors to the various activities (including processor splitting). Typically, the number of intervals in the solution will be around twice the number of buffers. It is instructive to examine the processes u, q in some of the intervals of the example. Recall that superscripts number the interval and subscripts the buffer or activity.

- In the first interval, machine 3 is working on buffer 9, machine 2 is working on buffer 5, and machine 1 is working on buffer 1. We have $u_1^1 = u_5^1 = u_9^1 = 1$, the remaining $u_j = 0$.
- In the second interval, machine 1 has switched from buffer 1 to buffer 7. This indicates that dual variable y_1 (dual to u_1) has hit the value 0 at time t_1. Now $u_1^2 = 0$ and $u_7^2 = 1$. At the end of this interval buffer 9 becomes empty, this determines the time of the breakpoint t_2.
- In the third interval, buffer 9 is empty, but there is flow through the empty buffer 9. Machine 3 is still working fully on buffer 9. Buffer 9 remains empty because machine 2 is now pumping fluid out of buffer

8 and into buffer 9 at the same rate that machine 3 is pumping it out of buffer 9. Machine 2 is still also processing buffer 5, but it is only using the part of its capacity, which is left over from its processing of buffer 8, so buffer 5 is now emptying at a slower rate than in interval 2. Machine 1 is still working on buffer 7. We now have $u_7^3 = 1$, $u_9^3 = 1$ and $0 < u_8^3 < 1$, $0 < u_5^3 < 1$, $u_8^3 + u_5^3 = 1$. At the end of this interval, buffer 7 becomes empty; this determines the time of the breakpoint t_3 (see Exercise 14.6).

Two major features of the fluid solution should be stressed here:

(a) In each interval of the fluid solution, every buffer is either empty in the whole interval, or it is non-empty in the entire open interval.

(b) Fluid buffers can have positive flow into the buffer and positive flow out of the buffer, while the buffer is empty; i.e. fluid buffers can have positive flow through them while they are empty. This is not possible in the discrete stochastic system.

The second phase of the method is to track the optimal fluid solution with the discrete stochastic system by controlling the deviations between the fluid solution and the actual discrete stochastic buffer levels. This will be done in a decentralized way, as we describe in the next section.

14.6 Modeling Deviations from the Fluid as Networks with IVQs

The fluid solution divides the time horizon into M intervals, in each of which the allocation of processors to activities is constant. The problem of tracking the fluid solution breaks down to the problem of tracking the fluid solution in each of the M intervals. In each interval, the fluid solution specifies the rate at which activities should be applied. In each interval, there is an important distinction between buffers with positive fluid and buffers with zero fluid. Buffers with positive fluid will not be empty during the interval (except perhaps close to the beginning or end of the interval). Therefore, they can be processed throughout the interval and processors assigned to them will never need to be idle. Thus, they behave like IVQs. On the other hand, buffers that have 0 fluid in the interval will behave as standard queues, where the allocated processors will attempt to serve all the arriving customers and stay close to empty.

As an example, we consider the sixth interval of the fluid solution, as shown in Figure 14.5. Here buffers 5, 7, 9 are empty, while the remaining buffers have positive fluid. So, if we are able to track the fluid solution, then during this interval, buffers 0–4, 6, 8, will be able to process items all the time, at the rates indicated by the fluid solution. To track the fluid

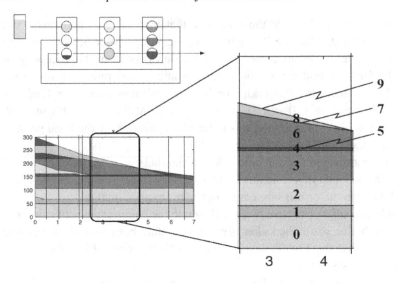

Figure 14.5 Modeling deviations in a time interval as IVQ.

solution, the output from these buffers should be at rate $\sum_{j \in \mathcal{J}} C_{j,k} \mu_j u_j$ for buffer k (for a re-entrant line, the sum consists of a single activity; in a general processing network, more than one activity may work on a single buffer). Note that the fluid buffer 8 empties at the end of the interval. This means that a tracking solution will become empty at approximately the end of the interval, and possibly toward the end of the interval the tracking solution will behave as a standard queue. The other buffers will not be empty either at the beginning or the end of the interval. At the same time, during the whole interval, buffers 5,7,9 will have input at a rate indicated by the fluid solution, and will need to process items at the same rate and to stay reasonably empty. We note that buffers 5, 7, 9 may be bottlenecks during this interval; they need to provide output at a rate equal to the input. This may require that the processors that serve them have a workload of 1.

Corresponding to the optimal fluid solution, we now define a process $\tilde{Q}(t)$ that we call the fluid tracking process, and it simply measures the deviation of the stochastic system from the fluid solution. $\tilde{Q}(t)$ is defined in each of the intervals (t_{m-1}, t_m) as the queue length process of a processing network in which the buffers with 0 fluid are standard queues and the buffers with positive fluid are IVQs. We let $\mathcal{K}_0^m$ denote the standard queues and $\mathcal{K}_\infty^m$ denote the IVQs of the tracking process in the interval (t_{m-1}, t_m). For the IVQs, we calculate the nominal outflow rate using the optimal fluid solution u^m in the mth interval, as:

$$\alpha_k^m = \sum_{j \in \mathcal{J}} C_{j,k} \, \mu_j u_j^m, \quad t \in (t_{m-1}, t_m), \ k \in \mathcal{K}_\infty^m. \tag{14.9}$$

We then have for the interval $t \in (t_{m-1}, t_m]$:

$$\tilde{Q}_k(t) = \tilde{Q}_k(t_{m-1}) - \sum_{j \in \mathcal{J}} C_{j,k} \left(S_j(T_j(t)) - S_j(T_j(t_{m-1})) \right) \tag{14.10}$$

$$+ \sum_{l \neq k} \sum_{j \in \mathcal{J}} C_{j,l} \left(\mathcal{R}_{l,k}^j (S_j(T_j(t))) - \mathcal{R}_{l,k}^j (S_j(T_j(t_{m-1}))) \right), \quad k \in \mathcal{K}_0^m,$$

$$\tilde{Q}_k(t) = \tilde{Q}_k(t_{m-1}) - \sum_{j \in \mathcal{J}} C_{j,k} \left(S_j(T_j(t)) - S_j(T_j(t_{m-1})) \right)$$

$$+ \alpha_k^m (t - t_{m-1}), \quad k \in \mathcal{K}_\infty^m.$$

We initialize the process with $\tilde{Q}(0) = 0$, and for all subsequent intervals we obtain the initial value $\tilde{Q}(t_{m-1})$ as carried over from the previous interval.

At this point, we have for each interval a processing network with IVQs, and we have nominal flow rates for all the IVQs, and wish to keep this system stable or rate stable, so that the IVQs are close to the nominal flow, and the standard queues are close to 0. Our policy will be determined by the choice of allocations $\frac{d}{dt} T_j(t)$ at all times. There may be several policies that can achieve stability of the tracking process.

We choose to use maximum pressure policy on the tracking process. We will show in the next section that maximum pressure is rate stable, and under appropriate scaling we will show that using the two-phase policy, solving for the optimal fluid solution, and tracking it with Maximum pressure policy, is asymptotically optimal.

We will need the following property that ensures weak stability of fluid solutions for the fluid tracking process under maximum pressure policy:

Proposition 14.1 *Let $Q(t)$ be a processing network satisfying EAA. Let $q(t)$ be the optimal fluid SCLP solution for $Q(t)$ as defined by (14.7)–(14.8), and let $\tilde{Q}(t)$ be the fluid tracking process (14.10). Then $\tilde{Q}(t)$ satisfies the conditions of Theorem 13.5, which imply that each of the processes $\tilde{Q}(t)$ is rate stable.*

Proof We need to show that the processing network with IVQs that is modeled by the fluid tracking process $\tilde{Q}(t)$ in each of the intervals (t_{m-1}, t_m) of the fluid solution satisfies EAA, and that the nominal flow rates α_k for $k \in \mathcal{K}_\infty^m$ given as (14.9) by the fluid solution provide a feasible solution to the static production planning problem (13.11).

We note that by definition of EAA property, if the original processing network has EAA property, then so do all the processing networks with IVQs that define the tracking process, since IVQs always have extreme actions available.

The optimal solution of the fluid control problem (14.7), (14.8), in the interval (t_{m-1}, t_m), is given by the controls u^m. By taking derivatives of the dynamics constraints, we have that $\dot{q}(t) + Ru^m = 0$ in the interval. For each $k \in \mathcal{K}_0^m$ in that interval $\dot{q}(t) = 0$. For each $k \in \mathcal{K}_\infty^m$, $\dot{q}(t)$ consists of inflow and of outflow such that the outflow equals $\sum_{j \in \mathcal{J}} C_{j,k} \, \mu_j u_j^m$ which is exactly how we defined α_k. It follows that for the process $\tilde{Q}(t)$, $t \in (t_{m-1}, t_m)$ the controls u^m satisfy:

$$R_{\mathcal{K}_\infty^m} u^m = \alpha,$$
$$R_{\mathcal{K}_0^m} u^m = 0,$$
$$A u^m \leq \mathbf{1},$$
$$u^m \geq 0,$$

which are exactly the conditions for feasibility in (13.11).   □

14.7 Asymptotic Optimality of the Two-Phase Procedure

We summarize the model and the policy. We consider a processing network with buffers $k = 1, \ldots, K$, activities $j = 1, \ldots, J$, resources $i = 1, \ldots, I$, with constituency matrix C and resource consumption matrix A. We assume that the service time $S_j(t)$ and routing $\mathcal{R}^j(\ell)$ processes satisfy SLLN conditions $S_j(t)/t \to \mu_j$, and $\mathcal{R}_{k,l}^j(\ell)/\ell \to p_{k,l}^j$. We consider this processing network with given initial buffer contents $Q(0)$, and wish to control it over a finite time horizon $[0, T]$.

The two-phase policy will be denoted by subscript $*$. It works as follows:
(i) We solve the SCLP fluid optimization problem, to obtain a fluid solution $\bar{q}(t)$, $0 \leq t \leq T$.
(ii) We define the fluid tracking process $\tilde{Q}(t)$, over the various intervals of the fluid solution, with the definition of standard queues and IVQs, and nominal input rates for the IVQs obtained from the fluid solution.
(iii) At any moment in time we then determine the resource allocation to activities as dictated by the maximum pressure policy for the fluid tracking process, with its IVQs.
(iv) We use this resource allocation to process the buffer contents $Q(t)$ whenever possible. If an IVQ buffer of the tracking process $\tilde{Q}(t)$ requires

processing but the actual buffer of $Q(t)$ is empty, no processing is done.
(v) We add the following proviso: If at any time there is a tie under maximum pressure between idling or starting an activity, we break it in favor of activity. This is needed for the case in which all of $\tilde{Q}_k(t)$ (standard and IVQ) are 0, and then if one idles (especially under non-preemptive non-splitting policy), there will be no event at which to start any future activity.

We denote the minimum objective of the optimal fluid solution by

$$V_* = \int_0^T c^\top \bar{q}(t)dt.$$

We will wish to compare it to the random objective for any general policy π,

$$V_\pi = \int_0^T c^\top Q(t)dt.$$

The following theorem states that the two-phase policy is asymptotically optimal. For the asymptotic statement, we consider the original network scaled up by a factor of N, so we have initial buffer contents $Q^N(0) = NQ(0)$, and time horizon $[0, NT]$.

We denote by $Q_*^N(t)$ the sequence of processes obtained under our two-phase policy, and by V_*^N its random objective. We denote by $Q_\pi^N(t)$, V_π^N, the corresponding quantities for policy π.

Theorem 14.2 *The two-phase policy is asymptotically optimal in the sense that:*

(i) For any policy π, a.s.

$$\liminf_{N \to \infty} V_\pi^N/N^2 \geq V_*. \tag{14.11}$$

(ii) For the two-phase policy,

$$\lim_{N \to \infty} \frac{1}{N} Q_*^N(Nt) = \bar{q}(t), \quad 0 < t < T, \tag{14.12}$$

and

$$\lim_{N \to \infty} V_*^N/N^2 = V_*. \tag{14.13}$$

Proof (i) Consider any policy π. Let ω be a sample path that satisfies SLLN, and assume that for some subsequence N_k, $V_\pi^{N_k}(\omega)$ converges. Using the arguments of Theorem 11.2 fluid limits of $\mathcal{T}_\pi^{N_k}(t, \omega)$ exist, and therefore fluid limits of $Q_\pi^{N_k}(t, \omega)$ exist, so for some subsequence r of N_k, $Q_\pi^r(t, \omega)/r \to q_\pi(t)$, and $V_\pi^r(\omega)/r^2 \to V_\pi$. But in that case, $q_\pi(t)$ must satisfy the constraints of the fluid SCLP (14.6), which proves that $V_\pi \geq V_*$. This argument holds almost surely, which proves (14.11).

(ii) Obviously (14.13) follows from (14.12). We will show that (14.12) holds in every interval (t_{m-1}, t_m). By definition, it holds at time 0. We note that the fluid SCLP problem is completely determined by the initial values, so for system N we have that the SCLP optimal fluid solution $q_*^N(t)$, $0 < t < NT$ satisfies $\frac{1}{N} q_*^N(Nt) = \bar{q}(t)$ (the solution of the SCLP), and so for system N the first tracking interval is simply $(0, Nt_1)$, and the fluid tracking maximum pressure policy has standard queues $\mathcal{K}_0^1$ and IVQs $\mathcal{K}_\infty^1$. By Proposition 14.1 the tracking policy is rate stable in each interval, so as we let $N \to \infty$, for any $t \in (0, t_1)$, $\frac{1}{N} \tilde{Q}_*^N(Nt) \to 0$. Furthermore, for any $t \in (0, t_1)$, for all N large enough $Q_*^N(t)$ will be larger than constant κ. This implies that the maximum pressure allocation calculated for $\tilde{Q}_*^N(t)$ can be implemented for $Q_*^N(t)$. We therefore have that for any $t \in (0, t_1)$, $\frac{1}{N} Q_*^N(Nt) \to \bar{q}(t)$. Similarly, by induction this holds for $t \in (t_{m-1}, t_m)$, $m = 2, \ldots, M$. This completes the proof.

$\square$

14.8 An Illustrative Example

Consider the three-buffer re-entrant line of previous Section 13.2.2, Figure 13.4, but assume now that $m_2 > m_1 + m_3$, so station 2 is a bottleneck. We start off with initial fluid in all three buffers, and no further input, and we wish to empty the system with minimum cumulative inventory. The fluid problem is:

$$\min V = \int_0^T (q_1(t) + q_2(t) + q_3(t))\, dt$$

s.t.

$$q_1(t) = q_1(0) - \int_0^t \mu_1 u_1(s)\, ds$$

$$q_2(t) = q_2(0) - \int_0^t \mu_2 u_2(s)\, ds + \int_0^t \mu_1 u_1(s)\, ds$$

$$q_3(t) = q_3(0) - \int_0^t \mu_3 u_3(s)\, ds + \int_0^t \mu_2 u_2(s)\, ds$$

$$u_1(t) \qquad\qquad + u_3(t) \leq 1$$

$$u_2(t) \qquad\quad \leq 1$$

$$u(t), q(t) \geq 0 \qquad\qquad t \in [0, T]$$

We take: $q_0 = (8, 1, 15)$, $\mu = (1, 0.25, 1)$, and $T = 40$.

The following Figure 14.6 shows three feasible solutions: minimum time solution, which keeps buffer 2 fully utilized all the time, LBFS which

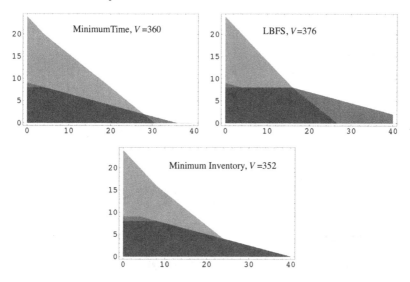

Figure 14.6 Fluid solutions: minimum time, LBFS, and optimal.

empties buffer 3 first and only then starts flow from buffer 1, and the optimal fluid solution of minimum area, obtained by solving the SCLP. It is seen that the optimal fluid solution starts with the greedy allocation of emptying buffer 3, but at time t_1, it slows the flow from buffer 3 and starts flow from buffer 1. The flow from buffer 1 allows positive flow through buffer 2 so that the bottleneck station 2 is fully utilized from t_1 onward.

A simulation of the system shows how the procedure tracks the optimal fluid solution, and how the tracking improves as $N \to \infty$. This is shown in Figure 14.7, taken from Nazarathy and Weiss (2009). Exercises 14.9–14.12 give more details for this example.

14.9 Implementation and Model Predictive Control

We discuss implementation in the case of unitary networks, for which the asymptotic optimality under non-splitting, non-preemptive implementation of the two-phase method is rate stable and asymptotically optimal. These include all MCQN where, in addition to scheduling at each machine, there is a choice of where to direct input and how to route items inside the network.

Fluid optimization problems, for networks of 30 machines and 500 buffers, are easily solved centrally by the simplex-type algorithm for SCLP.

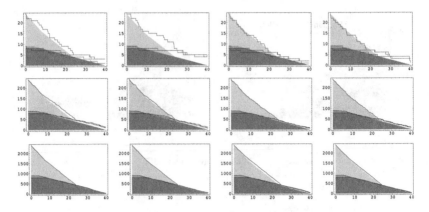

Figure 14.7 Example simulation: four samples (columns) with scalings $N = \{1, 10, 100\}$ for each sample, reproduced with permission from Nazarathy and Weiss (2009).

This solution is then translated to instructions for the operators of each of the machines in the network.

The instructions to each machine operator consist of the time breakpoints, and for each interval, the list of IVQ buffers that are processed by activities that use his machine, and the nominal flow rates for these IVQs, which may be positive or zero.

The machine operator of machine i is at all times responsible for activities $\{j : A_{i,j} = 1\}$, and buffers $\{k : A_{i,j}C_{j,k} = 1\}$. Given the nominal flow rates, he can track the deviations $\tilde{Q}_k(t)$ at the buffers under his control. He will also need the values of $\tilde{Q}_l(t)$ for the buffers directly down stream from k. Calculating the pressure from these, he will then obtain the next activity to be implemented. In the special case of a re-entrant line (with $j = k$), he will choose to activate buffer $k \in C_i$ for which $\mu_k(\tilde{Q}_k(t) - \tilde{Q}_{k+1}(t))$ is maximal, or idle if the pressure is negative.

The practical way to implement the two-phase finite horizon transient control is to use *rolling horizon* or, as it is now called, *model predictive control*. This method can control a processing network that is not necessarily stationary, over an unlimited time horizon, and it incorporates both inference and control.

Model predictive control (or rolling horizon) is implemented as follows: The SCLP is solved for a time horizon T, and it is then implemented by tracking the fluid solution over a time horizon $S < T$. For example, T may be six weeks, while S may be one week, or T is a week and S is one day.

After time S the SCLP is solved again, starting from the current state at the end of S, for time horizon T, and again implemented for S, etc.

The model predictive implementation allows also re-adjustments of the system parameters, i.e. μ_j, $p_{k,l}^j$, as new data is obtained and decisions are changed. It can incorporate control of systems with periodic changes. It can also incorporate online inference about the system parameters.

14.10 Sources

Wafer fabs motivated much research in the queueing community, including Kumar (1993), Chen et al. (1988), and Wein (1988). An engineering text on wafer fabs is Van Zant (2014). The job-shop scheduling problem is discussed in Lawler et al. (1993), and Fleischer and Sethuraman (2005). Various approaches to the control of MCQN are described in Kushner (2001), Harrison (1988), Harrison (1996), Meyn (2008), and Dai and Neuroth (2002). The simplex-type algorithm for solution of SCLP was developed by Weiss (2008), and its implementation for very large re-entrant lines and other fluid network problems is described in Shindin et al. (2021). The modeling of the deviations from the optimal fluid solution and the use of maximum pressure to track it, and the proof of asymptotic optimality, are due to Nazarathy and Weiss (2009).

Exercises

14.1 The objective (14.4) gives a reward of $(T - t)h_k$ for each departure from buffer k at time t. Find an objective that will give a reward of $w_{j,k}(T - t)$ for each completion of service of an item in buffer k by activity j, write the equation for this objective in terms of buffer contents, and write its fluid approximation.

14.2 Repeat the proof that fluid limits exist and are Lipschitz continuous, and that almost surely for all ω, the scaled queue lengths and allocations converge to fluid limits that satisfy equations (14.6).

14.3 Show that the problems SCLP and its symmetric dual SCLP* satisfy weak duality, i.e. if x, u and y, v are feasible solutions, then the objective value of SCLP* is greater or equal to the objective value of SCLP. Show that the objective values are equal if and only if the complementary slackness condition holds, in which case they are optimal solutions.

14.4 Show how to reconstruct the primal and dual solutions of SCLP, SCLP* from $x^0, y^M, B_1, \ldots, B_M$, and T.

14.5 Prove that solutions that satisfy all the conditions listed in Section 14.4 are optimal solutions of SCLP, SCLP*.

14.6 For the fluid solution example described in Section 14.5, describe what is happening in intervals 4–9.

14.7 For the example in Section 14.5, write the Boundary-LP/LP*, write the Rates-LP/LP*, list the primal and dual bases of the Rates-LP/LP* for the 9 intervals, and write the equations for the breakpoints.

14.8 (∗) Reverse engineer the SCLP problem with the solution described in Section 14.5.

14.9 Use the data for the 3-buffer re-entrant problem of Section 14.8 to calculate the minimum time to empty solution, and the last buffer first served solution.

14.10 Find the optimal fluid solution for the 3-buffer re-entrant problem of Section 14.8. Discuss its properties in terms of bases etc.

14.11 Write the equations for the tracking process for the 3-buffer re-entrant problem of Section 14.8, and describe the policies for tracking the fluid in each of the intervals of the fluid solution.

14.12 Simulate the control of the 3-buffer re-entrant problem of Section 14.8, for a scaling by 10 and by 100. Use exponential processing times for each activity.

Part V

Diffusion Scaled Balanced Heavy Traffic

In Part IV of the book we looked at MCQN, and studied the transient behavior by looking at their fluid model. Fluid models provide a rough approximation of the network, and they served us to determine stability by verifying that starting from any congested state, the fluid converged to zero, and further, we used the fluid to control networks over a finite time horizon, by finding an optimal fluid solution and tracking it.

When a network is operated for a longer time horizon under uniform conditions, we are interested in optimizing its steady state. For that the fluid is no longer informative; the fluid scaling of a stationary network converges to a constant at all times. To optimize the steady state, we will use diffusion scaling.

We now look at control of networks in balanced heavy traffic: In many networks one can ignore the nodes of the network that are not in heavy traffic, since the overwhelming majority of customers in the system will be at the nodes with heavy traffic. For nodes in heavy traffic, one is naturally led to diffusion approximation of the network. The paradigm that we will follow is to approximate the primitives of the network by Brownian processes, and then formulate control problems for the resulting Brownian network. This will induce a policy for the actual network. We consider in Chapter 15 routing to parallel servers; this illustrates our aim to always pool the system resources. We also observe state space collapse and briefly discuss diffusion limits for MCQN. In Chapter 16 we consider Brownian problems of scheduling and admission control, where we force congestion to be kept at the least costly nodes, and use admission control to regulate congestion. In Chapter 17 we extend the methods developed in Chapter 16 to routing control, and demonstrate significant savings that result from pooling efforts and balancing the contents of the nodes.

257

15

Join the Shortest Queue in Parallel Servers

In this chapter we consider the simplest problem of routing: Customers requesting a single service arrive, and need to choose which of several queues to join. This problem becomes acute when the arrival rate is close to the total service capacity. The aim is to get as close as possible to the performance of a single pooled server. In this chapter we discuss several policies and in particular the policy of join the shortest queue (JSQ). We will return to this problem when the number of servers becomes very large in Chapter 21.

In Section 15.1 we consider two identical single servers, Poisson arrivals, and exponential service, and obtain the stationary distribution of the queue lengths under join the shortest queue policy. In Section 15.2 we discuss the advantage of resource pooling in comparison to other policies. In Section 15.3 we show that the policy of join the shortest queue is asymptotically optimal in heavy traffic, by considering a diffusion approximation, and we highlight the property of state space collapse. In Section 15.4 we show that asymptotic optimality and state space collapse are also achieved by more general threshold policies. Finally, in Section 15.5 we briefly comment on diffusion limit results for general MCQN.

15.1 Exact Analysis of Markovian Join the Shortest Queue

We first look at the simplest two identical servers case. There are two identical $\cdot$/M/1 servers with service rates 1, customers arrive in a Poisson stream of rate $2\rho < 2$, and an arriving customer joins the shortest queue at his moment of arrival. When both queues have the same length, an arriving customer will choose either with probability $1/2$.

The system can be described by a continuous time countable state Markov chain that moves as a random walk on the non-negative quadrant integer grid. We define the state (m, n) where m is the length of the shortest queue, and n is the difference between the queue lengths. Figure 15.1 shows the system, as well as the states and transition rates. The analysis of this two-dimensional Markov chain has puzzled researchers for many years, and illustrates the difficulty of solving balance equations even for simple looking non-birth and death chains. It was finally resolved by Ivo Adan, Jaap Wessels and Henk Zijm, who showed that the stationary distribution can be expressed as an infinite sum of product-form distributions. We sketch here their derivation, using what they call the *compensation method*. We leave out the details of the various steps, which can be followed in Exercise 15.1.

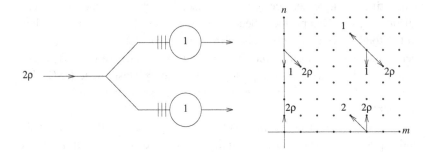

Figure 15.1 Two servers, join the shortest queue; m is the shortest queue, n is the difference.

It is immediately clear that the stationary distribution is not of product form. As a first step, one can easily write the global balance equations for the stationary probabilities $p(m, n)$, and we can then eliminate those for $p(m, 0)$ and obtain three sets of equations:

(i) Equations for $p(m, n)$, $m > 0$, $n > 1$ that need to hold in the interior of the quadrant.

(ii) Equations for $p(m, 1)$, $m > 0$ that need to hold for the horizontal boundary.

(iii) Equations for $p(0, n)$, $n > 0$ that need to hold for the vertical boundary.

We can try a product-form solution of the form $p(m, n) = \alpha^m \beta^n$. The equations (i) will all be satisfied if and only if α, β solve:

$$\beta^2(\alpha + 2\rho) - 2\beta\alpha(\rho + 1) + \alpha^2 = 0.$$

Any linear combination of such solutions will also satisfy (i). This equation is solved by points on a curve illustrated in Figure 15.2. Note that for fixed α it is a quadratic equation in β, and vice versa.

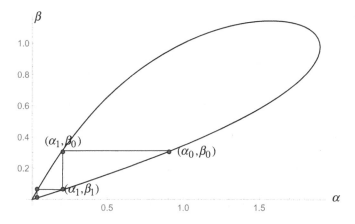

Figure 15.2 Curve of product-form candidates, and the converging set of compensations.

If we choose

$$\alpha_0 = \rho^2, \quad \beta_0 = \frac{\rho^2}{2 + \rho},$$

both (i) and (ii) will be satisfied, but not (iii).

We now take as α_1 the other root of the quadratic with β_0, and find a linear combination $\alpha_0^m \beta_0^n + c_1 \alpha_1^m \beta_0^n$ that satisfies (i) and (iii) but not (ii). This new term forms a compensation. The next compensation is to take β_1 that is the other root of the quadratic with α_1, and find a combination $c_1 \alpha_1^m \beta_0^n + c_1 d_1 \alpha_1^m \beta_1^n$ that will again satisfy (ii) but not (iii). Proceeding in this way, we get that at each odd step we satisfy (i), (ii), and at each even step we satisfy (i), (iii). We then see that the α_k, β_k converge geometrically to zero and that the sum with the c_k, d_k coefficients converges. The procedure is illustrated by the following Figure 15.2, in which we see how the successive values

$$\alpha_0 > \beta_0 > \alpha_1 > \beta_1 > \alpha_2 > \cdots$$

are determined as pairs of roots of the quadratics. The compensating coefficients are determined with $c_0 = d_0 = 1$ and

$$c_{k+1} = -\frac{\alpha_{k+1} - \beta_k}{\alpha_k - \beta_k} c_k, \quad d_{k+1} = -\frac{(\alpha_{k+1} + \rho)/\beta_{k+1} - (\rho + 1)}{(\alpha_{k+1} + \rho)/\beta_k - (\rho + 1)} d_k.$$

The stationary distribution is then the infinite sum of these product-form solutions:

$$\pi(m, n) = B \sum_{k=0}^{\infty} d_k \left(c_k \alpha_k^m + c_{k+1} \alpha_{k+1}^m \right) \beta_k^n$$

$$= B \beta_0^n \alpha_0^m + B \sum_{k=0}^{\infty} c_{k+1} \left(d_k \beta_k^n + d_{k+1} \beta_{k+1}^n \right) \alpha_{k+1}^m,$$

where the normalizing constant is

$$B = \frac{\rho(2 + \rho)}{2(1 - \rho^2)(2 - \rho)}.$$

We have encountered another example of the compensation method in Section 13.2.1.

It can be shown that join the shortest queue is optimal; in fact, it minimizes stochastically the discounted number of customers served up to any time t. This is also true for s identical $\cdot$/M/1 servers (see Exercise 15.4).

Join the shortest queue is also optimal if the distribution of service times is IHRA (increasing hazard rate on the average). It may not be optimal for other distributions of service time, as some examples show. A famous example is when processing time is 0 with high probability and a large constant with small probability. In that case, it may be better to join the longer queue, since it indicates that the customer in service has already been in service for a longer time than the customer in service in the other queue, and there is a good chance that all the waiting customers, even in the longer queue, have 0 processing times (see Exercise 15.5).

15.2 Variability and Resource Pooling

The following Table 15.1 illustrates the advantage of join the shortest queue. We compare five different policies. The five policies are: two independent M/M/1 queues, equivalently, two servers and customers choose which queue to join at random; two servers and customers are routed to them in alternating order; join the shortest queue; two servers serving a single queue; one server that works at twice the speed. The table lists average sojourn times for Poisson arrivals and exponential mean 1 service time, as a functions of ρ. The first thing to notice is that a single twice as fast server has half the queue of two separate servers, each with his own queue. This is the famous pooling effect: The queue length for an M/M/1 queue is a function of ρ alone, so we have the same queue in both systems, but for the fast server everything is happening at twice the rate, so the sojourn time is half. Hence,

Table 15.1 *Resource pooling under various policies – average sojourns.*

ρ	Speed-Up M/M/1	M/M/2	Join Short Queue	Alternate Routing	Two M/M/1's
0.3	0.7143	1.0989	1.1441	1.2166	1.4286
0.5	1.0000	1.3333	1.4262	1.6180	2.0000
0.7	1.6667	1.9608	2.1081	2.6006	3.3333
0.9	5.0000	5.2632	5.4748	7.5883	10.0000

pooling two separate queues into a single twice as fast queue has a saving factor of 2. By the same argument, pooling s separate single-server M/M/1 queues into a single s time faster server queue has a saving factor of s. The same holds for the first moment of the queue length for M/G/1, which is a function of the squared coefficient of variation of service time alone, and not of the average service time.

Next we see that pooling two separate M/M/1 queues into a single queue served by two servers, i.e. an M/M/2 queue, is almost as good. Comparing an M/M/2 queue with an M/M/1 queue with a twice faster server, we can see that the average waiting time before service is in fact shorter for M/M/2, but the sojourn time is slightly longer, and the difference is bounded by the length of a single service period, which becomes negligible in heavy traffic (see Exercise 15.6).

More surprising, we see that join the shortest queue is almost as good as the single fast server pooled queue, as the traffic intensity increases. In the next section we show that this observation is valid for general renewal arrivals and services. We make a similar comparison for many servers in Section 21.2.

We also observe that alternate routing achieves half the savings of pooling, by reducing the variability of interarrival times, since in alternate routing each queue has Erlang 2 interarrival times, as opposed to exponential with the same mean for two independent M/M/1 queues. See Exercises 15.6–15.8.

15.3 Diffusion Approximation and State Space Collapse

We now show that JSQ achieves resource pooling in heavy traffic under more general conditions, by obtaining a diffusion approximation.

We consider the following model of two servers and three arrival streams, where customer streams 1 and 2 are served by servers 1 and 2, respectively, and customers of arrival stream 3 are free to choose which queue to join. We assume that $\lambda_3 > \lambda_1 - \lambda_2 \geq 0$, so that stream 3 can be split to balance the arrival rates of the two servers. We assume general i.i.d. services of rate 1, and general independent renewal arrivals, and let $c_{a,i}^2$, $i = 1, 2, 3$, $c_{s,i}^2$, $i = 1, 2$ be the squared coefficients of variation of the various interarrival and service times, the system is shown in Figure 15.3.

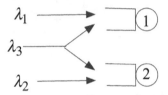

Figure 15.3 A two-server model with choice of routing.

To study the behavior of this system in heavy traffic, we consider a sequence of systems in which $\sqrt{n}(\lambda_1^n + \lambda_2^n + \lambda_3^n - 2) \to -\theta < 0$, as $n \to \infty$. In this sequence of systems, as $n \to \infty$ the system approaches $\rho = 1$. We let $W_i^n(t)$, $i = 1, 2$ be the workloads at the two servers, i.e. the total time needed to clear all the customers currently at each queue, and let $\hat{W}_i^n(t) = W_i^n(t)/\sqrt{n}$, $i = 1, 2$ be the diffusion scaled workloads. We then have the following :

Theorem 15.1 *(i) Under join the shortest queue, $|\hat{W}_1^n - \hat{W}_2^n| \to_p 0$ as $n \to \infty$.*

(ii) $\hat{W}_1^n + \hat{W}_2^n \to_w RBM(\theta, \sum \lambda_i c_{a,i}^2 + \sum c_{s,i}^2)$ as $n \to \infty$.

In words: Sending customers of arrival stream 3 to the shortest queue equalizes the two workload processes, and the system converges in heavy traffic to the same diffusion limit as a G/G/2 system.

The theorem will follow from the following two propositions. For simplicity, assume a single arrival stream of rate $\lambda < 2$ with all customers free to choose which queue to join.

Proposition 15.2 *Let $W_1(t)$, $W_2(t)$ be the workloads of the two queues under JSQ, and let $W(t)$ be the total workload of a G/G/2 system, with the* same sequence of customers (i.e. we couple the JSQ system with the G/G/2 system). Then, for the coupled system, for every sample path,

$$0 \le \mathcal{W}_1(t) + \mathcal{W}_2(t) - \mathcal{W}(t) \le \sup_{0 \le s \le t} |\mathcal{W}_1(s) - \mathcal{W}_2(s)|. \qquad (15.1)$$

Proposition 15.3 *Under join the shortest queue, if $\sqrt{n}(\lambda_1^n + \lambda_2^n + \lambda_3^n - 2) \to -\theta < 0$, as $n \to \infty$, then $\sup_{0 < t < T} |\hat{\mathcal{W}}_1^n(t) - \hat{\mathcal{W}}_2^n(t)| \to_p 0$ for all T, as $n \to \infty$.*

The upshot of this is that rather than having a two-dimensional process, $(\mathcal{W}_1, \mathcal{W}_2)$, in heavy traffic the system behaves as a one-dimensional process, of $\mathcal{W}_1 + \mathcal{W}_2$, where $\mathcal{W}_1 \approx \mathcal{W}_2$. This phenomenon is called *state space collapse* – a two-dimensional process behaves like a one-dimensional process. Furthermore, in heavy traffic, because of this state space collapse and the inequality (15.1), the process $\mathcal{W}_1 + \mathcal{W}_2$ behaves like the process $\mathcal{W}(t)$, the workload of the G/G/2 system. This proves that JSQ is asymptotically optimal.

Proof of Proposition 15.2   Denote $L(t) = \mathcal{W}_1(t) + \mathcal{W}_2(t) - \mathcal{W}(t)$ and $K(t) = |\mathcal{W}_1(t) - \mathcal{W}_2(t)|$; we wish to show $0 \le L(t) \le \sup_{0 \le s \le t} K(s)$. Consider a single busy period of the JSQ system, starting at time 0, so that $L(0) = K(0) = 0$. Let t_0 be the first time that G/G/2 has both servers busy, while JSQ has one server idle, if such a time exists, or else we let t_0 be the end of the busy period. Then $L(t) = 0$ for $0 \le t < t_0$. At the time t_0 w.l.g., assume server 2 is idle and server 1 is busy. Then, $L(t_0) \ge 0$ and $K(t_0) = \mathcal{W}_1(t_0) \ge \mathcal{W}(t_0)$. In the time following t_0, while G/G/2 has two busy servers and JSQ only one, $L(t)$ increases at rate 1, while $K(t)$ decreases at rate 1. However, this can only continue for at most a duration $\mathcal{W}(t_0)/2$, and so in this entire period that $L(t)$ is increasing, $0 \le L(t) \le L(t_0) + \mathcal{W}(t_0)/2 \le K(t_0) - \mathcal{W}(t_0)/2 \le K(t_0)$.

Before the time $t_0 + \mathcal{W}(t_0)/2$, three things can happen: (i) there may be an arrival, in which case both systems will again have two servers busy, or (ii) one of the servers in the G/G/2 system will become idle, so both systems will have a single busy server, and in both cases $L(t)$ will continue with no change. Finally (iii), we may reach a time t when $\mathcal{W}(t) = 0$ while $L(t) = \mathcal{W}_1(t) \ge 0$. At that time, the JSQ system will have one busy server while the G/G/2 system will be idle, and $L(t)$ will be decreasing until it reaches 0 and the busy period ends, or there is a new arrival and then at first the G/G/2 system will have one server busy, and the JSQ will have both servers busy, and $L(t)$ will continue to decrease, again until the end of the busy period, or until another arrival makes both systems fully busy.

Throughout, $0 \leq L(t) \leq K(t_0)$ will be maintained, until the busy period is over, or until at some time $t_1 > t_0$, again for the first time, system G/G/2 has two busy servers while the JSQ system has one idle server.

Then at t_1, $L(t_1) \geq 0$ and $K(t_1) = \mathcal{W}_i(t_1)$, with $\mathcal{W}_{3-i}(t_1) = 0$ for $i = 1$ or $i = 2$, and in the next period, by the same argument, $0 \leq L(t) \leq K(t_1)$. This can be repeated with $t_0 < t_1 < \cdots t_k < \cdots$ being successive times at which G/G/2 has both servers busy and in the JSQ system one of the servers becomes idle. This shows that (15.1) holds throughout the busy period.

To summarize: $L(t)$ never becomes negative; it only increases when G/G/2 has both servers busy and JSQ has one idle server, and within a period of increase it stays below $K(t_k)$ where t_k is the last time before t that one server became idle in the JSQ system.   □

Proof of Proposition 15.3   The proof is similar to the proof of Theorem 6.3, and is left as Exercise 15.10.   □

15.4 Threshold Policies for Routing to Parallel Servers

Join the shortest queue achieves two things: It is optimal for each joining customer, and in heavy traffic it minimizes the total number in the system, i.e. minimizes average waiting time. It achieves this by equalizing the queues, so we rarely have one server idle while the other has a long queue. It manages in heavy traffic to keep both servers working almost all the time that there are customers in the system, i.e. it behaves as if both servers share a single queue. However, the objective of keeping both servers busy at most times can also be achieved by a threshold policy. A threshold policy allows customers to join any queue they choose, as long as the number in each of the queues is above a threshold. When the number of customers at a queue falls below that threshold, arriving customers are directed to that queue.

Analysis of the Markovian case and simulations show that threshold policies also achieve complete resource pooling in heavy traffic.

15.5 A Note about Diffusion Limits for MCQN

We have found in Chapter 6 that by centering and scaling the G/G/1, G/G/s, and G/G/∞, we can obtain convergence to a diffusion or a Gaussian process. Similarly, in Chapter 9 we saw that by centering and scaling a generalized Jackson network, we can also obtain convergence to a multi-variate diffusion process. A natural question is whether this can be done for MCQN. It turns out that this is a hard question, and existence of diffusion limits is, similar to stability of MCQN, dependent on the policy. There are examples of networks that under given policies do not converge to a diffusion limit under diffusion scaling.

The question of finding diffusion limits for MCQN operated by a given policy is as yet unresolved in general, but a deep theory was developed by Maury Bramson and Ruth Williams to answer this question. It involves two major steps. First, one needs to verify that under heavy traffic state space collapse occurs: as $\rho_i \to 1$ for stations $i = 1, \ldots, I$, the K-dimensional queue lengths process behaves like a process of dimension $\leq I$. This indeed is what we showed for the JSQ system, where the two-dimensional process collapsed to a single dimension. The second step is then to verify some conditions under which the collapsed process converges to a diffusion. This two step procedure was carried out for FCFS Kelly networks, and for general MCQN under the policy of HLPPS (head of the line proportional processor sharing). Another example where a diffusion limit was established is a MCQN under maximum pressure policy, under a condition of complete resource pooling. We shall say more about this in Section 16.7, when we discuss asymptotic optimality of policies.

15.6 Sources

The exact analysis of symmetric join the shortest queue, using the compensation method to obtain an infinite sum of product forms was derived by Adan et al. (1990, 1993); see also Kingman (1961). Optimality of JSQ was established by Winston (1977) and Weber (1978). For a counter-example to JSQ optimality, see Whitt (1986). The advantage of alternate routing and a proof of optimality of extremal splitting for more parallel servers is proved by Hajek (1985). The term state space collapse was coined by Reiman (1984b), where he analyzed JSQ in heavy traffic, for general input and service times and proved asymptotic optimality. Comparison of join the shortest queue and threshold policies is illustrated in Kelly and Laws (1993), and further results on threshold models are obtained by Van Houtum et al. (1998).

An example of non-existence of diffusion limits is presented in Dai and Wang (1993). Verifying state space collapse and proving diffusion limits are outlined in the two seminal papers by Bramson (1998) and Williams (1998). Diffusion limit under maximum pressure policy, when the state space collapses to a single bottleneck station is derived in Dai and Lin (2008).

Exercises

15.1 (∗) For the case of two symmetric ·/M/1 queues, under join the shortest queue (JSQ) policy, complete the derivation of the stationary distribution as an infinite sum of product forms. Perform the following steps [Adan et al. (1990)]:

   (i) Write the balance equations, eliminate those for $n = 0$, and set up the three sets of equations, for states in the interior of the quadrant, for states on the horizontal $n = 1$ boundary, and for states on the vertical $m = 0$ boundary.

   (ii) Derive the quadratic equation for α, β that give product-form solutions for the states in the interior.

   (iii) Verify that α_0, β_0 satisfy balance equations for states (i),(ii).

   (iv) Do the first compensation, to satisfy (i),(iii).

   (v) Derive the sequence of α_k, β_k and the coefficients c_k, d_k.

   (vi) Show that the coefficients and the roots converge geometrically.

15.2 (∗) The following model is called *join the shortest queue with jockeying*: in the two symmetric ·/M/1 queues, customers always join the shortest queue; however, when the difference in queue length at the two servers exceeds a threshold d, a waiting customer is moved from the longer to the shorter queue. For $d = 2$, draw the states and transitions diagram, write the balance equations, and suggest how to solve them [Adan et al. (1991, 1994)].

15.3 For the case of two symmetric ·/M/1 queues, under join the shortest queue policy, obtain the stationary marginal distribution of the difference in length between the two queues, and find an M/M/1-type approximation that is an upper bound for it.

15.4 For two identical ·/M/1 servers and any arrival stream, prove that join the shortest queue maximizes the probability that k items will complete service by time s, for all s and k (hint: use induction on k [Winston (1977); Weber (1978)]).

15.5 Consider Poisson arrivals, and service time that is 0 with probability $1 - \epsilon$, and n with probability ϵ. Consider the following alternative to JSQ: If there is an empty server or if the difference in queue length is 0 or 1, use JSQ. Otherwise, join the longer queue. Show that for fixed λ and n, this policy outperforms JSQ as $\epsilon \to 0$ (hint: show that under any policy, the probability that a busy cycle contains k long jobs is of order ϵ^k [Whitt (1986)]).

15.6 Calculate the expected waiting time and the expected sojourn time for an M/M/2 queue, and compare it with the expected waiting time and expected sojourn time of an M/M/1 queue with a server that has twice the speed.

15.7 Calculate the expected waiting time for a system with s ·/M/1 queues, when Poisson customers are routed on a round-robin policy to the different queues.

15.8 Verify the calculated values of the resource pooling effect on expected sojourn time as listed in Table 15.1 .

15.9 An alternative to the proof of the bound on the difference between JSQ and G/G/2, is to compare JSQ to a G/G/1 queue, with half the processing times. With $\tilde{\mathcal{W}}(t)$ the unfinished workload of the G/G/1 queue, show that

$$\tilde{\mathcal{W}}(t) \le W_1(t) + W_2(t) \le \tilde{W}(t) + \sup_{0 < s < t} |\mathcal{W}_1(t) - W_2(t)|.$$

15.10 Complete the proof of Proposition 15.3.

15.11 Complete the proof of Theorem 15.1 without assuming $\lambda_1 = \lambda_2 = 0$.

15.12 Prove that similar to Propositions 15.2, 15.3 about workloads, the following theorem holds for the queue lengths:

(i) Under join the shortest queue, $|\hat{Q}_1^n - \hat{Q}_2^n| \to_p 0$ as $n \to \infty$.

(ii) $\hat{Q}_1^n + \hat{Q}_2^n \to_w \text{RBM}(\theta, \sum_i \lambda_i c_{a,i}^2 + \sum c_{s,i}^2)$ as $n \to \infty$.

16

Control in Balanced Heavy Traffic

16.1 MCQN in Balanced Heavy Traffic

In Part IV we studied the transient behavior of MCQN via fluid models. We now want to study long-term behavior of stable but congested systems. We use diffusion scaling for approximating the network: We look at time in units of n, and space consisting of number of customers or immediate workload in units of $\sqrt{n}$, or equivalently we consider customers in units of N, and time in units of N^2. We will in fact use the N, N^2 scaling (rather than $n, \sqrt{n}$), since we are interested in those workstations that in steady state contain a large number N of customers, with corresponding workload of order N, and wish to study their behavior over a time horizon of length N^2. In this situation, workstations with a light or moderate workload can be ignored and will be eliminated from discussion, and we are left with what we call a MCQN in balanced heavy traffic.

The dynamics of MCQN were derived in Sections 9.5 and 10.1; we rewrite (10.1), the dynamics for buffer k. Because we are interested in the long-run, we can take $Q(0) = 0$:

$$Q_k(t) = \mathcal{A}_k(t) - \mathcal{S}_k(T_k(t)) + \sum_{l=1}^{K} \mathcal{R}_{l,k}(\mathcal{S}_l(T_l(t))).$$

Given stationary primitives of the system, with arrival rates α, service rates μ and routing probabilities P (with spectral radius < 1), stability of the system under any policy imposes certain constraints on the system. We get in steady state that the rate of flow in and out of each buffer, λ_k, is given by the unique solution of the traffic equations:

$$\lambda = \alpha + P^{\mathsf{T}}\lambda \implies \lambda = (I - P^{\mathsf{T}})^{-1}\alpha,$$

and we then have that the offered load at buffer k, denoted β_k, is given by $\beta_k = \lambda_k/\mu_k$. Using the input-output matrix $R = (I - P^{\mathsf{T}})\text{diag}(\mu)$ and its inverse $R^{-1} = \text{diag}(m)(I - P^{\mathsf{T}})^{-1}$, we can express β in vector form as the unique solution of the equation,

$$\alpha - R\beta = 0 \Longrightarrow \beta = R^{-1}\alpha = \text{diag}(m)(I - P^{\mathsf{T}})^{-1}\alpha = \text{diag}(m)\lambda.$$

Recall that $R_{k,l}$ is the rate of decrease in buffer k due to processing of buffer l. Then $\alpha - R\beta = 0$ spells out that input rate equals output rate, which is necessary for stability. β_k is the fraction of time that server $s(k)$ needs to devote to the processing of buffer k, to obtain balance of input and output. There is no flexibility in this; the long-run average time that needs to be devoted to buffer k needs to satisfy $\lim_{t \to \infty} \frac{1}{t}\mathcal{T}_k(t) = \beta_k$, $k = 1, \ldots, K$ for any stable policy.

We let $\rho_i = \sum_{k \in C_i} \beta_k$, which is the offered load to workstation i. We define $\nu_k = \beta_k/\rho_{s(k)}$. This is the fraction of busy time that workstation i needs to devote to buffer $k \in C_i$. We note that $\sum_{k \in C_i} \nu_k = 1$. We call ν_k the *nominal allocations*, they will play a major role in the sequel.

We now decompose the dynamics of the queues:

$$
\begin{aligned}
Q_k(t) &= (\alpha_k - \mu_k \nu_k + \sum_{l=1}^{K} p_{l,k}\mu_l \nu_l)t + (\mathcal{A}_k(t) - \alpha_k t) \\
&\quad - (S_k(\mathcal{T}_k(t)) - \mu_k \mathcal{T}_k(t)) \\
&\quad + \sum_{l=1}^{K} (\mathcal{R}_{l,k}(S_l(\mathcal{T}_l(t))) - p_{l,k}\mu_l \mathcal{T}_l(t)) \\
&\quad + \mu_k(\nu_k t - \mathcal{T}_k(t)) - \sum_{l=1}^{K} p_{l,k}\mu_l(\nu_l t - \mathcal{T}_l(t)) \\
&= \mathcal{X}_k(t) + \mu_k(\nu_k t - \mathcal{T}_k(t)) - \sum_{l=1}^{K} p_{l,k}\mu_l(\nu_l t - \mathcal{T}_l(t)),
\end{aligned}
$$

which in vector form is written as:

$$Q(t) = X(t) + R(\nu t - \mathcal{T}(t)). \tag{16.1}$$

We compare this decomposition to the decomposition for single class networks (e.g. generalized Jackson networks) as given in Section 9.3, equations (9.7)–(9.10). In a single class network, the entire busy time of each server is allocated to the single buffer it serves, the idle time is $t - \mathcal{T}_i(t)$, and therefore under work-conserving HOL non-preemptive policies, the idle time and the queue length processes are uniquely determined. In contrast, for MCQN each workstation divides its busy time between the various buffers it serves, where in the long-run a fraction ν_k is allocated to buffer k, and the new quantity, $\mathcal{J}_k(t) = \nu_k t - \mathcal{T}_k(t)$, which we call the *free time of buffer k* replaces the idle time. The idle time of workstation i is now:

$$\mathcal{I}_i(t) = t - \sum_{k \in C_i} \mathcal{T}_k(t) = \sum_{k \in C_i} (v_k t - \mathcal{T}_k(t)) = \sum_{k \in C_i} \mathcal{J}_k(t),$$

and it has the properties $\mathcal{I}_i(0) = 0$, $\mathcal{I}_i(t)$ is Lipschitz continuous, and $\mathcal{I}_i(t)$ in non-decreasing, which is the same as for single class networks. Similarly, $|\mathcal{J}_k(t) - \mathcal{J}_k(s)| < t - s$, so $\mathcal{J}_k(t)$ is also Lipschitz continuous. However, unlike $\mathcal{I}_i(t)$, $\mathcal{J}_k(t)$ are no longer restricted to be non-decreasing or positive. Furthermore, the total idle time, for work-conserving HOL policy, is not uniquely determined – it will depend on our scheduling policy: Work conservation requires that while there are customers at workstation i we do not idle, but we still have a choice of which buffer to serve, and this will influence congestion and idling at other buffers and stations. Our goal is to find good $\mathcal{T}_k(t)$, and this can be reformulated as a policy that controls the deviations $\mathcal{J}_k(t)$ from the predetermined nominal allocation $v_k t$.

Our balanced heavy traffic assumption is that $Q_k(t)$ are typically $O(N)$ with large N, that ρ_i, the offered load of workstation i, is close to 1, and that $N(1 - \rho_i)$ is of moderate size. We defined $v_k = \beta_k / \rho_{s(k)}$, so we have that $N(v_k - \beta_k)$ is also of moderate size. We then consider $X(t)$ of (16.1), which as before we name the netput process. We follow the steps of deriving diffusion limits in Section 9.5, Theorem 9.4, where we use $\frac{1}{N}\mathcal{T}_k(N) \approx \beta_k \approx v_k$ for large N (see Exercise 16.1). We obtain that $\frac{1}{N}X(N^2 t)$ is approximated by the Brownian process:

$$\hat{X}(t) = \theta t + (\Gamma)^{1/2} BM(t),$$

where $BM(t)$ is a standard Brownian motion, and $\hat{X}(t)$ has drift and co-variances (see Exercise 16.2),

$$\theta = N(\alpha - Rv), \quad \Gamma = \Gamma^0 + \sum_{j=1}^{K} v_j \Gamma^j, \quad \Gamma_{k,l}^0 = \alpha_k c_{ak}^2 \delta_{k,l}, \quad (16.2)$$

$$\Gamma_{k,l}^j = \mu_j p_{j,k}(\delta_{k,l} - p_{j,l}) + \mu_j c_{sj}^2 (\delta_{j,k} - p_{j,k})(\delta_{j,l} - p_{j,l}),$$

where $\delta_{k,l}$ is the Kronecker delta.

It is important to note that because for any stable HOL work-conserving non-preemptive policy, β_k is independent of the policy and $\frac{1}{N}\mathcal{T}_k(Nt) \to \beta_k t$, the approximating Brownian scaled netput process $\hat{X}(t)$ is independent of the policy.

16.2 Brownian Control Problems

For a MCQN in balanced heavy traffic, any stable policy will have the same throughput (rate of input and output), but the choice of policy will affect the waiting times and the queue lengths. We wish to find policies that will minimize the queue lengths, with the objective

$$\min \limsup_{T \to \infty} \frac{1}{T} \mathbb{E} \int_0^T \sum_{k=1}^K h_k Q_k(t) dt, \qquad (16.3)$$

where $h_k > 0$ are holding costs, cost per unit time of an item waiting in buffer k. This problem is however intractable even under the most simplifying assumptions of Poisson arrivals and exponential service times. Furthermore, in more general systems, the exact optimal policy will depend on detailed parameters of the distributions of interarrival and service times, which will hardly ever be available.

We have seen that under balanced heavy traffic, $\frac{1}{N} X(N^2 t)$ can be approximated by a Brownian motion process, $\hat{X}(t)$. We emphasize that $\hat{X}(t)$ is independent of the policy we use. We now formulate the following problem, which we name the *Brownian control problem* (BCP): For a given Brownian motion $\hat{X}(t)$, find $\hat{Q}(t), \hat{\mathcal{J}}(t), \hat{I}(t)$ such that

$$\min \limsup_{T \to \infty} \frac{1}{T} \mathbb{E} \int_0^T \sum_{k=1}^K h_k \hat{Q}_k(t) dt$$

$$\hat{Q}(t) = \hat{X}(t) + R\hat{\mathcal{J}}(t),$$

$$\hat{I}(t) = C\hat{\mathcal{J}}(t), \qquad (16.4)$$

$$\hat{Q}(t) \geq 0, \quad \hat{I}(0) = 0, \ \hat{I}(t) \text{ non-decreasing},$$

$$\hat{Q}(t), \hat{\mathcal{J}}(t), \hat{I}(t) \text{ non-predictive with respect to } \hat{X}(t).$$

In this problem, $\hat{Q}(t), \hat{\mathcal{J}}(t), \hat{I}(t)$ are the scaled version analogs of the actual queue length, free time, and idle time $Q(t), \mathcal{J}(t), I(t)$. Because the BCP involves only a Brownian motion, its solution depends only on first- and second-order parameters of the queueing system, and Brownian motion is a well-understood process. As we shall see, the BCP can be analyzed and sometimes even solved explicitly, for a wide range of MCQN. A solution of the BCP may then be used to suggest a policy for the queueing network control problem.

We now give an overview of the solution of the Brownian control problems. It involves the following three steps:

- Formulating the BCP in terms of workloads, the *workload formulation*.
- Solving for $\hat{Q}_k(t)$ for given $\hat{I}(t)$.

- Solving for $\hat{I}(t)$.

  We define the expected current workload process $\mathcal{W}(t)$ as

$$\mathcal{W}(t) = MQ(t) = CR^{-1}Q(t) = MX(t) + I(t). \tag{16.5}$$

Here $M_{i,k}$ is the expected amount of work that is required at workstation i, for each unit in buffer k until it leaves the system, and so $\mathcal{W}_i(t)$, $i = 1, \ldots, I$, is the expected amount of work for workstation i, in order to complete the processing of all the customers that are in the system at the present time t.

We define the workload netput $\hat{\mathcal{Z}}(t) = M\hat{X}(t)$, which is a Brownian motion with drift $M\theta$ and covariance matrix $M^{\mathsf{T}}\Gamma M$. The workload formulation of the BCP is then: For a given Brownian motion process $\hat{\mathcal{Z}}(t)$, find $\hat{Q}(t)$, $\hat{I}(t)$ such that for sample path $\hat{\mathcal{Z}}(t, \omega)$,

$$\begin{aligned}
\min \ \limsup_{T \to \infty} \frac{1}{T} \mathbb{E} \int_0^T \sum_{k=1}^K & h_k \hat{Q}_k(t) dt \\
\hat{\mathcal{W}}(t) &= M\hat{Q}(t), \\
\hat{\mathcal{W}}(t) &= \hat{\mathcal{Z}}(t) + \hat{I}(t), \\
\hat{Q}(t) &\geq 0, \quad \hat{I}(0) = 0, \ \hat{I}(t) \ \text{non-decreasing}, \\
\hat{Q}(t), \ \hat{I}(t) & \ \text{non-predictive with respect to } \hat{\mathcal{Z}}(t, \omega).
\end{aligned} \tag{16.6}$$

The workload formulation has the advantage that it reduces the number of constraints from K to I, and eliminates the unknowns $\hat{\mathcal{J}}_k(t)$. Once it is solved, we can retrieve the optimal solution of (16.4), from $\hat{\mathcal{J}}_k(t) = R^{-1}(\hat{Q}_k(t) - \hat{X}_k(t))$.

The solution of (16.6) is done in two steps. In the first step, we assume that $\hat{\mathcal{W}}(t) = \hat{\mathcal{Z}}(t) + \hat{I}(t)$ is known. We then formulate: for every T, and every sample path $\hat{\mathcal{Z}}(t, \omega)$ and its corresponding $\hat{\mathcal{W}}(t, \omega)$, find $\hat{Q}(t)$ such that:

$$\begin{aligned}
\min \int_0^T \sum_{k=1}^K & h_k \hat{Q}_k(t) dt \\
M\hat{Q}(t) &= \hat{\mathcal{Z}}(t, \omega) + \hat{I}(t, \omega), \\
\hat{Q}(t) &\geq 0, \\
\hat{Q}(t) & \ \text{non-predictive with respect to } \hat{\mathcal{Z}}(t, \omega).
\end{aligned} \tag{16.7}$$

This problem is separable, and can be solved separately for every t, by the standard linear program (LP):

$$\min \sum_{k=1}^{K} h_k \hat{Q}_k(t)$$

$$M\hat{Q}(t) = \hat{W}(t, \omega), \qquad (16.8)$$

$$\hat{Q}(t) \geq 0.$$

We note that this LP is feasible and bounded, and the LP solution can be chosen to have no more than I non-negative buffers (see Exercise 16.3).

Once we have solved for $\hat{Q}(t)$, we can use the solution to obtain the value of the objective for the given $\hat{W}(t)$. This value, denoted by $H(\hat{W}(t))$, is a convex piecewise linear function of $\hat{W}(t)$.

In the second step of the solution, we need for given Brownian motion $\hat{Z}(t)$, to find $\hat{I}(t)$ that solves

$$\min \limsup_{T \to \infty} \frac{1}{T} \mathbb{E} \int_0^T H(\hat{W}(t)) dt$$

$$\hat{W}(t) = \hat{Z}(t) + \hat{I}(t), \qquad (16.9)$$

$$\hat{W}(t) \geq 0, \ \hat{I}(0) = 0, \ \hat{I}(t) \text{ non-decreasing,}$$

$$\hat{I}(t) \text{ non-predictive with respect to } \hat{Z}(t, \omega).$$

Here it appears that we wish to use minimum idling, so as to keep the queues non-negative, and the solution will be a singular control. This second step is the crucial hard step, and we will indicate how it can be solved for some special examples.

Given a solution to the BCP, we then return to the actual queueing system in balanced heavy traffic. The Brownian solution requires us to transfer customers between queues instantaneously. This is almost possible for the discrete system because of the scale: Moving N customers is achieved much faster than the time scale of N^2. If we stop working on one queue and work on all the other queues at a workstation, the single queue will very quickly grow and all the others reduce, so we can emulate the Brownian control approximately. We can also transfer workload from one workstation to another very quickly, so a station can be kept fully busy even when it contains only very few customers, and so we can emulate the Brownian requirement to keep a station empty and yet not idling. Similarly, by controlling the input we can keep inventory low and yet not idle.

In the following sections we will use these ideas to solve a succession of problems of increasing complexity. So far, we outlined only control of the scheduling at the workstations, by means of the free times $\mathcal{J}(t)$. In addition to scheduling, we can control admissions to the system via $\mathcal{A}(t)$ and routing via $\mathcal{R}(n)$. In Section 16.3 we consider the simple so-called criss-cross network, in this network there is only one scheduling decision to control the system. In Section 16.4 we discuss scheduling in a closed MCQN with two stations; this adds the constraint that the scaled queue lengths satisfy $\mathbf{1}^\mathsf{T}\hat{Q}(0) = \mathbf{1}^\mathsf{T}\hat{Q}(t) = 1$. In Section 16.5 we extend these results to scheduling and admission control in an open MCQN with two stations. Finally, in Section 16.6 we generalize these results to a multi-station MCQN, with scheduling and admission controls. Further results, on routing control will be postponed until Chapter 17. Finally, in Section 16.7 we discuss what is known about the relation between the solution of the BCP and the control of the MCQN.

16.3 The Criss-Cross Network

The earliest application of the method suggested in Section 16.2 was the *criss-cross network*. It is the simplest example for illustrating the ideas of Sections 16.1, 16.2. The criss-cross network consists of two stations and two types of customers. Customers of type A require a single service at station 1, while customers of type B require service at station 1 followed by service at station 2. The criss-cross network is illustrated in Figure 16.1. Arrival rates of customers of type A are $\alpha_1 = \rho$ and of customers of type B are $\alpha_2 = \rho$. Expected processing times at station 1 are $m_1 = 1/\mu_1 = 0.5$, and at station 2 are $m_2 = 1/\mu_2 = 1$. As $\rho \to 1$, the system will be in balanced heavy traffic, both stations will be close to critically loaded.

Let $Q_k(t)$, $k = 1, 2, 3$ be the queue lengths at time t, and let $\mathcal{W}(t) =$

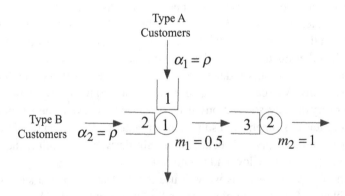

Figure 16.1 The criss-cross network.

$MQ(t)$ be the expected workload for these queues, with $M = \begin{bmatrix} 0.5 & 0.5 & 0 \\ 0 & 1 & 1 \end{bmatrix}$.

We note that for $\rho < 1$, as a feed-forward MCQN by Section 11.3.2, this system will be stable under any work-conserving policy. Our aim is to find a policy that will minimize the number of customers in the system. The objective then is to find a policy such that

$$\min \quad \limsup_{T \to \infty} \frac{1}{T} \mathbb{E} \int_0^T (Q_1(t) + Q_2(t) + Q_3(t)) dt. \qquad (16.10)$$

Since we are interested in the long-term average, we can for simplicity take $Q_k(0) = 0$, $k = 1, 2, 3$. Clearly, both stations should be work conserving, which determines the policy at station 2. Our only control is at station 1, to decide which type of customer to serve next. We could serve customers at station 1 on a FCFS basis (see Exercise 16.4), but we may do better if we use a decision rule that will be based on the state of the system at time t.

Both stations will be busy a fraction ρ of the time. Station 1 will devote a fraction $v_1 = 1/2$ of its busy time to customers of type A, and a fraction $v_2 = 1/2$ to customers of type B. v_1, v_2 are the nominal allocations, and $v_3 = 1$. Under any rate-stable policy, these will be the long-run allocations devoted to each class. Busy time, free time, and idle time are related by

$$\mathcal{I}_1(t) = t - (\mathcal{T}_1(t) + \mathcal{T}_2(t)) = v_1 t - \mathcal{T}_1(t) + v_2 t - \mathcal{T}_2(t) = \mathcal{J}_1(t) + \mathcal{J}_2(t).$$

Recall that $\mathcal{J}_k(t)$ is not restricted to be non-decreasing or non-negative. Our controls are $\mathcal{T}_k(t)$, or equivalently $\mathcal{J}_k(t)$.

We spell out the dynamics of the systems to illustrate the theory described in Sections 16.1 and 16.2:

$$
\begin{aligned}
Q_1(t) &= (\alpha_1 - v_1\mu_1)t + (\mathcal{A}_1(t) - \alpha_1 t) - (\mathcal{S}_1(\mathcal{T}_1(t) - \mu_1\mathcal{T}_1(t))) \\
&\quad + \mu_1(v_1 t - \mathcal{T}_1(t)) = \mathcal{X}_1(t) + \mu_1\mathcal{J}_1(t), \\
Q_2(t) &= (\alpha_2 - v_2\mu_1)t + (\mathcal{A}_2(t) - \alpha_2 t) - (\mathcal{S}_2(\mathcal{T}_1(t) - \mu_1\mathcal{T}_2(t))) \qquad (16.11) \\
&\quad + \mu_1(v_2 t - \mathcal{T}_2(t)) = \mathcal{X}_2(t) + \mu_1\mathcal{J}_2(t), \\
Q_3(t) &= (v_2\mu_1 - v_3\mu_2)t + (\mathcal{S}_2(\mathcal{T}_1(t) - \mu_1\mathcal{T}_2(t))) - (\mathcal{S}_3(\mathcal{T}_3(t) - \mu_2\mathcal{T}_3(t))) \\
&\quad - \mu_1(v_2 t - \mathcal{T}_2(t)) + \mu_2(v_3 t - \mathcal{T}_3(t)) = \mathcal{X}_3(t) - \mu_1\mathcal{J}_2(t) + \mu_2\mathcal{J}_3(t),
\end{aligned}
$$

and in terms of workload, the dynamics are

$$
\begin{aligned}
\mathcal{W}_1(t) &= m_1(Q_1(t) + Q_2(t)) \\
&= m_1 X_1(t) + m_1 X_2(t) + (v_1 t - \mathcal{T}_1(t)) + (v_2 t - \mathcal{T}_2(t)) \\
&= \mathcal{Z}_1(t) + I_1(t), \\
\mathcal{W}_2(t) &= m_2(Q_2(t) + Q_3(t)) \\
&= m_2 \left[(\alpha_2 - v_3\mu_2)t + (\mathcal{A}_2(t) - \alpha_2 t) - (\mathcal{S}_3(\mathcal{T}_3(t) - \mu_2\mathcal{T}_3(t))) \right] \\
&\quad + (v_3 t - \mathcal{T}_3(t)) \\
&= m_2(X_2(t) + X_3(t))) + m_2(v_3 t - \mathcal{T}_3(t)) \\
&= \mathcal{Z}_2(t) + I_2(t).
\end{aligned}
\tag{16.12}
$$

Notice that while $I_1(t)$ is forced to increase only when $\mathcal{W}_1(t) = 0$, this is not the case for $\mathcal{W}_2$, since I_2 may be forced to increase even in periods when $\mathcal{W}_2$ is positive, if Q_3 is empty while Q_2 is positive.

We now approximate the scaled $\frac{1}{N}X(N^2t)$ and $\frac{1}{N}Z(N^2t)$ by Brownian motions, $\hat{X}(t)$, $\hat{Z}(t)$, with $\hat{Q}(t), \hat{\mathcal{J}}(t), \hat{I}(t)$ the presumed approximations to the scaled $Q(t), \mathcal{J}(t), I(t)$. For the Brownian motions $\hat{X}(t), \hat{Z}(t)$, we can calculate the drift and covariances (see Exercise 16.7), but as it turns out, these do not play a role in the solution. Following the steps described in Section 16.2, the workload formulation of the BCP is: For each sample path $\hat{Z}(T, \omega)$ of the Brownian motion $\hat{Z}(t)$, and any T, find $\hat{Q}(t), \hat{I}(t)$ such that

$$
\begin{aligned}
\min \quad & \int_0^T \sum_{k=1}^3 \hat{Q}_k(t)\,dt \\
\text{s.t.} \quad & \frac{1}{2}\hat{Q}_1(t) + \frac{1}{2}\hat{Q}_2(t) = \hat{Z}_1(t, \omega) + \hat{I}_1(t), \\
& \hat{Q}_2(t) + \hat{Q}_3(t) = \hat{Z}_2(t, \omega) + \hat{I}_2(t), \\
& \hat{Q}(t) \ge 0, \quad \hat{I}(0) = 0, \quad \hat{I} \text{ non-decreasing}, \\
& \hat{Q}, \hat{I} \text{ non-anticipating with respect to } \hat{Z}(t).
\end{aligned}
\tag{16.13}
$$

We solve the BCP in two steps:
Step 1: Solve for optimal $\hat{Q}_k(t)$, when $\hat{\mathcal{W}}(t)$ are given.
Once $\hat{W}_i(t) = \hat{Z}_i(t) + \hat{I}_i(t)$, $i = 1, 2$ are given, the constraints on $\hat{Q}$ are

separate for each t, and we need to solve the linear program (LP):

$$\min \; \hat{V} = \sum_{k=1}^{3} \hat{Q}_k(t)$$

$$\text{s.t.} \; \frac{1}{2}\hat{Q}_1(t) + \frac{1}{2}\hat{Q}_2(t) = \hat{W}_1(t),$$

$$\hat{Q}_2(t) + \hat{Q}_3(t) = \hat{W}_2(t), \qquad\qquad (16.14)$$

$$\hat{Q}(t) \geq 0.$$

This LP can be solved by inspection: Clearly $\hat{Q}_2(t)$ must be made as big as possible, and from this we get

$$\hat{Q}_2(t) = 2\hat{W}_1(t) \wedge \hat{W}_2(t), \qquad \hat{V} = 2\hat{W}_1(t) \vee \hat{W}_2(t),$$
$$\hat{Q}_1(t) = \left(2\hat{W}_1(t) - \hat{W}_2(t)\right)^+, \quad \hat{Q}_3(t) = \left(\hat{W}_2(t) - 2\hat{W}_1(t)\right)^+. \qquad (16.15)$$

This means that for any given workloads $\hat{W}_i(t)$, we should keep as many of the customers as possible in $\hat{Q}_2(t)$, and the rest in one of the other two queues. This is enabled by the possibility of instantly moving diffusion scaled customers between queues.

Step 2: Solve for optimal $\hat{I}(t)$:
By step 1, q_1, q_2, q_3 are directly calculated from (16.15) for any w_1, w_2. Since the objective is $\hat{V} = 2w_1 \vee w_2$, the optimal workloads need to solve

$$\min \quad \hat{V} = 2\hat{W}_1(t) \vee \hat{W}_2(t),$$
$$\hat{W}_1(t) = \hat{Z}_1(t, \omega) + \hat{I}_1(t) \geq 0, \qquad (16.16)$$
$$\hat{W}_2(t) = \hat{Z}_2(t, \omega) + \hat{I}_2(t) \geq 0,$$

and this is achieved by the Skorohod reflection,

$$\hat{I}_i(t)) = \varphi(\hat{Z}_i(t)), \qquad \hat{W}_i(t)) = \psi(\hat{Z}_i(t)). \qquad (16.17)$$

This solution is non-anticipating; all that is needed at any time are the current values of $\hat{W}(t)$ and $\hat{Z}(t)$, and in fact, $\hat{I}_i(t) = \inf_{0 < s < t} \hat{Z}_i(s)$.

We now summarize how the policy works:

- Keep as much of the workload in class 2. This means that at any time class 2 contains all the workload for one of the stations, and the remaining workload is in class 1 if the workload of station 1 exceeds that of station 2, and in class 3 if the workload for station 2 exceeds that of station 1.
- While $\hat{Q}_2(t) > 0$, both stations will be working at full rate (see Exercise 16.6).
- Whenever $\hat{Q}_2(t) = 0$, one of two things happens: either also $\hat{Q}_3(t) = 0$, and then $\hat{W}_2(t) = 0$, and we use singular control $\hat{I}_2(t)$ to keep it non-negative, i.e. we idle workstation 2, or also $\hat{Q}_1(t) = 0$, and then $\hat{W}_1(t) =$

0, and we use singular control $\hat{\mathcal{I}}_1(t)$ to keep it non-negative, i.e. we idle workstation 1.

We have solved the BCP. We saw that it is non-anticipating. Furthermore, this policy minimizes the objective for every sample path, and it minimizes the objective for each sample path also pointwise, i.e. at every time t. This is a very strong optimality. It means in particular that it minimizes not just the long-term average, but also the objective of the transient Brownian motions for any time interval, for every realization of the process. In particular, it also minimizes the discounted objective, and various moments. We can calculate the distributions of the optimal $\hat{W}_1(t)$ and $\hat{W}_2(t)$, and obtain estimates of mean and variance of the Brownian objective (see Exercises 16.8 and 16.9).

We now return to the problem of the original discrete stochastic criss-cross queueing network. In the Brownian solution, we keep station 2 working at full rate on class 3, even when class 3 is empty, and all the customers are actually stored in class 2, and only flow through class 3 without queueing, leaving at the same rate that they enter. This is similar to what we observed with fluid approximations, where we could have positive flow through an empty buffer. It can be achieved for the diffusion, since the scaling is N in space and N^2 in time. It cannot be done in the discrete queueing network, but we can attempt to approximate it.

The following policy is suggested by the solution of the Brownian network problem. We wish to give priority to class 1 over class 2, as long as we can keep station 2 working. We therefore set a threshold κ, and use a threshold policy:
- While $Q_3(t) > \kappa$, station 1 will work on class 1.
- When $Q_3(t) \leq \kappa$, station 1 will work on class 2.

For every policy we can write the lower bound:

$$Q_1(t) + Q_2(t) + Q_3(t)$$
$$= (Q_1(t) + Q_2(t)) \vee (Q_2(t) + Q_3(t)) + Q_1(t) \wedge Q_3(t) \qquad (16.18)$$
$$\geq (Q_1(t) + Q_2(t)) \vee (Q_2(t) + Q_3(t)).$$

The Brownian solution achieves this lower bound by keeping one of $Q_1(t)$, $Q_3(t)$ always empty. In heavy traffic, we hope that the threshold policy will keep Q_1 small by giving priority to class 1, as long as Q_3 is not small, and allow Q_1 to grow only when Q_3 is small. When this is implemented, we will have both $Q_1, Q_3 = O(\kappa)$ while $Q_2 = O(N)$, and will need to adjust κ to achieve good balance between them.

We note that our Brownian control and presumably also the threshold policy of the queueing network again induce a state space collapse: First, we do not need to consider the three-dimensional process but only the two-dimensional process $\mathcal{W}_1(t), \mathcal{W}_2(t)$. Furthermore, instead of the two-dimensional $\mathcal{W}_1(t)$, $\mathcal{W}_2(t)$ we have the representation of the objective by just $(2\mathcal{W}_1(t) \vee \mathcal{W}_2(t)) and (Q_1(t) \wedge Q_3(t))$, and in this two-dimensional representation, under the Brownian control, and presumably also under the threshold policy in heavy traffic, $(Q_1(t) \wedge Q_3(t))$ becomes negligible so the system behaves as a one-dimensional process. Such a state space collapse that leaves a one-dimensional process to describe the system, so that it is acting as a single workstation, is called *complete resource pooling*; see Exercise 16.10.

16.4 Sequencing for a two-station Closed Queueing Network

In this section we consider a closed queueing network with two service stations. There is a fixed total number N of customers in the system, distributed among buffers $1, 2, \ldots, K$, located at the two service stations. Service rate for buffer k is μ_k with average service time m_k, $k = 1, \ldots, K$. Customers move between buffers, with routing probabilities $p_{k,l}$, where P is a stochastic irreducible matrix. This in particular means that in the long-run every customer visits all the stations.

An equivalent description is an open system where each time a customer leaves, a new customer enters, so that the number of customers is kept a constant N. To obtain the equivalence, we can assign buffer K as the exit buffer. Transitions are now according to the $p_{k,l}$, $k \neq K$, and on completion of service in buffer K a customer leaves, and a new customer enters class (buffer) l with probability $p_{K,l}$. This alternative description fits a Kanban system with N tokens as it is used in manufacturing; it is illustrated in Figure 16.2.

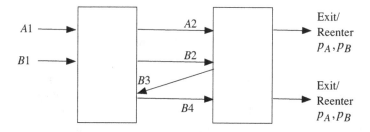

Figure 16.2 A two-station closed queueing network.

In a closed queueing system, there is no waiting time; instead, what is desired is to maximize the speed at which customers circulate through the system. Since customers circulate through all the buffers, we can measure the speed of circulation by counting how many customers exit from any one of the buffers, say buffer K, per unit time. Again our control is to decide at each time t which buffer to serve at each station, and it is clear that we should use stationary work-conserving policies (see Exercise 16.13).

A single customer walking from buffer to buffer according to P will visit buffer k with frequency π_k, the unique stationary probability vector solving $P^{\mathsf{T}}\pi = \pi$, and hence the amount of work required at buffer k will be $\beta_k \propto \pi_k m_k$, so that β is a solution of the equation $R\beta = 0$ (where $R = (I - P^{\mathsf{T}})\mathrm{diag}(\mu)$). We let $\rho_i = \sum_{k \in C_i} \beta_k$, and fix the values of β by letting $\max\{\rho_1, \rho_2\} = 1$. Similar to Section 16.2, we define the nominal allocation $\nu_k = \beta_k/\rho_{s(k)}$. Then in the long-run, under a stationary policy, the fraction of busy time of workstation i devoted to buffer $k \in C_i$ will be ν_k. Our *balanced heavy traffic assumption* for the closed queueing network is that the population N of the network is large, and that $\rho_1 \approx \rho_2$, so that $N(\rho_1 - \rho_2)$ is of moderate size.

We write the dynamics of the closed network:

$$
\begin{aligned}
Q_k(t) &= Q_k(0) - S_k(\mathcal{T}_k(t)) + \sum_{l=1}^{K} \mathcal{R}_{l,k}(S_l(\mathcal{T}_l(t))) \\
&= \Big[Q_k(0) - (\nu_k \mu_k - \sum_{l=1}^{K} p_{l,k} \nu_l \mu_l)t - (S_k(\mathcal{T}_k(t)) - \nu_k \mu_k \mathcal{T}_k(t)) \\
&\quad + \sum_{l=1}^{K} (\mathcal{R}_{l,k}(S_l(\mathcal{T}_l(t))) - p_{l,k} \nu_l \mu_l \mathcal{T}_l(t)) \Big] \\
&\quad + \Big[\mu_k(\nu_k t - \mathcal{T}_k(t)) - \sum_{l=1}^{K} p_{l,k} \mu_l(\nu_l t - \mathcal{T}_l(t)) \Big],
\end{aligned}
$$

which as before we write as: $Q(t) = X(t) + R\mathcal{J}(t)$, where in particular $\mathcal{J}_k(t) = (\nu_k t - \mathcal{T}_k(t))$. In addition, we have that $\mathbf{1}^{\mathsf{T}}Q(0) = \mathbf{1}^{\mathsf{T}}Q(t) = N$ at all times, for our closed network.

To maximize the throughput, we can control the scheduling policy: at each time t, which class is served at each server. If our policy is work conserving, then there will always be circulation, but we can choose policies that are patently bad: If each station will avoid serving customers that are routed to the other station and give priority to customers that stay at the same station, then workstations will often be starved and be idle. A natural policy is to treat the network as a Kelly network, i.e. handle all customers in a buffer equally, without regard to their destination, e.g. using FCFS or a symmetric policy (see Exercise 16.14). It is, however, possible to do better by observing the current state of all the buffers and basing the policy on that. Clearly, to maximize the throughput (circulation), we need to minimize the idling at the two stations. As we have already argued, in the long-run the busy time at each station will be proportional to $\rho_i(t)$, i.e. for large t, $\sum_{k \in C_1} \mathcal{T}_k(t) / \sum_{k \in C_2} \mathcal{T}_k(t) \approx \rho_1/\rho_2$. It follows that minimizing $\mathcal{I}_1(t) = t - \sum_{k \in C_1} \mathcal{T}_k(t)$ in the long-run will also minimize $\mathcal{I}_2(t)$. So our control problem is:

$$\min \ \limsup_{T \to \infty} \frac{1}{T} \mathbb{E}(\mathcal{I}_1(T))$$
$$Q(t) = X(t) + R\mathcal{T}(t),$$
$$\mathcal{I}(t) = C\mathcal{T}(t), \tag{16.19}$$
$$\mathbf{1}^\mathsf{T} Q(t) = N, \quad Q(t) \geq 0, \quad \mathcal{I}(0) = 0, \ \mathcal{I}(t) \text{ non-decreasing},$$
$$Q(t), \mathcal{T}(t), \mathcal{I}(t) \text{ non-anticipating},$$

where by non-anticipating we mean that our decisions are based only on events observed up to time t.

Again, under balanced heavy traffic the scaled netput $\frac{1}{N} X(N^2 t)$ can be approximated by a Brownian motion $\hat{X}(t)$, with drift θ and covariance matrix Γ that are given by (16.2) modified to exclude the input, and one can see that $\mathbf{1}^\mathsf{T}\theta = 0$, and $\mathbf{1}^\mathsf{T}\Gamma\mathbf{1} = 0$ (see Exercise 16.14). Using $\hat{Q}, \hat{\mathcal{T}}, \hat{\mathcal{I}}$, for the scaled queues and the free and idle times, the BCP is: For given Brownian motion $\hat{X}(t)$, find $\hat{Q}(t), \hat{\mathcal{T}}(t)$ such that:

$$\min \ \limsup_{T \to \infty} \frac{1}{T} \mathbb{E}[\hat{\mathcal{I}}_1(T)]$$
$$\text{s.t. } \hat{Q}(t) = \hat{X}(t) + R\,\hat{\mathcal{T}}(t),$$
$$\hat{\mathcal{I}}(t) = C\hat{\mathcal{T}}(t), \tag{16.20}$$
$$\mathbf{1}^\mathsf{T}\hat{Q}(t) = 1, \quad \hat{Q}(t) \geq 0, \quad \hat{\mathcal{I}}(0) = 0, \ \hat{\mathcal{I}}(t) \text{ non-decreasing},$$
$$\hat{Q}, \hat{\mathcal{T}} \text{ are non-anticipating with respect to } \hat{X}(t), \quad 0 \leq t \leq T.$$

We now wish to derive the workload formulation of the BCP, which is a bit more tricky than for open queueing networks. To do it, we resort to the Kanban representation of the network, with buffer K designated as exit buffer, and count the workload of all customers in the system until they reach and complete service at buffer K. We introduce the following notations:

$P_{\backslash K}$ is the matrix P in which the last row has been replaced by 0.

$$\begin{bmatrix} p_{K,1} \\ \vdots \\ p_{K,K} \end{bmatrix}, \text{ the last column of } P^\mathsf{T}, \text{ is the routing vector out of buffer } K.$$

$$R^- = \operatorname{diag}(m)(I - P^\mathsf{T}_{\backslash K})^{-1} = \operatorname{diag}(m)(I + P^\mathsf{T}_{\backslash K} + P^{2^\mathsf{T}}_{\backslash K} + \cdots).$$

$$M = CR^- = C\operatorname{diag}(m)(I - P^\mathsf{T}_{\backslash K})^{-1}.$$

Because P is irreducible, the matrix $P_{\backslash K}$ has spectral radius < 1 (see Exercise 16.15), so R^- is well defined and $R^-_{k,l}$ is the expected amount of work at buffer k that is needed for a customer currently at buffer l until that customer exits from buffer K. Hence, the vector $R^- Q(t)$ is the total expected amount of work required at the various buffers for all the customers currently in the system, to reach and exit buffer K, and the workload for the two stations is

$$\mathcal{W}(t) = MQ(t) = CR^- Q(t) = C\operatorname{diag}(m)(I - P^\mathsf{T}_{\backslash K})^{-1}Q(t)$$
$$= M\mathcal{X}(t) + MR\mathcal{J}(t). \tag{16.21}$$

We define $\mathcal{Z}(t) = M\mathcal{X}(t)$, and express $MR\mathcal{J}(t)$ as follows:

$$MR\mathcal{J}(t) = C\operatorname{diag}(m)(I - P^\mathsf{T}_{\backslash K})^{-1}(I - P^\mathsf{T}_{\backslash K}))\operatorname{diag}(\mu)\,\mathcal{J}(t)$$

$$- M\begin{bmatrix} p_{K,1} \\ \vdots \\ p_{K,K} \end{bmatrix} \mu_K \mathcal{J}_K(t)$$

$$= C\mathcal{J}(t) - \Upsilon\eta(t) = \mathcal{I}(t) - \Upsilon\eta(t),$$

where we define $\Upsilon = \begin{bmatrix} \Upsilon_1 \\ \Upsilon_2 \end{bmatrix} = M\begin{bmatrix} p_{K,1} \\ \vdots \\ p_{K,K} \end{bmatrix}$, and $\eta(t) = \mu_K \mathcal{J}_K(t)$. The

vector $R^-\begin{bmatrix} p_{K,1} \\ \vdots \\ p_{K,K} \end{bmatrix}$ is the expected amount of work per customer that has

just exited buffer K, until it will again reach and exit from buffer K, so it encompasses the expected amount of work at each buffer, per one circulation of an expected customer. Hence, (Υ_1, Υ_2) is the amount of work for each of the workstations per circulation of a customer, counting between exits from buffer K. The quantities (Υ_1, Υ_2) are again independent of the policy, and we wish to minimize the times for each of these circulations. It is important to note that (Υ_1, Υ_2) is proportional to (ρ_1, ρ_2) (see Exercise 16.16).

We now approximate $\frac{1}{N}\mathcal{Z}(N^2 t)$ by the Brownian motion $\hat{\mathcal{Z}}(t) = M\hat{\mathcal{X}}(t)$, and let $\hat{Q}, \hat{\mathcal{J}}, \hat{\mathcal{I}}, \hat{\eta}$ correspond to the scaled $Q, \mathcal{J}, \mathcal{I}, \eta$. We then obtain the workload formulation for the BCP: For a given Brownian motion $\hat{\mathcal{Z}}(t)$, for any sample path $\hat{\mathcal{Z}}(t, \omega)$ and any T, find $\hat{Q}(t), \hat{I}(t), \hat{\eta}(t)$ such that:

$$
\begin{aligned}
&\min\ \hat{I}_1(T) \\
&\quad M\hat{Q}(t) = \hat{W}(t), \\
&\quad \hat{W}(t) = \hat{\mathcal{Z}}(t) + \hat{I}(t) - \Upsilon\hat{\eta}(t), \\
&\quad \mathbf{1}^\top\hat{Q}(t) = 1, \\
&\quad \hat{Q}(t) \geq 0, \quad \hat{I}(0) = 0, \quad \hat{I} \text{ non-decreasing}, \\
&\quad \hat{Q}, \hat{I}, \hat{\eta} \text{ are non-anticipating with respect to } \hat{X}(t).
\end{aligned}
\tag{16.22}
$$

Note that this is more ambitious than the formulation (16.20), since we no longer look for minimizing long-range expectation, instead we wish to solve this for every sample path and every finite time horizon.

In trying to interpret $\hat{\eta}(t)$, we note that $\hat{\mathcal{Z}}(t) + \hat{I}(t)$ corresponds to the workload that we would have if each workstation had been working independently of the other, doing its own work, and never having to idle because the other station did not supply it with work; so $\hat{\eta}(t)$ is the deviation from independent operation, expressed in number customers whose processing is lost in one of the stations, because they were present at the other station. It can be positive or negative, indicating loss due to one or the other of the stations.

The workload formulation has reduced the problem of determining K free time controls to one where the controls are $\hat{I}_1(t), \hat{I}_2(t), \hat{\eta}(t)$. We now do another reformulation, that will further reduce the number of controls. Define the workload imbalance process $\hat{W}_\Delta(t) = \rho_2\hat{W}_1(t) - \rho_1\hat{W}_2(t)$. Define also the approximating scaled Brownian motion netput to the work imbalance, $\hat{\mathcal{Z}}_\Delta(t) = \rho_2\hat{\mathcal{Z}}_1(t) - \rho_1\hat{\mathcal{Z}}_2(t)$. The drift of this netput is $N(\rho_1 - \rho_2)$ which is of moderate size. Let $\hat{R}(t) = \rho_2\hat{I}_1(t)$ and $\hat{L}(t) = \rho_1\hat{I}_2(t)$. We get a workload imbalance formulation of the BCP: For a given Brownian motion $\hat{\mathcal{Z}}_\Delta(t)$, for any sample path $\hat{\mathcal{Z}}_\Delta(t, \omega)$ and any T, find $\hat{Q}, \hat{R}, \hat{L}$ such that:

$$
\begin{aligned}
&\min\ \hat{R}(T) \\
&\text{s.t.}\ \hat{W}_\Delta(t) = \hat{\mathcal{Z}}_\Delta(t) + \hat{R}(t) - \hat{L}(t), \\
&\quad \hat{W}_\Delta(t) = \sum_{k=1}^{K}(\rho_2 M_{1,k} - \rho_1 M_{2,k})\hat{Q}_k(t), \\
&\quad \mathbf{1}^\top\hat{Q}(t) = 1, \quad \hat{Q}(t) \geq 0, \quad \hat{R}(0) = \hat{L}(0) = 0, \quad \hat{R}, \hat{L} \text{ non-decreasing}, \\
&\quad \hat{Q}, \hat{R}, \hat{L}, \text{ are non-anticipating with respect to } \hat{\mathcal{Z}}_\Delta(t).
\end{aligned}
$$

We now solve this problem. In fact, we will obtain a solution that minimizes the objective pointwise for every t.

We rename the buffers so that they are ordered by their individual work imbalance $\delta_k = \rho_2 M_{1,k} - \rho_1 M_{2,k}$, so that $b = \delta_1 \geq \cdots \geq \delta_K = a$. Recall that $\sum_{k=1}^{K}(\rho_2 M_{1,k} - \rho_1 M_{2,k}) = 0$, so $a < b$, $a \leq 0 \leq b$. Also, by its maximality, class 1 is served at station 1, and similarly, class K is served at station 2. From the constraints that $\hat{Q}(t) \geq 0$ and $\mathbf{1}\hat{Q}_k = 1$ it follows that $\hat{W}_\Delta(t)$ can have any value in the interval $[a, b]$ (with more than one solution $\hat{Q}$ for each point in of (a, b)), but cannot have any value outside $[a, b]$.

Ignoring the actual values of $\hat{Q}_k(t)$, we now wish to find functions $\hat{R}, \hat{L}$ which start at 0 and are non-decreasing, non-anticipating with respect to a Brownian motion $\hat{Z}_\Delta(t)$, such that

$$\hat{W}_\Delta(t) = \hat{Z}_\Delta(t) + \hat{R}(t) - \hat{L}(t),$$
$$a \leq \hat{W}_\Delta(t) \leq b, \quad 0 < t < T,$$

and such that $\hat{R}(t)$ is minimized (which as we already argued will also minimize $\hat{L}(t)$). We have solved this problem in Section 7.4. It is given by the unique Skorohod reflection controls, defined through the recursive definition:

$$\hat{R}(t) = \sup_{0 < s \leq t} (\hat{Z}_\Delta(s) - a - \hat{L}(s))^-, \quad \hat{L}(t) = \sup_{0 < s \leq t} (b - \hat{Z}_\Delta(s) - \hat{R}(s))^-,$$

$$\tag{16.23}$$

and these $\hat{R}, \hat{L}$ are minimal for every sample path at every t. The mnemonic is that $\hat{R}$ is push to the right, $\hat{L}$ is push to the left, to keep $a < \hat{W}_\Delta < b$.

With these values of $\hat{R}(t), \hat{L}(t)$ we have that $\hat{I}_1(t) = \hat{R}(t)/\rho_2$, $\hat{I}_2(t) = \hat{L}(t)/\rho_1$ are non-anticipating, and minimal for every t. It remains to construct $\hat{Q}(t)$, to satisfy all the constraints (note, this is not a unique solution, but it will be feasible and achieve the optimal objective):

$$\hat{Q}_k(t) = \begin{cases} \frac{\hat{W}_\Delta(t)-a}{b-a} & k = 1, \\ \frac{b-\hat{W}_\Delta(t)}{b-a} & k = K, \\ 0 & k = 2, \ldots, K-1. \end{cases}$$

It is then immediate to see that $\hat{W}_\Delta(t) = \sum_{k=1}^{K}(\rho_2 M_{1,k} - \rho_1 M_{2,k})\hat{Q}_k(t)$. This completes the solution of the Brownian problem.

We explain how this Brownian solution works: We always give lowest priority to buffer 1 at workstation 1 and to buffer K at workstation K. While both buffers 1 and K are not empty both workstations are non-idling, and idling of workstation 1 increases only when buffer 1 is empty, idling of workstation 2 increases only when buffer K is empty. in between the idling periods, the work imbalance netput $\hat{Z}_\Delta(t)$ is regulating the levels of the two buffers 1 and K.

By using the results of Section 7.4 we can calculate the long-run average cost of the optimal policy (see Exercise 16.17).

Having solved this BCP, we come back to the original closed network, with N customers. We cannot implement the Brownian policy. To emulate it we will use a static priority policy, using the analog of δ_k as priority index. At station 1 we give priority to buffers with low δ_k, while at station 2 we give priority to buffers with high δ_k. In balanced heavy traffic this policy tends to give higher priority at one workstation to the buffers that require more work along their route at the other workstation.

The work imbalance formulation of the BCP can be generalized to closed networks with I workstations. In that case one can define an $I - 1$-dimensional imbalance process, and idleness will only occur on the boundary of this polytope (see Exercise 16.29).

16.5 Admission Control and Sequencing for a Two-Station MCQN

In this section we take the two-station network of the last Section 16.4, but instead of N customers in the system at all times, we now have control over the number of customers. Recall that the closed network could also be regarded as an open network in which each time a customer leaves, we immediately introduce a new customer, and in this way we keep the number in the system a constant N, and we tried to optimize the rate at which customers enter and leave (the circulation time). Now we are given an average rate at which customers should enter and leave, but we have complete control over the times at which we introduce new customers, and our aim is to minimize the number of customers or the holding costs of customers.

The system has two workstations, K buffers with constituency matrix C, processing rates μ_k, and routing matrix P with spectral radius < 1. Input is $\mathcal{A}(t)$ with long range rate $\bar{\alpha}$, but subject to that rate we have complete control on the times that customers are introduced. Customers that enter the system go to buffer k with probability $p_{0,k}$, so the rate of $\mathcal{A}_k(t)$, the input to buffer k, is $\alpha_k = p_{0,k}\bar{\alpha}$. We denote the input rates vector by α and the input routing fractions vector by p_0. Note that we have control over admission times, but not on the routing of admissions to the various buffers.

The offered loads ρ_1, ρ_2 are determined by the processing rates μ_k, the routing matrix P, the input rate $\bar{\alpha}$, and the admission fractions p_0. Our balanced heavy traffic assumption is that for some large N, at both stations $N(1 - \rho_i)$ is of moderate size.

The stochastic optimization problem then is

$$\min \ \limsup_{T \to \infty} \frac{1}{T} \mathbb{E}\Big[\int_0^T \sum_{k=1}^K h_k Q_k(t)\, dt \Big]$$

$$\text{s.t.} \ \ Q_k(t) = \mathcal{A}_k(t) - \mathcal{S}_k(\mathcal{T}_k(t)) + \sum_{l=1}^K \mathcal{R}_{l,k}(\mathcal{S}_l(\mathcal{T}_l(t))),$$

$$\mathcal{I}_i(t) = t - \sum_{k \in C_i} \mathcal{T}_k(t), \tag{16.24}$$

$$\lim_{T \to \infty} \frac{1}{T} \mathbb{E}(\mathcal{A}(T)) = \bar{\alpha},$$

$$Q_k(t) \geq 0, \quad \mathcal{A}(0), \mathcal{T}(0), \mathcal{I}(0) = 0, \ \mathcal{A}, \mathcal{T}, \mathcal{I} \text{ non-decreasing,}$$

$$\mathcal{A}, \mathcal{T} \text{ are non-anticipating.}$$

It differs from the closed network problem (16.19) in two ways: The constraint $\mathbf{1}Q(t) = N$ is replaced by $\mathcal{A}(t)/t \to \bar{\alpha}$, and instead of maximizing the circulation, we are minimizing inventory costs, the waiting times, and the queue lengths, with $h_k > 0$ the holding cost of a class k customer per unit time.

From the parameters of the problem we calculate (as before): $R = (I - P^{\mathsf{T}})\mathrm{diag}(\mu)$, $\lambda = \bar{\alpha}(I - P^{\mathsf{T}})^{-1}p_0$, $\beta = \mathrm{diag}(m)\lambda = \bar{\alpha}R^{-1}p_0$, $\rho_i = \sum_{k \in C_i} \beta_k$, $\nu_k = \beta_k/\rho_{s(k)}$, $M = CR^{-1}$. Note that p_0 now plays the role of the circulating probabilities given by the last column of P^{T} in Section 16.4, and the current P is what was $P_{\backslash K}$ in Section 16.4.

Similar to Sections 16.2–16.4, we decompose the queue length processes $Q(t)$ to netput that is independent of policy and two types of control processes: the free times $\mathcal{J}(t)$ for each class, and the input deviations $\eta(t)$:

$$Q(t) = X(t) + R\mathcal{J}(t) - p_0\eta(t), \tag{16.25}$$

$$X_k(t) = (p_{0,k}\bar{\alpha}t - \mu_k\nu_k t + \sum_{l=1}^K p_{l,k}\mu_l\nu_l t) + (\mathcal{A}_k(t) - p_{0k}\mathcal{A}(t)) \tag{16.26}$$

$$- (\mathcal{S}_k(\mathcal{T}_k(t)) - \mu_k\mathcal{T}_k(t)) + \sum_{l=1}^K (\mathcal{R}_{l,k}(\mathcal{S}_l(\mathcal{T}_l(t))) - p_{l,k}\mu_l\mathcal{T}_l(t)),$$

$$\mathcal{J}_k(t) = \nu_k t - \mathcal{T}_k(t),$$

$$\eta(t) = \bar{\alpha}t - \mathcal{A}(t),$$

where ν_k are the nominal allocations.

Under diffusion scaling, $\frac{1}{N}X(N^2t)$ can be approximated by a Brownian motion $\hat{X}(t)$ that (because of balanced heavy traffic) is independent of the policy, with drift vector θ and covariance matrix Γ. Here, similar to (16.2,

$$\theta_k = p_{0,k}\bar{\alpha}t - \mu_k \nu_k t + \sum_{l=1}^{K} p_{l,k}\mu_l \nu_l t, \qquad \Gamma = \Gamma^0 + \sum_{j=1}^{K} \nu_j \Gamma^j,$$

$$\Gamma_{k,l}^0 = \bar{\alpha}p_{0,k}(\delta_{k,l} - p_{0,l}), \tag{16.27}$$

$$\Gamma_{k,l}^j = \mu_j p_{j,k}(\delta_{k,l} - p_{j,l}) + \mu_j c_{sj}^2(\delta_{j,k} - p_{j,k})(\delta_{j,l} - p_{j,l}).$$

Unlike the models of Sections 16.3 and 16.4, where the parameters of the Brownian motion were irrelevant to the policy, for the current problem we will need the values of θ and Γ.

We now follow steps similar to 16.4 and formulate a BCP for $\hat{Q}(t)$ with the Brownian motion $\hat{X}(t)$. Using $M = C\text{diag}(m)(I - P^{\intercal})^{-1}$, we then formulate a Brownian workload control problem for the workload $\hat{W}(t) = M\hat{Q}(t)$, with the Brownian motion $\hat{Z}(t) = M\hat{X}(t)$. We then define the work imbalance $\hat{W}_\Delta(t) = \rho_2\hat{W}_1(t) - \rho_1\hat{W}_2(t)$, and the scalar Brownian motion $\hat{Z}_\Delta(t) = \rho_2\hat{Z}_1(t) - \rho_1\hat{Z}_2(t)$. The process $\hat{Z}_\Delta(t)$ has drift and covariance:

$$\text{drift} = N(\rho_1 - \rho_2), \quad \text{variance} = \sigma^2 = [\rho_2, -\rho_1]M^{\intercal}\Gamma M \begin{bmatrix} \rho_2 \\ -\rho_1 \end{bmatrix}. \tag{16.28}$$

All this leads (see Exercise 16.20) to the reduced dimension Brownian control problem:

$$\min \ \limsup_{T\to\infty} \frac{1}{T}\mathbb{E}\Big[\int_0^T \sum_{k=1}^{K} h_k \hat{Q}_k(t)dt \Big]$$

$$\text{s.t. } \hat{W}_\Delta(t) = \sum_{k=1}^{K} (\rho_2 M_{1,k} - \rho_1 M_{2,k})\hat{Q}_k(t),$$

$$\hat{W}_\Delta(t) = \hat{Z}_\Delta(t,\omega) + \rho_2\hat{I}_1(t) - \rho_1\hat{I}_2(t), \tag{16.29}$$

$$\limsup_{T\to\infty} \frac{1}{T}\mathbb{E}(\hat{I}_i(T)) = N(1 - \rho_i),$$

$$\hat{Q}(t) \geq 0, \quad \hat{I}(0) = 0, \ \hat{I} \text{ non-decreasing},$$

$$\hat{Q}, \hat{I} \text{ are non-anticipating w.r.t. } \hat{X}.$$

We first solve this problem for $\hat{Q}(t)$, assuming that $\hat{W}_i(t)$ are known.

The problem is then separated into a standard LP for each t:

$$\min \ \hat{V}(t) = \sum_{k=1}^{K} h_k \hat{Q}_k(t) dt$$

$$\text{s.t.} \ \sum_{k=1}^{K} (\rho_2 M_{1,k} - \rho_1 M_{2,k}) \hat{Q}_k(t) = \hat{W}_\Delta(t),$$

$$\hat{Q}_k(t) \geq 0.$$

This can be solved by inspection. Let $\delta_k = \frac{\rho_2 M_{1,k} - \rho_1 M_{2,k}}{h_k}$ and rename the buffers so that $\delta_1 \geq \cdots \geq \delta_K$. We note that $\delta_1 \geq 0 \geq \delta_K$ (see Exercise 16.21), and in fact, under most conditions $\delta_1 > 0 > \delta_K$ (see Exercise 16.22). The solution of the LP is then to keep all the buffers empty, except one: buffer 1 if $\hat{W}_\Delta(t) \geq 0$ and buffer K if $\hat{W}_\Delta(t) < 0$. So,

- If $\hat{W}_\Delta(t) \geq 0$, then $\hat{Q}_1(t) = \frac{\hat{W}_\Delta(t)}{\rho_2 M_{1,1} - \rho_1 M_{2,1}}$, all other $\hat{Q}_k(t) = 0$, and the objective value is $\hat{V}(t) = \hat{W}_\Delta(t) \frac{h_1}{\rho_2 M_{1,1} - \rho_1 M_{2,1}}$.

- If $\hat{W}_\Delta(t) < 0$, then $\hat{Q}_K(t) = \frac{\hat{W}_\Delta(t)}{\rho_2 M_{1,K} - \rho_1 M_{2,K}}$, all other $\hat{Q}_k(t) = 0$, and the objective value is $\hat{V}(t) = \hat{W}_\Delta(t) \frac{h_K}{\rho_2 M_{1,K} - \rho_1 M_{2,K}}$.

Next we need to solve for $\hat{I}_1, \hat{I}_2$, so as to satisfy the demands of non-negativity and flow $\bar{\alpha}$, and determine the optimal $\hat{W}_\Delta$. As we see, the value of the objective at time t is a function of $\hat{W}_\Delta(t)$. We introduce some notations:

$$d_R = \frac{1}{\delta_1} = \frac{h_1}{\rho_2 M_{1,1} - \rho_1 M_{2,1}}, \quad d_L = -\frac{1}{\delta_K} = \frac{h_K}{\rho_1 M_{2,K} - \rho_2 M_{1,K}},$$

$$H(x) = \begin{cases} d_R x, & x \geq 0, \\ -d_L x, & x < 0, \end{cases}$$

$$\hat{R}(t) = \rho_2 \hat{I}_1(t), \quad \hat{L}(t) = \rho_1 \hat{I}_2(t).$$

Note that $d_R, d_L > 0$. When we use the optimal values of $\hat{Q}_k(t)$, then

$$\sum_{k=1}^{K} h_k \hat{Q}_k(t) dt = H(\hat{W}_\Delta(t)).$$

We need to solve

$$\min \lim_{T \to \infty} \sup \frac{1}{T} \mathbb{E}\Big[\int_0^T H(\hat{\mathcal{W}}_\Delta(t)) dt \Big]$$

$$\text{s.t.} \quad \hat{\mathcal{W}}_\Delta(t) = \hat{\mathcal{Z}}_\Delta(t, \omega) + \hat{R}(t) - \hat{L}(t) \geq 0, \qquad (16.30)$$

$$\lim_{T \to \infty} \sup \frac{1}{T} \mathbb{E}(\hat{R}(T)) = N\rho_2(1 - \rho_1),$$

$$\lim_{T \to \infty} \sup \frac{1}{T} \mathbb{E}(\hat{L}(T)) = N\rho_1(1 - \rho_2).$$

16.5.1 Solution of (16.30)

The shape of $H(x)$ is a minimum of 0 at $x = 0$, convex, and linear for $x > 0$ as well as for $x < 0$ (it is V shaped). It can be shown that the solution to this optimization problem is to use left and right singular controls that keep $a < \hat{\mathcal{W}}_\Delta(t) < b$, as described in Section 7.4:

$$\hat{R}(t) = \sup_{0 < s \leq t} (\hat{\mathcal{Z}}_\Delta(s) - a - \hat{L}(s))^-, \qquad \hat{L}(t) = \sup_{0 < s \leq t} (b - \hat{\mathcal{Z}}_\Delta(s) - \hat{R}(s))^-.$$

It remains to determine the appropriate $a < 0 < b$ that will give us the correct amount of idling, and that will minimize the expected holding costs. We use equation (7.19), which gave the stationary behavior of $\hat{R}, \hat{L}, \hat{\mathcal{W}}_\Delta$. Recall that the Brownian motion $\hat{\mathcal{Z}}_\Delta(t)$ has drift $N(\rho_1 - \rho_2)$, and variance (diffusion coefficient) σ^2 given by (16.28). Let $\zeta = \frac{2N(\rho_1 - \rho_2)}{\sigma^2}$.

We first obtain the width of the interval in which $\hat{\mathcal{W}}_\Delta(t)$ is kept, $b - a$.

Proposition 16.1 *The width of the interval, $b - a$, is*

$$b - a = \begin{cases} \dfrac{\sigma^2}{2N\rho(1 - \rho)}, & \rho_1 = \rho_2 = \rho, \\[2ex] \dfrac{\sigma^2}{2N(\rho_1 - \rho_2)} \log\left(\dfrac{\rho_1(1 - \rho_2)}{\rho_2(1 - \rho_1)}\right), & \rho_1 \neq \rho_2. \end{cases} \qquad (16.31)$$

Proof When $\rho_1 = \rho_2 = \rho$ we need to equate

$$\lim_{T \to \infty} \frac{\hat{R}(T)}{T} = \lim_{T \to \infty} \frac{\hat{L}(T)}{T} = \frac{\sigma^2}{2(b - a)} = N\rho(1 - \rho).$$

When $\rho_1 \neq \rho_2$ we need to equate

$$\lim_{T \to \infty} \frac{\hat{R}(T)}{T} = \frac{N(\rho_1 - \rho_2)}{e^{\zeta(b-a)} - 1} = N\rho_2(1 - \rho_1),$$

or equivalently

$$\lim_{T \to \infty} \frac{\hat{L}(T)}{T} = \frac{N(\rho_1 - \rho_2)}{1 - e^{-\zeta(b-a)}} = N\rho_1(1 - \rho_2),$$

and (16.31) follows.   □

We now need to find the locations of a and b that will minimize the stationary expected cost per unit time. Again using (7.19), we can calculate $\mathbb{E}(H(\hat{W}_\Delta(\cdot)))$ as a function of a and b. We can then locate the optimal a and get $b = a + (b - a)$.

Proposition 16.2 *If $\rho_1 = \rho_2 = \rho$, then*

$$a = -\frac{d_R}{d_R + d_L} \frac{\sigma^2}{2N\rho(1 - \rho)}, \qquad b = \frac{d_L}{d_R + d_L} \frac{\sigma^2}{2N\rho(1 - \rho)}. \qquad (16.32)$$

If $\rho_1 \neq \rho_2$, then

$$a = \frac{\sigma^2}{2N(\rho_1 - \rho_2)} \log \Big(\frac{(d_L + d_R)\rho_2(1 - \rho_1)}{d_L\rho_2(1 - \rho_1) + d_R\rho_1(1 - \rho_2)} \Big),$$
$$b = \frac{\sigma^2}{2N(\rho_1 - \rho_2)} \log \Big(\frac{(d_L + d_R)\rho_1(1 - \rho_2)}{d_L\rho_2(1 - \rho_1) + d_R\rho_1(1 - \rho_2)} \Big). \qquad (16.33)$$

Proof The value of the objective over the time that $\hat{W}(t)$ is at a or b is determined by $\lim_{T \to \infty} \frac{1}{T}\mathbb{E}(\hat{R}(T)) = N\rho_2(1-\rho_1)$ and $\lim_{T \to \infty} \frac{1}{T}\mathbb{E}(\hat{L}(T)) = N\rho_1(1 - \rho_2)$.

The idea of the proof is to use the stationary density $f(z)$ of $\hat{W}_\Delta(\cdot)$ as given by (7.19) to evaluate

$$\hat{V}(a) = \mathbb{E}\Big[\int_a^{a+(b-a)} H(z)f(z)dz \Big],$$

and then to find a that minimizes this expression.

To prove (16.32):

$$\hat{V}(a) = \int_a^0 (-d_L z)\frac{2N\rho(1 - \rho)}{\sigma^2}dz + \int_0^{a + \frac{\sigma^2}{2N\rho(1-\rho)}} (d_R z)\frac{2N\rho(1 - \rho)}{\sigma^2}dz.$$

Taking the derivative and equating to 0:

$$0 = \hat{V}'(a) = d_L a \frac{2N\rho(1 - \rho)}{\sigma^2} + d_R\Big(a + \frac{\sigma^2}{2N\rho(1 - \rho)}\Big)\frac{2N\rho(1 - \rho)}{\sigma^2},$$

and (16.32) follows.

The proof of (16.33) is more laborious, and is left as Exercise 16.24.   □

Next we obtain the control $\hat{\eta}(t)$ by substituting our values of

$$\hat{\eta}(t) = \frac{1}{\Upsilon_1} \Big[\hat{\mathcal{Z}}_1(t) + \frac{\mathcal{R}(t)}{\rho_2} - \sum_{k=1}^{K} M_{1,k} \hat{Q}_k(t) \Big]$$

$$= \frac{1}{\Upsilon_2} \Big[\hat{\mathcal{Z}}_2(t) + \frac{\mathcal{L}(t)}{\rho_1} - \sum_{k=1}^{K} M_{2,k} \hat{Q}_k(t) \Big], \tag{16.34}$$

where $\Upsilon = Mp_0$.

We can also obtain expressions for $\hat{\mathcal{J}}(t)$, namely,

$$\hat{\mathcal{J}}(t) = R^{-1} \big(\hat{Q}(t) - \hat{X}(t) + \hat{\eta}(t)p_0 \big). \tag{16.35}$$

16.5.2 Summary of the Brownian Policy and Heuristic

We have derived the optimal solution to the BCP, by obtaining explicit expressions for $\hat{Q}(t), \hat{\mathcal{J}}(t), \hat{I}(t), \hat{\eta}(t)$. We now examine how this optimal control works for the control of the Brownian system; it is depicted in Figure 16.3, which shows the workload at the two stations.

We found in the solution that the workload imbalance behaves like a

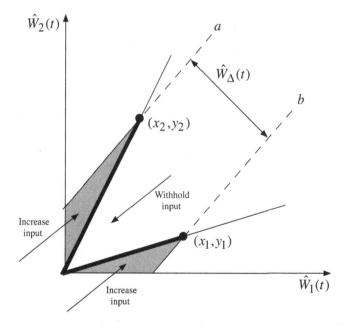

Figure 16.3 Two-station Brownian control policy.

reflected Brownian motion between a and b. Because $\rho_i \approx 1$, the lines $\hat{W}_\Delta \geq a$ and $\hat{W}_\Delta \leq b$ are approximately the lines $\hat{W}_2 + a \leq \hat{W}_1 \leq \hat{W}_2 + b$ (recall $a < 0 < b$). Whenever $\hat{W}_\Delta > b$, it means that $\hat{W}_1$ is too big, and we need to idle workstation 1, and whenever $\hat{W}_\Delta < a$, it means that $\hat{W}_2$ is too big, and we need to idle workstation 2. The will keep the workloads between the two parallel lines, with distance $b - a$ between them.

Next we found that when the workload imbalance is positive we give lowest priority to buffer 1, and $\hat{Q}_1 > 0$, while all other buffers are empty, and when the workload imbalance is negative buffer K has lowest priority and $\hat{Q}_K > 0$, while all other buffers are empty. Using $\hat{W}(t) = M\hat{Q}(t)$, this implies:

$$\hat{W}_\Delta(t) > 0 \implies \hat{W}_1(t) = M_{1,1}\hat{Q}_1(t) \text{ and } \hat{W}_2(t) = M_{2,1}\hat{Q}_1(t)$$

$$\implies \hat{W}_2(t) = \frac{M_{2,1}}{M_{1,1}}\hat{W}_1(t),$$

$$\hat{W}_\Delta(t) < 0 \implies \hat{W}_1(t) = M_{1,K}\hat{Q}_K(t) \text{ and } \hat{W}_2(t) = M_{2,K}\hat{Q}_K(t)$$

$$\implies \hat{W}_2(t) = \frac{M_{2,k}}{M_{1,K}}\hat{W}_1(t).$$

Note that $\frac{M_{2,1}}{M_{1,1}} < 1$ while $\frac{M_{2,k}}{M_{1,K}} > 1$, by $\delta_1 > 0 > \delta_K$. We then see that while $\hat{W}_\Delta(t)$ moves between a and b, $\hat{W}_1(t), \hat{W}_2(t)$ move along the two boundaries of the cone in Figure 16.3.

Finally, we examine the control of the input via $\eta(t)$. We saw in (16.25) that $\hat{Q}(t)$ includes the control $\eta(t)p_0$, and so the workload $\hat{W}(t)$ includes the control $\eta(t)Mp_0$, and we note that $\Upsilon = Mp_0$ is proportional to (ρ_1, ρ_2), so the control $\eta(t)$ moves the workloads of the two stations up or down along the line $\hat{W}_1 \approx \hat{W}_2$. This means that the input is withheld when the workloads are inside the cone, and input is admitted when outside the cone.

In summary: Only buffers 1 and K are ever non-empty, 1 is non-empty when the imbalance is positive, K when it is negative. The admissions control is keeping the workloads on the boundary of the cone, and idling is only implemented at the points $(x_1, y_1), (x_2, y_2)$ where the cone boundary meets the reflecting lines. On the two line segments of the cone the workloads move like Brownian motion.

This Brownian control policy cannot be applied directly to the original discrete network. A policy that will emulate the Brownian policy is as follows: Use static priority policy giving priority to buffers with larger δ_k at work station 1 and to larger $-\delta_k$ at station 2. Note the similarity to the "$c\mu$" rule for single-server queue: the priority is determined by cost over time, where instead of time we use the buffer work imbalance. Also, the admission rule for the Brownian problem allows admission only until the cone is reached. This is because on the diffusion scale processing and input are instantaneous. To avoid starvation on the original scale one needs to shift the cone from the origin at 0 to an origin at (κ, κ). Finally, a station is idled only when it is empty.

16.6 Admission Control and Sequencing in Multi-Station MCQN

In this section we generalize the results on sequencing and admission control of MCQN from two workstations to I workstations. Following exactly the same steps as in Section 16.5, the BCP for the scaled queueing network is: For a Brownian motion $\hat{X}(t)$, find $\hat{Q}, \hat{\mathcal{J}}, \hat{\eta}$,

$$\min \ \limsup_{T \to \infty} \frac{1}{T} \mathbb{E}\Big[\int_0^T \sum_{k=1}^K h_k \hat{Q}_k(t) dt \Big]$$

$$\text{s.t. } \hat{Q}(t) = \hat{X}(t) + R\,\hat{\mathcal{J}}(t) - p_0\,\hat{\eta}(t),$$
$$\hat{I}(t) = C\hat{\mathcal{J}}(t),$$
$$\limsup_{T \to \infty} \frac{1}{T} \mathbb{E}[\,\hat{I}_i(T)\,] = N(1 - \rho_i),$$
$$\hat{Q}(t) \geq 0, \qquad \hat{I}(0) = 0, \ \hat{I} \text{ non-decreasing,}$$
$$\hat{Q}, \hat{\mathcal{J}}, \hat{\eta} \text{ are non-anticipating with respect to } \hat{X}(t),$$

and the workload formulation is: For the Brownian motion $\hat{Z}(t) = M\hat{X}(t)$, find $\hat{Q}, \hat{I}, \hat{\eta}$,

$$\min \ \limsup_{T \to \infty} \frac{1}{T} \mathbb{E}\Big[\int_0^T \sum_{k=1}^K h_k \hat{Q}_k(t) dt \Big]$$

$$\text{s.t. } M\hat{Q}(t) = \hat{W}(t),$$
$$\hat{W}(t) = \hat{Z}(t) + \hat{I}(t) - \Upsilon \hat{\eta}(t),$$
$$\limsup_{T \to \infty} \frac{1}{T} \mathbb{E}[\,\hat{I}_i(T)\,] = N(1 - \rho_i),$$
$$\hat{Q}(t) \geq 0, \qquad \hat{I}(0) = 0, \ \hat{I} \text{ non-decreasing,}$$
$$\hat{Q}, \hat{I}, \hat{\eta} \text{ are non-anticipating with respect to } \hat{Z}(t),$$

where $\Upsilon = Mp_0$.

We note that apart from $\hat{I}(t)$, the remaining unknowns, $\hat{Q}(t)$, $\hat{\eta}(t)$ have separate constraints for each t, so assuming that $I(t)$ is given we can solve for them directly. Before we do that we will eliminate $\hat{\eta}$ by defining work imbalance.

Recall Υ_i are proportional to ρ_i, define the $I - 1$-dimensional workload imbalance $\hat{W}_{\Delta,i}(t) = \rho_I \hat{W}_i(t) - \rho_i \hat{W}_I(t)$, $i = 1, \ldots, I - 1$, and define similarly the Brownian motion $\hat{Z}_{\Delta,i}(t)$ and the idling $\hat{I}_\Delta(t)$. We can now eliminate $\hat{\eta}(t)$ and one constraint from the formulation to get

$$\min \ \limsup_{T \to \infty} \frac{1}{T} \mathbb{E}\Big[\int_0^T \sum_{k=1}^K h_k \hat{Q}_k(t) dt \Big]$$

$$\text{s.t.} \ \sum_{k=1}^K (\rho_I M_{i,k} - \rho_i M_{I,k}) \hat{Q}_k(t) = \hat{W}_{\Delta,i}(t),$$

$$\hat{W}_{\Delta,i}(t) = \hat{Z}_{\Delta,i}(t) + \rho_I \hat{I}_i(t) - \rho_i \hat{I}_I(t), \qquad (16.36)$$

$$\limsup_{T \to \infty} \frac{1}{T} \mathbb{E}[\hat{I}_i(T)] = N(1 - \rho_i),$$

$$\hat{Q}(t) \geq 0, \qquad \hat{I}(0) = 0, \ \hat{I} \text{ non-decreasing},$$

$$\hat{Q}, \hat{I} \text{ are non-anticipating with respect to } \hat{Z}_\Delta(t).$$

We first solve for $\hat{Q}$ for given $\hat{W}_\Delta$, and again obtain a separate LP at every point t. The following LP will always be bounded and feasible; see Exercise 16.26:

$$\min \ \sum_{k=1}^K h_k \hat{Q}_k(t)$$

$$\text{s.t.} \ \sum_{k=1}^K (\rho_I M_{i,k} - \rho_i M_{I,k}) \hat{Q}_k(t) = \hat{W}_{\Delta,i}(t), \quad i = 1, \ldots, I - 1,$$

$$\hat{Q}(t) \geq 0. \qquad (16.37)$$

Unlike the two workstation case that had a single constraint and could be solved by inspection for every $\hat{W}_\Delta(t)$, the LP (16.37) has $I - 1$ constraints so a basic optimal solution could have up to $I - 1$ buffers k for which $\hat{Q}_k(t) > 0$, with all the other buffers empty. The basic optimal solution will keep all the inventory in $I - 1$ buffers. However, the set of basic buffers will vary with t.

Given the solution for $\hat{Q}(t)$, we retrieve $\hat{\eta}(t)$ from

$$\hat{\eta}(t) = \big(M\hat{Q}(t) - \hat{Z}(t) - \hat{I}(t)\big)_i \big/ \Upsilon_i, \quad \text{for any } i. \qquad (16.38)$$

The next step (the harder step) is to solve for $\hat{I}$. Before we do that we describe the solution that we have for $\hat{Q}(t)$ and $\hat{\eta}(t)$, in the space of workloads, the I-dimensional orthant with coordinates $w_1, \ldots, w_I$.

The workloads for buffer k will be along a ray $(M_{1,k}, \ldots, M_{I,k})$. A basic solution for given $\hat{\mathcal{W}}_\Delta(t)$ will keep the inventory in a subset of $I-1$ buffers, so it will be in an $I-1$-dimensional polyhedral cone formed by the convex hull of the corresponding basic buffer rays. The union of these cones, for all possible optimal basic solutions, will form the boundary of an I-dimensional cone, not necessarily convex, pointed at 0. For any given point $\hat{\mathcal{Z}}(t) + \hat{I}(t)$ in the I-dimensional orthant, the inventory will be kept on the boundary of the $I-1$-dimensional facet of this cone that corresponds to the optimal basic solution for the work imbalance $\hat{\mathcal{W}}_\Delta(t) = \hat{\mathcal{Z}}_\Delta(t) + \hat{I}_\Delta(t)$. It will be kept on that facet by the admission control $\Upsilon\hat{\eta}(t) = \hat{\mathcal{Z}}(t) + \hat{I}(t) - M\hat{Q}(t)$: withholding admissions inside the cone, and injecting admissions outside the cone. Recall that in heavy traffic $\rho_i \approx 1$, so the admission control will move workload approximately parallel to the main diagonal of the orthant.

To tackle the solution for $\hat{I}$ we now look at the LP dual to (16.37):

$$\max \ \sum_{i=1}^{I-1} \hat{\mathcal{W}}_{\Delta,i}(t) y_i(t) \tag{16.39}$$

$$\text{s.t.} \ \sum_{i=1}^{I-1} (\rho_I M_{i,k} - \rho_i M_{I,k}) y_i(t) \le h_k, \quad k = 1, \ldots, K.$$

Denote by $\hat{Q}^*(t)$ and $y^*(t)$ basic optimal solutions of (16.37) and (16.39) at the time t. We express the value of the objective at time t as a function of the work imbalance, and use duality to write:

$$H(\hat{\mathcal{W}}_\Delta) = \sum_{i=1}^{I-1} \hat{\mathcal{W}}_{\Delta,i}(t) y_i^*(t) = \sum_{k=1}^{K} h_k \hat{Q}_k^*(t). \tag{16.40}$$

Then $H(\hat{\mathcal{W}}_\Delta)$ is a positive, piecewise linear, and convex function with minimum of 0 at the origin; see Exercise 16.27).

We now consider the solution for $\hat{I}(t)$. Looking at the workload imbalance formulation (16.36), we know the values of $\hat{Q}(t)$, and we replace the objective with the equivalent $H(\hat{\mathcal{W}}_\Delta)$. The drift and variance of $\hat{\mathcal{Z}}_\Delta$ are obtained from those of $\hat{X}$. The problem now is for given Brownian motion $\hat{\mathcal{Z}}_\Delta$, find $\hat{I}_i(\cdot)$, $i = 1, \ldots, I$ such that:

$$\min \ \limsup_{T \to \infty} \frac{1}{T} \mathbb{E}\left[\int_0^T H(\hat{\mathcal{W}}_\Delta(t)) dt \right] \tag{16.41}$$

s.t. $\hat{W}_{\Delta i}(t) = \hat{Z}_{\Delta i}(t) + \rho_I \hat{I}_i(t) - \rho_i \hat{I}_I(t),$

$$\limsup_{T \to \infty} \frac{1}{T} \mathbb{E}[\,\hat{I}_i(t)\,] \le N(1 - \rho_i), \qquad\qquad (16.42)$$

$\hat{I}(0) = 0, \quad \hat{I}$ non-decreasing,

$\hat{I}$ is non-anticipating with respect to $\hat{Z}_\Delta$.

We make one more transformation to the problem: Define $\bar{\rho}_i = \prod_{j \ne i, I} \rho_j$ (product of all except i and I), and let $\hat{W}_{\Delta,i}^\circ = \bar{\rho}_i \hat{W}_{\Delta,i}$, $\hat{Z}_{\Delta,i}^\circ = \bar{\rho}_i \hat{Z}_{\Delta,i}$, $\hat{I}_{\Delta,i}^\circ = \bar{\rho}_i \hat{I}_{\Delta,i}$, $H^\circ(\hat{W}_\Delta^\circ) = H(\hat{W}_\Delta)$. We then get a Brownian optimization problem for the balanced workload imbalance: For the given Brownian motion $\hat{Z}_\Delta^\circ(t)$, find $I_i^\circ(t)$, $i = 1, \dots, I$ such that

$$\min \; \limsup_{T \to \infty} \frac{1}{T} \mathbb{E}\Big[\int_0^T h^\circ(\hat{W}_\Delta^\circ(t)) dt \Big]$$

s.t. $\hat{W}_{\Delta i}^\circ(t) = \hat{Z}_{\Delta i}^\circ(t) + \hat{I}_i^\circ(t) - \hat{I}_I^\circ(t),$

$$\limsup_{T \to \infty} \frac{1}{T} \mathbb{E}[\,\hat{I}_i^\circ(t)\,] \le N\bar{\rho}_i(1 - \bar{\rho}_i), \qquad\qquad (16.43)$$

$\hat{I}^\circ(0) = 0, \quad \hat{I}^\circ$ non-decreasing,

$\hat{I}^\circ$ is non-anticipating with respect to $\hat{Z}_\Delta^\circ$.

This problem has the same structure as the two-workstation problem (16.30): We wish to find singular controls that will satisfy the right amount of idling, and will keep the workload imbalances in a region that will minimize the expected costs. In terms of the workload, this will define a convex polyhedral prism that has its $I - 1$-dimensional facets parallel to the line directed approximately as the vector $\mathbf{1}$, and $\hat{I}^\circ(t)$ will be singular controls that keep the workload inside this prism. The prism for $\hat{I}(t)$ will have the direction parallel to $(\rho_1, \dots, \rho_I)$. This sort of multidimensional Brownian singular control problem, in which we want to satisfy constraints on the average amount of idling, as well as minimize the expected cost inside the region, is intractable for $I > 2$, and has to be solved numerically.

To solve it numerically, one way is to use the finite difference approximation approach of Kushner. The idea is to replace the diffusion by a discrete-time discrete-space Markov chain and translate the control problem to a Markov decision problem. The number of actions at each state is finite, and it turns out that the process under optimal policy stays in a bounded region around the origin, so by proper truncation this becomes a finite-dimensional Markov decision problem with linear side constraints, for which there are several solution algorithms.

We illustrate the solution of this BCP by a problem taken from the original paper of Wein (1992). This example problem has 12 buffers and 3 workstations, and the solution is shown in Figure 16.4. The pathwise

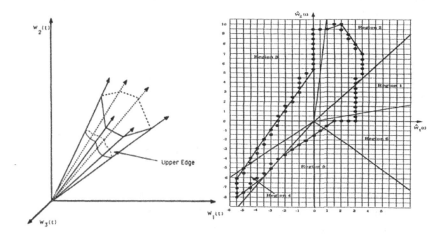

Figure 16.4 Workload control for a three station network, reproduced with permission from Wein (1992).

solution of the sequencing problem keeps the three-dimensional scaled workload process on the boundary of the polyhedral cone pointed at 0. Each of the extreme rays of the cone corresponds to a single buffer, and each two-dimensional face of the cone corresponds to a basic solution of the LP (16.37). The workload is kept on the face of the cone by the singular control $\hat{\eta}(t)$, which feeds input when the workload moves out of the cone and withholds input when it moves inside the cone. Idling happens at the intersection of the reflecting prism and the cone; this intersection defines an upper edge of the cone. The figure shows the three-dimensional cone with 6 two-dimensional faces, and next to it the approximate numerical solution that calculates the region in which the two-dimensional workload imbalance lives, and this determines the boundaries of the reflecting prism and the upper edge of the cone where idling occurs.

To implement a solution of the BCP for the actual MCQN, one can proceed as follows: First, the cone of workloads would be translated from the origin to a point $(\kappa, \ldots, \kappa)$. Next, input would be aimed to keep the workload within distance κ from the boundary of the cone, and idling will be exercised at the upper edge, sometimes idling non-empty workstations. Finally, at every time t, according to the workload and the corresponding workload imbalance, a priority ordering of the buffers at each workstation will be observed: Buffers in the basic solution will receive lowest priority. For the remaining buffers, priority could be determined by the reduced costs of the LP ((16.37), $\bar{h}_k(t) = h_k - \sum_{i=1}^{I} M_{i,k} y_i^*(t)$, with $\bar{h}_k / \sum_i M_{i,k}$ serving as a "$c\mu$" index .

16.7 Asymptotic Optimality of MCQN Controls

The idea of using the solution of a BCP to suggest control policies for the associated MCQN in balanced heavy traffic was pioneered by Michael Harrison as a heuristic. We used this idea in Sections 16.3–16.6, following Harrison and Larry Wein, to solve a series of BCPs of increasing complexity and obtain priority and threshold policies for sequencing and admission controls. We will pursue similar ideas also to obtain routing policies in Chapter 17. In all cases, simulation studies showed that the suggested policies performed extremely well. This left the intriguing question, whether these policies are in fact asymptotically optimal.

It took some years for Martin, Shreve, and Soner to prove that in the criss-cross network with Poisson arrivals and exponential service, the policy of priority to class 1 as long as class 3 is above a threshold is asymptotically optimal.

This was followed by two papers of Bell and Williams, the first proving asymptotic optimality of a threshold policy for a two server N-network, the second extending the result to parallel servers with overlapping skills. Next Stolyar showed asymptotic optimality of maximum weight policy for a generalized switch. In these three papers, customers require a single service, i.e. they are single pass systems, and the proofs are to show that the systems under the proposed policy have state space collapse to a one-dimensional workload process, which in heavy traffic converges to the one-dimensional solution of the BCP.

Dai and Lin then considered extending these results to processing networks, as discussed in Chapter 12, which are not single pass. They defined a *complete resource pooling* condition as follows: Recall the static planning problem LP (12.7), with the unknowns ρ, which is the traffic intensity for the network, and x, which is a nominal allocation. Write the dual to (12.7):

$$\max \sum_{i \in I_I} z_i$$

$$\text{s.t.} \sum_{k=1}^{K} y_k R_{k,j} \leq - \sum_{i \in I_I} z_i A_{i,j}, \text{ input activities } j,$$

$$\sum_{k=1}^{K} y_k R_{k,j} \leq \sum_{i \in I_S} z_i A_{i,j}, \text{ service activities } j, \qquad (16.44)$$

$$\sum_{i \in I_S} z_i = 1,$$

$$z_i \geq 0, \quad i \in I_S,$$

where I_I is the set of input processors, and I_S the set of service processors. Complete resource pooling holds if this dual LP has a unique optimal solution. In that case, the unique solution defines a bottleneck workload process $\mathcal{W}(t) = y \cdot Q(t)$. They showed that under maximum pressure policy, this bottleneck workload process converges to the solution of the BCP and is asymptotically optimal.

More recently, Budhiraja and Ghosh considered general unitary processing networks with a discounted holding cost objective. These include all the networks considered in Chapters 16 and 17. For such networks, one cannot expect state space collapse to a one-dimensional workload process. They showed first that no sequencing policy can outperform the associated BCP, i.e. the BCP solution provides a lower bound on the objective. Next they showed that the optimal objective value for the unitary network converges in the heavy traffic limit to the objective value of the BCP. To show the convergence, they constructed an ϵ-optimal solution of the BCP that is a multivariate reflected pure jump process, and then constructed a sequence of policies for the network, for which the convergence held.

Proofs of all these results are beyond the scope of this book.

16.8 Sources

The material in this chapter follows a series of papers by J. Michael Harrison and Larry Wein, which carry out the program of solving Brownian control programs to obtain heuristics for the control of multi-class queueing networks in balanced heavy traffic. The main ideas are outlined in Harrison (1988), and the criss-cross network is solved in Harrison and Wein (1989). This was followed by the two-station closed queueing network model, Harrison and Wein (1990). The two-station open queueing network with admission control is solved in Wein (1990a,b), and extended to multi-stations in Wein (1992). The finite difference dynamic programming approach is described in the book *Numerical methods for stochastic control problems in continuous time*, Kushner (1990). Additional material on control of MCQN is found in the textbook *Fundamentals of Queueing Networks: Performance, Asymptotics, and Optimization*, Chen and Yao (2001), and in the monographs *Control Techniques for Complex Networks*, Meyn (2008), *Heavy Traffic Analysis of Controlled Queueing and Communication Networks*, Kushner (2001). Extensions to processing network type models with an emphasis on workload formulation and state space collapse and canonical workload formulation, are discussed in Harrison and Van Mieghem (1997), Harrison (2000, 2002, 2003), and Bramson and Williams (2003). The asymptotic optimality of the threshold policy for the criss-cross network was proved by Martins et al. (1996), and further proofs for asymptotic optimality of threshold policies associated with BCP solutions were derived by Bell and Williams (2001, 2005). Optimality of maximum pressure policy for processing networks with complete resource pooling is shown by Dai and Lin (2005). Budhiraja and Ghosh (2006) prove for unitary networks that the BCP solution for a discounted objective is a lower bound on the MCQN problem, and in Budhiraja and Ghosh (2012) they show that the heavy traffic asymptotic optimal objective converges to the BCP objective.

Exercises

16.1 Derive the diffusion limits for the MCQN netput process $\frac{1}{N}\mathcal{X}(N^2 t)$, using the property that in balanced heavy traffic $\lim_{N\to\infty} \frac{1}{N}\mathcal{T}_k(Nt) = \nu_k t$ u.o.c. a.s.

16.2 Verify equation (16.2) for the drift and covariance of the netput process.

16.3 Show that the LP (16.8), which solves for $\hat{Q}(t)$ in terms of $\hat{W}(t)$, is feasible and bounded.

16.4 Consider the criss-cross network with Poisson arrivals and exponential service times. Assume that customers that arrive at station 1 are served on a FCFS basis. Analyze the performance of the network under that policy.

16.5 Consider the criss-cross network with Poisson arrivals and exponential service times. Analyze the performance of the network under the policy of priority to class 1 and under the policy of priority to class 2.

16.6 For the criss-cross network, plot a schematic path of the buffer contents, using drift and deviations from the drift to illustrate various possible states.

16.7 For the criss-cross network, calculate the drift and the covariances for the netput Brownian motion $\hat{X}(t)$ and for the netput Brownian workload $\hat{Z}(t)$.

16.8 Obtain the marginal distributions of the Brownian workloads $\hat{W}_i(t)$ for the criss-cross network under the optimal policy. Note, they are not independent and we cannot obtain their joint distribution. Try and estimate the mean objective.

16.9 Consider a modified criss-cross network, where customers of type B can start their service at station 2 without waiting to complete service at station 1, and customers of type B leave the system when both services are complete. Assume Poisson arrivals and exponential service times, and server 1 is giving non-preemptive priority to customers of type A. Calculate the expected number of customers in the system as $\rho \nearrow 1$.

16.10 Consider the criss-cross network under the policy of maximum pressure, and compare this to the proposed threshold policy. Evaluate both by simulation.

16.11 Consider the criss-cross network with general parameters, i.e. arrival rates α_i, service rates μ_k. Assume renewal arrivals and services, with coefficients of variation $c_{a,i}, c_{s,k}$. Follow the next steps:

(i) Write the dynamics of $Q_k(t)$, using nominal allocations ν_k.

(ii) Define netput processes $X_k(t)$, and write $Q_k(t)$ in terms of the netput $X_k(t)$ and the free times $\mathcal{J}_k(t)$.

(iii) Write the dynamics of the workload processes $W_i(t)$, define workload netput $Z_i(t)$, and write $W_i(t)$ in terms of the workload netput $Z_i(t)$ and the idle time processes $I_i(t)$.

(iv) Calculate the offered load for each of the servers, ρ_i.

(v) Define a sequence of systems indexed by N. Use fluid scaling with time and space scaled by N, and obtain the fluid limits of $Q^N, X^N, \mathcal{J}^N$ and of W^N, Z^N, I^N.

(vi) Formulate the conditions for balanced heavy traffic and determine the diffusion approximation to $X(t)$ and $Z(t)$.

(vii) Calculate the drift and and covariance matrix for the diffusion limits $\hat{X}$ and $\hat{Z}$.

16.12 Formulate the Brownian control problem of the general criss-cross network in heavy traffic, and solve it for general parameters, $\alpha_i, i = 1, 2, \mu_k, k = 1, 2, 3$ and cost coefficients $h_k, k = 1, 2, 3$. Note that there are several cases, according to the parameters.

16.13 For the closed two-station multi-class network, show that to maximize throughput one should use a work-conserving policy.

16.14 For the closed two-station multi-class network, write down the dynamics, center and scale all components and derive a decomposition with a netput process and control processes. Derive the drift and covariance matrix of the limiting Brownian netput process.

16.15 Prove that if the stochastic transition matrix P is irreducible, then the matrix $P_{\backslash K}$ has spectral radius < 1 so that $I - P_{\backslash K}^\top$ is invertible.

16.16 Show that (Υ_1, Υ_2) is proportional to (ρ_1, ρ_2).

16.17 For the closed two-station multi-class network, use the results of Section 7.4 to obtain the stationary distribution of the Brownian workload imbalance $\hat{W}_\Delta(t)$ when using the optimal policy, and find the stationary rate of idling at the two stations.

16.18 Analyze the closed two-station queueing network as a Kelly network under a symmetric policy, and find the stationary distribution of the queue lengths.

16.19 For the closed Kumar–Seidman Rybko–Stolyar two-station network, described in the following figure, with N customers and feed back to the top or bottom route with probabilities $\alpha, 1 - \alpha$, perform all the steps of the analysis,

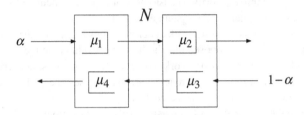

to reach the BCP and solve it.

16.20 For the open two-station network with admission controls, follow the steps necessary to derive the formulation of the work imbalance BCP (16.29).

16.21 For the open two-station network with admission controls, show that $\delta_1 \geq 0 \geq \delta_K$.

16.22 For the open two-station network with admission controls, analyze the cases when $\delta_1 = 0 > \delta_K$, $\delta_1 > 0 = \delta_K$, and $\delta_1 = \delta_K = 0$. When can that happen?

16.23 Analyze the two-station queueing network with admission control as a Kelly network under a symmetric policy. Assume Poisson arrivals with rate $\bar{\alpha}$, and find the stationary distribution of the queue lengths and the expected waiting times.

16.24 For the open two-station network with admission controls, prove equation (16.33) in Proposition 16.2 [Wein (1990a)].

16.25 For the open Kumar–Seidman Rybko–Stolyar two-station network, described in Section 10.2.3, with arrival rates α_1, α_2 to the top and bottom routes, respectively, perform all the steps of the analysis, to reach the BCP and solve it.

16.26 Provide an argument to show that if $\hat{I}(t)$ is given, then the LP (16.37) and its dual (16.39) are bounded and feasible.

16.27 For the multi-station network with admission control, show that the objective function $H(\hat{W}(t))$ is convex [Wein (1990a)].

16.28 For the multi-station network with admission control, show that the boundary forms a prism parallel to $(\rho_1, \ldots, \rho_I)$ [Wein (1990a)].

16.29 (∗) Analyze the control of a closed MCQN with I workstations, in analogy with Sections 16.4 and 16.6.

17

MCQN with Discretionary Routing

17.1 The General Balanced Heavy Traffic Control Problem

In this chapter we add control over routing to control over scheduling and admissions. To control admissions we may now have more than one input activity, and to control routing we may now have several service activities at each buffer. This leads us to the more general models of processing networks of Chapters 12 and 13. So we now assume the system consists of workstations $i = 1, \ldots, I$, buffers $k = 1, \ldots, K$, and activities $J = 1, \ldots, J$, where some of the buffers are input buffers with infinite supply $k \in \mathcal{K}_\infty$, and the rest are service buffers $k \in \mathcal{K}_0$, with standard queues. The topology of the network is now described by the constituency matrix C and consumption matrix A, and parameters are processing rates μ_j and routing matrices P^j.

Unlike the MCQN models of Chapter 16, for more general processing networks with more choice of activities, the overall throughput rate does not determine the workloads of the stations, and does not determine the nominal allocations of the activities. To determine those we resort to the *static production planning problem* (13.11) of Section 13.4. In this problem the unknowns are the admission rates α_k out of the infinite supply buffers $\mathcal{K}_\infty$, and the nominal allocations x:

$$\max_{v,\alpha} \sum_{k \in K_\infty} w_k \alpha_k \tag{17.1}$$

$$\text{s.t.} \quad R_{\mathcal{K}_\infty} v = \alpha, \quad R_{\mathcal{K}_0} v = 0, \quad A v \leq 1, \quad v, \alpha \geq 0.$$

Here R is the extended input-output matrix defined in (12.5), and $k \in \mathcal{K}_\infty$ are types of products of the network, with w_k the revenue per item. Any feasible solution of (17.1) presents a plan for operating the network, and an optimal solution will maximize its revenue. By the nature of the LP, some of the constraints $A v \leq 1$ will be satisfied as equalities, and those workstations will be in heavy traffic. We then discard all the lightly loaded workstations, and the remaining network is then in balanced heavy traffic.

We now have nominal input rates from the buffers $k \in \mathcal{K}_\infty$ and nominal allocations v_j for the service activities, with flows at rates $v_j \mu_j$. We know that there exist policies that are rate stable with these maximal revenue inputs and flows. Under EAA and in particular for unitary networks, the system will be rate stable under maximum pressure policy. Our balanced heavy traffic problem is to find a policy that is rate stable and minimizes the long-term congestion, measured by $\sum_{k \in \mathcal{K}_0} h_k Q_k(t)$. We look for a dynamic policy, that uses the deviations from nominal inputs and nominal allocations, to determine our admission, sequencing, and routing decisions.

In principle we can now approximate this general network by a Brownian network, formulate a BCP, and look for an optimal Brownian policy. Since we now have a wide variety of controls, we can expect an efficient solution. Unfortunately, in this general form the problem is often intractable.

Instead of attempting to obtain general results, we shall limit our treatment to unitary networks, where each activity consumes a single resource and processes a single buffer, so columns of A and C are unit columns. We will further restrict attention to networks with deterministic routes. We assume renewal arrivals and i.i.d. service times for each activity.

We first analyze a simple model, similar to the criss-cross network, in Section 17.2, and move on to more complex models in Sections 17.3 and 17.4. For the networks analyzed in these sections, we find that the optimal solutions to the BCPs are a.s. pathwise optimal, i.e. for almost every sample path the solution is optimal at every t. That this is not always the case is demonstrated by an example in Section 17.4. We then look at a more general case in Section 17.5, and identify those problems that do have a.s. pathwise optimal solutions. A plausibility argument indicates that these may include many practical problems.

17.2 A Simple Network with Routing and Sequencing

Consider the network illustrated in Figure 17.1, where customers of type A can on arrival be routed to station 1 or station 2, and customers of type B need both workstations. This model combines routing to parallel servers, as in Chapter 15, with the criss-cross network of Section 16.3. Arrivals of both types are rate λ with c.o.v. c_a, services in both stations are rate μ with c.o.v. c_s. Balanced heavy traffic is reached when $N(3\lambda - 2\mu)$ is of moderate size for large N. To control queue lengths $Q_k(t)$, we use the sequencing controls $\mathcal{J}_k(t) = \begin{cases} \frac{2}{3}t - \mathcal{T}_k(t), & k = 1, 2 \\ \frac{1}{3}t - \mathcal{T}_k(t), & k = 3, 4 \end{cases}$, where $\frac{2}{3}, \frac{1}{3}$ are the nominal allocations, and the routing controls $\mathcal{V}_k(t) = \mathcal{A}_k(t) - \frac{1}{2}\lambda t$, $k = 3, 4$.

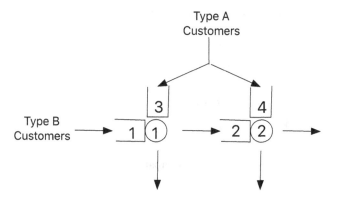

Figure 17.1 Network with sequencing and routing controls.

We wish to minimize the total number of customers in the system. We then formulate a scaled approximate Brownian optimal control problem: For given Brownian motion $\hat{X}(t)$, find $\hat{Q}(t)$, $\hat{\mathcal{J}}(t)$, $\hat{V}(t)$ such that

$$\min \ \limsup_{T \to \infty} \frac{1}{T} \mathbb{E}\Big[\int_0^T \sum_{k=1}^4 \hat{Q}_k(t)dt \Big]$$

$$\text{s.t. } \hat{Q}(t) = \hat{X}(t) + R\,\hat{\mathcal{J}}(t) - G\hat{V}(t), \quad t \geq 0.$$

$$\hat{I}(t) = C\hat{\mathcal{J}}(t), \qquad H\hat{V}(t) = 0, \quad t \geq 0.$$

$$\hat{Q}(t) \geq 0, \qquad \hat{I}(0) = 0, \quad \hat{I} \text{ non-decreasing}, \quad t \geq 0,$$

$$\hat{Q}, \hat{\mathcal{J}}, \hat{V} \text{ are non-anticipating with respect to } \hat{X}(t),$$

with

$$R = \mu(I - P^{\mathsf{T}}) = \mu \begin{bmatrix} 1 & 0 & 0 & 0 \\ -1 & 1 & 0 & 0 \\ 0 & 0 & 1 & 0 \\ 0 & 0 & 0 & 1 \end{bmatrix}, \quad G = \begin{bmatrix} 0 & 0 & 1 & 0 \\ 0 & 0 & 0 & 1 \end{bmatrix},$$

$$C = \begin{bmatrix} 1 & 0 & 1 & 0 \\ 0 & 1 & 0 & 1 \end{bmatrix}, \qquad H = [1 - 1].$$

Workloads are given by

$$\hat{W}(t) = CR^{-1}Q(t) = \mu^{-1} \begin{bmatrix} 1 & 0 & 1 & 0 \\ 1 & 1 & 0 & 1 \end{bmatrix} \hat{Q}(t).$$

Similar to Chapter 15, routing pools the resources of the two stations in the system. We therefore look at the sum of the workloads, $\hat{W}_\Sigma = \hat{W}_1 + \hat{W}_2 =$

$\mu^{-1}[2\ 1\ 1\ 1]\hat{Q}$. The pooled workload control problem is to find $\hat{Q}, \hat{I}$ such that:

$$\min \ \limsup_{T \to \infty} \frac{1}{T}\mathbb{E}\left[\int_0^T \sum_{k=1}^4 \hat{Q}_k(t)dt\right]$$

$$\text{s.t.} \ \ \mu^{-1}(2\hat{Q}_1(t) + \hat{Q}_2(t) + \hat{Q}_3(t) + \hat{Q}_4(t)) = \hat{W}_\Sigma(t),$$

$$\hat{W}_\Sigma(t) = \hat{Z}(t) + \hat{I}_1(t) + \hat{I}_2(t),$$

$$\hat{Q}(t) \geq 0, \quad \hat{I}(0) = 0, \quad \hat{I} \text{ non-decreasing}, \quad t \geq 0,$$

$$\hat{Q}(t), \hat{I}(t) \text{ are non-anticipating with respect to } \hat{Z}(t),$$

where the netput $\hat{Z}$ is a scalar Brownian motion with drift $N(3\lambda - 2\mu)$ and variance $\sigma^2 = 2\lambda c_a^2 + 4\mu c_s^2$. To minimize the objective, it is certainly optimal to minimize $\hat{W}_\Sigma(t)$ pathwise, which is achieved by the Skorohod reflection, taking

$$\hat{I}_1(t) + \hat{I}_2(t) = -\inf_{0 \leq s \leq t} \hat{Z}(s).$$

This control says that the idling at any of the stations occurs only when the entire system is empty.

Following the pathwise minimization of $\hat{W}_\Sigma(t)$, we see by inspection that we should have

$$\hat{Q}_1(t) = \frac{1}{2}\mu\hat{W}_\Sigma(t), \quad \hat{Q}_k(t) = 0, \ k = 2, 3, 4,$$

that is, all the inventory in the system is kept at buffer 1. This will minimize the objective for every sample path almost surely, and it will be minimal at all time points t. Values of $\hat{I}_i, i = 1, 2, \hat{V}, \hat{J}_k, k = 1, 2, 3, 4$ can then be determined. We take $\hat{I}_1(t) = \hat{I}_2(t)$, i.e. both stations idle when the system is empty. We use the control of the routing $\hat{V}(t)$ to keep station 2 empty. Finally, the scheduling at station 1 is to give priority to type A customers, thus keeping buffer 3 empty.

For the unscaled system, a policy that attempts to imitate this is to route type A jobs to buffer 3 only when buffer 3 has fewer jobs than the total in station 2, and to give priority to type A over type B at station 1 unless station 2 falls below a certain threshold.

To get an idea of the advantage of coordinated routing and sequencing, assume Poisson arrivals and exponential service. Because we achieve resource pooling, the system under the suggested control behaves like an M/M/1 queue, with expected total number of customers in system $2\lambda/(2\mu - 3\lambda)$. In comparison, if arrivals are routed randomly and service is FCFS, we have expected total number of customers in system $6\lambda/(2\mu - 3\lambda)$, so we have a factor 3 improvement. A factor 2 of savings is the result of the pooling of the two stations. Another factor of 3/2 is achieved because by keeping station 2 close to empty, customers of type B do not wait at station 2 (see Exercise 17.1).

17.3 The Network of Laws and Louth

We now consider a more complex network due to Laws and Louth, where the savings are even more impressive. Consider the following network shown in Figure 17.2, where horizontal customers arrive at rate λ_H and can choose upper or lower route and vertical customers arrive at rate λ_V and can choose left or right route, and service rates at all the stations are μ. The system

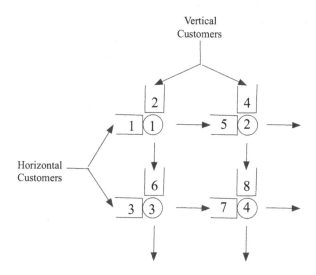

Figure 17.2 The Laws and Louth network.

will be in balanced heavy traffic if $N(2\lambda_H + 2\lambda_V - 4\mu)$ is of moderate size for large N. To keep the system stable, we require nominal choice of $1/2$ to each route, and nominal service allocation of $v_H = \lambda_H/(\lambda_H + \lambda_V)$, $v_V = \lambda_V/(\lambda_H + \lambda_V)$ at each station. So the queues $Q_k(t)$ are controlled by the deviations from nominal rate $\mathcal{J}_k(t) = v_H t - \mathcal{T}_k(t), k = 1, 3, 5, 7$ and $\mathcal{J}_k(t) = v_V t - \mathcal{T}_k(t), k = 2, 4, 6, 8$, and by the deviations of the actual routing from the nominal rate, $\mathcal{V}_k(t) = \mathcal{A}_k(t) - \lambda_H t/2, k = 1, 3, \mathcal{V}_k(t) = \mathcal{A}_k(t) - \lambda_V t/2, k = 2, 4$. We again have input-output matrix $R = \mu(I - P^\top)$, and resource consumption matrix $M = CR^{-1}$ where M_{ik} is how much work needs to be done by station i per customer of class k. The BCP for scaled queue lengths is to find $\hat{Q}, \hat{\mathcal{J}}, \hat{\mathcal{V}}$ such that (see Exercise 17.4)

$$\min \ \limsup_{T \to \infty} \frac{1}{T} E\Big[\int_0^T \sum_{k=1}^{8} \hat{Q}_k(t)\, dt \Big]$$

$$\text{s.t.} \ \ \hat{Q}(t) = \hat{X}(t) + R\,\hat{\mathcal{J}}(t) + G\hat{\mathcal{V}}(t),$$

$$\hat{I}(t) = C\hat{\mathcal{J}}(t), \qquad H\hat{\mathcal{V}}(t) = 0, \tag{17.2}$$

$$\hat{Q}(t) \geq 0, \quad \hat{I}(0) = 0, \ \hat{I} \text{ non-decreasing}, \quad t \geq 0,$$

$$\hat{Q}, \hat{\mathcal{J}}, \hat{\mathcal{V}} \text{ are non-anticipating with respect to } \hat{X}(t).$$

Here $H = \begin{bmatrix} 1 & 0 & 1 & 0 \\ 0 & 1 & 0 & 1 \end{bmatrix}$ and $G^{\mathsf{T}} = [I_4 \ 0_4]$, and $\mathcal{X}$ is a Brownian motion with known drift and covariance (see Exercise 17.5).

We now make the crucial observation here that every customer needs to go through either station 1 or station 4, and also every customer needs to go through either station 2 or station 3. Accordingly, we define a two-dimensional pooled workload process, $\mathcal{W}_P = (\mathcal{W}_{1,4}, \mathcal{W}_{2,3})$ where

$$\mathcal{W}_P = \mu^{-1} \begin{bmatrix} 1 & 1 & 1 & 1 & 0 & 0 & 1 & 1 \\ 1 & 1 & 1 & 1 & 1 & 1 & 0 & 0 \end{bmatrix} Q(t) := MQ(t).$$

The pooled workloads formulation of the BCP is to find $\hat{Q}, \hat{I}$ such that

$$\min \ \limsup_{T \to \infty} \frac{1}{T} E\Big[\int_0^T \sum_{k=1}^{8} \hat{Q}_k(t)\, dt \Big]$$

$$\text{s.t.} \ \ \hat{\mathcal{W}}_P = M\hat{Q}(t),$$

$$\hat{\mathcal{W}}_{1,4}(t) = \hat{\mathcal{Z}}_1(t) + \hat{I}_1(t) + \hat{I}_4(t),$$

$$\hat{\mathcal{W}}_{2,3}(t) = \hat{\mathcal{Z}}_2(t) + \hat{I}_2(t) + \hat{I}_3(t),$$

$$\hat{Q}(t) \geq 0, \quad \hat{I}(0) = 0, \ \hat{I} \text{ non-decreasing}, \quad t \geq 0,$$

$$\hat{Q}, \hat{I} \text{ are non-anticipating with respect to } \hat{\mathcal{Z}}(t),$$

where $\hat{\mathcal{Z}}$ is a two-dimensional Brownian motion with known drift and covariance (see Exercise 17.5).

Because the constraints on the stations in the two workload netputs of $\hat{\mathcal{Z}}_{1,4}$ and $\hat{\mathcal{Z}}_{2,3}$ do not overlap, we can now minimize $\hat{\mathcal{W}}_P(t)$ a.s. pathwise, at every point t, by choosing

$$\hat{\mathcal{W}}_{1,4}(t) = \hat{\mathcal{Z}}_1(t) - \inf_{0 \leq s \leq t} \hat{\mathcal{Z}}_1(s), \quad \hat{\mathcal{W}}_{2,3}(t) = \hat{\mathcal{Z}}_2(t) - \inf_{0 \leq s \leq t} \hat{\mathcal{Z}}_2(s).$$

For the minimal $\hat{\mathcal{W}}_P(t)$, we now solve for the optimal $\hat{Q}(t)$. By inspection we see that:

$$\sum_{k=1}^{8} \hat{Q}_k(t) \geq \hat{W}_{1,4}(t) \vee \hat{W}_{2,3}(t),$$

and this lower bound is reached pathwise by

$$\sum_{k=1}^{4} \hat{Q}_k(t) = \hat{W}_{1,4}(t) \wedge \hat{W}_{2,3}(t),$$

$$\hat{Q}_5(t) + \hat{Q}_6(t) = [\hat{W}_{2,3}(t) - \hat{W}_{1,4}(t)]^+ \quad \hat{Q}_7(t) + \hat{Q}_8(t) = [\hat{W}_{1,4}(t) - \hat{W}_{2,3}(t)]^+.$$

This completes the solution of the BCP, where we note that the solution is not unique, presumably because of the symmetry in the system (see Exercise 17.9).

The following policy for the unscaled system will try to imitate the Brownian control. The sequencing rules to keep Q_k, $k = 5, 6, 7, 8$ close to empty are:

At station 1, give priority to class 1 if $\mathcal{W}_2(t) < \mathcal{W}_3(t)$, and to class 2 otherwise.

At stations 2, 3, give priority to exit classes unless $\mathcal{W}_4(t) <$ threshold κ.

At station 4, use FCFS.

The routing rules need to keep the two pooled stations in each pair equal so that idling will occur only when both are close to empty:
Define

$$r_{1,4}(t) = \frac{\mathcal{W}_4(t) - \mathcal{W}_1(t)}{\mathcal{W}_1(t) \wedge \mathcal{W}_4(t)}, \qquad r_{2,3}(t) = \frac{\mathcal{W}_3(t) - \mathcal{W}_2(t)}{\mathcal{W}_2(t) \wedge \mathcal{W}_3(t)}.$$

These quantities are large positive or large negative if one of the pooled stations is nearly empty while the other one is not. Combining these indications we have: $r_{1,4} + r_{2,3} > 0$ indicates $\mathcal{W}_1$ and/or $\mathcal{W}_2$ are near empty, and $r_{1,4} - r_{2,3} > 0$ indicates $\mathcal{W}_1$ and/or $\mathcal{W}_3$ are near empty. Use the following routing rule:

If $r_{1,4} + r_{2,3} > 0$, route horizontal customers up to Q_1, otherwise route them down, to Q_3.

If $r_{1,4} - r_{2,3} > 0$, route vertical customers left to Q_2, otherwise route them right, to Q_4.

The Brownian network under optimal control behaves like the fork-join network shown in Figure 17.3: Each job that arrives splits into two jobs, one is served FCFS by the pooled stations 1,4 the other is served FCFS by the pooled stations 2,3; when both parts of the job are completed, it leaves. This fork-join system provides a lower bound on the behavior of the

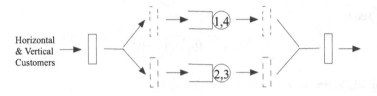

Horizontal
& Vertical
Customers

Figure 17.3 Fork-join analog of the four node network.

unscaled Laws and Louth network (see Exercise 17.6). Simulations show convergence to this lower bound in heavy traffic. The dynamic routing policy provides a substantial improvement over random routing and FCFS sequencing; see Exercises 17.7 and 17.8.

17.4 Further Examples of Pathwise Minimization

We now look first at two more examples of a similar nature to the Laws and Louth network, and we then consider an example that has a more complex behavior.

The following 2×2 cube network, illustrated in Figure 17.4, has eight workstations arranged in a cube and three types of customers, left, top and front, each of which can choose one of four routes starting at the corners of their entry face, going through two stations, and exiting from the opposite

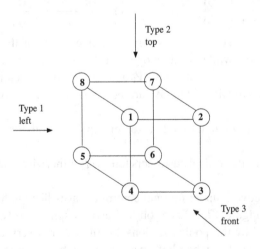

Figure 17.4 A $2 \times 2 \times 2$ cube network.

face. Every customer has to pass through one of the stations $\{1, 3, 5, 7\}$ as well as through one of the stations $\{2, 4, 6, 8\}$. By considering the combined workloads of these, $\mathcal{W}_{1,3,5,7}$, $\mathcal{W}_{2,4,6,8}$, one can again formulate a BCP to control these pooled workloads and minimize the number of customers in the system. The BCP again has a pathwise optimal solution at every time point t, which achieves the behavior of a two-station fork-join network with the pooled service stations (see Exercise 17.11). This can be generalized to a 2^n cube network, with the two pooled resources consisting of the odd and of the even stations, where odd (even) refers to the number of steps on a path from station 1.

A further example is the following ring network, illustrated in Figure 17.5, which consists of six workstations and three types of customers, each having two choices of moving through three of the stations, clockwise or counter-clocwise. Every customer has to go through one of the workstations $\{1, 4\}$, one of $\{2, 5\}$, and one of $\{3, 6\}$. Using the pooled workload of these three sets, one again obtains a Brownian network formulation that has a pathwise solution minimal at each time point t, where it behaves as a three-station fork-join network with the three pooled service stations (see Exercise 17.12).

Note: This can be generalized to an eight-station ring with four types of

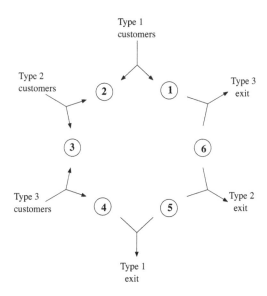

Figure 17.5 A six-station ring network.

customers. One obtains four sets of pooled workstations, and one can mini-mize idleness for these pooled workstaions pathwise. However, in that case the sum of the queue lengths is not minimized when the pooled workloads are minimized pathwise, and the BCP does not have a pathwise optimal solution (see Exercise 17.13).

In all the three previous examples, the pooled resources constituted cuts in the network; every customer needed to pass through each of these cuts. Furthermore, these cuts consisted of disjoint sets of workstations. This led to a formulation of a BCP in terms of the pooled workloads of the cuts, and because the cuts were disjoint, workload in each cut could be minimized a.s. pathwise by one-dimensional Skorohod reflection.

Almost sure pathwise optimal solutions, based on disjoint cuts, are not always possible, as can be seen from the six-station example shown in Fig-ure 17.6. Assume arrival rates λ_1, λ_2 and service rates μ_i, $i = 1, \ldots, 6$. To accommodate the flow, several cuts can be found, some of those are *gener-alized cuts*. An example of a generalized cut is the following requirement:

$$3\lambda_1 + 2\lambda_2 \leq 2\mu_1 + \mu_3 + 2\mu_5 + \mu_6.$$

The interpretation of this generalized cut constraint is as follows: Assume the workstations $j = 1, \ldots, 6$ charge $(2, 0, 1, 0, 2, 1)$ per customer that passes, and each customer of type $1, 2$ has a budget of $3, 2$ to pass. Then the

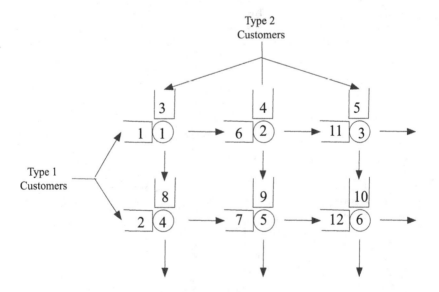

Figure 17.6 A 2×3 six-node network.

maximum charge collected can be $2\mu_1 + \mu_3 + 2\mu_5 + \mu_6$ per unit time, and this will be an upper bound on the combinations of customers $3\lambda_1 + 2\lambda_2$ that can pass the network.

Two generalized cuts of the same form are $(2, 1, 0, 0, 1, 2)$, $(0, 2, 1, 2, 0, 1)$, as can be observed by symmetry. Assume now that both arrival rates are λ and all the service rates are μ, then all three pooled resources of these three cuts will be in balanced heavy traffic when $5\lambda \approx 6\mu$. But because they are using overlapping resources, their Brownian approximations cannot be minimized pathwise.

In Section 17.5 we will study this example in more detail. We will define generalized cuts and formulate the BCP in terms of such cuts. We will then argue that in many cases we can find a.s. pathwise optimal solutions to these problems.

17.5 Routing and Sequencing with General Cuts

17.5.1 Multicommodity Flow Networks

We now consider general unitary processing networks with deterministic routes. Those can be described as multicommodity flow networks.

The network consists of nodes (workstations) $i = 1, \ldots, I$ with rates (capacities) μ_i, and of customer types $c \in C$, each with an arrival rate λ_c, and a set of routes $\mathcal{R}_c$, where route $r \in \mathcal{R}_c$ consists of a sequence of nodes, $i_{r,1}, \ldots, i_{r,s_r}$, and $a_{i,r}$ is the amount of resources from node i that route r requires (in our networks so far it is either 0 or 1, but more generally it counts the number of times that a route passes through a workstation). A compatible flow $\mathbf{f} = (f_r, r \in \mathcal{R}_c, c \in C)$ for λ, μ is found by solving the *compatibility problem LP*, to determine flows f_r and a value η:

$$
\begin{aligned}
&\text{min} \quad \eta \\
\text{LP}(\lambda, \mu) \quad &\text{s.t.} \sum_{c \in C} \sum_{r \in \mathcal{R}_c} a_{i,r} f_r - \eta \leq \mu_i, \qquad i = 1, \ldots, I, \\
&\qquad\quad \sum_{r \in \mathcal{R}_c} f_r = \lambda_c, \qquad c \in C, \\
&\qquad\quad f_r \geq 0.
\end{aligned}
$$

The set of flows and capacities (λ, μ) is compatible if $\eta \leq 0$.

The dual LP with unknowns h_c, π_i is

$$\max \sum_{c \in C} \lambda_c h_c - \sum_{i=1}^{I} \mu_i \pi_i,$$

$$\text{LP}^*(\lambda, \mu) \quad \text{s.t.} \quad h_c \le \sum_{i=1}^{I} a_{i,r} \pi_i, \qquad r \in \mathcal{R}_c, c \in C,$$

$$\sum_{i=1}^{I} \pi_i = 1, \qquad \pi_i \ge 0, \quad i = 1, \dots, I.$$

Note that the solution of the dual will always have $h_c = \min_{r \in \mathcal{R}_c} \sum_{i=1}^{I} a_{i,r} \pi_i \ge 0$.

The condition $\eta \le 0$ is by duality of LP equivalent to $\sum_{c \in C} \lambda_c h_c \le \sum_{i=1}^{I} \mu_i \pi_i$. The constraints of the dual LP are independent of λ, μ. Let $h^{(\ell)}, \pi^{(\ell)}, \ell = 1, \dots, L$ be the list of vertices (extreme points) of the dual constraint set. We obtain a set of L *generalized cut constraints* necessary for the existence of a compatible flow:

$$\sum_{c \in C} h_c^{(l)} \lambda_c \le \sum_{i=1}^{I} \pi_i^{(l)} \mu_i, \text{ where } h_c^{(l)} = \min_{r \in \mathcal{R}_c} \sum_{i=1}^{I} a_{i,r} \pi_i^{(l)}, \qquad l = 1, \dots, L.$$

For the six-station network of Figure 17.6, there are 29 cut constraints that are necessary for compatible flows, six of which are:

$$\lambda_1 \le \mu_1 + \mu_4, \quad \lambda_1 \le \mu_1 + \mu_5, \quad \lambda_2 \le \mu_1 + \mu_2 + \mu_3, \quad \lambda_1 + \lambda_2 \le \mu_1 + \mu_2 + \mu_6,$$

$$2\lambda_1 + \lambda_2 \le \mu_1 + \mu_2 + 2\mu_6, \quad 3\lambda_1 + 2\lambda_2 \le 2\mu_1 + \mu_3 + 2\mu_5 + \mu_6.$$

The first four constraints are cut constraints; the other two are generalized cuts. The remaining constraints are obtained by symmetry (see Exercise 17.16).

17.5.2 *Balanced Heavy Traffic with a Single Generalized Cut*

Consider any network with compatible flows, and increase all the input rates uniformly until the system gets into heavy traffic. In many cases, this will happen when a single generalized cut constraint becomes tight. We now assume that this is the case. Note that this is not the case for the Laws and Louth network of Section 17.3 with symmetric processing rates, nor for the six-node network of Section 17.4 with identical input rates and identical processing rates at all nodes. However, in many realistic networks it is unlikely that such a high degree of symmetry exists, and so this assumption may hold for many types of networks.

Denote by h^*, π^* the extreme point of the dual that corresponds to the cut that becomes tight. As before, we can think of π_i^* as the charge for using node i, and h_c^* as the total budget of customers of type c. Let also $\mathbf{f}^*$ be a compatible flow for this solution, which we choose so that it has a maximal number of positive coordinates. We have for this solution that: $\sum_{c \in C} \lambda_c h_c^* = \sum_{i=1}^{I} \mu_i \pi_i^*$.

We can now consider a reduced network, which consists of the heavy traffic part of the original network. We can discard all nodes i for which $\pi_i^* = 0$; we can discard all types c for which $h_c^* = 0$. We can also discard all routes for which $\sum_{i=1}^{I} a_{i,r} \pi_i^* > h_c^*$. The resulting network has all its stations in balanced heavy traffic ($\pi_i^* > 0$ implies that the capacity constraint of the node is tight), and all remaining routes for type c, $r \in \mathcal{R}_c$, have the same cost, h_c^*. The discarded stations are not bottlenecks, the discarded types have routes that do not pass through bottlenecks, and the discarded routes will not be used, since they are too expensive.

We now assume that in this reduced network with the given μ_i and with λ_c increased uniformly, we have the balanced heavy traffic condition that for some large N, which is the unscaled populations size,

$$N\left(\sum_{c \in C} h_c^* \lambda_c - \sum_{i=1}^{I} \pi_i^* \mu_i \right) < 0 \text{ is of moderate size.}$$

Our controls for the network are $\mathcal{J}_k(t) = \lambda_k t - \mathcal{T}_k(t)$ (where each step on a route defines a class k), and $\mathcal{V}_r(t) = \mathcal{A}_r(t) - f_r t$. The former are the deviations from nominal in the processing of each class; the latter are deviations from nominal in the routing. When we scale time by N^2 and space by N, we can formulate a BCP: Find $\hat{Q}, \hat{\mathcal{J}}, \hat{\mathcal{V}}$ such that

$$\min \ \limsup_{T \to \infty} \frac{1}{T} \mathbb{E}\left[\int_0^T \sum_{k=1}^{K} \hat{Q}_k(t) dt \right]$$

$$\text{s.t. } \hat{Q}(t) = \hat{X}(t) + R\hat{\mathcal{J}}(t) - G\hat{\mathcal{V}}(t),$$
$$\hat{I}(t) = C\hat{\mathcal{J}}(t), \qquad H\hat{\mathcal{V}}(t) = 0, \tag{17.3}$$
$$\hat{Q}(t) \geq 0, \qquad \hat{I}(0) = 0, \quad \hat{I} \text{ non-decreasing,}$$
$$\hat{Q}, \hat{\mathcal{J}}, \hat{\mathcal{V}} \text{ are non-anticipating with respect to } \hat{X}(t).$$

See Exercise 17.14 for exact expressions of R, G, C, H. We now define a pooled workload process based on the generalized cut. For each class k let $m_{i,k}$ be the number of visits of class k in station i along its route until it exits. Let also $\bar{m}_k = \sum_{i=1}^{I} m_{i,k} \pi_i^*$. Recall that $\sum_{i=1}^{I} \pi_i^* = 1$, so $\bar{m}_k$ is a weighted average of the workloads of customers of class k over the various workstations, where workstation i has weight π_i^*. We then define

$$\mathcal{W}_P = \sum_{k=1}^{K} \bar{m}_k Q_k.$$

Note that this pooled workload is in terms of number of services, rather than processing times. Let also $\bar{\bar{m}} = \max\{\bar{m}_k, \ k = 1, \ldots, k\}$, and define the set of maximal work classes, $K^* = \{k : \bar{m}_k = \bar{\bar{m}}\}$.

With this we can formulate the Brownian pooled workload control problem: For given $\hat{\mathcal{Z}}$, find $\hat{Q}, \hat{I}$ such that

$$\min \ \limsup_{T \to \infty} \frac{1}{T} \mathbb{E}[\int_0^T \sum_{k=1}^{K} \hat{Q}_k(t) dt \,]$$

$$\text{s.t.} \ \ \hat{W}(t) = \hat{\mathcal{Z}}(t) + \sum_{i=1}^{I} \pi_i^* \mu_i \hat{I}_i(t), \qquad (17.4)$$

$$\hat{Q}(t) \geq 0, \qquad \hat{I}(0) = 0, \qquad \hat{I} \text{ non-decreasing,}$$

$$\hat{Q}, \hat{I} \text{ are non-anticipating with respect to } \hat{\mathcal{Z}}(t),$$

where $\hat{\mathcal{Z}}(t) = \sum_{k=1}^{K} \bar{m}_k X_k$ is the pooled workload netput which is taken as a Brownian motion, and has drift $N(\sum_{c \in C} \lambda_c h_c^* - \sum_{i=1}^{I} \mu_i \pi_i^*)$ of moderate size, and variance $\sigma^2 = \sum_{c \in C} h_c^{*2} \lambda_c c_{a,c}^2 + \sum_{i=1}^{I} \pi_i^{*2} \mu_i c_{s,i}^2$. This problem has the following pathwise optimal solution:

$$\sum_{i=1}^{I} \pi_i^* \mu_i \hat{I}_i(t) = -\inf_{0 \leq s \leq t} \hat{\mathcal{Z}}(s),$$

$$\sum_{k \in K^*} \hat{Q}_k(t) = \frac{1}{\bar{\bar{m}}} \hat{W}(t), \qquad \hat{Q}_k(t) = 0, \quad k \notin K^*.$$

It can be shown that the regulated process $\mathcal{Z}(t)$ is a lower bound for the sum of queue lengths under any policy. For the Brownian problem this lower bound is achieved. There is no proof that the unscaled system under a policy that imitates the Brownian control converges to the Brownian system.

17.6 Sources

The criss-cross network with routing is due to Wein (1991), the four-station network was suggested and analyzed by Laws and Louth (1990), and the six-station network as well as the multicommodity flows approach are derived in Laws (1992). A reference on multicommodity flows is Gondran and Minoux (1984). Our presentation in this chapter is based on a survey paper of Kelly and Laws (1993).

Exercises

17.1 Derive the factor three saving for the two-station model with criss-cross customers and routing, over a random routing FCFS scheduling. Assume Poisson arrivals and exponential service.

17.2 For the two-station model with criss-cross customers and routing, repeat the Brownian problem formulation and solution for general parameters, $\lambda_A, \lambda_B, \mu_1, \mu_2$.

17.3 For the two-station model with criss-cross customers and routing, perform the Brownian problem formulation and solution for general parameters, λ_A, λ_B and individual processing rates, μ_k, $k = 1, 2, 3, 4$.

17.4 For the network of Laws and Louth, derive the formulation of the BCP, equation (17.2).

17.5 For the network of Laws and Louth, obtain the drift and variance of $\hat{X}(t)$ and of $\hat{Z}(t)$.

17.6 Explain why the Brownian Laws and Louth network under optimal control behaves exactly like the fork-join network, and explain why for the original network the fork-join network provides a lower bound.

17.7 In the network of Laws and Louth, assume Poisson arrivals and exponential service times, and use the policy of random routing and FCFS. Analyze this as a Jackson network, and find total expected number of customers in the system, expected total workloads, and expected sojourn times.

17.8 (∗) For the fork-join network that imitates the Laws and Louth network, under Poisson arrivals and exponential service times, find estimates for expected number in system, total workload, and sojourn time, when the system is in heavy traffic. Compare it to the uncontrolled Jackson network.

17.9 For the network of Laws and Louth, repeat the derivations when service rates of the four stations are not all equal. Notice that there are some conditions on the processing rates that are needed so that the behavior of the resulting balanced heavy traffic network will be similar to the symmetric network.

17.10 (∗) Formulate the control problem for the Laws and Louth network with additional admission control, and required total input rate of $\bar{\lambda}$ and describe its optimal solution.

17.11 For the cube network with three types of customers, formulate the BCP and derive the optimal Brownian solution. Show the analogy to a fork-join network.

17.12 For the ring network with six stations and three types of customers, formulate the BCP and derive the optimal Brownian solution. Show the analogy to a fork-join network.

17.13 (∗) For a ring network with eight stations, obtain a pooled workload Brownian problem, and show how to minimize the pooled workloads a.s. pathwise, but then show that this solution does not minimize the sum of queue lengths and that there is no a.s. pathwise optimal solution to the original BCP.

17.14 Obtain the exact expressions for R, G, C, H in equation 17.3.

17.15 Justify the formulation and the solution of the BCP 17.4.

17.16 (∗) For the six node network of Figure 17.6, write the compatibility linear program and its dual, and identify all the 29 cut constraints. Show that there are no more than 29.

17.17 Consider the six node network of Figure 17.6, with the following arrival and processing rates: $\lambda = (2, 4)$, $\mu = (2, 6, 3, 6, 2, 3)$. Locate the critical generalized cut constraint and formulate and solve the BCP. Obtain the distribution of the pooled workload of the solution.

Part VI

Many-Server Systems

The internet, cloud computing, call centers, and health systems are environments that handle huge volumes of customers, jobs, agents, or queries, with very large numbers of servers. To model these, we no longer use fluid or diffusion scaling. Instead, we let arrival rates and number of servers tend to infinity and scale space, but we do not scale time. This is the so-called many-server scaling.

Because time is not scaled, limiting results for many-server scaling retain dependence on service time distribution, as we already saw in scaling of $G/G/\infty$ in Section 6.9. In Chapter 18 we extend these infinite server results to general time dependent arrival streams.

Heavy traffic $M/M/s$ has full utilization of servers at the cost of congestion, while $M/M/\infty$ has no waiting but poor utilization. These are termed efficiency driven (ED) and quality driven (QD) regimes, respectively. A golden middle road of quality and efficiency driven (QED) is the Halfin–Whitt regime, studied and extended to $G/G/s$ in Chapter 19.

While in theory systems with $\rho > 1$ blow up, in reality they are stabilized by abandonments. We study limiting results for many-server systems with abandonments in Chapter 20.

With many servers, some new issues emerge: using join shortest queue, which we studied in Chapter 15, is no longer practical, since searching for the shortest queue becomes too laborious. We study the supermarket model in Chapter 21 and show that choosing the shortest of just a few randomly chosen queues is almost as good as JSQ. Also in Chapter 21 we see that JSQ performs as well as complete pooling by a single super-fast server, and discuss some other ways of achieving this goal. Another issue with many-server systems is specialization, with several customer and server types, and limited compatibility between them. We study matching in parallel specialized service systems in Chapter 22.

18

Infinite Servers Revisited

In Section 6.9 we looked at G/G/∞ systems and special cases G/D/∞, G/M/∞, and M/G/∞; we assumed arrival rates $n\lambda$ with increasing n and scaled the number in system by n without scaling time. We found that the number in system, $Q(t)$, converges under this scaling to a constant plus a diffusion scaled Gaussian deviations process that is not a diffusion (i.e. not Markovian), and the autocovariance function of the stationary limiting Gaussian process depends on the full distribution of the processing time. In this section we take a fresh look at infinite server systems: We assume i.i.d. service times, but we now look at general arrival processes $\mathcal{A}(t)$, and we assume $Q(0)$ initial customers with general i.i.d. remaining service times. We let $Q(0)$ and $\mathcal{A}(t)$, $t > 0$ increase, and scale the number in system but do not scale time. The main new idea is a detailed representation of $Q(t)$ in terms of all arrival and processing times, presented in Section 18.2. This leads to both fluid and diffusion limits in Sections 18.3 and 18.4. We define two more processes that we need in the representations of general infinite server queues and their limits in Section 18.1.

18.1 Sequential Empirical Processes and the Kiefer Process

We summarize some known facts about the empirical distribution, its Brownian bridge approximation, and the sequential empirical process and its Kiefer process approximation.

Let v_i, $i = 1, 2, \ldots, n$ be a sample of i.i.d. random variables from some distribution F. For a sample of size n, the empirical distribution is the function $F^n(x) = \frac{1}{n} \sum_{i=1}^{n} \mathbb{1}(v_i \leq x)$. So for each x it counts the number in the sample that are $\leq x$ and divides by n.

The Brownian bridge $\mathcal{B}(x)$, $0 \leq x \leq 1$ is a Gaussian process with continuous paths, mean zero, and autocovariance function $\mathbb{E}(\mathcal{B}(x)\mathcal{B}(y)) = x \wedge y - xy$. It is obtained from a Brownian motion $BM(x)$ via $\mathcal{B}(x) =$

$BM(x) - xBM(1)$; in words, it is a Brownian motion tied down to 0 both at $t = 0$ and at $t = 1$ (see Exercise 18.1).

Theorem 18.1 *Let $u_1, \ldots, u_n$ be a sample from a uniform distribution, $U(x) = \mathbb{P}(u_1 \leq x) = x$, $0 \leq x \leq 1$, with empirical distribution $U^n(x)$. Then: $U^n(x) - U(x) \to 0$ u.o.c. a.s., and $\sqrt{n}(U^n(x) - U(x)) \to_w \mathcal{B}(x)$, as $n \to \infty$.*

As a corollary from this, for a sample $v_1, \ldots, v_n$ from a continuous distribution F, $F^n(x) - F(x) \to 0$ u.o.c. a.s. (this is the Glivenko–Cantelli theorem), and $\sqrt{n}(F^n(x) - F(x)) \to_w \mathcal{B} \circ F(x)$, i.e. the centered diffusion scaled deviations converge weakly to a transformed Brownian bridge, calculated as $\mathcal{B}(F(x))$.

The sequential empirical process $F^n(t, x)$ defined on $(t, x) \in [0, 1] \times \mathbb{R}$ is defined as

$$F^n(t, x) = \frac{1}{n} \sum_{i=1}^{\lfloor nt \rfloor} \mathbb{1}(v_i \leq x).$$

It builds the empirical distribution by adding one observation at a time in the order that they occur in the sample. It is illustrated in Figure 18.1.

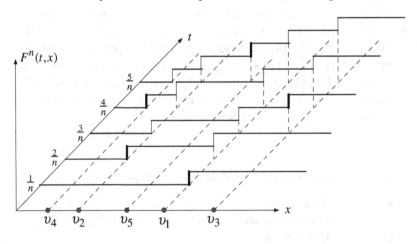

Figure 18.1 The sequential empirical process.

The Kiefer process $\mathcal{B}(t, x)$ is a two-parameter continuous path Gaussian process on $[0, 1] \times [0, 1]$ with mean 0 and autocovariance function

$$\mathbb{E}[\mathcal{B}(s, x)\mathcal{B}(t, y)] = (s \wedge t)(x \wedge y - xy).$$

In particular, this means that for x fixed, $\mathcal{B}(\cdot, x)$ is a Brownian motion (scaled by $\sqrt{x(1-x)}$), and for t fixed, $\mathcal{B}(t, \cdot)$ is a Brownian bridge (scaled by $\sqrt{t}$). In fact, it is the limiting form of the sequential empirical process, as the following theorem states:

Theorem 18.2 (Bickel–Wichura '71 and Krichgina–Puhalski '97) *Let* $u_i, i = 1, 2, \ldots$ *be i.i.d. uniform on* $[0, 1]$ *variables, and let* $U^n(t, x)$ *on* $[0, 1] \times [0, 1]$ *be the sequential empirical processes for the* u_i. *Then the centered scaled empirical processes converge in distribution to the Kiefer process, that is, as* $n \to \infty$:

$$\hat{U}^n(t, x) = \frac{1}{\sqrt{n}} \sum_{i=1}^{\lfloor nt \rfloor} [\mathbb{1}(u_i \le x) - x] \to_w \mathcal{B}(t, x).$$

As a corollary from this, for a sample $v_1, \ldots, v_n$ from a general continuous distribution F, $\sqrt{n}(F^n(t, x) - tF(x)) \to_w \mathcal{B}(t, F(x))$.

Although we defined the sequential empirical process $F^n(t, x)$ only over $t \in [0, 1]$, it is straight forward to see that it remains well defined for $t \in [0, \infty)$, and one can show that the Bickel–Wichura theorem and corollary continue to hold.

18.2 Stochastic System Equations for Infinite Servers

We consider a sequence of infinite server systems indexed by n. The primitives include the arrival stream $\mathcal{A}^n(t)$ with arrival times $a_1^n < a_2^n < \cdots$, the initial number of customers in the system, $Q_0^n = Q^n(0)$, and two sequences of processing times: remaining processing times of initial customers, $\tilde{v}_1, \tilde{v}_2, \ldots$, i.i.d. from distribution G_0, and processing times of arriving customers $v_1, v_2, \ldots$, i.i.d. from distribution G. We shall also use the notation $\bar{G}_0(t) = 1 - G_0(t)$, $\bar{G}(t) = 1 - G(t)$. Note that the processing time sequences are common to all n. On the other hand, we make minimal assumptions about the arrival time sequences. We will allow $Q^n(0) \to \infty$, and assume that $\mathcal{A}^n(t) \to \infty$ for all $t > 0$, as $n \to \infty$, to describe systems with a large volume of customers and many busy servers.

A very general way of representing the number of customers in system n at time t is to use the sequential empirical process. It will also be essential to Chapter 19. The number of customers in system n at time t can be written as

$$Q^n(t) = \sum_{i=1}^{Q^n(0)} \mathbb{1}(\tilde{v}_i > t) + \sum_{i=1}^{A^n(t)} \mathbb{1}(a_i + v_i > t) \tag{18.1}$$

$$= \sum_{i=1}^{Q^n(0)} \mathbb{1}(\tilde{v}_i > t) + n \int_{s=0}^{t} \int_{x=0}^{\infty} \mathbb{1}(s + x > t)\, dG^n(A^n(s)/n, x),$$

where $G^n(t, x)$ is the sequential empirical process for the sequence v_i, so that by the definition of the sequential empirical process,

$$G^n(\mathcal{A}^n(t)/n, x) = \frac{1}{n} \sum_{i=1}^{\mathcal{A}^n(t)} \mathbb{1}(v_i \leq x).$$

In the integral representation, $dG^n(A^n(s)/n, x)$ will contribute $1/n$ if there is an arrival at time $(s, s + ds)$, with processing time of length $(x, x + dx)$, and if $\mathbb{1}(s + x > t)$, i.e. the customer that arrived at time s with processing time x is still present in the system at time t.

It is already possible to guess from this representation of $Q^n(t)$ what the fluid and diffusion scale limits will look like.

18.3 Fluid Approximation of Infinite Server Queues

To obtain fluid limits for the infinite server systems we scale the number of customers by n, but we do not scale time. We are now looking at $\frac{1}{n}A^n(t)$, the scaled arrivals, and at the resulting scaled queues, $\frac{1}{n}Q^n(t)$. We will make minimal assumptions about the arrivals.

Theorem 18.3 *Assume that for some constant q_0 and a deterministic continuous increasing function $\bar{a}(t)$*

$$\frac{1}{n}Q_0^n \to q_0, \quad \frac{1}{n}A^n(t) \to \bar{a}(t) \quad u.o.c.\ a.s.\ as\ n \to \infty.$$

Then,

$$\frac{1}{n}Q^n(t) \to_p q_0\bar{G}_0(t) + \int_0^t \bar{G}(t-s)d\bar{a}(s) \quad u.o.c.\ as\ n \to \infty.$$

Proof We sketch the proof. By the strong law of large numbers, initial customers satisfy

$$\frac{1}{n} \sum_{i=1}^{Q^n(0)} \mathbb{1}(\tilde{v}_i > t) \to q_0\bar{G}_0(t)$$

We use the property that the sequential empirical process $G^n(t, x)$ converges in probability uniformly on bounded regions to $tG(x)$. Hence, for 0 initial customers, by $A^n(t)/n \to \bar{a}(t)$:

$$
\begin{aligned}
\frac{1}{n}(Q^n(t) \,|\, Q^n(0) = 0) &= \int_0^t \int_0^\infty 1(s + x > t)\, dG^n(A^n(s)/n, x) \\
&\to_p \int_0^t \int_0^\infty 1(s + x > t)\, dG(x) d\bar{a}(s) \\
&= \int_0^t \bar{G}(t - s)\, d\bar{a}(s).
\end{aligned}
$$

$\square$

18.4 Diffusion Scale Approximation of Infinite Server Queues

We now look for stochastic limits under many-server diffusion scaling, where we center around the fluid limit and scale the deviations from the fluid limit by $\sqrt{n}$, but we do not scale time. We look at the many-server diffusion scaled arrivals $\sqrt{n}(\frac{1}{n}\mathcal{A}^n(t) - \bar{a}(t))$ and at the resulting many-server diffusion scaled queue lengths $\sqrt{n}(\frac{1}{n}Q^n(t) - \text{fluid limit})$.

Assume in addition to the conditions for fluid approximation that for some random variable X_0 and some stochastic process $\mathcal{Y}(t)$ with $\mathcal{Y}(0) = 0$ and continuous paths, as $n \to \infty$,

$$
\sqrt{n}\left(\frac{1}{n}Q^n(0) - q_0\right) \to_w X_0, \qquad \sqrt{n}\left(\frac{1}{n}\mathcal{A}^n(t) - \bar{a}(t)\right) \to_w \mathcal{Y}(t),
$$

and consider the queue length process, centered around its fluid approximation and scaled,

$$
X^n(t) = \sqrt{n}\left(\frac{Q^n(t)}{n} - q_0\bar{G}_0(x) - \int_0^t \bar{G}(t - s)d\bar{a}(s)\right).
$$

Theorem 18.4 *As $n \to \infty$, the sequence of processes $X^n(t)$ converges weakly to $X(t)$ given by*

$$
X(t) = \bar{G}_0(t)X_0 + \sqrt{q_0}\mathcal{B}^0(G_0(t)) + \int_0^t \bar{G}(t - s)d\mathcal{Y}(s)
$$

$$
- \int_0^t \int_0^t 1(s + x \le t)d\mathcal{B}(\bar{a}(s), G(x)),
$$

where $\mathcal{B}^0$ is a Brownian bridge, $\mathcal{B}(t, x)$ is the Kiefer process, and X_0, $\mathcal{Y}$, $\mathcal{B}^0$, $\mathcal{B}$ are independent.

We do not provide a proof to this theorem. Instead, we just discuss the result and interpret the four parts of the limiting process.

For the first part, $\bar{G}_0(t)\mathcal{X}_0$ is the deviation in the initial number of customers, multiplied by the expected fraction remaining at time t.

For the second part, consider $\dfrac{\sum_{i=1}^{Q^n(0)} \mathbb{1}(\tilde{v}_i \leq t) - Q^n(0)\, G_0(t)}{\sqrt{Q^n(0)}}$. This converges in distribution to the transformed Brownian bridge $\mathcal{B}^0(G_0(t))$. By the same argument,

$$\frac{\sum_{i=1}^{Q^n(0)} \mathbb{1}(\tilde{v}_i > t) - Q^n(0)\bar{G}_0(t)}{\sqrt{Q^n(0)}} \to_w \mathcal{B}^0(\bar{G}_0(t)).$$

But $\mathcal{B}^0(s)$ and $\mathcal{B}^0(1-s)$ have the same distribution. Finally, the expression we need is obtained when multiplying by $\frac{\sqrt{Q^n(0)}}{\sqrt{n}}$.

For the third expression, note first that the integral should be read as integration by parts:

$$\int_0^t \bar{G}(t-s)d\mathcal{Y}(s) = \left[\bar{G}(t-s)\mathcal{Y}(s)\right]_0^t - \int_0^t \mathcal{Y}(s)dG(t-s)$$

$$= \mathcal{Y}(t) - \int_0^t \mathcal{Y}(t-s)dG(s),$$

which is well defined. $\mathcal{Y}(s)$ is the scaled deviation of $\mathcal{A}^n(s)$ from its fluid approximation, and of this deviation at time s, a fraction $\bar{G}(t-s)$ are present at time t. So this takes into account the effect of fluctuations in the arrival process on the number of customers at time t.

For the last expression, we first replace $G^n(\mathcal{A}^n(s)/n, x)$ by its approximation $G^n(\bar{a}(s), x)$, since we already accounted for the deviation of $\mathcal{A}^n(s)$ from $\bar{a}(s)$ in the third part.

We then center and scale the term $n\int_0^t \int_0^\infty \mathbb{1}(s+x>t)\,dG^n(\bar{a}(s), x)$, to obtain:

$$\sqrt{n}\left[\int_0^t \int_0^\infty \mathbb{1}(s+x>t)\,dG^n(\bar{a}(s), x) - \int_0^t \bar{G}(t-s)\,d\bar{a}(s)\right]$$

$$\to_w \int_0^t \int_0^\infty \mathbb{1}(s+x>t)\,d\mathcal{B}(\bar{a}(s), \bar{G}(x))$$

$$= -\int_0^t \int_0^t \mathbb{1}(s+x\leq t)\,d\mathcal{B}(\bar{a}(s), G(x)).$$

Of these four expressions, the first two relate to the initial customers, while the last two relate to the arriving customers. The first and third expressions relate to the deviations in the number of customers, the second and fourth relate to the deviations in the service times. A remarkable property of these results is that the four terms in the limit are independent. This is explained by the independence of the old and new customers, and of the arrivals and service times.

The proof is very technical, mainly because it requires proofs of tightness for several processes, based on semi-martingale decomposition of the sequential empirical process and the use of powerful martingale techniques.

18.5 Sources

The material in this chapter is based on the beautiful seminal paper of Krichagina and Puhalskii (1997). The sequential empirical process is surveyed by Gänssler and Stute (1979), and Bickel and Wichura (1971) prove its convergence to the Kiefer process.

Exercises

18.1 Calculate the autocovariance function of the Brownian bridge $BM(x) - xBM(1)$, $0 \leq x \leq 1$, and conversely, show that the unique Gaussian process with continuous paths and autocovariance $x \wedge y - xy$ is $BM(x) - xBM(1)$, $0 \leq x \leq 1$.

18.2 For a Brownian bridge $\mathcal{B}(y)$ and a distribution function $F(x)$ of a nonnegative random variable, derive the autocovariance function of the process $\mathcal{B} \circ F(x)$ (calculated as $\mathcal{B}(F(x))$, $x \geq 0$).

18.3 Specialize the results of Theorems 18.3 and 18.4 to the G/G/∞, with interarrival time distribution F and i.i.d. service time distribution G, starting from empty.

18.4 Specialize the results of Theorems 18.3 and 18.4 to the G/G/∞, with interarrival time distribution F and i.i.d. service time distribution G, starting with initial number of customers $Q^n(0)/n \to q_0$, $\sqrt{n}(Q^n(0)/n - q_0) \to Q_0$ that have i.i.d. residual service time distribution G_{eq}.

18.5 Specialize the results of Theorems 18.3 and 18.4 to the G/M/∞, with interarrival time distribution F and i.i.d. exponential service times, starting with initial number of customers with the same exponential service time.

18.6 Specialize the results of Theorems 18.3 and 18.4 to the M/G/∞, with service time distribution G, starting with initial number of customers $Q(0)$ where $Q(0)$ is a Poisson random variable, and the initial customers have i.i.d. service time distribution G_{eq}.

18.7 Specialize the results of Theorems 18.3 and 18.4 to the M/M/∞ stationary process.

19

Asymptotics under Halfin–Whitt Regime

In this chapter we consider a single queue with many servers. Assume arrival rate of λ and service rate μ, where $\lambda \gg \mu$, so that $n = \frac{\lambda}{\mu}$ is large. Assume that the number of servers is $s = n + \beta \sqrt{n}$ for some finite positive constant β; this is called square root staffing. Then the system will be stable, with almost full utilization of the servers, and 0 or very little waiting by the customers. Such systems are said to be operating under the Halfin–Whitt regime. They provide both high quality of service and high utilization of servers; this is sometimes called quality and efficiency driven regime – QED. In this chapter we will derive exact results for the M/M/s queue in Halfin–Whitt regime (Section 19.2), and then extend the results to general G/G/s queues (Section 19.3). We first discuss the general concepts of many-server heavy traffic regimes (Section 19.1).

19.1 Three Heavy Traffic Limits

Consider a G/G/s system, with fixed i.i.d. service times of rate μ. For arrival rate λ, the traffic intensity is $\rho = \frac{\lambda}{s\mu}$. We have seen two ways of letting this system move into heavy traffic.

Standard heavy traffic:
Fix s, let $\rho \nearrow 1$. This is the classic heavy traffic, with all s servers busy almost all the time, customers almost always have to wait, the number of customers waiting for service, $Q(t) - s$, is very large, and so is the waiting time. We derived the asymptotics for G/G/s in Section 6.6 and Exercise 6.13. Under these asymptotics, the stationary queue length can be approximated as $Q(\infty) = s + X$, where X has exponential distribution with mean and standard deviation $\frac{c_a^2 + c_s^2}{2(1-\rho)}$, and the stationary process of waiting customers behaves approximately as a reflected Brownian motion.

Many servers underloaded:
Fix $\rho = \frac{\lambda}{s\mu} < 1$, and let both $s, \lambda \nearrow \infty$. Now the number of customers in

330

the system is still becoming very large, but almost all of them are in service, and there are almost always some idle servers, so customers start service immediately on arrival. As $\frac{\lambda}{\mu} \to \infty$, the average number of idle servers will be $\mu s - \lambda \to \infty$. In that case, the asymptotic behavior of the system is indistinguishable from that of G/G/∞ (see Section 6.9 and Chapter 18). In the particular case of the G/M/s system, for fixed ρ and large s the queue length will fluctuate around $s\rho$, and the fluctuations will be approximated by an Ornstein–Uhlenbeck process.

A Realistic Compromise:
In a realistic queueing system, it may not be desirable to have all servers busy with extremely long queues and waiting times, as in the fixed s standard heavy traffic case, nor will it be desirable to have no waiting at all at the cost of a considerable fraction of the servers idling at all times, as in the fixed ρ large s case. In a realistic system, one should have both $\rho \nearrow 1$ and $s \nearrow \infty$ in such a way that the stationary probability that all servers are busy is α, so that a fraction α of the customers needs to wait, and a fraction $1 - \alpha$ of customers enters service with no delay. It turns out that this happens when $(1 - \rho) \sqrt{s}$ converges to a constant. This is the *Halfin–Whitt regime*.

In the case of G/M/s, the scaled queue length process converges to a diffusion process around s, which is a reflected Brownian motion with negative drift above s, and an Ornstein–Uhlenbeck process below s. We discuss this in the next Section 19.2. The behavior under Halfin–Whitt regime for general G/G/s is then described in Section 19.3.

19.2 M/M/s in Halfin–Whitt Regime

In this section we derive the asymptotic behavior of the stationary M/M/s system, under many-server scaling. We have Poisson arrivals of rate λ, exponential services of rate μ, and traffic intensity $\rho = \frac{\lambda}{s\mu}$, and we look at a sequence of systems with $n = s$. We will then let $\lambda_n \to \infty$ with n in a special way, and study the stationary scaled processes $\frac{1}{n} Q^n(t)$, and $\hat{Q}^n(t) = \frac{1}{\sqrt{n}} (Q^n(t) - \mathbb{E}(Q^n(t)))$. Our aim is to be in the Halfin–Whitt QED regime, with probability $0 < \alpha < 1$ that all servers are busy and an arrival needs to wait.

We recall the stationary distribution of queue length for an M/M/s system

(Section 1.3):

$$
p_k = \mathbb{P}(Q(\infty) = k) = \begin{cases} \dfrac{(s\rho)^k}{k!} p_0, & k < s, \\[3mm] \dfrac{s^s \rho^k}{s!} p_0, & k \geq s, \end{cases} \tag{19.1}
$$

with probability of waiting

$$
\alpha = \mathbb{P}(Q(\infty) \geq s) = \frac{(s\rho)^s}{s!} \frac{1}{1-\rho} p_0, \tag{19.2}
$$

and the normalizing constant

$$
p_0 = \mathbb{P}(Q(\infty) = 0) = \left(\frac{(s\rho)^s}{s!} \frac{1}{1-\rho} + \sum_{k=0}^{s-1} \frac{(s\rho)^k}{k!} \right)^{-1}. \tag{19.3}
$$

We now consider a sequence of systems, indexed by n, with $\mu_n = \mu$, $s_n = n$, $\rho_n = \lambda_n / n\mu < 1$.

Theorem 19.1 *The probability of delay has a non-degenerate limit $\alpha = \lim_{n\to\infty} P(Q_n(\infty) \geq n)$, with $0 < \alpha < 1$, if and only if*

$$
\lim_{n\to\infty} \sqrt{n}(1 - \rho_n) = \beta, \qquad 0 < \beta < \infty, \tag{19.4}
$$

in which case

$$
\alpha = \left(1 + \sqrt{2\pi}\, \beta\, \Phi(\beta)\, e^{\beta^2/2} \right)^{-1} = \left(1 + \frac{\beta \Phi(\beta)}{\phi(\beta)} \right)^{-1}, \tag{19.5}
$$

where $\Phi(x)$ is the standard normal cdf, and $\phi(x)$ is the standard normal pdf.

Proof We consider the expressions: $A_n = \displaystyle\sum_{k=0}^{n-1} \frac{(n\rho_n)^k}{k!}$ and $B_n = \dfrac{(n\rho_n)^n}{n!(1-\rho_n)}$.
We note that $A_n/B_n = \mathbb{P}(Q^n(t) < n)/\mathbb{P}(Q^n(t) \geq n)$, and we can then express $\alpha = (1 + A_n/B_n)^{-1}$. The point of this is that we do not need to calculate p_0 to evaluate α. Furthermore, we now multiply A_n, B_n by $e^{-n\rho_n}$, so

$$
\gamma_n = A_n e^{-n\rho_n} = \sum_{k=0}^{n-1} \frac{(n\rho_n)^k}{k!} e^{-n\rho_n}, \qquad \xi_n = B_n e^{-n\rho_n} = \frac{(n\rho_n)^n}{n!(1-\rho_n)} e^{-n\rho_n}.
$$

We now note that if X_n is a Poisson random variable with parameter (i.e. mean and variance) $n\rho_n = \frac{\lambda_n}{\mu}$, then $\gamma_n = \mathbb{P}(X_n \leq n-1)$.

By CLT applied to X_n, if $(1 - \rho_n)\sqrt{n} \to \beta$, then $\gamma_n \to \Phi(\beta)$ (see Exercise 19.1).

Using Stirling's formula for $n!$ in ξ_n we obtain that if $(1 - \rho_n) \sqrt{n} \to \beta$, then $\xi_n \to \sqrt{2\pi} \, \beta \, e^{\beta^2/2}$ (see Exercise 19.2). This proves the convergence and the expression for α.

To prove only if: it can be shown (see Exercise 19.3) that if $(1 - \rho_n) \sqrt{n} \to 0$ then $\alpha_n \to 1$, and if $(1 - \rho_n) \sqrt{n} \to \infty$ then $\alpha_n \to 0$. Furthermore, if $(1 - \rho_n) \sqrt{n}$ fails to converge, then it must have subsequences with limits $\beta_1 \neq \beta_2$. The monotonicity of the limiting expression for α as a function of $\beta > 0$ implies that in that case α_n will also fail to converge. □

Note that for large $\frac{\lambda}{\mu}$, the condition $\sqrt{n}(1 - \rho_n) \approx \beta$ is equivalent to setting the number of servers s to be:

$$s = \frac{\lambda}{\mu} + \sqrt{\frac{\lambda}{\mu}}\beta, \tag{19.6}$$

i.e. we add to the minimum necessary number of servers, $\frac{\lambda}{\mu}$ a constant multiple of $\sqrt{\frac{\lambda}{\mu}}$. This is called *square root staffing*. Theorem 19.1 shows that to achieve $0 < \alpha < 1$, it is necessary and sufficient to have square root staffing.

For the rest of this section we assume that our M/M/s system is using square root staffing, i.e. (19.4) or equivalently (19.6), so that $0 < \alpha < 1$, and is given by (19.5).

Proposition 19.2 *(i) If $\delta_n < n$ and $(n - \delta_n)/\sqrt{n} \to \delta > 0$ as $n \to \infty$, then*

$$\lim_{n \to \infty} \mathbb{P}(Q^n(\infty) \leq \delta_n \mid Q^n(\infty) \leq n) = \Phi(\beta - \delta)/\Phi(\beta). \tag{19.7}$$

(ii) If $\delta_n > n$ and $(\delta_n - n)/\sqrt{n} \to \delta > 0$ as $n \to \infty$, then

$$\lim_{n \to \infty} \mathbb{P}(Q^n(\infty) \geq \delta_n \mid Q^n(\infty) \geq n) = e^{-\delta\beta}. \tag{19.8}$$

See Exercise 19.5 for proof.

We can now obtain an approximation to the stationary queue length distribution in the M/M/s system under Halfin–Whitt regime. Let $\hat{Q}^n(\infty) = (Q^n(\infty) - n)/\sqrt{n}$. By choosing $\delta_n = n + x \sqrt{n}$ in (19.7) and (19.8) we have immediately:

Theorem 19.3 $\hat{Q}^n(\infty) \to_w \hat{Q}$, *where the random variable $\hat{Q}$ satisfies:* $\mathbb{P}(\hat{Q} \geq 0) = \alpha$, $\mathbb{P}(\hat{Q} > x) \mid \hat{Q} \geq 0) = e^{-\beta x}$, $x > 0$, *and* $\mathbb{P}(\hat{Q} \leq x \mid \hat{Q} \leq 0) = \Phi(\beta + x)/\Phi(\beta)$, $x \leq 0$.

We see that the stationary queue length is centered around s, where positive deviations are the number of waiting customers, and negative deviations are idle servers. The distribution of the diffusion scaled deviations is composed of a Gaussian density below 0 (idle servers), and exponential density above 0 (waiting customers). Figure 19.1 illustrates the stationary distribution, for $\beta = 0.5$ and $\beta = 1$.

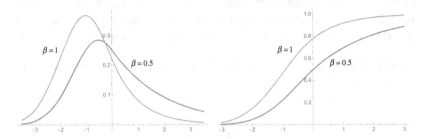

Figure 19.1 Halfin–Whitt stationary density and cdf

We now consider the entire sequence of processes $Q^n(t)$, under the Halfin–Whitt regime. We scale the process to have $\hat{Q}^n(t) = (Q^n(t)-n)/\sqrt{n}$, so that $\hat{Q}^n(t)^+$ will be the scaled number of waiting customers and $\hat{Q}^n(t)^-$ will be the scaled number of idle servers.

The queue length of M/M/s is a birth and death Markov chain, and with $\rho_n \to 1$, $n \to \infty$ the time between events is $O(1/n)$, while the change in the scaled queue length is $O(1/\sqrt{n})$, so that the process converges to a diffusion process (see Section 7.2). We see already from the behavior of the Markov chain that it behaves as a single-server queue with processing rate $n\mu$ when $Q^n(t) > n$, and like an infinite server queue when $Q^n(t) \le n$. This indicates that the process $\hat{Q}^n(t)$ will behave as Brownian motion when positive, and as an Ornstein–Uhlenbeck process when negative. This is stated in the next theorem.

Theorem 19.4 *If $\hat{Q}^n(0) \to \hat{Q}(0)$, then $\hat{Q}^n(t) \to_w \hat{Q}(t)$ in $\mathbb{D}(0, \infty)$, where $\hat{Q}(t)$ is a diffusion process defined by*

$$d\hat{Q}(t) = m(\hat{Q}(t))dt + \sigma(\hat{Q}(t))dBM(t),$$

where the drift and diffusion coefficient are

$$m(x) = \begin{cases} -\mu\beta & x \ge 0, \\ -\mu(x + \beta) & x \le 0, \end{cases} \qquad \sigma^2(x) = 2\mu.$$

Proof Recall that $Q^n(t)$ are birth and death processes. The theorem follows from Stone's theorem 7.2 about convergence of birth and death processes to diffusion processes. We assume that $(1 - \rho_n)\sqrt{n} = (n - \lambda_n/\mu)/\sqrt{n} \to \beta$ as $n \to \infty$. We calculate the infinitesimal mean and variance of changes in the birth and death process $\hat{Q}^n$.

For $x > 0$:

$$m_n(x) = \lim_{h \to 0} \frac{\mathbb{E}(\hat{Q}^n(t + h) - \hat{Q}^n(t) \mid \hat{Q}^n(t) = x)}{h}$$

$$= \lambda_n \frac{1}{\sqrt{n}} - n\mu \frac{1}{\sqrt{n}} \to -\mu\beta,$$

$$\sigma_n^2(x) = \lim_{h \to 0} \frac{\mathbb{V}\text{ar}(\hat{Q}^n(t + h) - \hat{Q}^n(t) \mid \hat{Q}^n(t) = x)}{h}$$

$$= (n\mu + \lambda_n)/n = \mu(1 + \rho_n) \to 2\mu,$$

and for $x < 0$:

$$m_n(x) = \lim_{h \to 0} \frac{\mathbb{E}(\hat{Q}^n(t + h) - \hat{Q}^n(t) \mid \hat{Q}^n(t) = x)}{h}$$

$$= \lambda_n \frac{1}{\sqrt{n}} - (n + \sqrt{n}x)\mu \frac{1}{\sqrt{n}} \to -\mu(\beta + x),$$

$$\sigma_n^2(x) = \lim_{h \to 0} \frac{\mathbb{V}\text{ar}(\hat{Q}^n(t + h) - \hat{Q}^n(t) \mid \hat{Q}^n(t) = x)}{h}$$

$$= ((n + \sqrt{n}x)\mu + \lambda_n)/n = \mu(1 + \rho_n + \frac{x}{\sqrt{n}}) \to 2\mu.$$

So, the states of $\hat{Q}^n(t)$ become dense on the real line, and the infinitesimal means and variances converge. The theorem follows. □

We observe that the scaled queue length behaves like a Brownian motion with constant negative drift when it is above n, the number of servers, and it behaves like an Ornstein–Uhlenbeck process when below n. This means that the unscaled number of waiting customers, when positive, behaves like a Brownian motion, and the number of idle servers, when positive, behaves like an Ornstein–Uhlenbeck process with a linear drift toward $\sqrt{n}\beta$.

The actual queue length (including all customers in the system) is stable, and fluctuates around the number of servers, with fluctuations of order $\sqrt{n}$. With probability $1 - \alpha$ there are some idle servers (and no waiting), but the fraction of idle servers is negligible (of order $1/\sqrt{n}$), and with probability α customers may need to wait, but the waiting time is negligible (of order $1/\sqrt{n}$).

19.2.1 Extension to G/M/s

We now assume that arrivals to system n are renewal, interarrivals are i.i.d. u_n, with mean $1/\lambda_n$, squared coefficient of variation c_a^2, and finite third moment. Under Halfin–Whitt regime $\hat{Q}_n$ will again converge to the same diffusion process, with the only change that $\sigma_n^2(x) \to \mu(1 + c_a^2)$, and in calculating the value of α, β is replaced by $2\beta/(1 + c_a^2)$. The proof is as follows: One considers the embedded Markov chain at times of arrival, and obtains the convergence of that to a diffusion (see Exercise 19.7). One then uses random time change and continuous mapping to obtain the result for the sequence of continuous Markov chains.

19.3 Fluid and Diffusion Limits for G/G/s under Halfin–Whitt Regime

We now extend the results for many-server systems under the more general assumptions of Chapter 18. For our G/G/n system we now assume that at time 0 there are Q_0 initial customers, of which $\min(Q_0, n)$ are already in service, with remaining i.i.d. service times $\tilde{v}_i$, distributed as G_0, and the remaining $(Q_0 - n)^+$ customers are waiting for service. Assume arrival stream $A(t)$, with customer i arriving at time a_i. We have processing times v_i, $i = 1 \ldots, (Q_0 - n)^+$ for the waiting initial customers, and $v_{(Q_0-n)^++i}$ for the ith new arrival, all i.i.d. distributed as G. We denote $\bar{G}_0(t) = 1 - G_0(t)$ and $\bar{G}(t) = 1 - G(t)$. Let $\tilde{w}_i$ be the waiting times of the $(Q_0 - n)^+$ initial customers, and w_i the waiting times of the new arrivals.

Then, similar to (18.1) in Chapter 18, the queue length at time t, consisting of all customers in the system (in service or waiting), is given by

$$Q(t) = \sum_{i=1}^{\min(Q_0,n)} \mathbb{1}(\tilde{v}_i > t) + \sum_{i=1}^{(Q_0-n)^+} \mathbb{1}(\tilde{w}_i + v_i > t)$$

$$+ \sum_{i=1}^{A(t)} \mathbb{1}(a_i + w_i + v_{(Q_0-n)^++i} > t). \tag{19.9}$$

We next make a lengthy derivations toward a decomposition of the queue length. We centralize the three terms by subtracting the expectation of the indicator with respect to the processing times, to obtain:

$$Q(t) = \min(Q_0, n)\bar{G}_0(t) + W_0(t) + M_2(t)$$

$$+ \sum_{i=1}^{(Q_0-n)^+} \bar{G}(t - \tilde{w}_i) + \sum_{i=1}^{A(t)} \bar{G}(t - a_i - w_i), \tag{19.10}$$

where the centralized sums of indicators are

$$W_0(t) = \sum_{i=1}^{\min(Q_0,n)} (\mathbb{1}(\tilde{v}_i > t) - \bar{G}_0(t)),$$

$$M_2(t) = \sum_{i=1}^{(Q_0-n)^+} (\mathbb{1}(\tilde{w}_i + v_i > t) - \bar{G}(t - \tilde{w}_i)) \tag{19.11}$$

$$+ \sum_{i=1}^{A(t)} (\mathbb{1}(a_i + w_i + v_{(Q_0-n)^++i} > t) - \bar{G}(t - a_i - w_i)).$$

Next, adding and subtracting $A_G(t) = \int_0^t \bar{G}(t-s)dA(s)$, and $(Q_0-n)^+\bar{G}(t)$, we have

$$Q(t) = I(t) + W_0(t) + M_2(t) + A_G(t)$$
$$+ \sum_{i=1}^{(Q_0-n)^+} (\bar{G}(t - \tilde{w}_i) - \bar{G}(t)) + \sum_{i=1}^{A(t)} (\bar{G}(t - a_i - w_i) - \bar{G}(t - a_i)),$$

where

$$I(t) = \min(Q_0, n)\bar{G}_0(t) + (Q_0 - n)^+\bar{G}(t).$$

Note that for given $A(s)$, $0 \le s \le t$, the term $I(t) + A_G(t)$ is the expected number of customers in a G/G/∞ system, which has the same customers and sequence of processing times as our G/G/n system.

The following proposition gives a very useful way of describing the dynamics of the queue length process:

Proposition 19.5

$$\sum_{i=1}^{(Q_0-n)^+} (\bar{G}(t - \tilde{w}_i) - \bar{G}(t)) + \sum_{i=1}^{A(t)} (\bar{G}(t - a_i - w_i) - \bar{G}(t - a_i))$$
$$= \int_0^t (Q(t - s) - n)^+ dG(s).$$

Proof The number of customers waiting to be served at time t can be written as:

$$(Q(t) - n)^+ = \sum_{i=1}^{(Q_0-n)^+} \mathbb{1}(t < \tilde{w}_i) + \sum_{i=1}^{A(t)} \mathbb{1}(a_i < t < a_i + w_i).$$

We then have:

$$\sum_{i=1}^{A(t)} (\bar{G}(t - a_i - w_i) - \bar{G}(t - a_i))$$

$$= \sum_{i=1}^{A(t)} \int_{t-a_i-w_i}^{t-a_i} dG(s)$$

$$= \sum_{i=1}^{A(t)} \int_0^\infty \mathbb{1}(t - a_i - w_i < s < t - a_i) dG(s)$$

$$= \sum_{i=1}^{A(t)} \int_0^\infty \mathbb{1}(a_i < t - s < a_i + w_i) dG(s)$$

$$= \int_0^\infty \sum_{i=1}^{A(t)} \mathbb{1}(a_i < t - s < a_i + w_i) dG(s)$$

$$= \int_0^t \left((Q(t - s) - n)^+ - \sum_{i=1}^{(Q_0-n)^+} \mathbb{1}(t - s < \tilde{w}_i) \right) dG(s)$$

$$= \int_0^t (Q(t - s) - n)^+ dG(s) - \int_0^t \sum_{i=1}^{(Q_0-n)^+} \mathbb{1}(t - s < \tilde{w}_i) dG(s).$$

Similarly,

$$\sum_{i=1}^{(Q_0-n)^+} (\bar{G}(t - \tilde{w}_i) - \bar{G}(t)) = \int_0^t \sum_{i=1}^{(Q_0-n)^+} \mathbb{1}(t - s < \tilde{w}_i) dG(s),$$

and the proposition follows. □

With this proposition we can now represent the queue length $Q(t)$ in terms of the following renewal type equation:

$$Q(t) = I(t) + W_0(t) + M_2(t) + A_G(t) + \int_0^t (Q(t - s) - n)^+ dG(s). \quad (19.12)$$

19.3.1 The Renewal Type Mapping

One can prove the following theorem, as a result of which the representation (19.12) of $Q(t)$ uniquely determines $Q(t)$ as a deterministic function of Q_0 and the sequences of arrivals and service times, as expressed by I, W_0, M_2, A_G.

Theorem 19.6 *Let B be a cdf on $\mathbb{R}_+$, and κ a constant. For a function $x \in \mathbb{D}(0, \infty)$ define $\psi_{\kappa, B}$ as the solution z of*

$$z(t) = x(t) + \int_0^t (z(t - s) + \kappa)^+ dB(s).$$

Then for each x, there is a unique $\psi_{\kappa, B}(x) \in \mathbb{D}(0, \infty)$. Furthermore, the map $\psi_{\kappa, B} : \mathbb{D}(0, \infty) \to \mathbb{D}(0, \infty)$ is Lipschitz continuous.

Based on this, to obtain limiting results on a sequence $Q^n(t)$ of G/G/n systems, it is enough to discuss the corresponding limits of I^n, W_0^n, M_2^n, A_G^n.

19.3.2 Fluid limit results

We consider a sequence of systems, all sharing the same sequences of processing times, but system n has n servers, and its own $Q_0^n, A^n(t)$. We use many-server scaling, so that for a function $z^n(t)$ we have the scaled function $\bar{z}^n(t) = \frac{1}{n} z^n(t)$, i.e. we scale space but do not scale time. With the scaled $\bar{Q}_0^n$ and $\bar{\mathcal{A}}^n(t)$, we get the stochastic equations for the scaled queue length, following (19.12):

$$\frac{1}{n} Q^n(t) = \frac{1}{n} I^n(t) + \frac{1}{n} W_0^n(t) + \frac{1}{n} M_2^n(t) + \frac{1}{n} \mathcal{A}_G^n(t)$$
$$+ \int_0^t (\frac{1}{n} Q^n(t - s) - 1)^+ dG(s). \qquad (19.13)$$

We now assume that for some constant q_0 and some deterministic continuous increasing function $\bar{a}(t)$,

$$\frac{1}{n} Q_0^n(t) \to q_0, \text{ and } \frac{1}{n} \mathcal{A}^n(t) \to \bar{a}(t), \text{ u.o.c. a.s.}$$

Proposition 19.7 $\frac{1}{n} W_0^n \to_w 0$ *and* $\frac{1}{n} M_2^n \to_w 0$ *as* $n \to T$.

Proof We prove first that $\overline{W}_0^n \to 0$ u.o.c. a.s. By the Glivenko–Cantelli theorem,

$$\sup_{0 \le t < T} \frac{1}{n} \Big| \sum_{i=1}^n (\mathbb{1}(\tilde{v}_i > t) - \bar{G}_0(t)) \Big| \to 0 \text{ a.s.}$$

It then follows, similar to the proof of the FSLLN, that

$$\sup_{0 \le x \le 1} \sup_{0 \le t < T} \frac{1}{n} \Big| \sum_{i=1}^{\lfloor nx \rfloor} (\mathbb{1}(\tilde{v}_i > t) - \bar{G}_0(t)) \Big| \to 0 \text{ a.s.,}$$

and we have

$$\sup_{0 \le t < T} |\overline{W}_0(t)| = \sup_{0 \le t < T} \frac{1}{n} \left| \sum_{i=1}^{\min(Q_0^n, n)} (\mathbb{1}(\tilde{v}_i > t) - \bar{G}_0(t)) \right|$$

$$\le \sup_{0 \le x \le 1} \sup_{0 \le t < T} \frac{1}{n} \left| \sum_{i=1}^{\lfloor nx \rfloor} (\mathbb{1}(\tilde{v}_i > t) - \bar{G}_0(t)) \right| \to 0 \text{ a.s.}$$

The proof that $\bar{M}_2^n \to_w 0$ is quite technical and we only sketch the main points. Consider the times that initial waiting or newly arriving customers start service:

$$\tilde{w}_1 < \tilde{w}_2 < \cdots < \tilde{w}_{(Q_0-n)^+} < a_1 + w_1 < a_2 + w_2 < \cdots,$$

and define the process $\breve{\mathcal{A}}(t)$ as the total number of customers that have started service in $[0, t]$.

Define the centralized sequential empirical process for the distribution G and the counting process $\breve{\mathcal{A}}(t)$,

$$\hat{G}^n(t, x) = \sum_{i=1}^{\breve{\mathcal{A}}(t)} (\mathbb{1}(v_i \le x) - G(x)).$$

Then, after some rearranging, and similar to (18.1), we have

$$M_2(t) = \int_0^t \int_0^t \mathbb{1}(s + x \le t) d\hat{G}^n(s, x).$$

Note that setting

$$U^n(t, x) = \sum_{i=1}^{\lfloor nt \rfloor} (\mathbb{1}(G(v_i) \le x) - x), \quad t \ge 0, \ 0 \le x \le 1,$$

the process $U^n(t, x)$ satisfies, by the Bickel–Wichura theorem 18.2, that

$$\frac{1}{n} U^n(\cdot, \cdot) \to 0 \text{ u.o.c. a.s. and} \frac{1}{\sqrt{n}} U^n(\cdot, \cdot) \to_w \text{ the Kiefer process, as } n \to \infty,$$

and we have that

$$\hat{G}^n(t, x) = U^n\left(\frac{1}{n} \breve{\mathcal{A}}^n(t), G(x)\right).$$

We note also that $\frac{1}{n} \breve{\mathcal{A}}^n(t) \le \frac{1}{n} \mathcal{A}^n(t)$, and by $\frac{1}{n} \mathcal{A}^n(t) \to \bar{a}(t)$ it is bounded for all $0 \le t \le T$. It follows that $\frac{1}{n} \hat{G}^n(\cdot, \cdot) \to 0$, and hence $M_2 \to 0$. We skip the technical details.   □

We can now prove

Theorem 19.8 (Fluid approximation)   *If* $(\bar{Q}_0^n, \bar{\mathcal{A}}^n) \to (q_0, \bar{a}(t))$, *then* $\frac{1}{n}Q_n(\cdot) \to_w \bar{Q}(\cdot)$, *where* $\bar{Q}(t)$ *is a solution of*

$$\bar{Q}(t) = \min(q_0, 1)\bar{G}_0(t) + (q_0 - 1)^+\bar{G}(t) + \int_0^t \bar{G}(t-s)d\bar{a}(s)$$

$$+ \int_0^t (\bar{Q}(t-s) - 1)^+ dG(s). \tag{19.14}$$

Proof   We have already seen in Proposition 19.7 that $\frac{1}{n}W_0^n \to_w 0$ and $\frac{1}{n}M_2^n \to_w 0$ as $n \to \infty$. We now consider $\frac{1}{n}I^n(t)$ and $\frac{1}{n}\mathcal{A}_G^n(t)$. Since $\frac{1}{n}Q_0^n \to q_0$, and $\frac{1}{n}\mathcal{A}^n(t) \to \bar{a}(t)$, we have immediately:

$$\frac{1}{n}I^n(t) \to_w \min(q_0, 1)\bar{G}_0(t) + (q_0 - 1)^+\bar{G}(t),$$

$$\frac{1}{n}\mathcal{A}_G^n(t) \to_w \int_0^t \bar{G}(t-s)d\bar{a}(s).$$

But,

$$\frac{1}{n}Q^n(\cdot) = \psi_{-1,G}\left(\frac{1}{n}I^n(\cdot) + \frac{1}{n}W_0^n(\cdot) + \frac{1}{n}M_2^n(\cdot) + \frac{1}{n}\mathcal{A}_G^n(\cdot)\right),$$

and by the continuity of the $\psi_{-1,G}$ map, we have

$$\frac{1}{n}Q^n(\cdot) \to_w \psi_{-1,G}\left(\min(q_0, 1)\bar{G}_0(\cdot) + (q_0 - 1)^+\bar{G}(\cdot) + \int_0^\cdot \bar{G}(\cdot - s)d\bar{a}(s)\right).$$

This completes the proof.   □

The following is an interpretation of equation (19.14). We see that $\bar{Q} = \psi_{-1,G}(\bar{Q}_\infty)$, where

$$\bar{Q}_\infty(t) = \min(\bar{Q}_0, 1)\bar{G}_0(t) + (\bar{Q}_0 - 1)^+\bar{G}(t) + \int_0^t \bar{G}(t-s)d\bar{a}(s).$$

But $\bar{Q}_\infty(t)$ is exactly the fluid limit obtained for the G/G/∞ system, which started off with $\min(Q^n(0), n)$ customers that require i.i.d. G_0 service, and $(Q^n(0) - n)^+$ customers that require i.i.d. G service. Hence, the fourth term in (19.14), $\int_0^t (\bar{Q}(t-s) - 1)^+ dG(s)$, is the positive added fluid queue length due to waiting times in G/G/n, which G/G/∞ does not have. Note that it includes the additional waiting created at time 0.

Note that the fluid limit is completely general; depending on the fluid limit of the arrival stream and on the initial conditions, the fluid could move up, or go to zero or fluctuate. The equation (19.14) cannot in general be solved, mainly because of the non-linear operation $(\bar{Q} - 1)^+$. We give two examples where it can be solved.

Example 19.9 Assume that both G and G_0 are deterministic with value 1, and $q_0 = 1$, $\bar{a}(t) = t$. Then the fluid limit is a saw tooth function, starting at 1, going up to 2, and repeating periodically, as shown in Figure 19.2 (see Exercise 19.9). As we see, the fluid limit can be periodic and discontinuous.

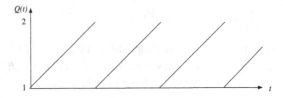

Figure 19.2 Periodic discontinuous fluid limit.

Next we consider the stationary case:

Corollary 19.10 *Assume that $Q_0/n \to 1$, $G_0 = G_{eq}$ is the equilibrium distribution of G, $\mu = 1$, and $\mathcal{A}^n(t)/n \to t$. Then $\frac{1}{n}Q^n(t) \to_w 1$.*

Proof The equation for the fluid limit $\bar{Q}(t)$ is in this case

$$\bar{Q}(t) = \min(1,1)\bar{G}_{eq}(t) + (1-1)^+\bar{G}(t) + \int_0^t \bar{G}(t-s)ds$$

$$+ \int_0^t (\bar{Q}(t-s) - 1)^+ dG(s)$$

$$= \bar{G}_{eq}(t) + \int_0^t \bar{G}(t-s)ds + \int_0^t (\bar{Q}(t-s) - 1)^+ dG(s)$$

$$= 1 + \int_0^t (\bar{Q}(t-s) - 1)^+ dG(s),$$

which is solved by $\bar{Q}(t) = 1$, and by the uniqueness of the $\psi_{\kappa,B}$ solutions, this is the unique fluid limit.   □

19.3.3 Diffusion limit results

So far in this section we made very general assumptions about initial states and arrivals. We now describe stochastic diffusion scale limits of the G/G/n system with regular input, under Halfin–Whitt regime.

We consider a sequence of systems and assume as follows: Customers arrive at rates $n\rho_n$ where $\rho_n \nearrow 1$, and the diffusion scaled arrivals satisfy:

$$\hat{\mathcal{A}}^n(t) = \frac{\mathcal{A}^n(t) - n\rho_n t}{\sqrt{n}} \to_w \mathcal{Y}(t), \qquad (19.15)$$

where $\mathcal{Y}(\cdot)$ is a stochastic process with a.s. continuous sample paths.

The arrival rates satisfy Halfin–Whitt square root staffing, that is

$$\sqrt{n}(1 - \rho_n) \to \beta, \text{ as } n \to \infty, \quad 0 < \beta < \infty. \tag{19.16}$$

The initial states satisfy the conditions of Corollary 19.10, i.e. $\frac{1}{n}Q^n(0) \to_w 1$, $G_0 = G_{eq}$, and without loss of generality $\mu = 1$. As a result, the fluid limit for the sequence is $\bar{Q}(t) = 1$. Assume also that for some random variable $\hat{Q}_0$,

$$\frac{1}{\sqrt{n}}(Q_0^n - n) \to \hat{Q}_0. \tag{19.17}$$

We centralize and scale the queue lengths, and using (19.9)–(19.12) and several manipulations we get:

$$\begin{aligned}
\hat{Q}^n(t) &= \frac{Q^n(t) - n}{\sqrt{n}} \\
&= \hat{M}_Q^n(t) + \hat{Q}_\infty^n(t) - \sqrt{n}(1 - \rho_n)G_{eq}(t) \\
&\quad + \int_0^t (\hat{Q}^n(t - s))^+ dG(s),
\end{aligned} \tag{19.18}$$

where

$$\hat{M}_Q^n(t) = \frac{1}{\sqrt{n}}(Q_0^n - n)^+(\bar{G}(t) - \bar{G}_{eq}(t)), \tag{19.19}$$

and $\hat{Q}_\infty^n(t)$ is composed of four parts,

$$\hat{Q}_\infty^n(t) = \hat{H}^n(t) + \hat{W}_0^n(t) + \hat{M}_1^n(t) + \hat{M}_2^n(t), \tag{19.20}$$

where

$$\begin{aligned}
\hat{H}^n(t) &= \frac{1}{\sqrt{n}}(Q_0^n - n)\bar{G}_{eq}(t), \\
\hat{W}_0^n(t) &= \frac{1}{\sqrt{n}}\sum_{i=1}^{\min(Q_0,n)} (\mathbb{1}(\tilde{v}_i > t) - \bar{G}_{eq}(t)), \\
\hat{M}_1^n(t) &= \frac{1}{\sqrt{n}}\int_0^t G(t - s)d(\mathcal{A}^n(s) - n\rho_n s), \\
\hat{M}_2^n(t) &= \frac{1}{\sqrt{n}}\Bigg[\sum_{i=1}^{(Q_0-n)^+} (\mathbb{1}(\tilde{w}_i + v_i > t) - \bar{G}(t - \tilde{w}_i)) \\
&\quad + \sum_{i=1}^{A(t)} (\mathbb{1}(a_i + w_i + v_{(Q_0-n)^++i} > t) - \bar{G}(t - a_i - w_i))\Bigg].
\end{aligned} \tag{19.21}$$

It is seen from (19.18 that:

$$\hat{Q}^n(t) = \psi_{0,G}(\hat{M}_Q^n(t) + \hat{Q}_\infty^n(t) - \sqrt{n}(1 - \rho_n)G_{eq}(t)). \qquad (19.22)$$

There follows the hardest part of the derivation, which is to show:

Proposition 19.11   *Under the conditions of Corollary 19.10 and assumptions (19.15)–(19.17),*

$$\hat{Q}_\infty^n(t) \to_w \hat{Q}_0\bar{G}_{eq} + \mathcal{B}(G_{eq}) + \int_0^t G(t-s)d\mathcal{Y}(s) + \hat{M}_2(t), \qquad (19.23)$$

where $\hat{M}_2(t)$ is a centered Gaussian process with autocovariance function:

$$\mathbb{E}(\hat{M}_2(t), \hat{M}_2(t+s)) = \int_0^t G(t-u)\bar{G}(t+s-u)du. \qquad (19.24)$$

Remarkably, this is exactly the stochastic limit for the scaled deviations of the queue lengths of the G/G/∞ system, which has the same initial customers, same arrival processes and same processing time sequences of initial and arriving customers, as our G/G/n system.

We now have the main theorem, which follows naturally from Proposition 19.11.

Theorem 19.12   *Under the conditions of Corollary 19.10 and assumptions (19.15)–(19.17), $\hat{Q}^n(\cdot) \to_w \hat{Q}(\cdot)$ where*

$$\hat{Q}(t) = (\hat{Q}_0)^+(\bar{G}(t) - \bar{G}_{eq}(t)) + \hat{Q}_\infty(t) - \beta G_{eq}(t)$$
$$+ \int_0^t (\hat{Q}(t-s))^+ dG(s), \qquad (19.25)$$

or equivalently,

$$\hat{Q}^n(\cdot) \to_w \psi_{0,G}\left((\hat{Q}_0)^+(\bar{G}(\cdot) - \bar{G}_{eq}(\cdot)) + \hat{Q}_\infty^n(\cdot) - \beta G_{eq}(\cdot)\right). \qquad (19.26)$$

Discussion   The proof is based on the uniqueness and continuity of the convolution type mapping $\psi_{\kappa,B}$, applied now to a sequence of stochastic processes that converge weakly, rather than to a sequence of functions that converge. The result is, similar to the fluid case: The limiting stochastic deviations of queue length from the constant fluid level consist of the G/G/∞ queue length deviations, given by $\hat{Q}_\infty(t)$, plus the effect of customers waiting at time $t - s$ integrated over $0 < s < t$. There are two additional terms: the term $(\hat{Q}_0)^+(\bar{G}(t) - \bar{G}_{eq}(t))$ accounts for the discrepancy that in G/G/∞ all initial customers have remaining processing times distributed as

$G_{eq}(t)$, whereas in G/G/n, only $\min(Q_0^n, n)$ have these processing times, while the remaining $(Q_0^n - n)^+$ have processing times distributed as $G(t)$. Finally, the term $-\beta G_{eq}$ is the only term that specifically includes the effect of the Halfin–Whitt staffing constant β. □

19.4 sources

The Halfin–Whitt result for the MM/s system appeared in 1981 (Halfin and Whitt (1981)) but it only gained prominence at the beginning of this century, with the growing emphasis on many-server systems. How to generalize it beyond M/M/s was a hard open problem, until it was solved in the pioneering paper of Reed (2009), who started from the results of Krichagina and Puhalskii (1997) described in Chapter 18 and derived the convolution type equation that governs both the fluid and the diffusion behavior of the many-server G/G/n system. A survey of applications of the Halfin–Whitt regime is Van Leeuwaarden et al. (2019)

Exercises

19.1 Assume $(1 - \rho_n)\sqrt{n} \to \beta$, and let X_n be a Poisson random variable with parameter $n\rho_n$. Use CLT to show that $\mathbb{P}(X_n \leq n - 1) \to \Phi(\beta)$.

19.2 Assume $(1 - \rho_n)\sqrt{n} \to \beta$, and let X_n be a Poisson random variable with parameter ρ_n. Use Stirling's approximation to $n!$ to show that $\mathbb{P}(X_n = n)/(1 - \rho_n) \to \phi(\beta)/\beta$.

19.3 In an M/M/n system show that if $(1 - \rho_n)\sqrt{n} \to 0$, then the probability of delay $\alpha_n \to 1$, and if $(1 - \rho_n)\sqrt{n} \to \infty$, then $\alpha_n \to 0$.

19.4 Show that α is monotone decreasing with β, for $0 < \beta < \infty$.

19.5 Verify equations (19.7), (19.8).

19.6 Verify the infinitesimal mean and variance (drift and diffusion coefficients) for the scaled M/M/n queue.

19.7 Consider the embedded Markov chain at arrival times of the G/M/n system. Derive the limit infinitesimal drift and diffusion under Halfin–Whitt regime for the sequence of centered and scaled queue length at the embedded times.

19.8 (*) For the Halfin–Whitt diffusion limit of the centered and scaled M/M/n process, in state $\hat{Q}(t) = x$, calculate the distribution of the times to return to 0 (no queue and all busy) from position $x > 0$ and from position $x < 0$ (hitting times).

19.9 Solve the renewal-type equation for Example 19.9.

20

Many Servers with Abandonment

It is important to determine whether a queueing system is stable or not; however, when the analysis shows that the system is not stable, in reality queues never grow without bound. One way to prevent that is to have a finite capacity, i.e. a finite waiting room, the simplest example of which is the M/M/K/K Erlang loss system. In this chapter we will consider another possibility that corresponds to reality, namely that when the queue becomes too long, and customer waiting times become excessive, customers will abandon the queue without service. A model for this is that each customer that joins the queue has a finite patience, and he will abandon when his patience runs out. We assume now that customers have random i.i.d. patience times, with some patience distribution H. We will use the notation M/M/n+M for the Markovian M/M/n queue with exponential patience distribution, G/G/n+G for the general G/G/n queue with general patience distribution.

In Section 20.1 we will present a heuristic derivation of the fluid behavior of the G/G/n+G queue with many servers, starting from empty. In Section 20.2 we will derive results on M/M/n+M under Halfin–Whitt regime, analogous to the M/M/n results of Section 19.2. In Section 20.3 we will analyze in detail the M/M/n+G system.

20.1 Fluid Approximation of G/G/n+G

We now consider systems of many servers with abandonments. Such systems will always be stable, by virtue of the abandonments. We consider a sequence of systems, indexed by n, where we let the number of servers n as well as the arrival rate λ_n go to infinity. We assume that arrivals in system n are $\mathcal{A}(\lambda_n t)$, where $\mathcal{A}(t)$ is a stationary point process, which satisfies $A(t)/t \to_w 1$ as $t \to \infty$. We assume i.i.d. processing times distributed as G, with mean 1, and i.i.d. patience times distributed as H, with densities g and h. Time is not scaled, but we scale the number of customers by n. We

346

wish to study the limiting behavior of this system when n, $\lambda_n \to \infty$ and we take thoughout $\lambda_n = \rho n$, so traffic intensity is ρ.

In this system, as n becomes large, we will always have a large number of customers in service, and if $\rho = \lambda_n/n > 1$, we will also have a large number of customers waiting, and many customers will abandon. The fluid behavior will then tell us approximately how many customers are waiting, how many are in service, and what is the age distribution of the waiting customers and the customers in service. This complete fluid description, for the stationary system, is illustrated in Figure 20.1.

We define the following two processes that describe the behavior of the system:

$$B^n(t, y) \quad \text{is the number of customers in service at time } t,$$
$$\text{that have been in service for time} \leq y,$$
$$Q^n(t, y) \quad \text{is the number of customers waiting at time } t,$$
$$\text{that have been waiting for time} \leq y.$$

Customers always join the queue, but abandon when their wait exceeds their patience.

We consider the fluid scaled processes:

$$\bar{B}^n(t, y) = \frac{B^n(t, y)}{n}, \qquad \bar{Q}^n(t, y) = \frac{Q^n(t, y)}{n}.$$

It is possible to show that these processes converge, when $n \to \infty$, to a deterministic fluid limit from some general initial conditions. We shall now assume that the system starts empty, and describe the fluid limits (without proof of convergence) by formulating fluid model equations. Denote these fluid limits by $\bar{B}(t, y), \bar{Q}(t, y)$.

In the limit $\bar{B}(t, y), \bar{Q}(t, y)$ record quantities of fluid, where each particle of fluid enters the system and gets into service immediately if there is no fluid in the queue, or it joins the queue. Once it joins the queue it will stay in the queue until it goes into service, or it may leave the system without service at age y at rate $r_a(y) = h(y)/(1 - H(y))$. If it goes into service, it will leave the system after being in service for a time y at rate $r_s(y) = g(y)/(1 - G(y))$. The hazard rate $r(y)$ is the rate of an event at age y, conditional on reaching age y. The total fluid in service is $\bar{B}(t) = \bar{B}(t, \infty)$, and the total fluid in queue is $\bar{Q}(t) = \bar{Q}(t, \infty)$. Assume $\bar{B}(t, y), \bar{Q}(t, y)$ for every t are absolutely continuous with respect to y, with densities $b(t, y), q(t, y)$.

The total rate of departures due to service completions at time t (referred to here as the service rate) is then

$$\sigma(t) = \int_0^\infty b(t, y) r_s(y) dy,$$

and the total rate of departures due to abandonments at time t (referred to here as the abandonment rate) is

$$\gamma(t) = \int_0^\infty q(t, y) r_a(y) dy.$$

The evolutions of b, q with change in t are

$$b(t + u, x + u) = b(t, x) \frac{1 - G(x + u)}{1 - G(x)},$$

since a fraction $\frac{1-G(x+u)}{1-G(x)}$ of those of age x remain, and are now aged $x + u$, and similarly

$$q(t + u, x + u) = q(t, x) \frac{1 - H(x + u)}{1 - H(x)}.$$

When the queue is not empty, fluid moves into service at rate $\sigma(t)$, which equals the departure rate and will keep all the servers busy, and the fluid that moves into service is the oldest fluid in the queue. We denote the age of the oldest fluid in the queue at time t by $w(t)$. Then

$$q(t, y) = 0, \text{ for all } y > w(t).$$

In the transient system $w(t)$ itself evolves in time. The input ρ enters the queue if all servers are busy and the queue is non-empty:

$$q(t, 0) = \rho, \text{ when } w(t) > 0 \text{ i.e. } \bar{Q}(t) > 0, \text{ and } \bar{B}(t) = 1 \text{ (by the scaling)}.$$

If the queue is empty, not all the servers are busy and then

$$b(t, 0) = \rho \text{ when } \bar{B}(t) < 1 \text{ and } w(t) = 0 \text{ when } \bar{Q}(t) = 0.$$

Finally, at the boundary between these states, if the queue is empty but all the servers are busy:

$$b(t, 0) = \sigma(t) \wedge \rho, \quad q(t, 0) = (\rho - \sigma(t))^+, \text{ when } \bar{B}(t) = 1 \text{ and } \bar{Q}(t) = 0.$$

It can be shown that these equations have a unique solution when $B(0) = Q(0) = 0$. In particular, one can derive the steady state behavior of the fluid model (see Figure 20.1).

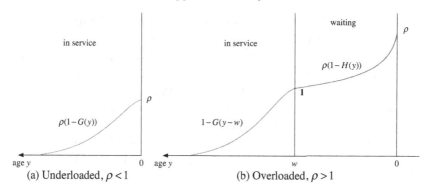

Figure 20.1 Distribution of ages in the stationary fluid limit.

Theorem 20.1 *(i) If $\rho \leq 1$, then*

$$\sigma = \bar{B} = \rho, \quad \gamma = \bar{Q} = 0, \quad b(y) = \rho(1 - G(y)).$$

(ii) If $\rho > 1$, then

$$\sigma = \bar{B} = 1, \quad \gamma = \rho - 1, \quad \bar{Q} > 0, \quad b(y) = 1 - G(y),$$
$$q(y) = \rho(1 - H(y)), \ 0 < y < w,$$

where w is determined by:

$$H(w) = \frac{\rho - 1}{\rho}, \quad and \quad \bar{Q} = \rho \int_0^w (1 - H(y)dy.$$

Proof In steady state, input rate to the servers equals output rate, so $b(0) = \sigma$. By the dynamics $b(y) = b(0)(1 - G(y)) = \sigma(1 - G(y))$. It then follows also that $\bar{B} = \sigma$.

Also, when $\rho > 1$, $q(y) = q(0)(1 - H(y)) = \rho(1 - H(y))$. So $\gamma = \int_0^w q(y)\frac{h(y)}{1-H(y)}dy = q(0) \int_0^w h(y)dy$, so $\gamma = \rho H(w)$.

Total input equals total output, so $\rho = \gamma + \sigma$. Because we must have if $B < 1$ then $Q = 0$ and if $Q > 0$ then $B = 1$, hence,
- For $\rho < 1$, we have $\sigma = \rho, \gamma = 0, w = 0$.
- For $\rho > 1$, we have $\sigma = 1, \gamma = \rho - 1$, and so $H(w) = \frac{\rho-1}{\rho}$.
Finally, in the case that $\rho = 1$, assume that $\gamma > 0$, then $\sigma < 1$, so $Q = 0$ which implies $\gamma = 0$, contradiction. Hence,
- For $\rho = 1$, we have $\gamma = 0, \sigma = 1$.

□

Figure 20.1 shows the distribution of ages of customers in the system, for the ρ customers that enter per unit time. It shows the density of age measured from the time they entered the system. Figure 20.1(a) is for the case of $\rho \leq 1$, Figure 20.1(b) is for the case of $\rho > 1$.

We summarize the stationary behavior of a many-server system with abandonments, as derived in Theorem 20.1. When $\rho < 1$ almost all the customers enter service immediately on arrival, the number of customers in the waiting queue is $o(n)$, and hardly any customers abandon. Since service is immediate, such a system is said to operate in a *quality driven* (QD) *service regime*. When $\rho > 1$, all the servers will be busy almost all the time and almost all the customers will need to wait. There is a value w such that almost all the customers that have patience $< w$ will abandon, and customers with patience $> w$ will all be served after a wait of approximately w. w is determined by $H(w) = \rho - 1$. Since servers are fully utilized, such a system is said to operate in an *efficiency driven* (ED) *service regime*. Finally, when $\rho = 1$, there will be almost no abandonments, the servers will be busy most of the time and waiting time will be minimal. In fact, a fraction α of the time all the servers will be busy and customers will wait (not for long), and a fraction $1 - \alpha$ of the time there will be some idle servers (not many) and customers will enter service immediately. This is the situation of the Halfin–Whitt regime, and such a system is said to operate in a *quality and efficiency driven* (QED) *service regime*. The fluid approximations is not sensitive enough to determine the value of α.

20.2 The M/M/n+M System under Halfin–Whitt Regime

In the previous section we derived the fluid behavior of many-server systems with abandonment for the G/G/n+G system. Those results were informative enough for the ED regime, where at stationarity there is a waiting time of approximately w, where $H(w) = \rho - 1$, and for the QD regime where at stationarity a fraction $1 - \rho$ of the servers are idle. However, at ρ close to 1, the fluid behavior is uninformative. We now consider the abandonment in the Halfin–Whitt regime, and in particular the M/M/n+M system, with many servers.

The M/M/n+M system has Poisson arrivals of rate λ, exponential service times of rate μ, and exponential patience distribution, with rate γ. We keep μ fixed, and assume a large volume of traffic, so that $\lambda \to \infty$, and assume square root staffing, according to Halfin–Whitt regime,

$$n\mu = \lambda + \beta \sqrt{\lambda}, \qquad -\infty < \beta < \infty.$$

Note that unlike the M/M/n system in Halfin–Whitt regime, which is only stable for $\beta > 0$, the M/M/n+M system is stable for all values of β, because of abandonment.

We can immediately write expressions for the stationary probabilities for the queue lengths (see Exercises 1.9 and 20.2):

$$
\pi_k = \begin{cases} \dfrac{1}{k!}\left(\dfrac{\lambda}{\mu}\right)^k \pi_0, & k = 0, 1, \ldots, n, \\[2ex] \dfrac{1}{n!}\left(\dfrac{\lambda}{\mu}\right)^n \left(\displaystyle\prod_{j=1}^{k-n} \dfrac{\lambda}{n\mu + j\gamma}\right)\pi_0, & k = n+1, n+2, \ldots, \end{cases}
$$

$$
\pi_0 = \left(1 + \sum_{k=1}^{n} \dfrac{1}{k!}\left(\dfrac{\lambda}{\mu}\right)^k + \dfrac{1}{n!}\left(\dfrac{\lambda}{\mu}\right)^n \sum_{k=n+1}^{\infty}\left(\prod_{j=1}^{k-n}\dfrac{\lambda}{n\mu + j\gamma}\right)\right)^{-1}.
$$

These expressions unfortunately do not yield simple formulas for performance measures. Instead, we obtain diffusion approximations in the many-server Halfin–Whitt regime.

Since the queue length is a birth and death process, we can use Theorem 7.2 and obtain diffusion approximations similar to the derivation of Theorem 19.4 for the M/M/n system without abandonments. We calculate the infinitesimal mean and variance of $Q^n(t)$:

$$
\begin{aligned}
m_{Q^n}(k) &= \lim_{h\to 0}\frac{1}{h}\mathbb{E}(Q^n(t+h) - Q_n(t)\,|\,Q_n(t) = k) \\
&= \lambda - \min(k, n)\mu - \gamma(k-n)^+ \\
&= \lambda - \mu n + \mu(n-k)^+ - \gamma(k-n)^+, \\
\sigma^2_{Q^n}(k) &= \lim_{h\to 0}\frac{1}{h}\mathrm{Var}(Q^n(t+h) - Q_n(t)\,|\,Q_n(t) = k) \\
&= \lambda + \min(k, n)\mu + \gamma(k-n)^+ \\
&= \lambda + \mu n - \mu(n-k)^+ + \gamma(k-n)^+.
\end{aligned}
$$

We now center and scale the queue lengths, and consider $\hat{Q}^n(t) = \frac{Q^n(t)-n}{\sqrt{\lambda}}$, for which the infinitesimal means and variances are

$$
\begin{aligned}
m_{\hat{Q}^n}(x) &= \frac{1}{\sqrt{\lambda}}m_{Q^n}\left(\sqrt{\lambda}x + n\right) \\
&= \frac{1}{\sqrt{\lambda}}\left(\lambda - \mu n + \mu(-\sqrt{\lambda}x)^+ - \gamma(\sqrt{\lambda}x)^+\right) \\
&\to \begin{cases} -\beta - \gamma x, & x > 0, \\ -\beta - \mu x, & x \le 0. \end{cases} \quad \text{as } \lambda \to \infty,
\end{aligned}
$$

$$\sigma_{\hat{Q}^n}^2(x) = \frac{1}{\lambda}\sigma_{Q^n}^2\left(\sqrt{\lambda}x + n\right)$$

$$= 1 + \frac{\mu n}{\lambda} - \mu\frac{(-x)^+}{\sqrt{\lambda}} + \gamma\frac{x^+}{\sqrt{\lambda}} \to 2, \quad \text{as } \lambda \to \infty.$$

By Stone's theorem 7.2 we have

Theorem 20.2 *If $\hat{Q}^n(0) \to_w \hat{Q}(0)$, then $\hat{Q}^n(t) \to_w \hat{Q}(t)$ in $\mathbb{D}(0, \infty)$, where $\hat{Q}(t)$ is the diffusion process defined by*

$$d\hat{Q}(t) = m(\hat{Q}(t))dt + \sigma(\hat{Q}(t))dBM(t),$$

where the drift and diffusion coefficient are

$$m(x) = \begin{cases} -\beta - \gamma x, & x > 0, \\ -\beta - \mu x, & x \le 0, \end{cases} \qquad \sigma^2(x) = 2.$$

This process is composed of two Ornstein–Uhlenbeck processes, on both sides of the line $x = 0$. For $x > 0$ it behaves as an Ornstein–Uhlenbeck process attracted to $-\beta/\gamma$, at rate γ, and for $x \le 0$ it behaves as an Ornstein–Uhlenbeck process attracted to $-\beta/\mu$, at rate μ.

Of course, $(Q(t) - n)^+$ is the number of waiting customers at time t, and $(n - Q(t))^+$ is the number of idle servers. One can then obtain stationary distributions for these, and obtain α, the probability of waiting, when $\lambda \to \infty$ (see Exercise 20.5):

$$\mathbb{P}\left(\frac{(Q(t) - n)^+}{\sqrt{\lambda}} \in (x, x + dx)\,\Big|\,\hat{Q}(t) > 0\right) = \sqrt{\gamma}\frac{\phi\left(\sqrt{\gamma}\left(x + \frac{\beta}{\gamma}\right)\right)}{\Phi\left(-\frac{\beta}{\sqrt{\gamma}}\right)}dx, \; x > 0,$$

$$\mathbb{P}\left(\frac{(n - Q(t))^+}{\sqrt{\lambda}} \in (x, x + dx)\,\Big|\,\hat{Q}(t) < 0\right) = \sqrt{\mu}\frac{\phi\left(\sqrt{\mu}\left(x - \frac{\beta}{\mu}\right)\right)}{\Phi\left(\frac{\beta}{\sqrt{\mu}}\right)}dx, \; x > 0,$$

$$\alpha = \mathbb{P}(\hat{Q}(t) > 0) = \left(1 + \sqrt{\frac{\gamma}{\mu}}\frac{\phi\left(\frac{\beta}{\sqrt{\gamma}}\right)}{\Phi\left(-\frac{\beta}{\sqrt{\gamma}}\right)}\frac{\Phi\left(\frac{\beta}{\sqrt{\mu}}\right)}{\phi\left(\frac{\beta}{\sqrt{\mu}}\right)}\right)^{-1}. \tag{20.1}$$

For the general case of G/G/n+G, a derivation similar to Section 19.3 yields a convolution-type equation for the queue length and for the number of abandoning customers, and can be used to obtain diffusion approximations.

20.3 The M/M/n+G System

We now study in more detail the M/M/n+G system, with n servers, Poisson arrivals at rate λ, exponential processing time with rate μ, and patience distribution H. Define the process: $(N(t), V(t))$, where $N(t)$ is the number of busy servers at time t taking values $1, \ldots, n$, and $V(t)$ is the offered waiting time at time t, which is defined as the time it would take a customer arriving at t with infinite patience until he will start service, where $V(t) = 0$ if $N(t) < n$. Then an arriving customer at time t will receive service if his patience is $\tau > V(t)$ and abandon (either at arrival if he can see $V(t)$ or at $t + \tau$ when his patience runs out) if his patience is $\tau \le V(t)$. $(N(t), V(t))$ is a Markov process (see Exercise 20.6). The Kolmogorov equations for this process, for $x > 0$, are (see Exercise 20.7)

$$\mathbb{P}(N(t + \delta) = n, V(t + \delta) > x) = \mathbb{P}(N(t) = n, V(t) > x + \delta)(1 - \lambda\delta)$$
$$+ \lambda\delta \Big[\mathbb{P}(N(t) = n, V(t) > x) + \mathbb{P}(N(t) = n - 1) \, e^{-n\mu x} \qquad (20.2)$$
$$+ \int_0^x \mathbb{P}(N(t) = n, V(t) = u) \, (1 - H(u)) \, e^{-n\mu(x-u)} du \Big] + o(\delta),$$

where we use the fact that with n servers and exponential processing times, each additional customer increases the virtual waiting time by an $\mathrm{Exp}(n\mu)$ distributed amount. Let

$$p_j = \lim_{t \to \infty} \mathbb{P}(N(t) = j), \quad j = 1, \ldots, n, \qquad (20.3)$$
$$v(x) = \lim_{t \to \infty} \lim_{dx \to 0} \mathbb{P}(N(t) = n, V(t) \in (x, x + dx))/dx, \quad x \ge 0.$$

Then for the stationary process, the balance equations for the states $N(t) = j, j < m$ are

$$\lambda p_0 = \mu p_1,$$
$$(\lambda + j\mu)p_j = \lambda p_{j-1} + (j + 1)np_{j+1}, \qquad j = 1, \ldots, n - 2, \quad (20.4)$$
$$(\lambda + (n - 1)\mu)p_{n-1} = \lambda p_{n-2} + v(0),$$

where in the last equation, when all n servers are busy, we have transition to $n - 1$ when the virtual waiting time reaches 0, as one of the customers leaves, i.e. from state $N(t) = n, V(t) = 0$ there is an immediate transition to $N(t) = n - 1$. Using equations (20.4), and the stationary version of (20.2), we obtain (see Exercise 20.8)

$$p_j = \left(\frac{\lambda}{\mu}\right)^j \frac{1}{j!} p_0, \quad j = 1, \ldots, n - 1,$$
$$v(0) = \lambda p_{n-1}, \qquad (20.5)$$
$$v(x) = \lambda p_{n-1} e^{-n\mu x} + \lambda \int_0^x v(u)(1 - H(u)) e^{-n\mu(x-u)} du, \quad x > 0.$$

To solve for $v(x)$, $x > 0$, note that

$$v(x)e^{n\mu x} = \lambda p_{n-1} + \lambda \int_0^x \{v(u)e^{n\mu u}\}(1 - H(u))du, \tag{20.6}$$

to obtain (see Exercise 20.9)

$$v(x) = \lambda p_{n-1} e^{\lambda \int_0^x (1-H(u))du - n\mu x}. \tag{20.7}$$

To further study the stationary behavior, define:

$$\tilde{H}(x) = \int_0^x (1 - H(u))du,$$

$$J = \int_0^\infty e^{\lambda \tilde{H}(x) - n\mu x} dx,$$

$$J_1 = \int_0^\infty x e^{\lambda \tilde{H}(x) - n\mu x} dx, \tag{20.8}$$

$$\mathcal{E} = \sum_{j=0}^{n-1} \frac{1}{j!}(\frac{\lambda}{\mu})^j \Big/ \frac{1}{(n-1)!}(\frac{\lambda}{\mu})^{n-1}.$$

These quantities have the following meanings:
- $\tilde{H}(\infty)$ is the expected patience time, and $\tilde{H}(x)/\tilde{H}(\infty) = H_{eq}(x)$ is the equilibrium distribution of the patience time,
- $\lambda p_{n-1} J = \int_0^\infty v(x)dx = \mathbb{P}(N(t) = n) = p_n$, is the probability that all servers are busy,
- $\frac{\lambda p_{n-1}}{p_n} J_1$ is the conditional expectation of $V(t)$, given that $V(t) > 0$,
- $\mathcal{E} = \frac{\mathbb{P}(N(t) \leq n-1)}{\mathbb{P}(N(t) = n-1)}$,

so that

$$p_0 = \Big(1 + \frac{\lambda}{\mu} + \cdots + (\frac{\lambda}{\mu})^{n-1} \frac{1}{(n-1)!}[1 + \lambda J]\Big)^{-1}.$$

Then the major performance measures of the system are given in the next theorem:

Theorem 20.3   *For the stationary M/M/n+G model,*
  (i) The probability of needing to wait is

$$\mathbb{P}(V(t) > 0) = p_n = \frac{\lambda J}{\mathcal{E} + \lambda J}.$$

  (ii) The probability of abandonment is

$$\mathbb{P}(ab) = \frac{1 + (\lambda - n\mu)J}{\mathcal{E} + \lambda J}.$$

(iii) The average wait of those needing to wait and receiving service is

$$\mathbb{E}(V(t) \mid V(t) > 0) = \frac{J_1}{J}.$$

We leave the verification to Exercise 20.10.

20.3.1 The M/M/n+G System under Many-Server Scaling

We now discuss how the M/M/n+G behaves for fixed μ when the number of servers (as well as the arrival rate) is large. Fix μ and fix $\rho = \frac{\lambda}{n\mu}$.

QD regime — Quality driven:

Under QD, $\rho = \lambda/n\mu < 1$, and we take $n \to \infty$. In this case, there are almost always free servers, almost no customers wait, and the number of abandonments is negligible. The probability of waiting approaches the probability of waiting without abandonment in an M/M/n system, as given in Section 19.2, equations (19.2), (19.3), so that (see Exercise 20.11)

$$\mathbb{P}(V > 0) \sim \frac{1}{\sqrt{2\pi n}} \frac{1}{1 - \rho} \left(\rho e^{1-\rho} \right)^n.$$

This probability declines to 0 exponentially fast with n. In summary, there is not much new here, the system behaves as if there were no abandonments, and with fixed ρ and $n \to \infty$ it behaves like an M/M/∞ queue.

ED regime — Efficiency driven:

Under ED, $\rho = \lambda/n\mu > 1$, and we take $n \to \infty$. In this case, all the servers are busy almost all the time, and almost all the customers have to wait, and the system is stabilized by abandonments. When all the servers are busy, there is no loss of processing by idle time, and so customers are processed at rate $n\mu$ and customers abandon at rate $\lambda - n\mu$. Assume H has density h. Let V^* be the unique value such that $H(V^*) = \frac{\rho-1}{\rho}$. Then customers with patience $< V^*$ will abandon, and customers with patience $\geq V^*$ will be served after a wait of V^*. We then have (as already indicated in Section 20.1):

$$\mathbb{P}(ab) \sim \frac{\rho - 1}{\rho}, \quad \mathbb{E}(V(t)) \sim V^*, \quad V(t) \to_p V^*, \quad \text{as } n \to \infty.$$

So there is a large $O(n)$ number of customers waiting, and the waiting time of patient customers is close to V^*. The deviations from V^* are of order $\sqrt{n}$ and are (see Exercise 20.11) asymptotically Gaussian:

$$\sqrt{n}(V(t) - V^*) \sim N(0, \mu \rho h(V^*)).$$

Also asymptotically for $n \to \infty$, by Little's law, the expected number of successful customers in the queue is $n\mu V^*$ with a total of $(\lambda - n\mu)V^*$ customers in queue that will at some point abandon, and n customers are in service at all times. The deviations in numbers of customers in queue from the mean is also asymptotically Gaussian. Equivalently, if customers can observe the workload, they can abandon on arrival (i.e. they balk). In practice, they can decide on balking by counting the number of customers ahead of them. In this case of many servers with abandonment, under ED regime, the fluid behavior is the dominant part in the system behavior, and can also be evaluated when arrival rates are subject to fluctuations.

QED regime — Quality and efficiency driven :

Under QED, the staffing is determined to be:

$$
n = \frac{\lambda}{\mu} + \beta \sqrt{\frac{\lambda}{\mu}}, \quad -\infty < \beta < \infty.
$$

This is the Halfin–Whitt regime with abandonments. As we saw in Section 19.2, without abandonments the centered and scaled number of customers in the system converges to a stationary diffusion process when $\beta > 0$. With abandonments, the system is stable also for $\beta \leq 0$. In general terms, we expect the system to offer customers short waits or no waiting at all, abandonments will be very rare when $\beta > 0$, and will consist of only very impatient customers even when $\beta \leq 0$; on the other hand, at all times either none or only a few of the servers will be idle. As is often the case for M/M/ models, exact calculation of the limiting performance measures of M/M/n+G is possible. We present some of the results without proof. Because waiting times are usually very short, only customers that have very little patience abandon, so abandonment is determined by probability of balking (zero patience) or by the density of the patience distribution at 0.

  Case (i)

Assume first that $h_0 = h(0) > 0$. Then:

  The probability of waiting is

$$
\alpha \sim \mathbb{P}(V(t) > 0) = \left(1 + \sqrt{\frac{h_0}{\mu}} \frac{\vartheta(\hat{\beta})}{\vartheta(-\beta)} \right)^{-1},
$$

where $\vartheta(x) = \frac{\phi(x)}{1-\Phi(x)}$ and $\hat{\beta} = \sqrt{\frac{\mu}{h_0}}\beta$.

Expected virtual waiting time when positive (waiting times of customers that get served) is

$$\mathbb{E}(V(t)\,|\,V(t) > 0) \sim \frac{1}{\sqrt{n}}\frac{1}{\sqrt{h_0\mu}}(\vartheta(\hat{\beta}) - \hat{\beta}).$$

The ratio of probability of abandoning and average virtual waiting time is

$$\frac{\mathbb{P}(ab)}{\mathbb{E}(V(t))} = \frac{\mathbb{P}(ab\,|\,V(t) > 0)}{\mathbb{E}(V(t)\,|\,V(t) > 0)} \sim h_0.$$

Case (ii)
Assume that customers that find all servers busy will *balk* (i.e. leave immediately, without service), with probability $\mathbb{P}(\text{balking})$.
The probability of waiting is

$$\mathbb{P}(V(t) > 0) \sim \frac{1}{\sqrt{n}}\frac{\vartheta(-\beta)}{\mathbb{P}(\text{balking})}.$$

Case (iii)
Assume that the density of the patience distribution vanishes at 0, and in fact all derivatives of order $0 \le i < k$ vanish, i.e. $h^{(i)}(0) = 0$, and $h_{0k} = h^{(k)}(0) > 0$.
For $\beta > 0$ the probability of waiting is

$$\mathbb{P}(V(t) > 0) \sim \left(1 + \frac{\beta}{\vartheta(-\beta)}\right)^{-1} = \left(1 + \frac{\beta\Phi(\beta)}{\phi(\beta)}\right)^{-1}$$

Otherwise, the probability of getting immediate service is

$$\mathbb{P}(V(t) = 0) \sim \begin{cases} \dfrac{1}{n^{k/(2k+4)}} \cdot \text{constant}, & \beta = 0, \\[2mm] \text{decays exponentially}, & \beta < 0. \end{cases}$$

The average waiting time is

$$\mathbb{E}(V(t)\,|\,V(t) > 0) \sim \begin{cases} \dfrac{1}{\beta\mu\sqrt{n}}, & \beta > 0, \\[2mm] \dfrac{1}{n^{1/(k+2)}} \cdot \text{constant}, & \beta = 0, \\[2mm] \dfrac{1}{n^{1/(2k+2)}} \cdot \text{constant}, & \beta < 0. \end{cases}$$

The probability of abandonment is

$$
\mathbb{P}(Ab \,|\, V(t) > 0) \sim
\begin{cases}
\dfrac{1}{n^{(k+1)/2}} \dfrac{h_{0k}}{(\beta\mu)^{k+1}}, & \beta > 0, \\[3ex]
\dfrac{1}{n^{(k+1)/(k+2)}} \cdot \text{constant}, & \beta = 0, \\[3ex]
\dfrac{|\beta|}{\sqrt{n}}, & \beta < 0.
\end{cases}
$$

20.4 Sources

The heuristic derivation of the fluid model for G/G/n+G is from Whitt (2006), who also proves the convergence for a discrete time version. The asymptotic analysis of the M/M/n+M system follows a survey of Ward (2012). That survey also includes further results on more general systems. Piecewise linear diffusion processes of the type obtained in Theorem 20.2 are studied by Browne et al. (1995). Dai et al. (2010) derive diffusion approximations for the many-server G/PH/n+G system, with phase-type service distributions; see also Dai and He (2010, 2013). Mandelbaum and Momcilovic (2012) derive a convolution type equation for the diffusion approximation to the general G/G/n+G system under Halfin–Whitt regime. The detailed painstaking analysis of the M/M/n+G system was carried out in Zeltyn and Mandelbaum (2005).

Exercises

20.1 Use Figure 20.1 to obtain the limiting expected sojourn time for an arriving customer if he abandons, or if he waits and and is patient enough to receive service.

20.2 For the M/M/n+M system, represent the queue length process as a birth and death process and derive its stationary distribution.

20.3 For the M/M/n+M system, assume that n is fixed, and that the value of μ increases according to the arrival rate λ, so that $\mu^{(\lambda)} = (\lambda + \beta \sqrt{\lambda})/n$. Calculate the infinitesimal mean and variance for the birth and death queue length process and obtain its limits as $\lambda \to \infty$. Obtain the diffusion approximation for the scaled queue length [Ward (2012)].

20.4 For the M/M/n+M system, assume that both n and μ increase according to the arrival rate λ, so that $\mu^{(\lambda)} = \mu(\lambda^{1-\alpha} + \beta\lambda^{\frac{1}{2}-\alpha})$, and $n^{(\lambda)} = \lambda^{\alpha}/\mu$ for $0 \le \alpha \le 1$. Show that the system goes into heavy traffic with $\rho_{\lambda} \to 1$ as

$\lambda \to \infty$. Calculate the infinitesimal mean and variance for the birth and death queue length process and obtain its limits as $\lambda \to \infty$, and obtain the diffusion approximation for the scaled queue length [Ward (2012)].

20.5 For the M/M/n+M system in Halfin–Whitt regime, derive the asymptotic distribution of the number waiting in queue when positive, the number of idle servers when positive, and the probability of waiting, as given in equation (20.1) [Browne et al. (1995)].

20.6 For the M/M/n+G system, explain why $(N(t), V(t))$ is a Markov process [Baccelli and Hebuterne (1981)].

20.7 For the M/M/n+G system, explain the Kolmogorov transition equations (20.2) for the process $(N(t), V(t))$ [Baccelli and Hebuterne (1981)].

20.8 For the M/M/n+G system, show that for the stationary system, p_j and $v(x)$ satisfy the equations (20.5) [Baccelli and Hebuterne (1981)].

20.9 For the M/M/n+G system, verify the solution of the integral equation (20.5) for $v(x)$ [Baccelli and Hebuterne (1981)].

20.10 For the M/M/n+G system, verify Theorem 20.3 [Zeltyn and Mandelbaum (2005)].

20.11 (∗) For the M/M/n+G system, verify the asymptotics for QD, ED, and QED [Zeltyn and Mandelbaum (2005)].

21

Load Balancing in the Supermarket Model

So far in this part of the book we considered a single queue at a single workstation with many servers. In this chapter we consider the situation in which there are many servers but they are separated, and each has his own queue. When a customer arrives, he needs to join one of the servers, and wait for his service there. We investigate strategies for dispatching the customers to the queues so as to improve performance.

Throughout the chapter we consider a system with n identical servers, a single stream of Poisson arrivals with arrival rate λ_n, and exponential service of rate 1 at each server. Hence, the traffic intensity is $\rho_n = \lambda_n/n$. We will analyze this as $n \to \infty$.

Consider first the situation when there is no centralized dispatching, and each customer chooses a server randomly, with probability $1/n$. Then each server will serve an M/M/1 queue, with immediate service probability $1 - \rho$, and mean queueing time, conditional on waiting, of $\frac{1}{1-\rho}$.

In contrast, dispatching each arriving customer to the shortest queue, the JSQ policy of Section 15.1, is optimal (Winston (1977); Weber (1978)). It stochastically minimizes the waiting time, which for fixed $\rho < 1$ is $O(1/n)$. JSQ evolves naturally in supermarkets with a small number of cashiers. The disadvantage of JSQ for many servers, e.g. in cloud data processing, is that finding the shortest queue becomes impractical, since it requires n communication messages for each customer.

We will discuss strategies to mitigate the costs of waiting times, communication costs, and memory requirements – these are load-balancing techniques.

We remark that with no communication at all, a centralized dispatcher can improve the waiting time by a factor of ~ 2, if he uses round-robin dispatching (Hajek (1985), see Exercise 21.1).

We will first show in Section 21.1 that for fixed ρ the policy of *join shortest of d*, that samples d servers chosen at random and dispatches an arriving customer to the shortest queue of the d, achieves exponential

360

improvement over random dispatching. Next in Section 21.2 we will derive the behavior of JSQ for many servers under Halfin–Whitt heavy traffic regime. We will see that the performance of JSQ is comparable to that of the single queue with n servers, with one major difference: Under JSQ the probability of waiting is $O(1/\sqrt{n})$ and the conditional expected waiting time is ~ 1, in contrast to the single queue (as in Section 19.2), where the probability of a wait is a fixed $\alpha > 0$ and waiting time is $O(1/\sqrt{n})$. Finally, in Section 21.3 we discuss briefly two strategies that are almost as good as JSQ in heavy traffic: One is to use choose shortest of d, where $d = d(n)$ depends on n, and diverges slowly. The other is *join idle queue*, JIQ, where the dispatcher, instead of sampling queues, receives messages from idle servers and stores those as tokens, to be used to dispatch customers.

The limits for the supermarket model under varying policies, as the number of servers $n \to \infty$, are given as solutions to some differential or integral equations. We will present those equations and study the resulting limiting processes first, and will discuss the proofs that the finite stochastic systems converge to these limits. We will not present the complete proofs, which are beyond the scope of this book.

21.1 Join Shortest of d Policy

We now consider the following policy: For a fixed $d \geq 2$, at each arrival, d servers are sampled (with replacement) and the customer is sent to the one with the shortest queue, breaking ties randomly. We assume that $\lambda_n = \lambda n$, so the traffic intensity $\rho = \lambda$ is fixed. This policy was analyzed independently in a thesis by Mitzenmacher and in a paper by Vvedenskaya, Dobrushin, and Karpelevich. We will follow mainly the more intuitive approach of the former, leaving out some important steps that are proved using different techniques by the latter. The main result for this policy, for fixed ρ, is exponential improvement over random dispatching.

Theorem 21.1 *Let $^dQ^n$ be the queue length at any one of the servers, and let $^dW^n$ be the sojourn time of an arriving customer, in the stationary n server system under choose shortest of d policy. Then, as $n \to \infty$:*

$$\mathbb{P}(^dQ^n \geq k) \to \lambda^{\frac{d^k-1}{d-1}}, \qquad \mathbb{E}(^dW^n) \to \sum_{k=1}^{\infty} \lambda^{\frac{d^k-d}{d-1}}. \qquad (21.1)$$

Our strategy toward proving this theorem is the following:

- Represent the state of the n server system by a Markov process $(S_k^n(t))_{k=0}^{\infty}$ that counts the number of servers with k or more customers, and show that it leads to a sequence of differential equations for a deterministic system $(s_k(t))_{k=0}^{\infty}$.
- Introduce *density dependent Markov chains* and show that $\frac{1}{n}S_k^n(t)$ converges to $s_k(t)$. This proof is incomplete.
- Investigate $s_k(t)$ to show that it decreases at doubly exponential rate with k, and it converges exponentially fast to a fixed point $s_k^* = \lambda^{\frac{d^k-1}{d-1}}$.
- Review the results, where we will list the details missing in the proof, and discuss the methods and conclusions in the works of Mitzenmacher and of Vvedenskaya, Dobrushin, and Karpelevich.

21.1.1 A Markov Representation and a Dynamic System

The analysis of the supermarket model is helped considerably by adopting the following description of the state of the system at any time t: We let $S_k^n(t)$, $k = 0, 1, 2, \ldots$ be the number of servers with a queue of k or more customers (including customer in service) at time t. We then have $n = S_0^n \geq S_1^n \geq S_2^n \geq \cdots$. By the symmetry of the system, this state description is fully informative. In particular, the number of servers with k customers is $S_k^n - S_{k-1}^n$, and the fraction of servers with k customers is $(S_k^n - S_{k-1}^n)/n$ (see Figure 21.1). This state description is often referred to as a *mean field representation*. It will serve us for all the policies in this chapter.

We assume a finite number of customers at all times, i.e. there is a minimal i for which $S_i^n(t) = 0$, and $S_1^n(t) = \sum_{k=1}^{\infty}(S_k^n(t) - S_{k+1}^n(t)) < \infty$. Clearly, $S^n(t) = (S_k^n(t))_{k=0}^{\infty}$ is a continuous time countable state space Markov chain for every n. We claim first:

Proposition 21.2 *Consider the n server system with $\rho < 1$, under choose shortest of d policy. Then the Markov chain $S^n(t)$ over all finite states is ergodic.*

Proof We say a random variable x is greater than random variable y in the convex majorization sense, denoted by $x \geq_c y$, if $\mathbb{E}(x - s)^+ \geq \mathbb{E}(y - s)^+$ for all s.

Let $^dQ_j^n(t)$ denote the queue length of server j in the n server system under choose shortest of $d \geq 2$ policy, and let $^1Q_j^n(t)$ be the queue length under random dispatching. The proposition then follows from $^1Q_j^n(t) \geq_c {}^dQ_j^n(t)$. We leave the proof of this majorization to Exercise 21.2.   □

Queue
Lengths

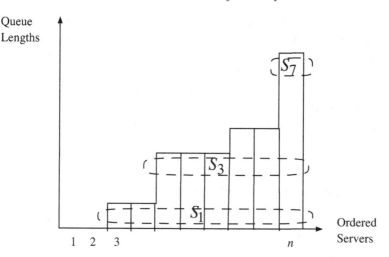

Figure 21.1 Description of the state by counting queues of each length.

We denote by $\bar{S}_k^n(t) = \frac{1}{n} S_k^n(t)$. Note that $1 \geq \bar{S}_1 \geq \bar{S}_2 \geq \cdots \geq 0$, is the empirical distribution of the occupancy of the n servers. The crucial connection between the process $\bar{S}^n(t) = (S_k^n(t))_{k=0}^{\infty}$ and the queue length process $({}^d Q_j^n(t))_{j=1}^{n}$ is the following:

Lemma 21.3 *Let j be an arbitrarily chosen server. Then*

$$\mathbb{P}({}^d Q_j^n(t) \geq k) = \mathbb{E}(\bar{S}_k^n(t)). \tag{21.2}$$

Proof Note that $\bar{S}^n(t)$ has the same values for $({}^d Q_1^n(t), \ldots, {}^d Q_n^n(t)) = (x_1, \ldots, x_n)$ as for $({}^d Q_{i_1}^n(t), \ldots, {}^d Q_{i_n}^n(t)) = (x_1, \ldots, x_n)$, for all the permutaions $i_1, \ldots, i_n$.

We therefore have by definition, given $\bar{S}^n(t)$, that for an arbitrarily chosen j,

$$\mathbb{P}\left({}^d Q_j^n(t) \geq k \mid \bar{S}^n(t)\right) = \bar{S}_k^n(t),$$

and taking expectation, we obtain

$$\mathbb{P}({}^d Q_j^n(t) \geq k) = \mathbb{E}\left(\mathbb{P}({}^d Q_j^n(t) \geq k \mid \bar{S}^n(t))\right) = \mathbb{E}(\bar{S}_k^n(t)).$$

$\square$

We use the memoryless property of the Poisson-exponential structure to write the short-term dynamics of each $S_k^n(t)$, $k = 1, 2, \ldots$. In a short interval of length h there is an arrival with probability $\lambda n h$. This will cause an increase in $S_k^n(t)$ if the arrival will join a queue of length exactly $k - 1$ and this will happen if all d servers chosen at random have at least $k - 1$ customers, but not all of them have more than $k - 1$ customers. So the probability of an increase in $S_k^n(t)$ is $\lambda n h[(S_{k-1}^n(t)/n)^d - (S_k^n(t)/n)^d]$. A decrease in $S_k^n(t)$ will happen if one of the servers with a queue of length exactly k completes a service, and since each of those servers works at rate 1, the probability of this is $h(S_k^n(t) - S_{k+1}^n(t))$, so the expected change in $S_k^n(t)$ is

$$\mathbb{E}(S_k^n(t + h) - S_k^n(t)|S(t)) \tag{21.3}$$
$$= \lambda n h[(S_{k-1}^n(t)/n)^d - (S_k^n(t)/n)^d] - h(S_k^n(t) - S_{k+1}^n(t)) + o(h).$$

We are interested in the many-server scaling of the process: We divide by n but leave time unscaled, so we have for $\bar{S}^n(t)$

$$\mathbb{E}(\bar{S}_k^n(t + h) - \bar{S}_k^n(t)|\bar{S}(t)) \tag{21.4}$$
$$= \lambda h(\bar{S}_{k-1}^n(t)^d - \bar{S}_k^n(t)^d) - h(\bar{S}_k^n(t) - \bar{S}_{k+1}^n(t)) + o(h).$$

It is important to note that n no longer appears in the sequence of equations. This suggests the limiting deterministic dynamic system that solves the sequence of differential equations:

$$\begin{cases} \dfrac{ds_k(t)}{dt} = \lambda(s_{k-1}(t)^d - s_k(t)^d) - (s_k(t) - s_{k+1}(t)), & k \geq 1, \\ \\ s_0(t) = 1. \end{cases} \tag{21.5}$$

For the n system the values of $\bar{S}_k^n(t)$ must be an integer multiple of $1/n$, but for the dynamic system, $s_k(t)$ can take any real value. We study this system in Section 21.1.3.

As we shall see, the supermarket model under shortest of d policy is an example of a density dependent jump Markov chain. We discuss such processes and their convergence next.

21.1.2 Density Dependent Markov Chains and Kurtz's Theorem

In this section we consider Markov chains that describe the dynamics of populations. We have a population with size given by n, which can be number of individuals, area, or volume. The population is classified into types indexed by k, with a set $\mathcal{K}$ of types, finite or countably infinite. We let $X_k^n(t)$ be the size of class k in population n, so that X_k^n describes the distribution of types in the population. We let $X^n(t) = (X_k^n(t))_{k \in \mathcal{K}}$ be the state of the population at time t, with $X^n(t) \in E$, where E is a subset of $\mathbb{Z}^{|\mathcal{K}|}$. Some examples with finite $\mathcal{K}$ are epidemic models where $k \in \{S, I, R\}$ stands for susceptible, infected, or removed (cured), and chemical reaction models, where k lists the different reagents. In the supermarket model n is the number of servers, servers are classified by the length of their queue $k = 1, 2, \ldots$, and $X_k^n(t)$ counts the number of servers with a queue of length k or more (here $k \in \mathcal{K} = \mathbb{N}$, and the state space of $X^n(t)$ is $E \subseteq \mathbb{Z}^\infty$).

One models $X^n(t)$ as a jump Markov process. There is a set of transitions, $\ell \in L \subset E$, and transitions are from state $x \in E$ to state $x + \ell \in E$, which marks the change in the population counts. A transition ℓ occurs according to a rate function $\beta_\ell(x/n)$, so that the transition rates of the jump Markov chain $X^n(t)$ from state $x \in E$ to state $y \in E$ is given by $q_{x,y}(t) = n\beta_{y-x}(x/n)$. The significant assumption here is that the transition rate functions depend on the density within the population, and are independent of the population size.

In the supermarket model, transitions are $\pm e_k$, where e_k is the infinite kth unit vector, and the transition $x \to x + e_k$ is an arrival of a customer to a queue of length $k - 1$ (increase of the number of queues of length k or more), while $x \to x - e_k$ is a departure of a customer from a queue of length k. The transition rates for the supermarket model are

$$q_{x,x+e_k}(t) = n\beta_{e_k}(x/n) = n\lambda[(S_{k-1}/n)^d - (S_k/n)^d],$$
$$q_{x,x-e_k}(t) = n\beta_{-e_k}(x/n) = n[(S_k/n) - (S_{k+1}/n)]. \tag{21.6}$$

We now use an elegant way to describe the dynamics of density dependent population processes through a set of stochastic integral equations. We let $D_\ell, \ell \in L$ be independent rate 1 Poisson processes, which govern the occurrences of ℓ transitions. The actual rates at which transitions of type ℓ occur are a function of the state $X^n(t)/n$, so the counting process of transitions of type ℓ is a doubly stochastic Poisson process (Cox process). We have:

$$X^n(t) = X^n(0) + \sum_{\ell \in L} \ell \, D_\ell \left(n \int_0^t \beta_\ell(X^n(s)/n)ds \right).$$

We will normalize these by dividing by the population size to get $\bar{X}^n(t) = X^n(t)/n$ which represents the frequencies of types in the population:

$$\bar{X}^n(t) = \bar{X}^n(0) + \sum_{\ell \in L} \frac{\ell}{n} D_\ell \left(n \int_0^t \beta_\ell(\bar{X}^n(s))ds \right). \tag{21.7}$$

We now centralize the Poisson processes $\tilde{D}_\ell(y) = D_\ell(y) - y$, and define

$F(x) = \sum_{\ell \in L} \ell \, \beta_\ell(x)$, where F is a function from E to E. We then have:

$$\bar{X}^n(t) = \bar{X}^n(0) + \sum_{\ell \in L} \frac{\ell}{n} \tilde{D}_\ell \left(n \int_0^t \beta_\ell(\bar{X}^n(s)) ds \right) + \int_0^t \sum_{\ell \in L} \ell \beta_\ell(\bar{X}^n(s)) ds$$

$$= \bar{X}^n(0) + \sum_{\ell \in L} \frac{\ell}{n} \tilde{D}_\ell \left(n \int_0^t \beta_\ell(\bar{X}^n(s)) ds \right) + \int_0^t F(\bar{X}^n(s)) ds. \qquad (21.8)$$

We note that by the FSLLN, $\frac{1}{n} \tilde{D}(nt) \to 0$ as $n \to \infty$ u.o.c. a.s. The next theorem states that under some conditions, $\bar{X}^n(t)$ converges to the solution of the integral equation (of the dimension of E):

$$\bar{X}(t) = x_0 + \int_0^t F(\bar{X}(s)) ds. \qquad (21.9)$$

Theorem 21.4 (Kurtz 1978) *Assume:*
 (i) $nX(0) \to x_0$,
 (ii) $|F(x) - F(y)| \le M|x - y|$ *for some M*
 (iii) $\sum_{\ell \in L} |\ell| \sup_{x \in E} \beta_\ell(x) < \infty$
Then:
$\lim_{n \to \infty} \sup_{0 \le s \le t} |\bar{X}^n(s) - \bar{X}(s)| = 0$ *a.s.*
Here $|\cdot|$ is the norm of the state space E.

Proof Let $\bar{\beta}_\ell = \sup_{x \in E} \beta_\ell(x)$. Then

$$\epsilon_n(t) := \left| \bar{X}^n(t) - \bar{X}^n(0) - \int_0^t F(\bar{X}^n(s)) ds \right|$$

$$\le \sum_{\ell \in L} \frac{|\ell|}{n} \left| \tilde{D}_\ell \left(n \int_0^t \beta_\ell(\bar{X}^n(s)) ds \right) \right|$$

$$\le \sum_{\ell \in L} \frac{|\ell|}{n} \sup_{u \le t} \left| \tilde{D}_\ell \left(n \bar{\beta}_\ell u \right) \right| \qquad (21.10)$$

$$\le \sum_{\ell \in L} \frac{|\ell|}{n} \left[D_\ell \left(n \bar{\beta}_\ell t \right) + n \bar{\beta}_\ell t \right].$$

The last two inequalities hold term by term, and are justified in Exercise 21.3. The terms inside the sum in the third row converge to 0 as $n \to \infty$, while the terms in the sum in the last row converge to $2|\ell|\bar{\beta}_\ell t$.

The process $\sum_{\ell \in L} \left(|\ell| D_\ell(\bar{\beta}_\ell t) + |\ell| \bar{\beta}_\ell t \right)$ is a process of independent increments, and so taking it at time nt and dividing by n, and letting $n \to \infty$, by the law of large numbers it converges to $\sum_{\ell \in L} 2|\ell|\bar{\beta}_\ell t$, which is finite by assumption. This justifies changing the order of limit and sum for the expression in the first inequality. Thus we have shown that $\lim_{n \to \infty} \epsilon_n(t) = 0$.

We now have, comparing the expressions for $\bar{X}^n(t)$ and $\bar{X}(t)$, and using the Lipschitz continuity of $F(\cdot)$, that

$$\left|\bar{X}^n(t) - \bar{X}(t)\right| = \left|\bar{X}^n(0) - x_0 + \epsilon_n(t) + \int_0^t [F(\bar{X}^n(s)) - F(\bar{X}(s))]\,ds\right|$$

$$\leq \left|\bar{X}^n(0) - x_0\right| + \epsilon_n(t) + \int_0^t M\left|\bar{X}^n(s) - \bar{X}(s)\right|\,ds.$$

We then obtain from Gronwalls inequality that

$$\left|\bar{X}^n(t) - \bar{X}(t)\right| \leq \left(\left|\bar{X}^n(0) - x_0\right| + \epsilon_n(t)\right) e^{Mt},$$

which proves the theorem. □

For the supermarket model under join the shortest of d policy, the integral equation (21.9) takes the form:

$$\begin{cases} s_k(t) = s_k(0) + \int_0^t \left[\lambda(s_{k-1}(s)^d - s_k(u)^d) - (s_k(u) - s_{k+1}(s))\right]\,ds, \\ \\ s_0(t) = 1. \end{cases}$$

(21.11)

Taking the derivative with respect to t, we obtain the sequence of equations (21.5) for the deterministic dynamic system.

Corollary 21.5 *For the supermarket model under choose shortest of d policy, assume $\bar{S}_k^n(0) \to s_k(0)$ a.s. as $n \to \infty$. Then:*

$$\lim_{n\to\infty} \sup_{0 \leq s \leq t} \sup_{k=1,2,\ldots} \left|\bar{S}_k^n(s) - s_k(s)\right| = 0, \quad a.s. \qquad (21.12)$$

that is, $\bar{S}^n(t)$ converges u.o.c. a.s. in the supremum norm of sequences to the deterministic sequence of functions $s_k(t)$.

Incomplete Proof This appears to follow directly from Kurtz's theorem 21.4 once we check the conditions on F. It is easy enough to show that F is Lipschitz continuous; see Exercise 21.4. However, the second condition fails: $\bar{\beta} = \sum_{\ell \in L} |\ell| \sup_{x \in E} \beta_\ell(x) < \infty$ fails, since for $\ell = -e_k$ we get $\sup_{x \in E} \beta_\ell(x) = 1$ for all $k > 0$.

One way around this problem is to modify the proof that $\epsilon_n(t) \to 0$ by bounding the quantity $\left|\tilde{D}_\ell\left(n \int_0^t \beta_\ell(\bar{S}^n(s))ds\right)\right|$ in (21.10) in terms of $\left|\tilde{D}_\ell\left(n\beta_\ell(\bar{S}^n(t))\right)\right|$ rather than by $\sup_{u \leq t} \left|\tilde{D}_\ell\left(n\bar{\beta}_\ell u\right)\right|$.

An alternative way is to truncate the states of the process to finite $k = 0, 1, \ldots, K$, where Kurtz theorem can be applied directly, and then let $K \to \infty$. □

21.1.3 The Deterministic Infinite Server Process

We now consider the deterministic dynamic system $s_k(t)$ of the ordered functions $1 = s_0(t) \geq s_1(t) \geq \cdots \geq 0$, that satisfy the system of equations (21.5). We also make the assumption that $\sum_{k=0}^{\infty} s_k(t) < \infty$ for all $t \geq 0$. We call this the *infinite server model*. We investigate the properties of such solutions in a series of propositions. To facilitate continuity of reading, we present the results first, and present the proofs at the end of this sub-section or in exercises.

First we find a fixed point of the system.

Proposition 21.6   *For fixed $\lambda > 0$ and integer $d \geq 2$, the set of differential equations (21.5) has a unique fixed point given by:*

$$s_k^* = \lambda^{\frac{d^k-1}{d-1}}, \quad k = 0, 1, 2, \ldots, \tag{21.13}$$

Note that the condition $\sum_{k=0}^{\infty} s_k(t) < \infty$ for all $t \geq 0$ is necessary. Without it, we get that $s(t) = (1, 1, \ldots)$ is also a fixed point.

The fixed point sequence s_k^* is decreasing in k at a doubly exponential rate, i.e. there exist constants $\alpha < 1$, $\beta > 1$, γ such that $s_k^* \leq \gamma \alpha^{\beta^k}$. The following proposition shows that the same is also the case for $s_k(t)$. We let $\pi_k = s_k^*$.

Proposition 21.7   *Assume there exists a minimal i for which $s_i(0) = 0$. Then the sequence $s_k(t)$, $k = 0, 1, 2, \ldots$ is decreasing at a doubly exponential rate for all t. In particular, if the system starts empty, i.e. $s_1(0) = 0$, then $s_k(t) \leq \pi_k$.*

Next we show that $s_k(t)$, $k = 0, 1, 2, \ldots$ converges to the fixed point exponentially fast as t increases.

Proposition 21.8   *Let $\Theta(t) = \sum_{k=1}^{\infty} w_k |s_k(t) - \pi_k|$. Then there exists a sequence of constants $w_k \geq 1$ such that if $\Theta(0) < \infty$, then $\Theta(t)$ converges to 0 exponentially fast, i.e. we can find $w_k \geq 1$ and $\delta > 0$ such that $\Theta(t) < \Theta(0)e^{-\delta t}$. It follows immediately that the L_1 distance of $s(t)$ and π, $g(t) = \sum_{k=1}^{\infty} |s_k(t) - \pi_k|$, converges to 0 exponentially fast, i.e. $g(t) \leq \Theta(0)e^{-\delta t}$.*

Denote by $\overline{W}_d(\lambda)$ the expected sojourn time of a customer in the infinite server system.

Proposition 21.9   *The expected sojourn time of a customer that joins the* *infinite server system converges as $t \to \infty$ to*

$$\overline{W}_d(\lambda) = \sum_{k=1}^{\infty} \lambda^{\frac{d^k - d}{d-1}}. \tag{21.14}$$

Furthermore, if system starts empty, then this is an upper bound for all t.

We now compare this expected value to that of the sojourn time under random dispatching, $d = 1$. Recall that for $d = 1$, i.e. random dispatching, the average sojourn time is $\overline{W}_1(\lambda) = \frac{1}{1-\lambda}$.

Proposition 21.10 *We compare the expected sojourn time for $d \geq 2$ to that for random dispatching, $d = 1$. As $\lambda \nearrow 1$, the expected time is smaller by an exponential factor, given by:*

$$\lim_{\lambda \to 1} \frac{\overline{W}_d(\lambda)}{\log \overline{W}_1(\lambda)} = \frac{1}{\log d}.$$

Proof of Proposition 21.6 Briefly, we first check that substituting s_k^* gives $\frac{d}{dt} s_k(t) = 0$, so s_k^* is a fixed point. Next, for uniqueness we need to show that $\frac{d}{dt} s_k(t) = 0$ implies $s_k(t) = s_k^*$. By summing all the equations and equating to 0 we can check that $s_1(t) = \lambda$. The proof is completed by induction on k. See Exercise 21.5 for the details. $\qquad\square$

Proof of Proposition 21.7 Briefly, the proof has the following steps:

(i) We note first a property of monotonicity of the solutions of (21.5), which is as follows: Increasing any initial value $s_j(0)$, also increases or leaves unchanged $s_k(t)$ for all k and t (see Exercise 21.6).

(ii) Define $M(t) = \sup_k (s_k(t)/\pi_k)^{1/d^k}$. We next show that $M(t) \leq M(0)$ for all $t \geq 0$. To do that we use the monotonicity. We change the initial state by raising all the values of $s_k(0)$, $k = 0, 1, \ldots$ to $\check{s}_k(0) = M(0)^{d^k} \pi_k$. It is then seen that this new initial sequence $\check{s}_k(0)$ is a fixed point of the system (with possibly $\check{s}_0(t) > 1$), and so the sequence of processes $\check{s}(t)$ has constant $\check{M}(t) = M(0)$. So, by the monotonicity, starting from the original $s(0)$, $M(t) \leq M(0)$.

(iii) Notice that for the specified minimal i for which $s_i(0) = 0$ we have (see Exercise 21.7):

$$M(0) \leq [1/\pi_{i-1}]^{1/d^{i-1}} < 1/\lambda^{1/(d-1)}.$$

We then have:

$$s_k(t) \leq \pi_k M(t)^{d^k} \leq \pi_k M(0)^{d^k} = \lambda^{-1/(d-1)} \left(\lambda^{1/(d-1)} M(0) \right)^{d^k},$$

which is decreasing with k at a doubly exponential rate: $s_k(t) \leq \gamma \alpha^{\beta^k}$ with $\alpha = \lambda^{1/(d-1)} M(0) < 1$, $\beta = d \geq 2$, $\gamma = \lambda^{-1/(d-1)}$.

$\square$

Proof of Proposition 21.8   We outline the proof for $d = 2$. The proof for $d > 2$ is similar. Let $\epsilon_k = s_k(t) - \pi_k$ (where we drop dependence on t for ease of writing). For $d = 2$, we write the derivative of ϵ_k,

$$\frac{d\epsilon_k}{dt} = \lambda \left((\epsilon_{k-1} + \pi_{k-1})^2 - (\epsilon_k + \pi_k)^2 \right) - \left((\epsilon_k + \pi_k) - (\epsilon_{k+1} + \pi_{k+1}) \right)$$

$$= \lambda \left(2\pi_{k-1}\epsilon_{k-1} + \epsilon_{k-1}^2 - 2\pi_k\epsilon_k - \epsilon_k^2 \right) - (\epsilon_k - \epsilon_{k+1}),$$

where we use the fact that π_k are a fixed point.

We then write the derivative of $\Theta(t)$:

$$\frac{d\Theta(t)}{dt} = \sum_{\epsilon_k > 0} w_k \left[\lambda (2\epsilon_{k-1}\pi_{k-1} + \epsilon_{k-1}^2 - 2\epsilon_k\pi_k - \epsilon_k^2) - (\epsilon_k - \epsilon_{k+1}) \right]$$

$$- \sum_{\epsilon_k < 0} w_k \left[\lambda (2\epsilon_{k-1}\pi_{k-1} + \epsilon_{k-1}^2 - 2\epsilon_k\pi_k - \epsilon_k^2) - (\epsilon_k - \epsilon_{k+1}) \right].$$

Here we treat the case of $\epsilon_j = 0$ by taking the right derivative, and including the term for ϵ_j in the summation over positive ϵ_k if the derivative is ≥ 0 and in the summation over negative ϵ_k if the derivative is < 0.

We now look at the terms involving ϵ_k in $\frac{d\Theta(t)}{dt}$. They include w_{k-1}, w_k, w_{k+1}, and they will depend on the signs of ϵ_{k-1}, ϵ_k, ϵ_{k+1}. For the case that all three are negative (by Proposition 21.7 this is the case if the system starts empty), the terms involving ϵ_k are

$$k \text{ term} = -w_{k-1}\epsilon_k + w_k[\lambda(2\epsilon_k\pi_k + \epsilon_k^2) + \epsilon_k] - w_{k+1}\lambda(2\epsilon_k\pi_k + \epsilon_k^2)$$

$$= (w_k - w_{k-1})\epsilon_k + (w_k - w_{k+1})\lambda(2\epsilon_k\pi_k + \epsilon_k^2).$$

We wish to find $w_k \geq 1$ and $\delta > 0$ so that each of these terms will be $\leq -\delta w_k |\epsilon_k|$. This will be the case if

$$(w_k - w_{k-1}) + (w_k - w_{k+1})(2\lambda\pi_k + \lambda\epsilon_k) \geq \delta w_k,$$

which, using $|\epsilon_k| \leq 1$, will hold if we choose

$$w_{k+1} \leq w_k + \frac{(1 - \delta)w_k - w_{k-1}}{\lambda(2\pi_k + 1)}.$$

One can check that this choice also works for all other sign combinations of ϵ_{k-1}, ϵ_k, ϵ_{k+1}.

One can indeed define an increasing sequence of w_k that will satisfy these conditions for small enough $\delta > 0$. See Exercise 21.8 for the details.

This sequence is chosen so that it is bounded by a geometric sequence, and because $s_k(t)$ and π_k decrease at a doubly exponential rate, with this choice $\Theta(t) < \infty$.

Finally, comparing $\Theta(t)$ and $\frac{d\Theta(t)}{dt}$, we see

$$\Theta(t) = \sum_{k=1}^{\infty} w_k |\epsilon_k|, \qquad \frac{d\Theta(t)}{dt} \leq -\sum_{k=1}^{\infty} \delta w_k |\epsilon_k|,$$

so we have $\frac{d\Theta(t)}{dt} \leq -\delta\Theta(t)$, and therefore $\Theta(t) < \Theta(0)e^{-\delta t}$.

For $g(t)$ we use simply the fact that $g(t) \leq \Theta(t)$. □

Proof of Proposition 21.9 An arrival at time t becomes the kth customer in a queue in the infinite server system with probability $s_{k-1}(t)^d - s_k(t)^d$. So his expected sojourn time (for i.i.d. mean 1 service time) is

$$\sum_{k=1}^{\infty} k\left(s_{k-1}(t)^d - s_k(t)^d\right) = \sum_{k=0}^{\infty} s_k(t)^d.$$

As $t \to \infty$, this converges to $\sum_{k=0}^{\infty} \pi_k^d = \sum_{k=1}^{\infty} \lambda^{\frac{d^k-d}{d-1}}$ by Proposition 21.8. Furthermore, if the system starts empty, this is an upper bound by Proposition 21.7. □

Proof of Proposition 21.10 The proof for this proposition is based on the result that for integer $d \geq 2$,

$$\lim_{\lambda \to 1} \frac{\sum_{k=0}^{\infty} \lambda^{d^k}}{\log \sum_{k=0}^{\infty} \lambda^k} = \lim_{\lambda \to 1} \frac{\sum_{k=0}^{\infty} \lambda^{d^k}}{\log \frac{1}{1-\lambda}} = \frac{1}{\log d}.$$

The derivation of this limit is quite involved; the interested reader can try to obtain it in Exercise 21.9. □

21.1.4 Proof of the Many-Server Limits

In this subsection we return to the proof of Theorem 21.1. We consider what can be said about the process $\bar{S}^n(t) = (\bar{S}_k^n(t))_{k=0}^{\infty}$ for $\lambda < 1$. On the one hand, it is an ergodic Markov chain, and so it has a stationary distribution that we will denote by $\bar{S}^n(\infty)$. On the other hand, because it is a density dependent Markov chain, by Kurtz's theorem (modified as outlined in the discussion of Corollary 21.5's proof), its sample paths converge as $n \to \infty$, to the deterministic sample path of the unique solution of the system of integral equations (21.11), or equivalently the system of differential equations (21.5) that we now denote $\bar{S}^\infty(t)$. Furthermore, the deterministic sample path $\bar{S}^\infty(t)$

converges exponentially fast to the stationary point of (21.5), which we now denote $\bar{S}^*$. To summarize this, we know

$$\bar{S}^n(t) \to_w \bar{S}^n(\infty) \text{ as } t \to \infty,$$
$$\bar{S}^n(\cdot) \to_w \bar{S}^\infty(\cdot) \text{ as } n \to \infty, \qquad (21.15)$$
$$\bar{S}^\infty(t) \to_w \bar{S}^* \text{ as } t \to \infty.$$

So we have that $\lim_{t\to\infty} \lim_{n\to\infty} \bar{S}^n(t) = \bar{S}^*$. However, what we wish to show is $\lim_{n\to\infty} \lim_{t\to\infty} \bar{S}^n(t) = \bar{S}^*$. This exchange of the order of limits needs justification. Mitzenmacher assumes that the system starts empty, and so the convergence $\bar{S}^\infty(t) \to_w \bar{S}^*$ is monotone, in which case one can change the order.

To prove Theorem 21.1 without additional conditions, we turn to the approach of Vvedenskaya, Dobrushin, and Karpelevich. For the process $\bar{S}^n(t)$, they consider its state space, which is included in the space $\mathcal{U}$ of sequences $1 \geq x_1 \geq x_2 \geq \cdots \geq 0$ with metric $\sup_{k \geq 1} \frac{|x_k - y_k|}{k}$ under which the space is compact. They define the generator of the process $\bar{S}^n(t)$ and its truncation to finite K, and use functional analysis techniques to prove the following:

Theorem 21.11 (Vvedenskaya, Dobrushin, and Karpelevich) *For the stationary distribution of $\bar{S}^n(t)$ when $\lambda < 1$, as $n \to \infty$*

$$\mathbb{E}(\bar{S}_k^n) \to \lambda^{\frac{d^k-1}{d-1}}.$$

We can now supply the proof of Theorem 21.1:

Proof of Theorem 21.1   By Lemma 21.3, $\mathbb{P}(^dQ^n \geq k) = \mathbb{E}(\bar{S}_k^n)$, so the first part of (21.1), that as $n \to \infty$, $\mathbb{P}(^dQ^n \geq k) \to \lambda^{\frac{d^k-1}{d-1}}$ follows from Theorem 21.11. The second part of (21.1), that $\mathbb{E}(^dW^n) \to \sum_{k=1}^\infty \lambda^{\frac{d^k-d}{d-1}}$ follows by the same argument as the proof of Proposition 21.9. $\qquad\square$

Intuitively, the sample paths of the n server system will be close to the path of the deterministic infinite server system. This means that from any starting point the n server supermarket system, similar to the infinite server system, converges exponentially fast to its stationary distribution. At that point, for large enough n the fraction of queues of length k or more is close to π_k, and the expected waiting time is close to $\overline{W}_d$ given by (21.14).

Figure 21.2 shows simulation results taken from the original paper by Mitzenmacher (1996, 2001). The figure shows four plots of average sojourn times against the arrival rate λ. They are the simulated sojourn times for choose shortest of 2 with 8 servers and with 100 servers, compared to a plot of the asymptotic values for the infinite server system, $\overline{W}_d(\lambda)$, given in

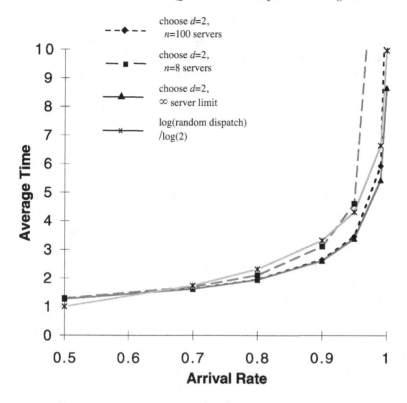

Figure 21.2 Comparing simulations for $d = 2$ to the limiting system and to $d = 1$, reproduced with permission from Mitzenmacher (2001).

(21.14), and compared to the plot of $\log\left(\overline{W}_1(\lambda)\right)/\log 2$. They demonstrate that for as few as 8 servers the expected sojourn time is close to the limiting system, and exponentially shorter than random dispatching.

21.2 Join the Shortest Queue under Halfin–Whitt Regime

We now consider the supermarket model in heavy traffic under the policy of *join the shortest queue* (JSQ). We know that for identical servers, Poisson arrivals, and exponential service times, this policy is optimal; in fact it stochastically minimizes the stationary waiting time of arriving customers. It is nevertheless not as good as a single queue M/M/n system, since in the supermarket model, under any policy, we may have some customers waiting for service while some servers are idle, which never happens in the M/M/n system. In this section we will show that in heavy traffic, under Halfin–Whitt square root staffing, average waiting time for the supermarket model under JSQ (denoted M/M/n-JSQ) is roughly the same as for M/M/n. Recall that this is also the case for just two servers under JSQ in heavy traffic, as shown in Section 15.3.

We take service rate 1. We will assume heavy traffic Halfin–Whitt staffing, i.e. for arrival rate λ_n a constant $\beta > 0$ is used to have

$$n = \lambda_n + \beta \sqrt{n}, \quad \text{so} \quad \rho_n = \frac{\lambda_n}{n} = 1 - \frac{\beta}{\sqrt{n}}. \tag{21.16}$$

We will show that under JSQ the number of idle servers is $O(\sqrt{n})$ and the number of customers waiting for service is also $O(\sqrt{n})$, but the number of queues of more than two is $o(\sqrt{n})$. These results will be summarized in Theorem 21.12, which provides a limiting scaled process for the system. Because there are almost always some idle servers, most customers get served immediately, while the ones that are not served immediately will almost always only need to wait for just one service completion, a single exponential random time with mean 1. Recall in comparison (see Chapter 19) that for the M/M/n system under Halfin–Whitt heavy traffic, customers need to wait with probability $\alpha > 0$, the queue length is $O(\sqrt{n})$, and average waiting time of customers that need to wait is $O(1/\sqrt{n})$.

We use the notations of Section 21.1, and use the ideas of Section 21.1.2, noting that the process is again a density dependent jump Markov chain (see Exercise 21.10). We let $\mathcal{A}(t)$, $D_k(t)$, $k = 1, 2, \ldots$, be independent rate 1 Poisson processes. We will now write the dynamics of the M/M/n-JSQ system. We will make the simplifying assumption that arrivals are rejected if there is no server with only 0 or 1 customer, i.e. there is a customer in service and at least one customer waiting at every server. We assume on the other hand that initially the longest queue in the system is of length $i > 2$. This means that while we may have $S_k^n(t), 2 \le k \le i$ positive, they will never increase. We will later relax this assumption. The dynamics of M/M/n-JSQ, similar to the derivations in Sections 21.1.1 and 21.1.2, will then be

$$S_1^n(t) = S_1^n(0) + \mathcal{A}(\lambda_n t) - D_1 \left(\int_0^t (S_1^n(s) - S_2^n(s))\, ds \right) - U_1^n(t),$$

$$S_2^n(t) = S_2^n(0) + U_1^n(t) - D_2 \left(\int_0^t (S_2^n(s) - S_3^n(s))\, ds \right) - U_2^n(t),$$

$$S_k^n(t) = S_k^n(0) - D_k \left(\int_0^t \left(S_k^n(s) - S_{k+1}^n(s) \right) ds \right), \quad 2 < k < i, \tag{21.17}$$

$$S_i^n(t) = S_i^n(0) - D_i\left(\int_0^t S_i^n(s)ds\right),$$

where U_1, U_2 are overflow processes that satisfy

$$U_1^n(t) = \int_0^t \mathbb{1}(S_1^n(s) = n)d\mathcal{A}(\lambda_n s),$$

$$U_2^n(t) = \int_0^t \mathbb{1}(S_1^n(s) = n, \ S_2^n(s) = n)d\mathcal{A}(\lambda_n s). \quad (21.18)$$

Here we have arrivals at rate λ_n and $U_1^n(t)$ directs arrivals to a busy server if $S_1^n(t) = n$ (no idle servers), while $U_2^n(t)$ counts rejected customers when $S_1^n(t) = S_2^n(t) = n$ (there is at least one waiting customer at every server), and $D_k(\cdot)$ count departures from queues of length k, which decrease S_k.

Similar to the M/M/n system under Halfin–Whitt regime, we expect the number of idle servers as well as the number of waiting customers in the M/Mn-JSQ system to be $O(\sqrt{n})$, and so we expect that $\bar{S}_1^n(t) \to 1$, $\bar{S}_k^n(t) \to 0$, $k > 1$. We will therefore study the diffusion scaled processes (without scaling time): $\hat{S}_1^n(t) = \dfrac{S_1^n(t) - n}{\sqrt{n}}$, and $\hat{S}_k^n(t) = \dfrac{S_k^n(t)}{\sqrt{n}}$, $k > 1$.

We assume the following initial conditions:

$$\hat{S}_k^n(0) \to \hat{S}_k(0), \ k = 1, \ldots, i, \qquad \hat{S}_k^n(0) = 0, \ k > i. \quad (21.19)$$

In particular, this implies that we start with $O(\sqrt{n})$ idle servers and waiting customers, and maximal queue length i.

We centralize and scale A, D_k, and denote

$$M_0^n(t) = \frac{1}{\sqrt{n}}\mathcal{A}(\lambda_n t) - \frac{\lambda_n t}{\sqrt{n}},$$

$$M_k^n(t) = \frac{1}{\sqrt{n}}D_k\left(\int_0^t \left(S_k^n(s) - S_{k+1}^n(s)\right)ds\right) \quad (21.20)$$

$$\qquad - \frac{1}{\sqrt{n}}\int_0^t \left(S_k^n(s) - S_{k+1}^n(s)\right)ds, \quad 1 \le k < i,$$

$$M_i^n(t) = \frac{1}{\sqrt{n}}D_i\left(\int_0^t S_i^n(s)ds\right) - \frac{1}{\sqrt{n}}\int_0^t S_i^n(s)ds. \quad (21.21)$$

We also let $\hat{U}_1^n(t) = \dfrac{U_1(t)}{\sqrt{n}}$, $\hat{U}_2^n(t) = \dfrac{U_2(t)}{\sqrt{n}}$, and recall $\beta = (1 - \rho_n)\sqrt{n} = (n - \lambda_n)/\sqrt{n}$.

We then rewrite, after simple manipulations, the dynamics of $\hat{S}^n(t)$:

$$\hat{S}_1^n(t) = \hat{S}_1^n(0) + M_0^n(t) - M_1^n(t) - \beta t - \int_0^t \left(\hat{S}_1^n(s) - \hat{S}_2^n(s)\right) ds - \hat{U}_1^n(t),$$

$$\hat{S}_2^n(t) = \hat{S}_2^n(0) + \hat{U}_1^n(t) - M_2^n(t) - \int_0^t \left(\hat{S}_2^n(s) - \hat{S}_3^n(s)\right) ds - \hat{U}_2^n(t),$$

$$\hat{S}_k^n(t) = \hat{S}_k^n(0) - M_k^n(t) - \int_0^t \left(\hat{S}_k^n(s) - \hat{S}_{k+1}^n(s)\right) ds, \quad 2 < k < i,$$

$$\hat{S}_i^n(t) = \hat{S}_i^n(0) - M_i^n(t) - \int_0^t \hat{S}_i^n(s) ds, \qquad\qquad (21.22)$$

$$0 = \int_0^t \mathbb{1}(\hat{S}_1^n(s) < 0) d\hat{U}_1^n(s),$$

$$0 = \int_0^t \mathbb{1}(\hat{S}_2^n(s) < \sqrt{n}) d\hat{U}_2^n(s).$$

We now present the main result of this section, followed by a discussion and outline of the proof.

Theorem 21.12 *Assume Halfin–Whitt staffing (21.16) and initial conditions (21.19). Then the fluid scaled process satisfies $S_1^n(t)/n \to 1$, $S_k^n(t)/n \to 0$, $k > 1$, as $n \to \infty$, and the diffusion scaled processes $\hat{S}_k^n(t)$, $k = 1, \ldots, i$ converge in distribution to $\hat{S}_k(t)$ that satisfy:*

$$\hat{S}_1(t) = \hat{S}_1(0) + \sqrt{2}BM(t) - \beta t - \int_0^t \left(\hat{S}_1(s) - \hat{S}_2(s)\right) ds - \hat{U}_1(t),$$

$$\hat{S}_2(t) = \hat{S}_2(0) + \hat{U}_1(t) - \int_0^t \left(\hat{S}_2(s) - \hat{S}_3(s)\right) ds,$$

$$\hat{S}_k(t) = \hat{S}_k(0) - \int_0^t \left(\hat{S}_k(s) - \hat{S}_{k+1}(s)\right) ds, \quad 2 < k < i,$$

$$\hat{S}_i(t) = \hat{S}_i(0) - \int_0^t \hat{S}_i(s) ds, \qquad\qquad (21.23)$$

$$0 = \int_0^t \mathbb{1}(\hat{S}_1(s) < 0) d\hat{U}_1(s),$$

where $BM(t)$ is a standard Brownian motion.

Because the equations for $3 \leq k \leq i$ are not stochastic, their solution is

(see Exercise 21.11)

$$\hat{S}_k(t) = \left(\hat{S}_k(0) + \sum_{j=1}^{i-k} \frac{t^j}{j!} \hat{S}_{k+j}(0)\right) e^{-t}, \qquad 3 \le k < i,$$

$$\hat{S}_i(t) = \hat{S}_i(0) e^{-t}. \tag{21.24}$$

We can see from this description that the limiting process will have the following properties:

- The limiting counts of queues with more than two customers decrease from the initial value to 0 at an exponential rate and are never replenished.
- The limiting count of waiting customers, $\hat{S}_2(t)$, is fed by the non-decreasing $\hat{U}_1(t)$ at times when all servers are busy. At times when some servers are idle, it decreases at an exponential rate.
- The limiting quantity of idle servers given by $-\hat{S}_1(t)$ behaves somewhat like an Ornstein–Uhlenbeck process, reflected at 0.

To explain the last paragraph, if we ignore $\hat{U}_1(t)$ we have

$$d\left[-\hat{S}_1(t)\right] = -\left(-\hat{S}_1(t)\right) + \left(\sqrt{2}\, dBM(t) + (\beta - \hat{S}_2(t))\right),$$

which is a firstst-order autoregression process driven by $\sqrt{2}BM(t)$ and by $(\beta - \hat{S}_2(t))$. This second term expresses the Halfin–Whitt β rate of processing provided by the extra servers minus waiting customers that claim idle servers. This Ornstein–Uhlenbeck type process is then reflected from 0 by $+\hat{U}(t)$.

In the n server system, the $S_0 - S_1$, S_k, $k \ge 2$ will behave approximately in this way as n becomes large, scaled up by $\sqrt{n}$. This is illustrated in Figure 21.3, taken from Eschenfeldt and Gamarnik (2018), which shows a simulated sample path, with $n = 10^5$, and $\beta = 2$.

The proof of Theorem 21.12 consists of the following steps:

(i) Existence and uniqueness of the solution to the pre-limit and limiting sets of equations (21.22), (21.23).

(ii) Finding the weak limits of $M_k^n(t)$ by using the FCLT for Poisson processes, and time change.

(iii) Obtaining the limits of the time change by finding the fluid limits of $S_k^n(t)$.

(iv) Relaxing the assumption of no arrivals to queues of length exceeding two.

We will state these in the following propositions, discussing the proofs briefly but leaving most of the details out or in the exercises.

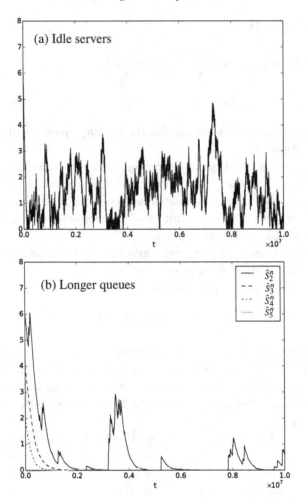

Figure 21.3 Simulation of Halfin–Whitt heavy traffic M/M/n queues under JSQ, reproduced with permission from Eschenfeldt and Gamarnik (2018).

Proposition 21.13 *The system of equations:*

$$x_1(t) = b_1 + y_1(t) - \int_0^t (x_1(s) - x_2(s))\, ds - u_1(t), \qquad (21.25)$$

$$x_2(t) = b_2 + y_2(t) - \int_0^t (x_2(s) - x_3(s))\, ds + u_1(t) - u_2(t),$$

$$x_k(t) = b_k + y_k(t) - \int_0^t (x_k(s) - x_{k+1}(s))\, ds, \quad 2 < k < i,$$

$$x_i(t) = b_i + y_i(t) - \int_0^t x_i(s)ds,$$

$$0 = \int_0^t \mathbb{1}(x_1(s) < 0)du_1(s),$$

$$0 = \int_0^t \mathbb{1}(x_2(s) < B)du_2(s),$$

$$x_1(t) \le 0, \quad 0 \le x_2(t) \le B, \quad x_k(t) \ge 0, \quad t \ge 0,$$

defines a function $f(B, b, y) = (x, u)$ from $\bar{\mathbb{R}} \times \mathbb{R}^i \times \mathbb{D}^i$ to $\mathbb{D}^i \times \mathbb{D}^2$, where $\bar{\mathbb{R}}$ is the real line including ∞, and $\mathbb{D}$ is the space of functions right continuous with left limits. Then the solution defining f is unique, f is continuous, and f transforms continuous y to continuous x, u. Furthermore,

$$\sup_{0 \le s \le t} x_k(s) \le 8e^{6t}|b_k + \sup_{0 \le s \le t} y_k(s)|. \tag{21.26}$$

Proof   We will only outline the proof of existence and uniqueness. Recall Skorohod's reflection mapping: For a function $y \in \mathbb{D}$, and constant B, let z be non-decreasing $z(0) = 0$, $x(t) = y(t) - z(t) \le B$, and $\mathbb{1}(x < B)dz = 0$, and denote $x = \psi_B(y)$, $z = \varphi_B(y)$, then φ, ψ are uniquely defined continuous mappings.

We saw already that x_k, $k \ge 3$ are uniquely defined, and substituting x_3 in the equation for x_2 yields the same equation with a new $\tilde{y}_2(y) = y_2(t) + \int_0^t x_3(s)ds$. So we need to show that the first two equations, without x_3 in the second equation, have a unique solution. We consider the equations:

$$w_1(t) = b_1 + y_1(t) - \int_0^t \left(\psi_0(w_1(s)) - \psi_B(w_2(s)) \right) ds,$$

$$w_2(t) = b_2 + \tilde{y}_2(t) + \varphi_0(w_1(t)) - \int_0^t \psi_B(w_2(s))ds. \tag{21.27}$$

One then sees that $x_1 = \psi_0(w_1)$, $u_1 = \varphi_0(w_1)$, $x_2 = \psi_B(w_2)$, $u_2 = \varphi_B(w_2)$ are the desired solutions to (21.25) (see Exercise 21.12). Next we consider instead

$$\tilde{w}_1(t) = b_1 + y_1(t) - \int_0^t \left(\psi_0(\tilde{w}_1(s)) - \psi_B(\tilde{w}_2(s) + \varphi_0(\tilde{w}_1(s))) \right) ds,$$

$$\tilde{w}_2(t) = b_2 + \tilde{y}_2(t) - \int_0^t \psi_B(\tilde{w}_2(s) + \varphi_0(\tilde{w}_1(s)))ds. \tag{21.28}$$

Then one can see that $(w_1, w_2) = (\tilde{w}_1, \tilde{w}_2 + \varphi_0(\tilde{w}_1))$ solves the equations for w_1, w_2 (see Exercise 21.12).

The proof of existence and uniqueness is completed by showing that the r.h.s. defines a contraction mapping on $(\tilde{w}_1, \tilde{w}_2)$ (see Exercise 21.13).   □

Proposition 21.14 *The sequence* $(M_0^n(\cdot), M_1^n(\cdot), \ldots, M_i^n(\cdot))$ *converges weakly to* $(BM_1(\cdot), BM_2(\cdot), 0, \ldots, 0)$ *where* BM_1, BM_2 *are two independent standard Brownian motions.*

Proof By the FCLT, for the rate 1 Poisson process $\mathcal{A}(t)$: $\frac{\mathcal{A}(nt) - \lambda nt}{\sqrt{n}} \to_w BM(t)$, and similarly for $D_k(t)$, $k = 1, 2, \ldots$. It follows immediately that

$$M_0^n(t) = \frac{\mathcal{A}(\lambda_n t) - \lambda_n t}{\sqrt{n}}$$

$$= \frac{\mathcal{A}(n(1 - \beta/\sqrt{n})t) - n(1 - \beta/\sqrt{n})t}{\sqrt{n}} \to_w BM_1(t),$$

where we used the time change $g_n(t) = (1 - \beta/\sqrt{n})t \to t$. For $M_k^n(t)$, we are looking at

$$M_k^n(t) = \frac{1}{\sqrt{n}} D_k\left(n \int_0^t \left(\frac{S_k^n(s)}{n} - \frac{S_{k+1}^n(s)}{n}\right) ds\right)$$
$$- \frac{1}{\sqrt{n}} \int_0^t n\left(\frac{S_k^n(s)}{n} - \frac{S_{k+1}^n(s)}{n}\right) ds$$

for $1 \le k < i$ with a similar expression for $M_i^n(t)$. Here the time change is $g_{k,n}(t) = \int_0^t \left(\bar{S}_k^n(s) - \bar{S}_{k+1}^n(s)\right) ds$. We note that because $S_k(s) \ge S_{k+1}(s)$, $g_{k,n}(t)$ are continuous increasing random functions of t. We will show that $g_{1,n}(t) \to_w t$, while $g_{k,n}(t) \to_w 0$, $2 \le k \le i$. Then by the time change theorem 5.6, $M_1^n(t) \to BM_2(t)$, while $M_k^n(t) \to 0$. Because the Poisson processes are independent, the independence of the limit follows. $\square$

Proposition 21.15 *The fluid scaled sequence* $\left(\frac{S_1^n(\cdot)}{n}, \frac{S_2^n(\cdot)}{n}, \ldots, \frac{S_i^n(\cdot)}{n}\right)$ *converges to* $(\omega(\cdot), 0, \ldots, 0)$, *where* $\omega(t) = 1$.

Proof By Proposition 21.14, for fixed t, each $M_k^n(t)$ is stochastically bounded (i.e. for any ϵ exists κ so $\mathbb{P}(M_k^n(t) > \kappa) < \epsilon$). Then by the bound (21.26) of Proposition 21.13 each process $\hat{S}_k^n(t)$ is stochastically bounded. So $\hat{S}_k^n(t)/\sqrt{n} \to 0$. Since $\hat{S}_1^n(t) = \frac{S_1^n(t) - n}{\sqrt{n}}$, we get $\frac{S_1^n(t)}{n} \to 1$, while $\frac{S_k^n(t)}{n} \to 0$, $k > 1$. $\square$

Propositions 21.13–21.15 prove Theorem 21.12 for the processes $\hat{S}_k^n(t)$ truncated by the rejection of arrivals when $S_1(t) = S_2(t) = n$. It remains to show that Theorem 21.12 holds also without the truncation.

Proposition 21.16 *Consider the sequence of truncated processes* $\hat{S}_k^n(t)$ *obtained by solving* (21.23). *Let* $t_n^* = \inf\{t : S_2^n(t) = n\}$. *Then for any*

t, $\mathbb{P}(t_n^* \leq t) \rightarrow 0$ as $n \rightarrow \infty$, and therefore the limit for the truncated processes $\hat{S}_k^n(t)$ and for the un-truncated processes is the same.

Proof The derivation so far assumed that no customers are sent to queues of length > 2. Those customers that should have gone to such queues are rejected. They are counted by $U_2^n(t)$. Then up to the time $t_n^* = \inf\{t : S_2^n(t) = n)$ the truncated and the un-truncated systems agree path by path. Hence, if we show that for any t, $\mathbb{P}(t_n^* \leq t) \rightarrow 0$ as $n \rightarrow \infty$ then for large n the truncated and un-truncated systems agree path by path with very high probability, and in the limit they are the same.

We have for the truncated system,

$$\mathbb{P}((t_n^* \leq t) = \mathbb{P}\left(\sup_{0 \leq s \leq t} \hat{S}_2^n(s) \geq \sqrt{n}\right) \leq \mathbb{P}\left(\sup_{0 \leq s \leq t} \hat{S}_2^n(s) \geq C\right)$$

for a constant $0 < C < \sqrt{n}$. We showed $\hat{S}_2^n \rightarrow_w \hat{S}_2$, so

$$\limsup_n \mathbb{P}\left(\sup_{0 \leq s \leq t} \hat{S}_2^n(s) \geq C\right) \leq \mathbb{P}\left(\sup_{0 \leq s \leq t} \hat{S}_2(s) \geq C\right)$$

and by continuity of probability

$$\lim_{C \rightarrow \infty} \mathbb{P}\left(\sup_{0 \leq s \leq t} \hat{S}_2(s) \geq C\right) = \mathbb{P}\left(\sup_{0 \leq s \leq t} \hat{S}_2(s) = \infty\right) = 0.$$

The proposition follows. □

21.3 Approaching JSQ: Shortest of $d(n)$ and Join Idle Queue

As we have seen, JSQ, which stochastically minimizes queue lengths and waiting times, is asymptotically as good as a single M/M/n queue. However, it requires n communication messages for each arriving customer, to determine the shortest queue to which the arrival is sent, and this introduces an unacceptable cost when $n \rightarrow \infty$.

We have also seen that sampling d servers and sending arrivals to the shortest of these queues is a significant improvement on random dispatching. Nevertheless, as the traffic intensity approaches 1, this policy will still lead to congestion.

Proposition 21.17 *Under Halfin–Whitt heavy traffic with $\lambda_n = n - \beta\sqrt{n}$, and using shortest of d policy, the stationary average sojourn time grows like $\frac{\log n}{2 \log d}$.*

This follows from Proposition 21.10; see Exercise 21.14.

It is still possible to approach the performance of JSQ in Halfin–Whitt heavy traffic by choosing as policy the shortest of $d(n)$ servers drawn at random, i.e. make d depend on n. The following theorem is proved by coupling the process under JSQ and under choose shortest of $d(n)$:

Theorem 21.18   *Under Halfin–Whitt heavy traffic with $\lambda_n = n - \beta\sqrt{n}$, under the policy of choose shortest of $d(n)$ where $d(n) \to \infty$, the fluid limit of the queue length processes is the same as for JSQ. If $d(n)/(\sqrt{n}\log n) \to \infty$ then the diffusion scaled queue lengths process converges weakly to the same limiting process as under JSQ.*

For a different approach we now consider the policy of *join idle queue* (JIQ): Instead of the dispatcher sampling the queues, the servers notify the dispatcher when they become idle, and the dispatcher maintains a list of idle servers. When a customer arrives, the dispatcher will send him to an idle server (and remove it from the list), or to a random server if no idle server is available.

Assume n servers and arrival rate λ_n with $\lambda_n/n < 1$. To describe the sample paths of $S^n(t)$, let $\mathcal{A}(t), D_k(t), k \geq 1$ be rate 1 Poisson processes, and let $\xi(s) = e_j$ with probability $1/n$, for $j = 1, \ldots, n$ where e_j is the jth unit vector. We will need a sequence of these for a sequence of time points s at which there is an arrival and no idle servers, and they are all independent. We can then write the sample paths of $S^n(t)$:

$$S_1^n(t) = S_1^n(0) + \mathcal{A}(\lambda_n t) - D_1\left(\int_0^t (S_1^n(s) - S_2^n(s))\,ds\right) - U_1^n(t),$$

$$S_k^n(t) = S_k^n(0) + \int_0^t \left(\sum_{j=S_k^n(s)+1}^{S_{k-1}^n(s)} \xi_j(s)\right) dU_1(s) \qquad (21.29)$$

$$- D_k\left(\int_0^t \left(S_k^n(s) - S_{k+1}^n(s)\right)ds\right), \qquad k > 1.$$

It can be shown then

Theorem 21.19   *Under Halfin–Whitt heavy traffic regime with $\lambda_n = n - \beta\sqrt{n}$, under the policy of join idle queue, JIQ, the fluid scaled as well as the diffusion scaled queue lengths processes converge weakly to the same limiting process as under JSQ.*

21.4 Sources

The behavior of shortest of d choice for fixed ρ was derived simultaneously in two indepnedent papers. We follow Mitzenmacher (1996, 2001) in most of our presentation, since the paper by Vvedenskaya et al. (1996), although more complete, uses techniques that are outside the scope of this book. Kurtz's theorem on density dependent Markov chains is from chapter 8 of his monograph *Approximation of Population Processes*, Kurtz (1981). The behavior of JSQ in the Halfin–Whitt square root staffing regime is derived in Eschenfeldt and Gamarnik (2018). A follow-up paper, Eschenfeldt and Gamarnik (2016) discusses choose shortest of d in Halfin–Whitt regime, and concludes that the queues increase as $\log(\frac{1}{1-\rho})$. That JSQ asymptotics are achieved by $d(n) \to \infty$ is from Mukherjee et al. (2018), and JIQ is discussed by Gamarnik et al. (2018) and Stolyar (2015). The analysis of shortest of d choice for general service distributions, using measure-valued processes is introduced by Aghajani et al. (2015) and Aghajani and Ramanan (2019). A survey of recent developments is Van der Boor et al. (2019).

Exercises

21.1 Consider the n server system with Poisson arrivals and exponential service time, and assume that customers are dispatched to the servers in round-robin order. Show that in heavy traffic the expected waiting time of each customer approaches $\frac{1}{2(1-\rho)}$ as $n \to \infty$, i.e. half of the time under random dispatching.

21.2 (*) Show that the stationary queue length for choose shortest of d, $^dQ^n$ is smaller in the sense of convex majorization than the stationary queue length under random dispatching $^1Q^n$, and prove that the choose shortest of d is a stable ergodic system [Vvedenskaya et al. (1996)].

21.3 Justify the inequalities:

$$\tilde{D}_l\left(n \int_0^t \beta_l(\bar{X}^n(s))ds\right)$$
$$\leq \sup_{u \leq t} \tilde{D}_l(n\bar{\beta}_l u)$$
$$\leq \left[\tilde{D}_l(n\bar{\beta}_l t) + n\bar{\beta}_l t\right].$$

21.4 Show that with

$$F(s(t)) = \left[\lambda(s_{k-1}(t)^d - s_k(t)^d) - (s_k(t) - s_{k+1}(t))\right]_{k=1,2,\ldots},$$

F is Lipschitz continuous, i.e. there exists M such that

$$|F(x) - F(y)| \leq M|x - y|.$$

21.5 Complete the steps of the proof of Proposition 21.6.

21.6 For the supermarket model, under choose shortest of d, show that in the limiting infinite server system (21.5) if we increase $s_j(0)$ for some j, this will increase or leave unchanged $s_k(t)$ for all $t > 0$ and all k.

21.7 In the proof of Proposition 21.7 show that $M(0) < 1/\lambda^{1/(d-1)}$.

21.8 (*) Show that one can find an increasing sequence $w_k \geq 1$ and $\delta > 0$ that satisfy

$$w_{k+1} \leq w_k + \frac{(1-\delta)w_k - w_{k-1}}{\lambda(2\pi_k + 1)}$$

so that this sequence is bounded by a geometric sequence.

21.9 (*) Prove that

$$\lim_{\lambda \to 1} \frac{\sum_{k=0}^{\infty} \lambda^{d^k}}{\log \frac{1}{1-\lambda}} = \frac{1}{\log d}$$

and use this to prove Proposition 21.10 [Mitzenmacher (1996)].

21.10 Show that the M/M/n-JSQ model can be described by a density dependent Markov chain.

21.11 For the M/M/n-JSQ truncated model, prove that for the limiting system, the scaled counts of the queues $\hat{S}_k(t)$, $k \geq 3$ are given by

$$\hat{S}_k(t) = \left(\hat{S}_k(0) + \sum_{j=1}^{i-k} \frac{t^j}{j!} \hat{S}_{k+j}(0) \right) e^{-t}, \qquad 3 \leq k < i,$$

$$\hat{S}_i(t) = \hat{S}_i(0)e^{-t}.$$

21.12 Show that if $\tilde{w}_1(t)$, $\tilde{w}_2(t)$ solve equations (21.28), then $(w_1, w_2) = (\tilde{w}_1, \tilde{w}_2 + \varphi_0(\tilde{w}_1))$ solve equations (21.27), and show that if (w_1, w_2) solve equations (21.27), then $x_1 = \psi_0(w_1)$, $u_1 = \varphi_0(w_1)$, $x_2 = \psi_B(w_2)$, $u_2 = \varphi_B(w_2)$ solve equations (21.25).

21.13 For the M/M/n-JSQ model, show that the transformation:

$$T_1(\tilde{w}_1, \tilde{w}_2)) = b_1 + y_1(t) - \int_0^t (\psi_0(\tilde{w}_1(s)) - \psi_B(\tilde{w}_2(s)) - \varphi_0(\tilde{w}_1(s))) \, ds,$$

$$T_2(\tilde{w}_1, \tilde{w}_2)) = b_2 + \bar{y}_2(t) - \int_0^t (\psi_B(\tilde{w}_2(s)) - \varphi_0(\tilde{w}_1(s))) \, ds,$$

is a contraction mapping from $\mathbb{D}^2$ to $\mathbb{D}^2$.

21.14 Show that under Halfin–Whitt heavy traffic staffing, with choose shortest of d policy, the stationary average sojourn time grows like $\dfrac{\log n}{2 \log d}$.

22

Parallel Servers with Skill-Based Routing

In this chapter we discuss systems with customers of various types that are served by pools of servers with different skills. Two extreme models for this are a single pool of servers that can serve all types of customers, and individual pools of servers each trained to serve a single type of customers. The first option is usually impractical as it requires each server to be trained in all skills; the second option is inefficient as it ignores all possibilities of pooling resources. A practical solution is to have some cross training of servers to acquire more than one skill, so that some pooling is achieved. In our parallel skilled service systems (PSBS) we assume customers of types $\mathscr{C} = \{c_1, \ldots, c_I\}$ and servers of types $\mathscr{S} = \{s_1, \ldots, s_J\}$ and a bipartite compatibility graph $\mathcal{G}$, where $(s_j, c_i) \in \mathcal{G}$ means that servers of type s_j are able to serve customers of type c_i.

Such systems, with large volumes of customers that are served either by a few high capacity servers or by pools of many servers, occur in many walks of life. Examples are call centers, health systems, internet shopping, and ride sharing.

We will make the plausible assumptions that arrivals are Poisson, of rate λ, and types of customers are i.i.d. with arrival rates $\lambda_{c_i} = \lambda \alpha_{c_i}$. For tractability we assume that services are exponential, with rates μ_{s_j, c_i}, with the special case of server dependent (SD) service rates $\mu_{s_j, c_i} = \mu_{s_j}$.

In Section 22.1 we derive exact results for the policy of FCFS-ALIS, where customers are assigned to the longest idle compatible server, and servers choose the longest waiting compatible customer. We assume SD service rates, and obtain stationary distributions of the queues and of the waiting times. In Section 22.2 we consider a simpler but more general model, of FCFS matching of an infinite sequence of i.i.d. customers to an infinite sequence of i.i.d. servers. This simple system only takes account of the order of types in each sequence, and is tractable. In particular it yields exact expressions for the fraction of matches of pairs $(s_j, c_i) \in \mathcal{G}$, so-called matching rates r_{s_j, c_i}. A similar model in which we have a single sequence of customers and servers, and servers are matched to earlier occurring customers, is discussed in Section 22.3; it provides a model for ride-sharing systems. In Section 22.4 we discard the assumption of SD exponential service times, and describe a design heuristic for the staffing of a FCFS-ALIS large-scale many-server system, based on the matching rates obtained in Section 22.2. Finally, in Section 22.5 a queue and idleness ratio (QIR) policy is discussed, that uses a dynamic priority policy for matching customers and servers, which in some cases achieves state space collapse and leads to diffusion limit approximations.

Many of the techniques used here have been explored in previous parts of the book, and we will therefore for the sake of brevity only give short discussions of proofs and derivations. The interested reader will be directed to further details in extensive exercises. The proofs of state space collapse in Section 22.5, as already mentioned in Section 15.5, are beyond the scope of this book and will not be outlined.

Some further notation will be used: $\mathcal{S}(c_i) = \{s_j : (s_j, c_i) \in \mathcal{G}\}$ are servers of c_i; similarly, $\mathcal{C}(s_j)$ are customers of s_j. For $C \subseteq \mathcal{C}$, $\mathcal{S}(C) = \bigcup_{c_i \in C} \mathcal{S}(c_i)$, and $\lambda_C = \sum_{c_i \in C} \lambda_{c_i}$, with similar definitions of α_C, and of $\mathcal{C}(S)$, μ_S, for $S \subseteq \mathcal{S}$ (assuming SD service). We also let $\mathcal{U}(S) = \mathcal{C}(\overline{S})$ be the unique customers of S.

Throughout this chapter we refer to queues, $Q(t)$ as the number of waiting customers, and exclude customers in service.

22.1 Parallel Skill-Based Service under FCFS

We now consider the parallel skill-based system for a fixed number of servers, under the policy of first come first served (FCFS) and assign to longest idle server (ALIS). When a customer arrives and there are several available idle servers, he is assigned to the compatible server that has been idle for the longest time, and when a server becomes available he will serve the longest waiting compatible customer. This policy is not trying to be efficient, but it has the virtue of being fair to customers and fair to the servers. We assume that arrivals are Poisson, services are exponential, and service rates are server dependent, i.e. $\mu_{s_j, c_i} = \mu_{s_j}$.

A major difficulty in analyzing the performance of the system under FCFS-ALIS is that we cannot easily determine the matching rates r_{s_j, c_i}, and as a result it is not even clear when the system will be stable. The natural necessary conditions for stability are that

$$\lambda_C < \mu_{\mathscr{S}(C)}, \quad \text{equivalently} \quad \mu_S > \lambda_{\mathscr{U}(S)}, \quad \text{for all subsets } S, C.$$
(22.1)

As we shall see, this condition is in fact necessary and sufficient for stability.

Example 22.1 *Three servers with three types of customers:* In this system customers of type c_1 can be served by all three servers, while customers of type c_2 are served only by server s_2, and customers of type c_3 are served only by server s_3. We then have: $\mathscr{S}(c_1) = \{s_1, s_2, s_3\}$, $\mathscr{S}(c_2) = s_2$, $\mathscr{S}(c_3) = s_3$, or equivalently $\mathscr{C}(s_1) = c_1$, $\mathscr{C}(s_2) = \{c_1, c_2\}$, $\mathscr{C}(s_3) = \{c_1, c_3\}$, and $\mathscr{U}(s_2) = c_2$, $\mathscr{U}(s_3) = c_3$, $\mathscr{U}(s_1) = \emptyset$. This is best seen from the bipartite compatibility graph shown on the right in Figure 22.1. There is a neat description of the state and dynamics of the system under FCFS-ALIS, illustrated on the left of Figure 22.1, which is a snapshot of the system at some point in time: Here the queue of customers is shown in order of arrivals from left (oldest) to right (newest), classified by types, and customers in service are shown together with the server that is serving them, so the oldest customer in the system, of type c_1, is currently in service by server s_2, then the second and third oldest customers are of type c_2 and are waiting for server s_2, and they were skipped by servers s_1, s_3. Server s_3 is serving the fourth oldest customer, of type c_3, and customers of types c_2, c_3, c_2 are next in line, waiting for a server, and they were skipped by server s_1 who is now idle.

Figure 22.1 3 servers 3 customer types under FCFS-ALIS,

More generally, at any time we will have the queue of all customers in the system, ordered in order of arrival from left (oldest) to right and classified by type, we will have customers in service attached to the server that is serving them, and in front of all the customers, a queue of all the idle servers ordered by idle time from right (oldest) to left (youngest) and classified by type. Customers waiting between servers have been skipped by incompatible servers ahead of them to the right. The dynamics of FCFS-ALIS are then seen from this description as follows: When a customer arrives, he scans the idle servers from right to left, and joins the end of the queue with the first compatible server that he finds, or without a server if there is no compatible idle server. When a server completes a service, he moves to the right, scanning the waiting customers until he finds the first compatible customer which he then serves; if there is no compatible waiting customer, he joins the left end of the idle servers queue.

We can now describe the dynamics of the FCFS-ALIS system by a Markov chain $X(t)$, $t \geq 0$ with state $X(t) = \mathfrak{s}$, where $\mathfrak{s} = (S_1, n_1, \ldots, S_i, n_i, S_{i+1}, \ldots, S_J)$. Here $S_1, \ldots, S_J$ is a random permutation of the J servers, $S_1, \ldots, S_i$ are busy serving compatible customers, ordered so that S_1 serves the oldest customer, and $S_{i+1}, \ldots, S_J$ are the idle servers, ordered with S_J the longest idle, S_{i+1} the latest to become idle, and n_j, $j = 1, \ldots, i$, counts the number of the customers between server S_j and S_{j+1}. Accordingly, the current state of the system described in Figure 22.1 is $X(t) = (s_2, 2, s_3, 3, s_1)$.

The process $X(t)$ is less detailed than the description in Figure 22.1, because it leaves out the identity of the customers in service, and of the customers that are waiting between the servers. The reason that this is a Markov chain is that customers waiting between server S_j and S_{j+1} must be of types in $\mathscr{U}(S_1, \ldots, S_j)$, and furthermore, their classes are i.i.d. with class c_i occurring with probability $\lambda_{c_i}/\lambda_{\mathscr{U}(S_1,\ldots,S_j)}$. We can therefore show that $X(t)$ is a jump Markov chain, and write down its transition probabilities (see Exercise 22.1). The state space is $\mathcal{P}(J) \times \mathbb{N}^J$, where $\mathcal{P}(J)$ is the set of all the permutations of the J servers.

Theorem 22.2   *The Markov chain $X(t)$ is ergodic if and only if (22.1) holds, and its stationary distribution is given by*

$$\pi_X(\mathfrak{s}) = B \prod_{j=1}^{i} \frac{\lambda_{\mathscr{U}(\{S_1,\ldots,S_j\})}^{n_j}}{\mu_{\{S_1,\ldots,S_j\}}^{n_j+1}} \prod_{j=i+1}^{J} \lambda_{\mathscr{C}(\{S_j\ldots,S_J\})}^{-1}, \qquad (22.2)$$

with normalizing constant

$$B = \left[\sum_{\mathcal{P}(J)} \sum_{i=0}^{J} \left(\prod_{j=1}^{i} \left(\mu_{\{S_1,\ldots,S_j\}} - \lambda_{\mathscr{U}(\{S_1,\ldots,S_j\})} \right)^{-1} \prod_{j=i+1}^{J} \left(\lambda_{\mathscr{C}(\{S_j,\ldots,S_J\})} \right)^{-1} \right) \right]^{-1}, \qquad (22.3)$$

where the summation is over all permutations $(S_1, \ldots, S_J) \in \mathcal{P}(J)$, and by convention, empty products are 1.

Proof   We outline the proof. This system actually satisfies *partial balance equations*, which are

(i) The total probability flux out of state $\mathfrak{s}$ due to an arrival that activates a server, equals the total probability flux into state $\mathfrak{s}$ due to a departure which idles a server.

(ii) The total probability flux out of state s, due to an arrival that joins the queue, equals the total probability flux into state s, due to a departure that is followed by another start of service (so that the set of idle servers is unchanged).

(iii) The total probability flux out of state s in which $n_i = 0$, due to a departure, equals the total probability flux into state s, due to an arrival of a customer that activates server S_i.

(iv) The total probability flux out of state s in which $n_i > 0$, due to a departure, equals the total probability flux into state s, due to an arrival of a customer that joins the queue.

It can now be checked that (22.2) satisfies these partial balance equations, and then calculate B. It is seen in the calculation of B that B^{-1} is finite only if each of the geometric terms in the product form converges, which is the case if and only if condition (22.1) holds.

Exercises 22.1–22.5 follow the necessary steps to complete the proof. □

Having found the stationary distribution, we can in principle obtain expressions for most performance measures. We note however that calculating B involves a summation over all permutations of the J servers, and it is very likely that this calculation is in fact #P-hard. Similarly, most expressions for performance measures will involve summation over all permutaions of the servers. This limits the exact calculations to relatively small J.

We can also obtain the distribution of waiting times, by using the distributional form of Little's law given in Theorem 2.27. The proof of the following Theorem is left as Exercise 22.6:

Theorem 22.3 *The LST of the steady-state waiting time W_c of a job of type c is equal to*

$$\mathbb{E}(e^{-sW_c}) = \sum_{\mathcal{P}(J)} \sum_{i=0}^{J} \left(\pi(S_1, \cdot, \ldots, S_i, \cdot, S_{i+1}, \ldots, S_J) \right.$$

$$\left. \prod_{\substack{j=1 \\ c \in \mathcal{U}(\{S_1, \ldots, S_j\})}}^{i} \frac{\mu_{\{S_1, \ldots, S_j\}} - \lambda_{\mathcal{U}(\{S_1, \ldots, S_j\})}}{\mu_{\{S_1, \ldots, S_j\}} - \lambda_{\mathcal{U}(\{S_1, \ldots, S_j\})} + s} \right), \quad (22.4)$$

where

$$\pi(S_1, \cdot, \ldots, S_i, \cdot, S_{i+1}, \ldots, S_J)$$

$$= \prod_{j=1}^{i} \left(\mu_{\{S_1, \ldots, S_j\}} - \lambda_{\mathcal{U}(\{S_1, \ldots, S_j\})} \right)^{-1} \prod_{j=i+1}^{J} \left(\lambda_{\mathcal{C}(\{S_j, \ldots, S_J\})} \right)^{-1}.$$

We explain this result: Here $\pi(S_1, \cdot, \ldots, S_i, \cdot, S_{i+1}, \ldots, S_J)$ is the stationary probability of the permutation $(S_1, \ldots, S_J)$ of the servers, with servers $(S_{i+1}, \ldots, S_J)$ idle, which is also the probability that an arriving customer will find the servers in that formation, by PASTA. If $c \in \mathscr{C}(S_{i+1}, \ldots, S_J)$, then the waiting time for a customer of type c is 0, expressed by an empty product in (22.4). Otherwise, the product represents the LST of a sum of independent exponential random variables. So the waiting time is given as a mixture of sums of independent exponential random variables, and a positive probability of no wait. The equation (22.4) suggests the following interpretation: A customer of type c that arrives will either enter service immediately, or it will wait as if it needs to go through a tandem series of M/M/1 queues, with arrival rates $\lambda_{\mathscr{U}(\{S_1,\ldots,S_j\})}$ and service rates $\mu_{\{S_1,\ldots,S_j\}}$, for $j = 1, \ldots, k$, where S_k is the first server compatible with c in the permutation. Of course, this is not how it really happens, as during the wait of each customer of type c servers complete services at rates μ_{S_j}, and overtake each other, until one of them (not necessarily S_k) will reach the customer. This type of *false interpretation* is typical of Markov chains with partial balance, such as Jackson networks, and other Markov chains showing insensitivity, such as symmetric queues (recall Sections 8.1, 8.8).

One can use the LST of the waiting time to obtain expressions for mean and variance of the waiting time; see Exercise 22.7.

22.2 Infinite Bipartite Matching under FCFS

Under the policy of FCFS-ALIS, the roles of servers and customers are almost symmetric: There is a queue of customers, and a queue of servers, and both are served on the basis of FCFS. The symmetry is broken by the fact that there is only a finite number of servers, and the analysis becomes harder because servers alternate between being busy and being idle. In this section we present a completely symmetric model. There is an infinite ordered sequence of customers, classified by types, and an infinite ordered sequence of servers classified by types, and customers and servers are matched according to a compatibility graph, using FCFS. We can think of this as a sequence of customers, and a sequence of services, rather than servers. All elements of time and of individual identity of the items are excluded from the model. FCFS matching is illustrated in Figure 22.2.

We use the notations at the beginning of the chapter, and the data for this model consists of the compatibility graph, $\mathcal{G}$, and the frequencies of the customer and server types, α_{c_i}, $i = 1, \ldots, I$, and β_{s_j}, $j = 1, \ldots, J$. The probability assumptions are that the types of customers and the types of

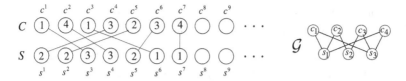

Figure 22.2 FCFS matching of two sequences (left), for the compatibility graph $\mathcal{G}$ (right).

servers are i.i.d. and the two sequences are independent. We will use the notation $\ldots, c^n, c^{n+1}, c^{n+2}, \ldots$, and $\ldots, s^m, s^{m+1}, s^{m+2}, \ldots$, with superscripts to denote the realized types in a sample path of the random i.i.d. system. The *exact definition of FCFS matching* is: s^m is matched to c^n if $(s^m, c^n) \in \mathcal{G}$, and all earlier customers in the sequence compatible with s^m and all earlier servers in the sequence compatible with c^n are already matched.

Throughout this section we will be interested in the complete system that consists of the *two sequences and all the FCFS matches between them*, as well as in the dynamics of establishing the matching from the earliest parts of the sequences onward. We describe three versions of these dynamics:

Matching server by server Starting from s^0, each successive server is matched to the first unmatched compatible customer.

Matching customer by customer Starting from c^0, each successive customer is matched to the first unmatched compatible server.

Matching pair by pair Starting from s^0, c^0 for each successive pair, s^m, c^m, complete all possible FCFS matches up to position m.

It is easy to see by induction that the matching is unique, and reached by either method, and if the number of items of each type is infinite (which it almost surely is for i.i.d. types), then every server and customer will be matched (see Exercise 22.8).

For each of the three matching dynamics, we define a discrete time countable states Markov chain. For server by server matching, $X(n) = (x^1, \ldots, x^L)$ lists the ordered unmatched customers that were skipped when all servers up to s^n have been matched. Similarly, for customer by customer matching, $Y(n) = (y^1, \ldots, y^K)$ lists the ordered servers that were skipped when all customers up to c^n were matched, and $Z(n) = (x^1, \ldots, x^L, y^1, \ldots, y^K)$ lists all the ordered customers and all the ordered servers left unmatched, when all possible FCFS matches up to s^n, c^n were performed. In all three cases, if all the customers and servers up to n were matched, we say that there is a *perfect match* at n and then $X(n) = Y(n) = Z(n) = 0$.

The countable state space consists of finite ordered words from $\mathscr{C}$ and $\mathscr{S}$. We shall see later that ergodicity exists if and only if:

Definition 22.4 (Complete resource pooling)   We say that the system satisfies *complete resource pooling* (CRP) if the following equivalent conditions hold for every subset of customer types $C \neq \mathscr{C}, \emptyset$ and server types $S \neq \mathscr{S}, \emptyset$ (see Exercise 22.9):

$$\alpha_C < \beta_{\mathscr{S}(C)}, \quad \beta_S < \alpha_{\mathscr{C}(S)}, \quad \beta_S > \alpha_{\mathscr{U}(S)}. \tag{22.5}$$

We now present three theorems that completely describe the behavior of this system and demonstrate its simplicity and elegance. We delay discussion of the proofs to the end of the section.

Theorem 22.5 (Uniqueness)   *Assume that complete resource pooling (22.5) holds. Then there is almost surely a unique matching of the two sequences for $-\infty < n < \infty$.*

It is clear that if we start matching from s^{-1}, c^{-1} for $n = -1, 0, 1, \ldots$ the unique matching is different from the matching starting from s^0, c^0. Moving back and starting from $-2, -3$, and so on to $-n$, the matching keeps changing. The theorem states that under the CRP condition this will stabilize almost surely to a unique matching as $n \to -\infty$. Moving back step-by-step is the famous Loynes scheme that we encountered in Section 4.1. The proof of the theorem is along similar lines.

Definition 22.6 (Exchange transformation)   The exchange transformation permutes the two sequences $(s^m)_{-\infty < m < \infty}$, $(c^n)_{-\infty < n < \infty}$ to obtain two new sequences $(\tilde{s}^m)_{-\infty < m < \infty}$, $(\tilde{c}^n)_{-\infty < n < \infty}$, by exchanging positions of matched pairs, so that if s^m, c^n are matched, then $\tilde{c}^m = c^n$, and $\tilde{s}^n = s^m$. This is illustrated in Figure 22.3.

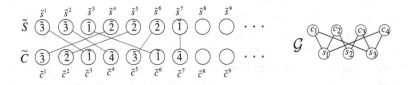

Figure 22.3 The matched sequences of Figure 22.2, after the exchange transformation.

Theorem 22.7 (Time reversal)   *The exchanged sequences are i.i.d. and* *independent of each other, and the unique FCFS matching of the exchanged sequences in reversed time retrieves the matched pairs before the exchange.*

The discovery of this time reversal is in agreement with our observation that the queueing process of Section 22.1 satisfies partial balance equations. It also suggests another approach to obtain stationary distributions.

We define a more detailed process: We now go from the past, and we match customers to servers on FCFS basis, and at every match we exchange the positions of customer and server. We now record at each step both skipped customers and servers as well as matched and exchanged customers and servers. For the server by server matching and exchanging process, we define a process $\mathring{X}(n), n = \ldots, 1, 2, \ldots$ as follows: $\mathring{X}(n)$ is the state after matching and exchanging all servers up to and including n, and $\mathring{X}(n) = (x_1, \ldots, x_L)$, where each x_l is either a skipped customer or a matched and exchanged server. In particular, $x_1 = c^{\underline{N}}$ is the first unmatched customer in position $\underline{N}$, and $x_L = \tilde{s}^{\overline{N}}$ where $\overline{N}$ is latest position of the last matched and exchanged customer, with $L = \overline{N} - \underline{N} + 1$.

Example 22.8 Consider Figure 22.2. The first server $s^1 = s_2$ (recall, this means that the server in position 1 is a server of type s_2) is matched with customer $c^4 = c_3$, and the state is $\mathring{X}(1) = (c_1, c_4, c_1, \tilde{s}_2)$, and it is followed by $\mathring{X}(2) = (c_1, c_4, c_1, \tilde{s}_2, \tilde{s}_2)$; $\mathring{X}(3) = (c_4, c_1, \tilde{s}_2, \tilde{s}_2)$; $\mathring{X}(4) = (c_1, \tilde{s}_2, \tilde{s}_2)$; $\mathring{X}(5) = (c_1, \tilde{s}_2, \tilde{s}_2, \tilde{s}_2)$; $\mathring{X}(6) = 0$; $\mathring{X}(7) = 0$.

Similarly, for matching and exchanging customer by customer, $\mathring{Y}(n) = (y_1, \ldots, y_K)$ where $y_1 = s^{\underline{M}}$ is the first unmatched server and $y_K = \tilde{c}^{\overline{M}}$ is the furthest matched and exchanged customer, with $K = \overline{M} - \underline{M} + 1$, and finally, for pair by pair matching, $\mathring{Z}(n) = (x_1, \ldots, x_L; y_1, \ldots, y_K)$, where $x_1 = c^{\underline{N}}$ is the first unmatched customer, $y_1 = s^{\underline{M}}$ is the first unmatched server, and $L = n - \underline{N} + 1, K = n - \underline{M} + 1$. Note that $\mathring{X}(n) = \mathring{Y}(n) = \mathring{Z}(n) = 0$ for a perfect match at n (see Exercise 22.10).

Theorem 22.9 (Bernoulli stationary distribution) *The stationary distributions of $\mathring{X}, \mathring{Y}, \mathring{Z}$ are given by*

$$\pi_{\mathring{X}}(\mathfrak{s}) = \pi_{\mathring{Y}}(\mathfrak{s}) = \pi_{\mathring{Z}}(\mathfrak{s}) = B \prod_{i=1}^{I} \prod_{j=1}^{J} \alpha_{c_i}^{\#c_i + \#\tilde{c}_i} \beta_{s_j}^{\#s_j + \#\tilde{s}_j}, \tag{22.6}$$

where we count the number of appearances of each type, original or exchanged, in the state $\mathfrak{s}$, and B is the same normalizing constant for all three Markov chains.

This result is of course expected: The unmatched customers are in their original position and got there with probability α_{c_i}, and, by the time-reversal theorem, the matched and exchanged $\tilde{c}_i$ are in their original position in the exchanged sequence, which is also with probability α_{c_i}, by the time-reversal Theorem 22.7. So the stationary distribution appears to be multi-Bernoulli, which seems simple, but this is only half the story, since the actual states that are possible are limited by the FCFS policy. For a characterization of all possible states, see Exercise 22.11.

We next obtain an expression for the normalizing constant, and verify the condition for ergodicity. We suspect that its calculation is #P-hard

Theorem 22.10 *The normalizing constant, obtained by summing over (22.6), is*

$$B = \left(\sum_{S_1,\ldots,S_J \in \mathcal{P}(J)} \prod_{j=1}^{J} \left(\beta_{\{S_1,\ldots,S_j\}} - \alpha_{\mathcal{U}(\{S_1,\ldots,S_j\})} \right)^{-1} \right)^{-1}. \tag{22.7}$$

It follows that complete resource pooling is necessary and sufficient for ergodicity of X, Y, Z.

Proof We note first that all of X, Y, Z as well as $\mathring{X}, \mathring{Y}, \mathring{Z}$ are ergodic or transient together. This is because all of them share the state 0 of perfect matching, which clearly is universally accessible. This also proves that the normalizing constant of all of them is the same, given by $B = \pi(0)$. To calculate B, we consider the parallel skilled service system of Section 22.1 under heavy traffic. In particular, assume that $\lambda_C < \mu_{\mathcal{S}(C)}$ for all non-empty subsets, and let the total arrival rate approach the total service rate. Then in heavy traffic the queue in front of all the servers will grow to infinity, but the servers will stay together, with states $(S_1, n_1, \ldots, S_J, n_J)$, and $\mathbb{P}(n_J > 0) \to 1$ (in fact, $\mathbb{P}(n_J > k) \to 1$ for every k). The jump chain for this Markov chain has the same stationary distribution as the continuous time process, since the time between events is exponential with parameter $\lambda + \mu$ for all states. But the jump chain for the parallel server process when $n_J > 0$ is exactly the process of bipartite FCFS infinite matching. So B is equal to the constant in (22.3) when we eliminate the cases of $i < J$. $\square$

Two important performance measures of the infinite FCFS bipartite matching model are the *matching rates*, r_{s_j,c_i}, the fraction of all the matches that are (s_j, c_i) matches, and the *link lengths*, L_{s_j,c_i}, the random distance between matched pairs (s_j, c_i) given by $n - m$ if $s^m = s_j$ matches with $c^n = c_i$. We present expressions for them with no proof.

Matching rates: for permutation $S_1, \ldots, S_J$ we denote:

$$\alpha_{(k)} = \alpha_{\mathscr{U}(\{S_1, \ldots, S_k\})}, \qquad \beta_{(k)} = \beta_{\{S_1, \ldots, S_k\}} = \beta_{S_1} + \cdots + \beta_{S_k}.$$

Note that if $\mathscr{U}(\{S_1, \ldots, S_k\}) = \emptyset$ then $\alpha_{(k)} = 0$. Further,

$$\phi_k = \frac{\alpha_{\mathscr{U}(\{S_1, \ldots, S_k\}) \cap \{c_i\}}}{\alpha_{\mathscr{U}(\{S_1, \ldots, S_k\})}}, \quad \psi_k = \frac{\alpha_{\mathscr{U}(\{S_1, \ldots, S_k\}) \cap (C(s_j) \setminus \{c_i\})}}{\alpha_{\mathscr{U}(\{S_1, \ldots, S_k\})}}, \quad \chi_k = 1 - \phi_k - \psi_k,$$

where by convention $0/0 = 0$.

The expression for the matching rate is:

$$r_{s_j, c_i} = \beta_{s_j} \sum_{(S_1, \ldots, S_J) \in \mathcal{P}(J)} \pi(S_1, \ldots, S_J) \tag{22.8}$$

$$\left(\sum_{k=1}^{J-1} \phi_k \frac{\alpha_{(k)}}{\beta_{(k)} - \alpha_{(k)} \chi_k} \prod_{l=1}^{k-1} \frac{\beta_{(l)} - \alpha_{(l)}}{\beta_{(l)} - \alpha_{(l)} \chi_l} + \frac{\phi_J}{\phi_J + \psi_J} \prod_{l=1}^{J-1} \frac{\beta_{(l)} - \alpha_{(l)}}{\beta_{(l)} - \alpha_{(l)} \chi_l} \right).$$

Link lengths: The generating functions of the distributions of L_{s_j, c_i} are:

$$\mathbb{E}(Z^{L_{s_j, c_i}}) = \sum_{(S_1, \ldots, S_J) \in \mathcal{P}(J)} \pi(S_1, \ldots, S_J) \sum_{l=1}^{J} \frac{\alpha_{(l)} \phi_l}{\beta_{(l)} - \alpha_{(l)} (\psi_l + \chi_l)}$$

$$\left(\prod_{k=1}^{l-1} \frac{\beta_{(k)} - \alpha_{(k)}}{\beta_{(k)} - \alpha_{(k)} (\psi_k + \chi_k)} \times \prod_{k=1}^{l} \frac{\beta_{(k)} - \alpha_{(k)} (\psi_k + \chi_k)}{\beta_{(k)} - \alpha_{(k)} \psi_k - \alpha_{(k)} \chi_k Z} \right.$$

$$\left. \times \prod_{k=l}^{J} \frac{\beta_{(k)} - \alpha_{(k)}}{1 - \alpha_{(k)} - (1 - \beta_{(k)}) Z^{-1}} \times \frac{1}{Z^{J-l}} \right) \tag{22.9}$$

where:

$$\pi(S_1, \ldots, S_J) = B \prod_{j=1}^{J-1} \left(\beta_{\{S_1, \ldots, S_j\}} - \alpha_{\mathscr{U}(\{S_1, \ldots, S_j\})} \right)^{-1}. \tag{22.10}$$

22.2.1 *Proof of the Uniqueness Theorem*

Assume that the ergodicity conditions (22.5) hold for the independent i.i.d. doubly infinite customer and server sequences. We define $Z^*(n)$, $-\infty < n < \infty$ as the stationary pair by pair FCFS matching process, and we define $Z^{[k]}(n)$, $k \le n < \infty$ as the process starting with $Z^{[k]}(k) = 0$. The proof of the uniqueness consists of the following steps: subadditivity, monotonicity, forward coupling, and backward coupling. We will skip some details left as exercises. Since the proof is of an almost sure result, it is helpful to think of a single sample path of the process, consisting of the sequences of customer and server types.

Proposition 22.11 (Subadditivity)   *Let $A' = (c^1, \ldots, c^m)$, $A'' = (c^{m+1}, \ldots, c^M)$ and $B' = (s^1, \ldots, s^n)$, $B'' = (s^{n+1}, \ldots, s^N)$ and let $A = A' \cup A'' = (c^1, \ldots, c^M)$, $B = B' \cup B'' = (s^1, \ldots, s^N)$. Consider the complete FCFS matching of A', B', of A'', B'', and of A, B and let K', K'', K be the number of unmatched customers and L', L'', L be the number of unmatched servers in these three matchings. Then $K \leq K' + K''$ and $L \leq L' + L''$.*

The proof is left as Exercise 22.13.

Proposition 22.12 (Monotonicity)   *Consider the processes $Z^*(n)$ and $Z^{[0]}(n)$. Let $0 \leq M_0 < M_1 < M_2, \cdots$, be the sequence of perfect match times of $Z^*(n)$, $n \geq 0$, and let N_j, $j = 0, 1, 2, \ldots$ be the number of unmatched customers and servers at $Z^{[0]}(M_j)$. Then $N_1 \geq N_2 \geq \cdots$.*

The proof is left as Exercise 22.14.

Proposition 22.13 (Perfect match probability)   *Consider an incompatible pair (c^0, s^0). Then there exists an h and a sequence $c^1, \ldots, c^h$, $s^1, \ldots, s^h$ with $h < \min(I, J)$, where (s^i, c^i), $i = 1, \ldots, h$ are compatible, such that the FCFS matching of $c^0, c^1, \ldots, c^h, s^0, s^1, \ldots, s^h$ is perfect. The probability of the occurrence of such a sequence is greater or equal to $\delta = \prod_{(s_j, c_i) \in \mathcal{G}} \alpha_{c_i} \beta_{s_j} > 0$.*

The proof is left as Exercise 22.15.

Proposition 22.14 (Forward coupling)   *Let $T = \min\{n : Z^*(n) = Z^{[0]}(n)\}$, be the coupling time of the two processes. Then T is finite and $\mathbb{E}(T) < \infty$.*

Proof   Let M_j, N_j be as in Proposition 22.12. By the ergodicity of Z, $\mathbb{E}(M_{j+1} - M_j) = \kappa < \infty$, and $\mathbb{E}(M_0) = \kappa_0 < \infty$. Clearly, $N_0 \leq M_0$. Consider $Z^{[0]}(M_l)$, and note that the nmber of unmatched customers and unmatched servers is equal. Assume $Z^{[0]}(M_l)$ contains an earliest unmatched pair (c, s). Then, by Proposition 22.13, there is a probability $\geq \delta$ that M_l will be followed by h compatible pairs that will form h perfect matched blocks of Z^* and that will give a perfect matching when combined with (c, s). In that case $N_{l+h} \leq N_l - 1$. So we need no more than a geometrically distributed, with parameter δ, number of perfectly matched blocks of Z^* to reduce the number of unmatched pairs by at least 1, which shows that
$$\mathbb{E}(T) \leq \frac{\kappa_0 \min(I, J) \kappa}{\delta}. \qquad \square$$

Proposition 22.15 (Backward coupling) $\lim_{k \to \infty} Z^{[-k]}(n) = Z^*(n)$ *for all* $-\infty < n < \infty$ *almost surely.*

Proof Define $T_k = \inf\{n \geq k : Z^{[-k]}(n) = Z^*(n)\}$. By the forward coupling Proposition 22.14, it is almost surely finite with finite expectation. Let $\hat{T}_K = \max_{0 \leq k \leq K} T_k$. It is ≥ 0, and is also almost surely finite with finite expectation, for any K. At time $\hat{T}_K$ any process starting empty at a time $-K \leq k \leq 0$ will be merged with Z^* and remain merged thereafter. Define the event $E_K = \{\omega : \forall n \geq 0, Z^{[-n]}(\hat{T}_K) = Z^*(\hat{T}_K)\}$, in words, those ω for which the process starting empty at any time before 0, will merge with Z^* by time $\hat{T}_K$. We claim that $\mathbb{P}(E_K) > 0$. We evaluate $\mathbb{P}(\overline{E}_K)$. For any fixed $n \geq 0$, call $E_{n,K}$ the event that $Z^{[-n]}$ couples with Z^* by time $\hat{T}_K$. We have $E_K = \bigcap_{n \geq 0} E_{n,K} = \bigcap_{n > K} E_{n,K}$ (by definition of $\hat{T}_K$, $E_{n,K}$ is always true for $n \leq K$), so $\overline{E}_K = \bigcup_{n > K} \overline{E_{n,K}}$.

The event $\overline{E_{n,K}}$ will happen if starting at the last time prior to $-K$ at which the process $Z^{[-n]}$ was empty, the next time that it is empty is after time 0. The reason is that otherwise the process $Z^{[-n]}$ reaches state $\emptyset$ at some time $k \in [-K, 0]$, and from that time onward it is coupled with $Z^{[-k]}$, and will couple with Z^* by time $\hat{T}_K$.

Define for $n > K$, $D_n = \{\omega : Z^{[-n]}(m) \neq \emptyset \text{ for all } -n < m \leq 0\}$. Clearly by the above, $\bigcup_{n > K} \overline{E_{n,K}} \subseteq \bigcup_{n > K} D_n$. Let τ denote the recurrence time of the empty state. Then:

$$\mathbb{P}(\overline{E_K}) = \mathbb{P}\left(\bigcup_{n > K} \overline{E_{n,K}}\right) \leq \mathbb{P}\left(\bigcup_{n > K} D_n\right) \leq \sum_{n > K} \mathbb{P}(\tau > n).$$

By the ergodicity $\sum_{l=0}^{\infty} \mathbb{P}(\tau > l) = \mathbb{E}(\tau) < \infty$. Hence, we have that $\mathbb{P}(\overline{E_K}) \to 0$ as $K \to \infty$, and therefore $\mathbb{P}(E_K) > 0$ for large enough K, and $\mathbb{P}(E_K) \to 1$ as $K \to \infty$. Note also that $E_K \subseteq E_{K+1}$.

Define now $\hat{T} = \sup_{k \geq 0} T_k$. We claim that $\hat{T}$ is finite a.s. Consider any ω. Then by $\mathbb{P}(E_K) \to 1$ as $K \to \infty$ and by the monotonicity of E_K, almost surely for this ω there exists a value l such that $\omega \in E_l$. But if $\omega \in E_l$, then $\hat{T}(\omega) \leq \hat{T}_l < \infty$.

So, all processes starting empty before time 0 will couple with Z^* by time $\hat{T}$. By the stationarity of the sequences $(s^n, c^n)_{n \in \mathbb{Z}}$ and of Z^*, we then also have that all processes $Z^{[-n]}$ starting empty before $-k$ will couple with Z^* by time 0, if $k \geq \hat{T}$. Therefore, using the Loynes scheme of starting empty at $-k$ and letting k increase, the process $Z^{[-k]}$ will merge with Z^* once k surpasses $\hat{T}$, and therefore letting $k \to \infty$ the constructed process will merge with Z^* at time 0. But the same argument holds not just for 0 but also for any negative time $-n$. Hence, $Z^{[-k]}$ and Z^* couple at $-n$ (and stay coupled) for any $k > n + \hat{T}$. This completes the proof. □

We have shown that almost surely the Loynes scheme converges to a unique limit, which is the FCFS matching over $\mathbb{Z}$, and this unique version is the realization of the stationary distribution of X, Y, Z.

22.2.2 Proof of the Time-Reversal Theorem

Proposition 22.16 *Assume there is a perfect FCFS matching of* $(s^1, \ldots, s^M)$ *and* $(c^1, \ldots, c^M)$. *Let* $(\tilde{c}^1, \ldots, \tilde{c}^M)$, $(\tilde{s}^1, \ldots, \tilde{s}^M)$ *be the sequences obtained by the exchange transformation, retaining the same links of the matched pairs. The resulting matching of* $(\tilde{c}^1, \ldots, \tilde{c}^M)$, $(\tilde{s}^1, \ldots, \tilde{s}^M)$ *is the unique FCFS matching in reversed time.*

The proof is left as Exercise 22.16.

Proposition 22.17 *The probability of observing a perfect matched block of length M is:*

$$\kappa_M \prod_{i=1}^{I} \alpha_{c_i}^{\#c_i} \prod_{j=1}^{J} \beta_{s_j}^{\#s_j},$$

where $\#s_j$, $\#c_i$ count the number of items of those types in the block, and κ_M is a constant that may depend on M.

Proof The conditional probability is calculated using Bayes formula:

$$\mathbb{P}(\text{seeing } c^{m+1}, \ldots, c^{m+M}, s^{m+1}, \ldots, s^{m+M} \mid \text{having a perfect match})$$
$$= \frac{\mathbb{P}(\text{having a perfect match} \mid \text{seeing } c^{m+1}, \ldots, c^{m+M}, s^{m+1}, \ldots, s^{m+M})}{\mathbb{P}(\text{having a perfect match of length } M)}$$
$$\times \mathbb{P}(\text{seeing } c^{m+1}, \ldots, c^{m+M}, s^{m+1}, \ldots, s^{m+M})$$
$$= \kappa_M \times \mathbb{1}\{c^{m+1}, \ldots, c^{m+M}, s^{m+1}, \ldots, s^{m+M} \text{ is a perfect match}\}$$
$$\times \prod_{i=1}^{I} \alpha_{c_i}^{\#c_i} \prod_{j=1}^{J} \beta_{s_j}^{\#s_j}$$

where $\kappa_M = 1/\mathbb{P}(\text{having a perfect match of length } M)$. $\qquad\square$

Proof of Theorem 22.7 We have a probability measure $\mathfrak{P}$ of the complete process of sequences $(s^n, c^n)_{n \in \mathbb{Z}}$ and the unique matches between them, with sample points $\mathfrak{p}$. We now make the exchange transformation, retain all the links, and consider it in reversed time. We obtain from $\mathfrak{p}$ a new sequence $(\tilde{s}^n, \tilde{c}^n)_{n \in -\mathbb{Z}}$ and links, denoted $\psi \mathfrak{p}$, and denote the probability measure associated with $\psi \mathfrak{p}$ by $\psi \mathfrak{P}$. We need to show that $\mathfrak{P} = \psi \mathfrak{P}$.

Let $\mathfrak{P}^0$ be the Palm measure associated with $\mathfrak{P}$, i.e. it is the probability measure $\mathfrak{P}$ conditional on the event $Z(0) = 0$ of perfect match at 0. We can construct sample paths of $\mathfrak{P}^0$, by constructing a doubly infinite sequence of perfectly matched blocks. We can construct those in the order $0, 1, -1, 2, -2$, etc. We now do the exchange transformation and time reversal on the points of $\mathfrak{P}^0$, to obtain the measure $\psi\mathfrak{P}^0$, which is the Palm version of $\psi\mathfrak{P}$. By Propositions 22.16, 22.17, $\mathfrak{P}^0 = \psi\mathfrak{P}^0$. But in that case, by the uniqueness of the relation of stationary measure and Palm measure, $\mathfrak{P} = \psi\mathfrak{P}$.   □

22.2.3 Proof of the Detailed Stationary Distribution

We will be using Kelly's lemma 8.11, which, as we saw, is helpful in solving balance equations.

Proof of Theorem 22.9   We will prove the theorem for $\mathring{X}(n)$. The proof for $\mathring{Y}(n)$ is analogous. Instead of proving the result for $\mathring{Z}(n)$, we find a representation of $\mathring{Z}(n)$ in terms of $\mathring{X}(n)$, $\mathring{Y}(n)$; see Exercise 22.12.

Consider matching server by server for $(s^m, c^m)_{m \in \mathbb{Z}}$, after all servers up to s^n have been matched and exchanged. At this point, on what used to be the server line, $\tilde{c}^m$ occupy positions $m \le n$, and positions $m > n$ still contain the original servers. On what used to be the customers line, all customers in positions $< \underline{N}$ have been matched and exchanged, while all positions $> \overline{N}$ still contain the original customers. In between, in positions $\underline{N}, \ldots, \overline{N}$, we have some unmatched customers, including $c^{\underline{N}}$, and some matched and exchanged servers, including $\tilde{s}^{\overline{N}}$; $\mathring{X}(n) = (c^{\underline{N}}, \ldots, \tilde{s}^{\overline{N}})$ consists of this section of the sequence. Consider now the entire matched and exchanged sequence, $(\tilde{s}^m, \tilde{c}^m)_{m \in \mathbb{Z}}$, and then perform all the matching and exchanging back to original values, matching exchanged customer by exchanged customer, in reversed time, until all exchanged customers in positions $\ge n + 1$ have been matched and exchanged back to servers. The result is exactly the same as when we matched servers in the original sequences up to n. Hence, the state $\mathring{X}(n)$ is also the state of $\mathring{Y}(n + 1)$ in the reversed and exchanged sequence, at time $n + 1$. We will use the notation $\overleftrightarrow{}$ to denote a reversed and exchanged state.

By Kelly's lemma, all we need to do to prove the theorem is to verify for any two states u, u' that:

$$\pi_{\mathring{X}}(u)\mathbb{P}(\mathring{X}(n + 1) = u' \mid \mathring{X}(n) = u) = \pi_{\mathring{X}}(u')\mathbb{P}(\mathring{X}(n) = u \mid \mathring{X}(n + 1) = u')$$

$$= \pi_{\mathring{Y}}(\overleftrightarrow{u'})\mathbb{P}(\mathring{Y}(m + 1) = \overleftrightarrow{u} \mid \mathring{Y}(m) = \overleftrightarrow{u'}). \quad (22.11)$$

There are three types of transitions of $\mathring{X}$, from n to $n + 1$, and they are also the transitions of $\mathring{Y}$ for the exchanged sequence, in reversed time, from

$n + 1$ to n. Figure 22.4 illustrates the initial state u, and the three possible transitions to u': In the initial state (top of the figure) clear circles are for customers and shaded circles for servers. Below are the transitions between n and $n + 1$. In the first (simple exchange), server s^{n+1} is matched to a customer c^m where $\underline{N} < m < \overline{N}$. In the second (deletion of elements), server s^{n+1} is matched to $c^{\underline{N}}$. In the third (addition of elements), server s^{n+1} is matched to a c^m where $m > \overline{N}$.

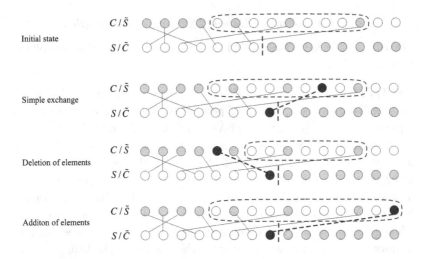

Figure 22.4 Transitions of server by server matching $\mathring{X}(n)$.

We check (22.11) for each type of transition, by substituting π as given in (22.6), and the transition rates.

(a) *Simple exchange:* If $u^l = c_i$ and $u'^l = s_j$, then according to (22.6) we have

$$\pi_{\mathring{Y}}(\overleftarrow{u'}) = \pi_{\mathring{X}}(u') = \pi_{\mathring{X}}(u) \frac{\beta_{s_j}}{\alpha_{c_i}},$$

and the direct and reversed transition rates are

$$P(\mathring{X}(n+1) = u' | \mathring{X}(n) = u) = \beta_{s_j},$$
$$P(\mathring{Y}(m+1) = \overleftarrow{\overline{u}} | \mathring{Y}(m) = \overleftarrow{u'}) = \alpha_{c_i}.$$

(b) *Deletion of elements from start of u:* If $u = (u^1, \ldots, u^L)$, and $u' = (u^{k+1}, \ldots, u^L)$, and the deleted part is $(c_i, s_{j_1}, \ldots, s_{j_{k-1}})$, then according to (22.6) we have

$$\pi_{\overleftarrow{Y}}(\overleftarrow{u'}) = \pi_{\overset{\circ}{X}}(u') = \pi_{\overset{\circ}{X}}(u)\frac{1}{\alpha_{c_i}\beta_{s_{j_1}}\cdots\beta_{s_{j_{k-1}}}},$$

and the direct and reversed transition rates are:

$$\mathbb{P}(\overset{\circ}{X}(n+1) = u'|\overset{\circ}{X}(n) = u) = \beta_{\mathscr{S}(c_i)},$$

$$\mathbb{P}(\overset{\circ}{Y}(m+1) = \overline{\tilde{u}}|\overset{\circ}{Y}(m) = \overleftarrow{u'}) = \alpha_{c_i}\beta_{s_{j_1}}\cdots\beta_{s_{j_{k-1}}}\beta_{\mathscr{S}(c_i)}.$$

(c) *Addition of elements to u:* If $u = (u^1,\ldots,u^l)$, and $u' = (u^1,\ldots,u^l,c_{i_1}, \ldots,c_{i_k},s_j)$, then according to (22.6) we have

$$\pi_{\overleftarrow{Y}}(\overleftarrow{u'}) = \pi_{\overset{\circ}{X}}(u') = \pi_{\overset{\circ}{X}}(u)\alpha_{c_{i_1}}\cdots\alpha_{c_{i_k}}\beta_{s_j}$$

and the direct and reversed transition rates are:

$$\mathbb{P}(\overset{\circ}{X}(n+1) = u'|\overset{\circ}{X}(n) = u) = \beta_{s_j}\alpha_{c_{i_1}}\cdots\alpha_{c_{i_k}}\alpha_{\mathscr{C}(s_j)},$$

$$\mathbb{P}(\overset{\circ}{Y}(m+1) = \overline{\tilde{u}}|\overset{\circ}{Y}(m) = \overleftarrow{u'}) = \alpha_{\mathscr{C}(s_j)}.$$

It is now immediate to check that the balance condition of (22.11) holds. □

22.3 A FCFS Ride-Sharing Model

Many modern service systems do not have a fixed set of servers; instead, there is a stream of arriving customers and a stream of arriving servers. In theory we can think of an infinite population of customers that require service from time to time, as well as an infinite population of servers that become available from time to time. Assuming that items in these populations are independent, we will have customers of types $c_i \in \mathscr{C}$, arriving in independent Poisson streams of rates λ_{c_i}, with total λ, and servers of types $s_j \in \mathscr{S}$, arriving in independent Poisson streams of rates μ_{s_j}, with total μ. This model may describe Uber passengers and drivers, buyers and sellers on Craigslist, etc. We assume a bipartite compatibility graph $\mathcal{G}$.

We will assume that customers wait for service, while an arriving server does not wait: He will on arrival match with a compatible customer and the two depart immediately or else he will depart without a customer. We again assume FCFS: An arriving server will match with the longest waiting compatible customer.

We can, as in Section 22.1, define a continuous time Markov chain to describe the state of the system, but instead we will consider, similar to

Section 22.2, the discrete time Markov chain of the jumps at times of events, i.e. arrivals of customers or servers. Since the rate of these events is constant $\lambda + \mu$, the stationary distribution of the discrete process and the continuous time process is the same.

The advantage of looking at the discrete time Markov chain is that it is still the correct model for a ride-sharing system if we discard the Poisson assumption: All we need is that the arrival processes of customers and of servers are independent of each other, and that the arriving types are i.i.d. Arrival times may be non-renewal, and in particular they may have time-varying arrival rates.

To study the process, we now consider a single infinite sequence of customers and servers ordered by order of arrival, and classified according to customer types and server types. Let $\alpha_{c_i} = \lambda_{c_i}/\lambda$, $\beta_{s_j} = \mu_{s_j}/\mu$, $\rho = \lambda/\mu$. Then items in the sequence are i.i.d., with customers occurring with probability $\frac{\rho}{1+\rho}$, and a customer is of type c_i with probability α_{c_i}, and servers occurring with probability $\frac{1}{1+\rho}$, and a server is of type s_j with probability β_{s_j}. In this sequence, by FCFS matching, each server is matched to the earliest unmatched compatible customer that occurs earlier in the sequence, or it is left unmatched. This is illustrated in Figure 22.5, where server s_2 matches with c_1, server s_3 finds no match, and server s_1 matches with customer c_1, and at that point c_2, c_3 are still unmatched and waiting.

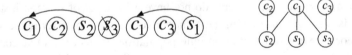

Figure 22.5 Ride-share matching.

The following definitions and theorems are analogous to the bipartite FCFS matching model of Section 22.2. For the process of matching all customers and servers up to n, we define the Markov chain $X(n) = (x^1, \ldots, x^L)$, where $(x^1, \ldots, x^L)$ are the ordered unmatched customers.

Theorem 22.18 *If $X(n)$ is ergodic, then there is a unique FCFS matching of servers to earlier customers over $n \in \mathbb{Z}$.*

Theorem 22.19 *Define the exchange transformation in which we switch the positions of each matched pair, with unmatched servers exchanged by themselves. The resulting sequence is i.i.d. and the FCFS matching in reversed time retrieves the same matched pairs.*

Theorem 22.20 Let $\mathring{X}(n) = (\mathring{x}^1, \ldots, \mathring{x}^K)$ include all matched and exchanged customers and servers, including also unmatched servers up to position n. Then the stationary distribution of $\mathring{X}(n)$ is

$$\pi_{\mathring{X}}(\mathring{x}^1, \ldots, \mathring{x}^K) = B \prod_{c_i \in \mathscr{C}} \left(\frac{\lambda_{c_i}}{\lambda + \mu}\right)^{\sharp c_i + \sharp \bar{c}_i} \prod_{s_j \in \mathscr{S}} \left(\frac{\mu_{s_j}}{\lambda + \mu}\right)^{\sharp s_j + \sharp \bar{s}_j}, \quad (22.12)$$

where the normalizing constant B is equal to the probability that all the customers are matched, and is given by equation (22.17)

The proofs are similar to those of Section 22.2 and are left as Exercises 22.17–22.20.

It is perhaps more useful to obtain the stationary distribution of $X(t)$:

Theorem 22.21 The process $X(t)$ is ergodic if and only if for every subset of customer types

$$\lambda_C < \mu_{\mathscr{S}(C)} \quad (22.13)$$

and the stationary distribution of $X(t)$ is given by

$$\pi_X(x^1, \ldots, x^L) = B \prod_{j=1}^{L} \frac{\lambda_{x^j}}{\mu_{\mathscr{S}(x^1, \ldots, x^j)}}. \quad (22.14)$$

In analogy with the skilled parallel service system of Section 22.1, we can define the Markov chain $Y(n)$ that for $X(n) = (c^1, \ldots, c^L)$ lists the types of customers that appear in $(c^1, \ldots, c^L)$ in the order in which they occur, and the number of additional waiting customers in between, i.e. $Y(n) = (C_1, n_1, \ldots, C_k, n_k)$, with stationary distribution

$$\pi_Y(C_1, n_1, \ldots, C_k, n_k) = B \prod_{\ell=1}^{k} \frac{\lambda_{C_\ell}}{\mu_{\mathscr{S}(C_1, \ldots, C_\ell)}} \left(\frac{\lambda_{C_1, \ldots, C_\ell}}{\mu_{\mathscr{S}(C_1, \ldots, C_\ell)}}\right)^{n_\ell}. \quad (22.15)$$

Summing over $n_j = 0, 1, \ldots$, we also obtain

$$\pi_Y(C_1, \cdot, \ldots, C_k, \cdot) = B \prod_{\ell=1}^{k} \frac{\lambda_{C_\ell}}{\mu_{\mathscr{S}(C_1, \ldots, C_\ell)} - \lambda_{C_1, \ldots, C_\ell}}, \quad (22.16)$$

and we can then obtain the normalizing constant:

$$B = \left(\sum_{k=0}^{I} \sum_{C \subseteq \mathscr{C}, |C|=k} \sum_{(C_1, \ldots, C_k) \in P(C)} \prod_{\ell=1}^{k} \frac{\lambda_{C_\ell}}{\mu_{\mathscr{S}(\{C_1, \ldots, C_\ell\})} - \lambda_{\{C_1, \ldots, C_\ell\}}}\right)^{-1}. \quad (22.17)$$

One can use equation (22.16) to obtain expressions analogous to (22.8) for matching rates, and for the fraction of servers of each type that is not utilized.

Also, analogous to (22.9) one can obtain expressions for link lengths. If one can assume stationary Poisson arrivals, then the expressions for link lengths can be used to obtain expressions for the Laplace transform of the waiting time distributions for each type of customer, and from those one can obtain explicit formulas for the mean and variance of waiting times.

22.4 A Design Heuristic for General Parallel Skill-Based Service

In this section we describe a heuristic for the design of a high-volume service center. The design starts with the classification of customers by types $\mathscr{C} = \{c_1, \ldots, c_I\}$. Customers arrive at rate λ and we assume that their types are i.i.d. with probabilities α_{c_i}. We make no further assumptions on the arrival process. We will denote by n the total number of servers to accommodate λ. At the service side of the system the design specifies the types and skills of the servers, where skills are neither universal nor exclusive. The designer needs to determine server types $\mathscr{S} = \{s_1, \ldots, s_J\}$, a compatibility graph $\mathcal{G}$, and for each link in $\mathcal{G}$ a service rate μ_{s_j, c_i}. We make no further assumptions on processing time distribution, except to say that they are all independent.

Note that service rates may depend both on the customer and the server type, which may incur inefficiency. A special case may be when service rates are of product form, i.e. $\mu_{s_j, c_i} = \mu_{s_j} \eta_{c_i}$. The interpretation of this is that $1/\eta_{c_i}$ is the average work required by a customer of type c_i, and μ_{s_j} is the speed at which server type s_j works. With product-form processing rates there will be no loss of efficiency, and the throughput (rate of departure of customers) when all servers are busy will be independent of the policy. In practice a designer may try to only include links for which $\mu_{s_j, c_i} \approx \mu_{s_j} \eta_{c_i}$; on the other hand, the various costs for different (s_j, c_i) may also influence the design.

Our design will use FCFS-ALIS policy throughout, and our decision will be how many servers of each type are needed to serve the customers. Our aim is that under FCFS-ALIS there will be resource pooling and all customers will receive the same level of service. We will consider three operational regimes: efficiency driven (ED), quality driven (QD), and efficiency as well as quality driven (QED). In ED regime we have all the servers busy at all times (in the limit as λ and n are large), and customers abandon when their patience runs out. For that we assume we know the patience distributions, given by H_{c_i}. In QD regime customers never wait, and under ALIS there is alway a queue of idle servers. In QED regime we are in Halfin–Whitt heavy traffic, with very little idleness and very little waiting, and there are hardly any abandonments.

Our heuristic is based on the conjecture that with large λ and many servers, the matching of customers and servers will occur as in the infinite bipartite matching model. In that case, we will be able to calculate matching rates (using equation (22.8)), and use those to obtain the desired design.

The reason we believe this conjecture may be true is that types of arrivals are i.i.d. and this will be preserved at least approximately if there are abandonments under FCFS-ALIS. At the same time, if the system is stationary, then each server will follow a stationary point process of starting jobs (of various types, and with zero or positive idle periods in between), and with many servers these stationary processes will be almost independent. But the superposition of many stationary independent point processes tends after scaling to a Poisson process. So in the limit this may approach the matching process of the bipartite model of Section 22.2. Currently, no conditions are known for this conjecture to hold, (except for exponential server dependent service times and Poisson arrivals), but similar results for other models exist. Simulation results support our conjecture.

The number of servers of each type necessary to achieve pooling and the same level of service for all customers is not uniquely determined, and we need to decide how to divide the total amount of service between the various server types. We will make the design decision that a fraction β_{s_j} of all the services will be supplied by servers of type s_j. We now describe our heuristic for the three regimes.

Efficiency driven regime:
We have specified $\lambda, \alpha_{c_i}, \beta_{s_j}$ and H_{c_i}. We now decide on *quality of service requirement*, which is that all customers that do not abandon will wait an amount of time $\approx W$.

(i) Abandonment: A wait of W will entail the fraction of abandonments $\theta_{c_i} = H_{c_i}(W)$.

(ii) Modified α: The fraction of customers of type c_i that will actually be served is $\tilde{\alpha}_{c_i} = \alpha_{c_i}(1 - \theta_{c_i})/\sum_{c_k \in \mathcal{C}} \alpha_{c_k}(1 - \theta_{c_k})$.

(iii) Verify complete resource pooling: Use (22.5) to verify, and change β_{s_j} if it fails.

(iv) Calculate matching rates: Use $\mathcal{G}, \tilde{\alpha}, \beta$ to calculate matching rates r_{s_j,c_i} from (22.8).

(v) Determine required numbers of servers:

$$n_{s_j} = \lambda \sum_{c_i \in \mathcal{C}(s_j)} \frac{r_{s_j,c_i}}{\mu_{s_j,c_i}}.$$

Quality driven regime:
We have specified $\lambda, \alpha_{c_i}, \beta_{s_j}$. We now decide on T, the amount of idleness that each server will experience following each service completion. T will be the approximate time that each server will require to spend in the queue of idle servers, under ALIS.

(i) Verify complete resource pooling: Use (22.5) to verify, and change β_{s_j} if it fails.

(ii) Calculate matching rates: Use $\mathcal{G}, \alpha, \beta$ to calculate matching rates r_{s_j,c_i} from (22.8).

(iii) Modify processing rates: $\tilde{\mu}_{s_j,c_i} = (T + \mu_{s_j,c_i}^{-1})^{-1}$.

(iv) Determine required servers:

$$n_{s_j} = \lambda \sum_{c_i \in \mathscr{C}(s_j)} \frac{r_{s_j,c_i}}{\tilde{\mu}_{s_j,c_i}}.$$

Quality and efficiency driven regime:
Same as above, with $W = T = 0$.

In QED regime, we can add or subtract multiples of $\sqrt{n_{s_j}}$ to control the fraction of customers receiving immediate service.

We can also design differentiated service, by reducing the number of servers for a subset of customers that have lower priority, so that resource pooling no longer holds. Under FCFS-ALIS, the customers of lower priority will experience longer waits. We can even design a system where some customer types are served in QD regime, some in QED regime, and some in ED regime when we use FCFS-ALIS, by designing the right server pool sizes, which automatically will decompose the bipartite graph.

Clearly this is a heuristic, which may or may not be effective under varying conditions. It has, however, been confirmed by extensive simulation studies, and simulation shows that it is useful also for moderate λ and n. See Exercise 22.21 for an example.

22.5 Queue and Idleness Ratio Routing

In this section we describe another approach to the design of a high-volume parallel skill-based service system, which operates in the Halfin–Whitt regime. Instead of using FCFS policy, it dynamically uses the current state of the queues of waiting customers and of the pools of idle servers to determine its actions. Let Q_{c_i} and I_{s_j} be the number of waiting customers and of idle servers, with Q_Σ, I_Σ their sums. The *queue and idleness ratio policy* (QIR) attempts to keep them in constant ratios $Q_{c_i} = p_{c_i} Q_\Sigma$ and

$\mathcal{I}_{s_j} = r_{s_j}\mathcal{I}_\Sigma$, by choosing for action customer or server of types furthest away from those ratios. We shall make this more precise next.

While the heuristic of Section 22.4 attempted to control the fluid level asymptotics, the current policy is aimed at controlling the diffusion level asymptotics. As a result, it achieves closer adherence to the desired performance, though this is at the cost of being more dependent on the accuracy of the assumptions.

We now describe the design. We let the total arrival rate λ be the scale of the system, which will tend to infinity. Customers of type c_i arrive at rate λ_{c_i}, and there are n_{s_j} servers of type s_j, where we have

$$\frac{\lambda_{c_i} - \alpha_{c_i}\lambda}{\sqrt{\lambda}} \to \xi_{c_i}, \qquad \frac{n_{s_j} - \beta_{s_j}\lambda}{\sqrt{\lambda}} \to \zeta_{s_j},$$

and service rates are μ_{s_j,c_i} for $(s_j, c_i) \in \mathcal{G}$. As before, we assume Poisson arrivals and exponential services. We also assume that customers have exponential patience, and abandon at rate $\theta_{c_i} \geq 0$.

The quantities α_{c_i}, β_{s_j} need to be chosen carefully so that the system will be able to operate in Halfin–Whitt regime: We formulate a static planning problem for the system, with unknowns ρ, η_{s_j,c_i}:

$$\min \quad \rho$$

$$\text{s.t.} \quad \sum_{s_j \in \mathcal{S}(c_i)} \mu_{s_j,c_i}\beta_{s_j}\eta_{s_j,c_i} = \alpha_{c_i}, \quad c_i \in \mathcal{C},$$

$$\sum_{c_i \in \mathcal{C}(s_j)} \eta_{s_j,c_i} \leq \rho, \quad s_j \in \mathcal{S},$$

$$\rho, \eta_{s_j,c_i} \geq 0.$$

It is necessary to choose β_{s_j} so that this LP will have optimal value $\rho = 1$, and in addition, for every optimal solution all the inequalities need to be tight. At that point, we need to choose exactly one of those optimal solutions, say η_{s_j,c_i}, and erase all the links in the solution for which $\eta_{s_j,c_i} = 0$. It is also necessary that this solution will remain connected.

We note that every basic solution to such an LP is always either a tree or a union of disjoint trees. Then our choice could be either a basic solution that is a connected tree, or a convex combinations of basic solutions that is connected. With the chosen $\eta_{s_j,c_i} > 0$, we will have matching rates $r_{s_j,c_i} = \mu_{s_j,c_i}\beta_{s_j}\eta_{s_j,c_i}$.

In addition to $Q_{c_i}(t)$, $\mathcal{I}_{s_j}(t)$, let $\mathcal{Z}_{s_j,c_i}(t)$ be the number of customers of type c_i currently being served by servers of type s_j, and let

$$X_{c_i}(t) = Q_{c_i}(t) + \sum_{s_j \in \mathcal{S}(c_i)} Z_{s_j,c_i}(t), \qquad X_\Sigma(t) = \sum_{c_i \in \mathcal{C}} X_{c_i}(t),$$

be the total number of type c_i customers in the system, and the overall total number of customers in the system.

We now introduce a centering and scaling that depends on the chosen solution of the LP,

$$\hat{X}_\Sigma^\lambda(t) = \frac{X_\Sigma^\lambda(t) - n_\Sigma^\lambda}{\sqrt{\lambda}}, \qquad \hat{I}_\Sigma^\lambda(t) = \frac{I_\Sigma^\lambda(t)}{\sqrt{\lambda}},$$

$$\hat{X}_{c_i}^\lambda(t) = \frac{X_{c_i}^\lambda(t) - \sum_{s_j \in \mathcal{S}(c_i)} \eta_{s_j,c_i} n_{s_j}^\lambda}{\sqrt{\lambda}},$$

$$\hat{Q}_{c_i}^\lambda(t) = \frac{Q_{c_i}^\lambda(t)}{\sqrt{\lambda}}, \quad \hat{I}_{s_j}^\lambda(t) = \frac{I_{s_j}^\lambda(t)}{\sqrt{\lambda}}, \quad \hat{Z}_{s_j,c_i}^\lambda(t) = \frac{Z_{s_j,c_i}^\lambda(t) - \eta_{s_j,c_i} n_{s_j}^\lambda}{\sqrt{\lambda}}.$$

Note that with this choice of $\alpha_{c_i}, \beta_{s_j}$ and the solution of the LP, if matching rates are kept as r_{s_j,c_i} the system will be in heavy traffic with $1 - \rho \approx O(1/\sqrt{\lambda})$, i.e. it should be in the Halfin–Whitt regime. In that case, we should have $X_\Sigma^\lambda(t) = O(\lambda)$, while $Q_\Sigma^\lambda(t)$ and $I_\Sigma^\lambda(t)$ are of order $O(\sqrt{\lambda})$. Hence $\left[\hat{X}_\Sigma^\lambda(t) \right]^+$ is approximately the total number of waiting customers, and $\left[\hat{X}_\Sigma^\lambda(t) \right]^-$ is approximately the total number of idle servers (it is not exactly that since we can have some waiting customers and at the same time some idle servers that are incompatible).

We now define our policy:

Definition 22.22 The queue and idleness ratio policy (QIR) is defined with weight vectors p, r (positive, summing to 1), as follows:

(i) Upon arrival of a c_i customer, choose from the compatible idle servers one that is of type

$$s_j^* = s_j^*(t) \in \arg\max_{s_j \in \mathcal{S}(c_i),\, I_{s_j}(t) > 0} \left[\hat{I}_{s_j}^\lambda(t) - r_{s_j} \left[\hat{X}_\Sigma^\lambda(t) \right]^- \right].$$

(ii) Upon completion of a service by a server of type s_j, choose from the compatible waiting customers one that is of type

$$c_i^* = c_i^*(t) \in \arg\max_{c_i \in \mathcal{C}(s_j),\, Q_{c_i}(t) > 0} \left[\hat{Q}_{c_i}^\lambda(t) - p_{c_i} \left[\hat{X}_\Sigma^\lambda(t) \right]^+ \right].$$

We then have

Theorem 22.23 *Under some restrictive conditions, as $\lambda \to \infty$, under QIR policy for $t > 0$:*

$$\hat{Q}^{\lambda}_{c_i}(t) - p_{c_i}\hat{Q}^{\lambda}_{\Sigma}(t) \to_w 0, \qquad \hat{I}^{\lambda}_{s_j}(t) - r_{s_j}\hat{I}^{\lambda}_{\Sigma}(t) \to_w 0.$$

The theorem holds under one of the following conditions:
- The service rates are server dependent: $\mu_{s_j,c_i} = \mu_{s_j}$.
- The service rates are customer dependent: $\mu_{s_j,c_i} = \mu_{c_i}$.
- The bipartite graph $\mathcal{G}$ is a tree.

This theorem establishes state space collapse: The multivariate processes of queues and idle pools become constant multiples of the single process $\hat{X}^{\lambda}_{\Sigma}(t)$. Furthermore, one can derive diffusion limits of $\hat{X}^{\lambda}_{\Sigma}(t)$ under each of the three conditions.

22.6 Sources

The description of the state for parallel service systems in Figure 22.1 is suggested in Visschers (2000). The results on parallel skill-based service under FCFS-ALIS are derived in Adan and Weiss (2014), and earlier related work is Visschers et al. (2012) and Adan et al. (2010). The model of infinite bipartite matching was proposed in Caldentey et al. (2009) and analyzed in Adan and Weiss (2012a) and Adan et al. (2018b). More general matching models, for general compatibility graphs are analyzed in Moyal et al. (2021), with further results in Mairesse and Moyal (2016) and Moyal and Perry (2017). The ride-sharing model is introduced in Adan et al. (2018a) and the performance measures for this model are obtained in Weiss (2020). The ride-sharing model is related to redundancy service policies introduced in Gardner et al. (2016), and generalized in Ayesta et al. (2019). For a survey on related product-form parallel service models, see Gardner and Righter (2020). The design heuristic with simulation results is in Adan et al. (2019). The queue and idleness ratio policy is derived and analyzed in Gurvich and Whitt (2009, 2010). Bell and Williams (2001) derive a dynamic asymptotically optimal policy for the N-system, and extend this to general parallel server dynamic matching in Bell and Williams (2005).

Exercises

22.1 For state $\mathfrak{s} = (S_1, n_1, \ldots, S_i, n_i, S_{i+1}, \ldots, S_J)$ of the PSBS system under FCFS-ALIS, write all the transitions out of state $\mathfrak{s}$ and their transition rates [Adan and Weiss (2014); Visschers et al. (2012)].

22.2 (continued) For state $s = (S_1, n_1, \ldots, S_i, n_i, S_{i+1}, \ldots, S_J)$, write all the possible transitions into this state, and find their transition rates.

22.3 (continued) Write down the partial balance equations for the four types of balanced transitions.

22.4 (continued) Verify that the proposed stationary distribution (22.2) satisfies the four partial balance equations.

22.5 (continued) Calculate the normalizing constant B given by (22.3).

22.6 Use the distributional form of Little's law to prove Theorem 22.3 [Visschers et al. (2012)].

22.7 Use equation (22.4) to obtain expressions for the first and second moment of the waiting time.

22.8 For the infinite bipartite matching model, show that matching of $s^1, s^2, \ldots,$ and $c^1, c^2, \ldots$ is unique, and if each type occurs infinitely often, it matches all customers and servers [Adan and Weiss (2012a)].

22.9 Show that the three conditions of the CRP definition (22.5) are equivalent.

22.10 Write down the first seven states of the processes $\mathring{Y}(n)$ and $\mathring{Z}(n)$ for the matching in Figure 22.2.

22.11 For the infinite bipartite matching model, show that $\mathring{X}(n) = (x_1, \ldots, x_L)$ is a possible state of the process $\mathring{X}$ if and only if: for any $1 \leq k < l \leq L$, if $x_k = c_i$ and $x_l = \tilde{s}_j$, then $(s_j, c_i) \notin \mathcal{G}$ with a similar characterization for possible states of $\mathring{Y}$ and $\mathring{Z}$ [Adan et al. (2018b)].

22.12 Show how to construct $\mathring{Z}(n)$ from $\mathring{X}(n)$ and $\mathring{Y}(n)$, and show that if $\mathring{Z}(n) = ((x_1, \ldots, x_L), (y_1, \ldots, y_K))$, then $(x_1, \ldots, x_L, y_K, \ldots, y_1)$ is a possible state of $\mathring{X}(n)$.

22.13 Prove the subadditivity property of FCFS matching, Proposition 22.11.

22.14 Prove the monotonicity result of Proposition 22.12.

22.15 For an incompatible pair (c^0, s^0), construct $(c^0, c^1, \ldots, c^h, s^0, s^1, \ldots, s^h)$ that are perfectly matched by FCFS, and find a lower bound for the probability of such a sequence, to prove Proposition 22.13.

22.16 Show that the exchange transformation on a perfectly matched block will retain the same links if we now do FCFS in reversed time, as stated in Proposition (22.16).

22.17 Prove the uniqueness theorem, 22.18, for the ride-sharing model [Adan et al. (2018a)].

22.18 Prove the time-reversal theorem, 22.19, for the ride-sharing model.

22.19 Verify the Bernoulli type stationary distributions, Theorem 22.20, for the ride-sharing model [Weiss (2020)].

22.20 Prove the ergodicity condition of Theorem 22.21 and verify equations (22.14)–(22.17) [Weiss (2020)].

22.21 The following table has data for a system with three types of customers and three types of servers. Calculate matching rates, and provide designs for ED with $W = 1$, for QD with $T = 0.5$, and for QED, for $\lambda = 20, 50, 100, 200$ and simulate the systems to evaluate the performance [Adan et al. (2019)].

Example – System and Data

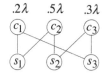

Patience distributions		Service time distributions			
H_{c_i}		G_{s_j, c_i}	c_1	c_2	c_3
c_1	Exp(0.1)	s_1	Pareto(2, 3)	Exp(0.125)	
c_2	U(0,10)	s_2		Exp(0.2)	U(2, 6)
c_3	Exp(0.2)	s_3	Pareto(3, 3)		U(1, 5)

Only the *mean* service times are used by the design algorithms. The full distributions are used in the simulations.

Resource allocation design parameters are: $\beta_{s_1} = 0.3$, $\beta_{s_2} = 0.3$, $\beta_{s_3} = 0.4$.

References

Adan, I., and Weiss, G. 2005. A two-node Jackson network with infinite supply of work. *Probability in the Engineering and Informational Sciences*, **19**(02), 191–212.

Adan, I., and Weiss, G. 2006. Analysis of a simple Markovian re-entrant line with infinite supply of work under the LBFS policy. *Queueing Systems*, **54**(3), 169–183.

Adan, I., and Weiss, G. 2012a. Exact FCFS matching rates for two infinite multitype sequences. *Operations Research*, **60**(2), 475–489.

Adan, I., and Weiss, G. 2012b. A loss system with skill-based servers under assign to longest idle server policy. *Probability in the Engineering and Informational Sciences*, **26**(3), 307–321.

Adan, I., and Weiss, G. 2014. A skill based parallel service system under FCFS-ALIS – steady state, overloads, and abandonments. *Stochastic Systems*, **4**(1), 250–299.

Adan, I., Wessels, J., and Zijm, H. 1990. Analysis of the symmetric shortest queue problem. *Stochastic Models*, **6**, 691–713.

Adan, I., Wessels, J., and Zijm, H. 1991. Analysis of the asymmetric shortest queue problem with threshold jockeying. *Communications in Statistics. Stochastic Models*, **7**(4), 615–627.

Adan, I., Wessels, J., and Zijm, H. 1993. A compensation approach for two-dimensional Markov processes. *Advances in Applied Probability*, 783–817.

Adan, I., VanHoutum, G. J., and Vander Wal, J. 1994. Upper and lower bounds for the waiting time in the symmetric shortest queue system. *Annals of Operations Research*, **48**(2), 197–217.

Adan, I., Hurkens, C., and Weiss, G. 2010. A reversible Erlang loss system with multitype customers and multitype servers. *Probability in the Engineering and Informational Sciences*, **24**(4), 535–548.

Adan, I., Kleiner, I., Righter, R., and Weiss, G. 2018a. FCFS parallel service systems and matching models. *Performance Evaluation*, **127**, 253–272.

Adan, I., Busic, A., Mairesse, J., and Weiss, G. 2018b. Reversibility and further properties of FCFS infinite bipartite matching. *Mathematics of Operations Research*, **43**(2), 598–621.

Adan, I., Boon, M., and Weiss, G. 2019. Design heuristic for parallel many server systems. *European Journal of Operational Research*, **273**(1), 259–277.

Adan, I., Foss, S., Shneer, S., and Weiss, G. 2020. Local stability in a transient Markov chain. *Statistics and Probability Letters*, **165**, 108855.

Aghajani, R., and Ramanan, K. 2019. The hydrodynamic limit of a randomized load balancing network. *Annals of Applied Probability*, **29**(4), 2114–2174.

Aghajani, R., Li, X., and Ramanan, K. 2015. *Mean-field dynamics of load-balancing networks with general service distributions*. arXiv:1512.05056.

Al Hanbali, A., Mandjes, M., Nazarathy, Y., and Whitt, W. 2011. The asymptotic variance of departures in critically loaded queues. *Advances in Applied Probability*, **43**(1), 243–263.

Asmussen, S. 2003. *Applied Probability and Queues*. Springer-Verlag.

Asmussen, S., and Glynn, P.W. 2007. *Stochastic Simulation: Algorithms and Analysis*. Vol. 57. Springer Science & Business Media.

Ayesta, U., Bodas, T., Dorsman, J.L., and Verloop, I.M. 2019. *A token-based central queue with order-independent service rates*. arXiv:1902.02137.

Baccelli, F., and Brémaud, P. 2013. *Elements of Queueing Theory: Palm Martingale Calculus and Stochastic Recurrences*. Vol. 26. Springer Science & Business Media.

Baccelli, F., and Foss, S. 1994. Ergodicity of Jackson-type queueing networks. *Queueing Systems*, **17**(1-2), 5–72.

Baccelli, F., and Hebuterne, G. 1981. *On Queues with Impatient Customers*. Tech. rept. RR-0094 inria-00076467. INRIA Rapports de Recherthe.

Baccelli, F., and Massey, W.A. 1989. A sample path analysis of the M/M/1 queue. *Journal of Applied Probability*, **26**(2), 418–422.

Baskett, F., Chandy, K.M., Muntz, R.R., and Palacios, F.G. 1975. Open, closed, and mixed networks of queues with different classes of customers. *Journal of the ACM (JACM)*, **22**(2), 248–260.

Bell, S., and Williams, R.J. 2005. Dynamic scheduling of a parallel server system in heavy traffic with complete resource pooling: Asymptotic optimality of a threshold policy. *Electronic Journal of Probability*, **10**, 1044–1115.

Bell, S.L., and Williams, R.J. 2001. Dynamic scheduling of a system with two parallel servers in heavy traffic with resource pooling: asymptotic optimality of a threshold policy. *Annals of Applied Probability*, **11**(3), 608–649.

Bellman, R.E. 1957. *Dynamic Programming*. Princeton University Press.

Berman, A., and Plemmons, R.J. 1994. *Nonnegative Matrices in the Mathematical Sciences*. SIAM.

Bertsimas, D., Gamarnik, D., and Tsitsiklis, J.N. 1996. Stability conditions for multiclass fluid queueing networks. *IEEE Transactions on Automatic Control*, **41**(11), 1618–1631.

Bertsimas, D., Gamarnik, D., and Tsitsiklis, J.N. 2001. Performance of multiclass Markovian queueing networks via piecewise linear Lyapunov functions. *Annals of Applied Probability*, **11**(4), 1384–1428.

Bickel, P.J., and Wichura, M.J. 1971. Convergence criteria for multiparameter stochastic processes and some applications. *Annals of Mathematical Statistics*, 1656–1670.

Billingsley, P. 1999. *Convergence of Probability Measures*. Wiley.

Borovkov, A.A. 1964. Some limit theorems in the theory of mass service. *Theory of Probability & Its Applications*, **9**(4), 550–565.

Borovkov, A.A. 1965. Some limit theorems in the theory of mass service, II multiple channels systems. *Theory of Probability & Its Applications*, **10**(3), 375–400.

Borovkov, A.A. 1967. On limit laws for service processes in multi-channel systems. *Siberian Mathematical Journal*, **8**(5), 746–763.

Borovkov, A.A. 1984. *Asymptotic Methods in Queuing Theory*. Wiley.

Borovkov, A.A., and Foss, S. 1992. Stochastically recursive sequences and their generalizations. *Siberian Advances in Mathematics*, **2**(1), 16–81.

Botvich, D. D., and Zamyatin, A. A. 1992. Ergodicity of conservative communication networks. *Rapport de Recherche, INRIA*, **1772**.

Boxma, O.J., and Daduna, H. 1990. Sojourn times in queueing networks. Pages 401–450 of: Takagi, H. (ed), *Stochastic Analysis of Computer and Communication Systems*. North-Holland.

Bramson, M. 1994. Instability of FIFO queueing networks. *Annals of Applied Probability*, **4**(2), 414–431.

Bramson, M. 1996. Convergence to equilibria for fluid models of head-of-the-line proportional processor sharing queueing networks. *Queueing Systems*, **23**(1-4), 1–26.

Bramson, M. 1998. State space collapse with application to heavy traffic limits for multiclass queueing networks. *Queueing Systems*, **30**(1-2), 89–140.

Bramson, M. 1999. A stable queueing network with unstable fluid model. *Annals of Applied Probability*, 818–853.

Bramson, M. 2001. Stability of earliest-due-date, first-served queueing networks. *Queueing Systems*, **39**(1), 79–102.

Bramson, M. 2008. *Stability of Queueing Networks*. Springer.

Bramson, M., and Williams, R.J. 2003. Two workload properties for Brownian networks. *Queueing Systems*, **45**(3), 191–221.

Brandt, A., Franken, P., and Lisek, B. 1990. *Stationary Stochastic Models*. Vol. 227. Wiley.

Breiman, L. 1992. *Probability*. SIAM.

Browne, S., Whitt, W., and Dshalalow, J.H. 1995. Piecewise-linear diffusion processes. *Advances in Queueing: Theory, Methods, and Open Problems*, **4**, 463–480.

Budhiraja, A., and Ghosh, A.P. 2006. Diffusion approximations for controlled stochastic networks: an asymptotic bound for the value function. *Annals of Applied Probability*, **16**(4), 1962–2006.

Budhiraja, A., and Ghosh, A.P. 2012. Controlled stochastic networks in heavy traffic: convergence of value functions. *Annals of Applied Probability*, **22**(2), 734–791.

Caldentey, R., Kaplan, E.H., and Weiss, G. 2009. FCFS infinite bipartite matching of servers and customers. *Advances in Applied Probability*, **41**(03), 695–730.

Chao, X., Pinedo, M., and Miyazawa, M. 1999. *Queueing Networks: Customers, Signals and Product Form Solutions*. Wiley.

Chen, H. 1995. Fluid approximations and stability of multiclass queueing networks: work-conserving disciplines. *Annals of Applied Probability*, **5**(3), 637–665.

Chen, H., and Mandelbaum, A. 1991. Stochastic discrete flow networks: Diffusion approximations and bottlenecks. *Annals of Probability*, **19**(4), 1463–1519.

Chen, H., and Mandelbaum, A. 1994a. Hierarchical modeling of stochastic networks, Part I: Fluid models. Pages 47–105 of: *Stochastic Modeling and Analysis of Manufacturing Systems*. Springer.

Chen, H., and Mandelbaum, A. 1994b. Hierarchical modeling of stochastic networks, Part II: Strong approximations. Pages 107–131 of: *Stochastic Modeling and Analysis of Manufacturing Systems*. Springer.

Chen, H., and Yao, D.D. 2001. *Fundamentals of Queueing Networks: Performance, Asymptotics, and Optimization*. Springer.

Chen, H., and Zhang, H. 1997. Stability of multiclass queueing networks under FIFO service discipline. *Mathematics of Operations Research*, **22**(3), 691–725.

Chen, H., and Zhang, H. 2000. Stability of multiclass queueing networks under priority service disciplines. *Operations Research*, **48**(1), 26–37.

Chen, H., Harrison, J.M., Mandelbaum, A., Van Ackere, A., and Wein, L.M. 1988. Empirical evaluation of a queueing network model for semiconductor wafer fabrication. *Operations Research*, **36**(2), 202–215.

Cohen, Jacob Willem. 2012. *The Single Server Queue*. Elsevier.

Cottle, R.W., Pang, J.S., and Stone, R.E. 2009. *The Linear Complementarity Problem*. SIAM.

Cox, D.R., and Smith, W.L. 1961. *Queues*. Methuen.

Daduna, H. 1991. On network flow equations and splitting formulas for sojourn times in queueing networks. *Journal of Applied Mathematics and Stochastic Analysis*, **4**(2), 111–116.

Dai, J.G. 1995. On positive Harris recurrence of multiclass queueing networks: a unified approach via fluid limit models. *Annals of Applied Probability*, **5**(1), 49–77.

Dai, J.G. 1996. A fluid limit model criterion for instability of multiclass queueing networks. *Annals of Applied Probability*, **6**(3), 751–757.

Dai, J.G., and Harrison, J.M. 1991. Steady-state analysis of RBM in a rectangle: numerical methods and a queueing application. *Annals of Applied Probability*, **1**(1), 16–35.

Dai, J.G., and Harrison, J.M. 1992. Reflected Brownian motion in an orthant: numerical methods for steady-state analysis. *Annals of Applied Probability*, **2**(1), 65–86.

Dai, J.G., and He, S. 2010. Customer abandonment in many-server queues. *Mathematics of Operations Research*, **35**(2), 347–362.

Dai, J.G., and He, S. 2013. Many-server queues with customer abandonment: Numerical analysis of their diffusion model. *Stochastic Systems*, **3**(1), 96–146.

Dai, J.G., and Lin, W. 2005. Maximum pressure policies in stochastic processing networks. *Operations Research*, **53**(2), 197–218.

Dai, J.G., and Lin, W. 2008. Asymptotic optimality of maximum pressure policies in stochastic processing networks. *Annals of Applied Probability*, **18**(6), 2239–2299.

Dai, J.G., and Meyn, S.P. 1995. Stability and convergence of moments for multiclass queueing networks via fluid limit models. *IEEE Transactions on Automatic Control*, **40**(11), 1889–1904.

Dai, J.G., and Miyazawa, M. 2011. Reflecting Brownian motion in two dimensions: Exact asymptotics for the stationary distribution. *Stochastic Systems*, **1**(1), 146–208.

Dai, J.G., and Neuroth, S. 2002. DPPS scheduling policies in semiconductor wafer fabs. Pages 194–199 of: *Proceedings of the 2002 International Conference on Modeling and Analysis of Semiconductor Manufacturing MASM*.

Dai, J.G., and Prabhakar, B. 2000. The throughput of data switches with and without speedup. Pages 556–564 of: *INFOCOM 2000. Nineteenth Annual Joint Conference of the IEEE Computer and Communications Societies. Proceedings Volume 2*. IEEE.

Dai, J.G., and Vande Vate, J.H. 2000. The stability of two-station multitype fluid networks. *Operations Research*, **48**(5), 721–744.

Dai, J.G., and Wang, Y. 1993. Nonexistence of Brownian models for certain multiclass queueing networks. *Queueing Systems*, **13**(1-3), 41–46.

Dai, J.G., and Weiss, G. 1996. Stability and instability of fluid models for reentrant lines. *Mathematics of Operations Research*, **21**(1), 115–134.

Dai, J.G., Hasenbein, J.J., and Vande Vate, J.H. 1999. Stability of a three-station fluid network. *Queueing Systems*, **33**(4), 293–325.

Dai, J.G., Hasenbein, J.J., and Vande Vate, J.H. 2004. Stability and instability of a two-station queueing network. *Annals of Applied Probability*, **14**(1), 326–377.

Dai, J.G., He, S., and Tezcan, T. 2010. Many-server diffusion limits for G/Ph/n+ GI queues. *Annals of Applied Probability*, **20**(5), 1854–1890.

Davis, M. 1984. Piecewise-deterministic Markov processes: a general class of non-diffusion stochastic models. *Journal of the Royal Statistical Society. Series B (Methodological)*, **46**(3), 353–388.

Donsker, M.D. 1951. An invariance principle for certain probability limit theorems. *Memoirs of the American Mathematical Society*, **6**, 1–10.

Down, D.G, Gromoll, H.C., and Puha, A.L. 2009. Fluid limits for shortest remaining processing time queues. *Mathematics of Operations Research*, **34**(4), 880–911.

Dubins, L.E. 1968. On a theorem of Skorohod. *Annals of Mathematical Statistics*, **39**(6), 2094–2097.

Erlang, A.K. 1909. The theory of probabilities and telephone conversations. *Nyt Tidsskrift for Matematik B*, **20**(33-39), 16.

Erlang, A.K. 1917. Solution of some problems in the theory of probabilities of significance in automatic telephone exchanges. *Elektrotkeknikeren*, **13**, 5–13.

Eschenfeldt, P., and Gamarnik, D. 2016. *Supermarket queueing system in the heavy traffic regime. Short queue dynamics.* arXiv:1610.03522.

Eschenfeldt, P., and Gamarnik, D. 2018. Join the shortest queue with many servers. The heavy-traffic asymptotics. *Mathematics of Operations Research*, **43**(3), 867–886.

Feller, W. 1971. *An Introduction to Probability Theory and Its Applications. (Vol 2), 2nd ed.* Wiley.

Fleischer, L., and Sethuraman, J. 2005. Efficient algorithms for separated continuous linear programs: the multicommodity flow problem with holding costs and extensions. *Mathematics of Operations Research*, **30**(4), 916–938.

Foss, S., and Konstantopoulos, T. 2004. An overview of some stochastic stability methods. *Journal of the Operations Research Society of Japan-Keiei Kagaku*, **47**(4), 275–303.

Foster, F.G. 1953. On the stochastic matrices associated with certain queuing processes. *Annals of Mathematical Statistics*, **24**(3), 355–360.

Frostig, E., and Weiss, G. 2016. Four proofs of Gittins multiarmed bandit theorem. *Annals of Operations Research*, **241**(1-2), 127–165.

Gamarnik, D., and Hasenbein, J.J. 2005. Instability in stochastic and fluid queueing networks. *Annals of Applied Probability*, **15**(3), 1652–1690.

Gamarnik, D., Tsitsiklis, J.N., and Zubeldia, M. 2018. Delay, memory, and messaging tradeoffs in distributed service systems. *Stochastic Systems*, **8**(1), 45–74.

Gänssler, P., and Stute, W. 1979. Empirical processes: a survey of results for independent and identically distributed random variables. *Annals of Probability*, **7**(2), 193–243.

Gardner, K., and Righter, R. 2020. Product forms for FCFS queueing models with arbitrary server-job compatibilities: an overview. *Queueing Systems*, **96**(1), 3–51.

Gardner, K., Zbarsky, S., Doroudi, S., Harchol-Balter, M., Hyytiä, E., and Scheller-Wolf, A. 2016. Queueing with redundant requests: exact analysis. *Queueing Systems*, **83**(3-4), 227–259.

Gelenbe, E. 1991. Product-form queueing networks with negative and positive customers. *Journal of Applied Probability*, **28**(3), 656–663.

Gittins, J., Glazebrook, K., and Weber, R.R. 2011. *Multi-Armed Bandit Allocation Indices*. Wiley.

Gittins, J.C., and Jones, D.M. 1974. A dynamic allocation index for the sequential design of experiments. In: Gani, J., Sarkadi, K., and Vince, I. (eds), *Progress in Statistics European Meeting of Statisticians (Hungary) 1972. Volume 1*. North-Holland.

Glynn, P.W. 1990. Diffusion approximations. *Handbooks in Operations Research and Management Science*, **2**, 145–198.

Glynn, P.W., and Whitt, W. 1991. A new view of the heavy-traffic limit theorem for infinite-server queues. *Advances in Applied Probability*, **23**(1), 188–209.

Gondran, M., and Minoux, M. 1984. *Graphs and Algorithms*. Wiley.

Goodman, J.B., and Massey, W.A. 1984. The non-ergodic Jackson network. *Journal of Applied Probability*, **21**(4), 860–869.

Gordon, W.J., and Newell, G.F. 1967. Closed queuing systems with exponential servers. *Operations Research*, **15**(2), 254–265.

Gromoll, H.C. 2004. Diffusion approximation for a processor sharing queue in heavy traffic. *Annals of Applied Probability*, **14**(2), 555–611.

Gromoll, H.C., Puha, A.L., and Williams, R.J. 2002. The fluid limit of a heavily loaded processor sharing queue. *Annals of Applied Probability*, **12**(3), 797–859.

Guo, Y., Lefeber, E., Nazarathy, Y., Weiss, G., and Zhang, H. 2014. Stability of multiclass queueing networks with infinite virtual queues. *Queueing Systems*, **76**(3), 309–342.

Gurvich, I., and Whitt, W. 2009. Queue-and-idleness-ratio controls in many-server service systems. *Mathematics of Operations Research*, **34**(2), 363–396.

Gurvich, I., and Whitt, W. 2010. Service-level differentiation in many-server service systems via queue-ratio routing. *Operations Research*, **58**(2), 316–328.

Hajek, B. 1985. Extremal splittings of point processes. *Mathematics of Operations Research*, **10**(4), 543–556.

Halfin, S., and Whitt, W. 1981. Heavy-traffic limits for queues with many exponential servers. *Operations Research*, **29**(3), 567–588.

Harchol-Balter, M. 2013. *Performance Modeling and Design of Computer Systems Queueing Theory in Action*. Cambridge University Press.

Harrison, J.M. 1985. *Brownian Motion and Stochastic Flow Systems*. Wiley.

Harrison, J.M. 1988. Brownian models of queueing networks with heterogeneous customer populations. Pages 147–186 of: *Stochastic Differential Systems, Stochastic Control Theory and Applications*. Springer.

Harrison, J.M. 1996. The BIGSTEP approach to flow management in stochastic processing networks. Chap. 4, pages 147–186 of: Kelly, F.P., Zachary, S., and Ziedins, I. (eds), *Stochastic Networks: Theory and Applications*. Clarendon.

Harrison, J.M. 2000. Brownian models of open processing networks: canonical representation of workload. *Annals of Applied Probability*, **10**(1), 75–103.

Harrison, J.M. 2002. Stochastic networks and activity analysis. Pages 53–76 of: Suhov, Yu.M. (ed), *Analytic Methods in Applied Probability: In Memory of Fridrikh Karpelevich. Vol 207*. Translations of the American Mathematical Society-Series.

Harrison, J.M. 2003. A broader view of Brownian networks. *Annals of Applied Probability*, **13**(3), 1119–1150.

Harrison, J.M. 2013. *Brownian Models of Performance and Control*. Cambridge University Press.

Harrison, J.M., and Reiman, M.I. 1981. On the distribution of multidimensional reflected Brownian motion. *SIAM Journal on Applied Mathematics*, **41**(2), 345–361.

Harrison, J.M., and Taksar, M.I. 1983. Instantaneous control of Brownian motion. *Mathematics of Operations Research*, **8**(3), 439–453.

Harrison, J.M., and Taylor, A.J. 1978. Optimal control of a Brownian storage system. *Stochastic Processes and Their Applications*, **6**(2), 179–194.

Harrison, J.M., and Van Mieghem, J.A. 1997. Dynamic control of Brownian networks: state space collapse and equivalent workload formulations. *Annals of Applied Probability*, **7**(3), 747–771.

Harrison, J.M., and Wein, L.M. 1989. Scheduling networks of queues: heavy traffic analysis of a simple open network. *Queueing Systems*, **5**(4), 265–279.

Harrison, J.M., and Wein, L.M. 1990. Scheduling networks of queues: heavy traffic analysis of a two-station closed network. *Operations Research*, **38**(6), 1052–1064.

Harrison, J.M., and Williams, R.J. 1987a. Brownian models of open queueing networks with homogeneous customer populations. *Stochastics*, **22**(2), 77–115.

Harrison, J.M., and Williams, R.J. 1987b. Multidimensional reflected Brownian motions having exponential stationary distributions. *Annals of Probability*, **15**(1), 115–137.

Haviv, M. 2013. *Queues: A Course in Queueing Theory*. Springer.

Henderson, W. 1983. Insensitivity and reversed Markov processes. *Advances in Applied Probability*, **15**(4), 752–768.

Iglehart, D.L. 1965. Limiting diffusion approximations for the many server queue and the repairman problem. *Journal of Applied Probability*, **2**(02), 429–441.

Iglehart, D.L. 1973. Weak convergence in queueing theory. *Advances in Applied Probability*, **5**(3), 570–594.

Iglehart, D.L., and Whitt, W. 1970a. Multiple channel queues in heavy traffic. I. *Advances in Applied Probability*, **2**(1), 150–177.

Iglehart, D.L., and Whitt, W. 1970b. Multiple channel queues in heavy traffic. II. *Advances in Applied Probability*, **2**(2), 355–369.

Jackson, J.R. 1963. Jobshop-like queueing systems. *Management Science*, **10**(1), 131–142.

Kang, W.N., and Williams, R.J. 2012. Diffusion approximation for an input-queued packet switch operating under a maximum weight algorithm. *Stochastic Systems*, **2**(2), 277–321.

Keilson, J., and Servi, L.D. 1988. A distributional form of Little's law. *Operations Research Letters*, **7**(5), 223–227.

Kelly, F.P. 1979. *Reversibility and Stochastic Networks*. Wiley, reprinted Cambridge University Press, 2011.

Kelly, F.P., and Laws, C.N. 1993. Dynamic routing in open queueing networks: Brownian models, cut constraints and resource pooling. *Queueing Systems*, **13**(1-3), 47–86.

Kendall, D.G. 1953. Stochastic processes occurring in the theory of queues and their analysis by the method of the imbedded Markov chain. *Annals of Mathematical Statistics*, **24**(3), 338–354.

Kiefer, J., and Wolfowitz, J. 1955. On the theory of queues with many servers. *Transactions of the American Mathematical Society*, **78**(1), 1–18.

Kingman, J.F.C. 1961. Two similar queues in parallel. *Annals of Mathematical Statistics*, **32**(4), 1314–1323.

Kingman, J.F.C. 1962. On queues in heavy traffic. *Journal of the Royal Statistical Society: Series B (Methodological)*, **24**(2), 383–392.

Kleinrock, L. 1975. *Queuing Systems, Volume 1: Theory*. Wiley.

Kleinrock, L. 1976. *Queueing Systems, Volume 2: Computer Applications*. Wiley.

Klimov, G.P. 1975. Time-sharing service systems. I. *Theory of Probability & Its Applications*, **19**(3), 532–551.

Komlós, J., Major, P., and Tusnády, G. 1975. An approximation of partial sums of independent RV'-s, and the sample DF. I. *Zeitschrift für Wahrscheinlichkeitstheorie und verwandte Gebiete*, **32**(1), 111–131.

Komlós, J., Major, P., and Tusnády, G. 1976. An approximation of partial sums of independent RV's, and the sample DF. II. *Zeitschrift für Wahrscheinlichkeitstheorie und verwandte Gebiete*, **34**(1), 33–58.

Kopzon, A., Nazarathy, Y., and Weiss, G. 2009. A push–pull network with infinite supply of work. *Queueing Systems*, **62**(1-2), 75–111.

Krichagina, E.V., and Puhalskii, A.A. 1997. A heavy-traffic analysis of a closed queueing system with a GI/∞ service center. *Queueing Systems*, **25**(1-4), 235–280.

Kruk, L., Lehoczky, J., Ramanan, K., and Shreve, S.E. 2007. An explicit formula for the Skorokhod map on [0, a]. *Annals of Probability*, **35**(5), 1740–1768.

Kumar, P.R. 1993. Re-entrant lines. *Queueing Systems*, **13**(1-3), 87–110.

Kumar, P.R., and Meyn, S.P. 1996. Duality and linear programs for stability and performance analysis of queuing networks and scheduling policies. *IEEE Transactions on Automatic Control*, **41**(1), 4–17.

Kumar, P.R., and Seidman, T.I. 1990. Dynamic instabilities and stabilization methods in distributed real-time scheduling of manufacturing systems. *IEEE Transactions on Automatic Control*, **35**(3), 289–298.

Kurtz, T.G. 1981. *Approximation of Population Processes*. SIAM.

Kushner, H.J. 1990. Numerical methods for stochastic control problems in continuous time. *SIAM Journal on Control and Optimization*, **28**(5), 999–1048.

Kushner, H.J. 2001. *Heavy Traffic Analysis of Controlled Queueing and Communication Networks*. Springer.

Latouche, G., and Taylor, P. 2000. *Advances in Algorithmic Methods for Stochastic Models, Proceedings of the 3rd International Conference on Matrix Analytic Methods*. Notable Publications Inc.

Latouche, G., and Taylor, P. 2002. *Matrix-Analytic Methods: Theory and Applications, Proceedings of the 4th International Conference on Matrix Analytic Methods*. World Scientific.

Latouche, G., Ramaswami, V., Sethuraman, J., Sigman, K., Squillante, M.S., and Yao, D. 2012. *Matrix-Analytic Methods in Stochastic Models*. Vol. 27. Springer Science & Business Media.

Lawler, E.L., Lenstra, J.K., Rinnooy Kan, A.H.G., and Shmoys, D.B. 1993. Sequencing and scheduling: algorithms and complexity. *Handbooks in Operations Research and Management Science*, **4**, 445–522.

Laws, C.N. 1992. Resource pooling in queueing networks with dynamic routing. *Advances in Applied Probability*, **24**(3), 699–726.

Laws, C.N., and Louth, G.M. 1990. Dynamic scheduling of a four-station queueing network. *Probability in the Engineering and Informational Sciences*, **4**(01), 131–156.

Lemoine, A.J. 1987. On sojourn time in Jackson networks of queues. *Journal of Applied Probability*, **24**(2), 495–510.

Lindley, D.V. 1952. The theory of queues with a single server. *Mathematical Proceedings of the Cambridge Philosophical Society*, **48**(02), 277–289.

Loynes, R.M. 1962. The stability of a queue with non-independent inter-arrival and service times. *Mathematical Proceedings of the Cambridge Philosophical Society*, **58**(03), 497–520.

Lu, S.H., and Kumar, P.R. 1991. Distributed scheduling based on due dates and buffer priorities. *IEEE Transactions on Automatic Control*, **36**(12), 1406–1416.

Maguluri, S.T., and Srikant, R. 2016. Heavy traffic queue length behavior in a switch under the MaxWeight algorithm. *Stochastic Systems*, **6**(1), 211–250.

Mairesse, J., and Moyal, P. 2016. Stability of the stochastic matching model. *Journal of Applied Probability*, **53**(4), 1064–1077.

Major, P. 1976. The approximation of partial sums of independent rv's. *Zeitschrift für Wahrscheinlichkeitstheorie und verwandte Gebiete*, **35**(3), 213–220.

Mandelbaum, A., and Momcilovic, P. 2012. Queues with many servers and impatient customers. *Mathematics of Operations Research*, **37**(1), 41–65.

Martins, L.F., Shreve, S.E., and Soner, H.M. 1996. Heavy traffic convergence of a controlled, multiclass queueing system. *SIAM Journal on Control and Optimization*, **34**(6), 2133–2171.

Matloff, N.S. 1989. On the value of predictive information in a scheduling problem. *Performance Evaluation*, **10**(4), 309–315.

Meyer, P.A. 1971. Sur un article de Dubins. Pages 170–176 of: *Séminaire de Probabilités V Université de Strasbourg*. Springer.

Meyn, S.P. 2008. *Control Techniques for Complex Networks*. Cambridge University Press.

Meyn, S.P., and Down, D.G. 1994. Stability of generalized Jackson networks. *Annals of Applied Probability*, **4**(1), 124–148.

Meyn, S.P., and Tweedie, R.L. 1993. *Markov Chains and Stochastic Stability*. Springer-Verlag.

Mitzenmacher, M. 1996. *The Power of Two Choices in Randomized Load Balancing*. Ph.D. thesis, University of California at Berkeley.

Mitzenmacher, M. 2001. The power of two choices in randomized load balancing. *IEEE Transactions on Parallel and Distributed Systems*, **12**(10), 1094–1104.

Moyal, P., and Perry, O. 2017. On the instability of matching queues. *Annals of Applied Probability*, **27**(6), 3385–3434.

Moyal, P., Busic, A., and Mairesse, J. 2021. A product form for the general stochastic matching model. *Journal of Applied Probability*, **58**(2), 449–468.

Mukherjee, D., Borst, S., Van Leeuwaarden, J., and Whiting, P. 2018. Universality of power-of-d load balancing in many-server systems. *Stochastic Systems*, **8**(4), 265–292.

Nazarathy, Y., and Weiss, G. 2009. Near optimal control of queueing networks over a finite time horizon. *Annals of Operations Research*, **170**(1), 233–249.

Nazarathy, Y., and Weiss, G. 2010. Positive Harris recurrence and diffusion scale analysis of a push pull queueing network. *Performance Evaluation*, **67**(4), 201–217.

Nazarathy, Y., Rojas-Nandayapa, L., and Salisbury, T.S. 2013. Non-existence of stabilizing policies for the critical push–pull network and generalizations. *Operations Research Letters*, **41**(3), 265–270.

Neuts, M.F. 1974. *Probability Distributions of Phase Type*. Purdue University. Department of Statistics.

Neuts, M.F. 1981. *Matrix-Geometric Solutions in Stochastic Models: an Algorithmic Approach*. Courier Corporation, reprinted Dover Publications 1994.

Obłój, J. 2004. The Skorokhod embedding problem and its offspring. *Probability Surveys*, **1**, 321–392.

Oksendal, B. 2013. *Stochastic Differential Equations: an Introduction with Applications*. Springer Science & Business Media.

Perel, N., and Yechiali, U. 2014. The Israeli queue with infinite number of groups. *Probability in the Engineering and Informational Sciences*, **28**(1), 1.

Pinedo, M.L. 2012. *Scheduling: Theory, Algorithms, and Systems*. Springer.

Ramanan, K. 2006. Reflected diffusions defined via the extended Skorokhod map. *Electronic Journal of Probability*, **11**, 934–992.

Reed, J. 2009. The G/GI/N queue in the Halfin–Whitt regime. *Annals of Applied Probability*, **19**(6), 2211–2269.

Reiman, M.I. 1984a. Open queueing networks in heavy traffic. *Mathematics of Operations Research*, **9**(3), 441–458.

Reiman, M.I. 1984b. Some diffusion approximations with state space collapse. Pages 207–240 of: *Modelling and Performance Evaluation Methodology*. Springer.

Ross, S.M. 1983. *Introduction to Stochastic Dynamic Programming*. Academic Press.

Ross, S.M. 2009. *Introduction to Probability Models*. 10 edn. Elsevier Academic Press.

Rybko, A.N., and Stolyar, A.L. 1992. Ergodicity of stochastic processes describing the operation of open queueing networks. *Problemy Peredachi Informatsii, Translation Problems of Information Transmission*, **28**(3), 3–26.

Schassberger, R. 1986. Two remarks on insensitive stochastic models. *Advances in Applied Probability*, **18**(3), 791–814.

Schrage, L.E. 1968. Proof of the optimality of the shortest remaining processing time discipline. *Operations Research*, **16**(3), 687–690.

Schrage, L.E., and Miller, L.W. 1966. The queue M/G/1 with the shortest remaining processing time discipline. *Operations Research*, **14**(4), 670–684.

Seneta, E. 1981. *Non-negative Matrices and Markov Chains*. Springer Science & Business Media.

Serfozo, R. 1999. *Introduction to Stochastic Networks*. Vol. 44. Springer Science & Business Media.

Servi, L.D., and Finn, S.G. 2002. M/M/1 queues with working vacations (m/m/1/wv). *Performance Evaluation*, **50**(1), 41–52.

Sevast'Yanov, B.A. 1957. An ergodic theorem for Markov processes and its application to telephone systems with refusals. *Theory of Probability & Its Applications*, **2**(1), 104–112.

Sevcik, K.C., and Mitrani, I. 1981. The distribution of queuing network states at input and output instants. *Journal of the ACM (JACM)*, **28**(2), 358–371.

Shah, D., and Wischik, D. 2012. Switched networks with maximum weight policies: fluid approximation and multiplicative state space collapse. *Annals of Applied Probability*, **22**(1), 70–127.

Shindin, E., Masin, M., Zadorojniy, A., and Weiss, G. 2021. *Revised Simplex-Type Algorithm for SCLP with Application to Large-scale Fluid Processing Networks*. arXiv:2103.04405.

Sigman, K. 1990. The stability of open queueing networks. *Stochastic Processes and Their Applications*, **35**(1), 11–25.

Skorokhod, A.V. 1961. Stochastic equations for diffusion processes in a bounded region. *Theory of Probability & Its Applications*, **6**(3), 264–274.

Smith, W.L. 1953. On the distribution of queueing times. *Mathematical Proceedings of the Cambridge Philosophical Society*, **49**(03), 449–461.

Smith, W.L. 1958. Renewal theory and its ramifications. *Journal of the Royal Statistical Society: Series B (Methodological)*, **20**(2), 243–284.

Stolyar, A.L. 2004. Maxweight scheduling in a generalized switch: State space collapse and workload minimization in heavy traffic. *Annals of Applied Probability*, **14**(1), 1–53.

Stolyar, A.L. 2015. Pull-based load distribution in large-scale heterogeneous service systems. *Queueing Systems*, **80**(4), 341–361.

Stone, C. 1963. Limit theorems for random walks, birth and death processes, and diffusion processes. *Illinois Journal of Mathematics*, **7**(4), 638–660.

Tassiulas, L. 1995. Adaptive back-pressure congestion control based on local information. *IEEE Transactions on Automatic Control*, **40**(2), 236–250.

Tassiulas, L., and Ephremides, A. 1992. Stability properties of constrained queueing systems and scheduling policies for maximum throughput in multihop radio networks. *IEEE Transactions on Automatic Control*, **37**(12), 1936–1948.

Van den Berg, J.L., and Boxma, O.J. 1989. Sojourn times in feedback queues. Pages 247–257 of: *Operations Research Proceedings 1988*. Springer.

Van den Berg, J.L., and Boxma, O.J. 1993. *Sojourn Times in Feedback and Processor Sharing Queues*. Vol. 97. Centrum voor Wiskunde en Informatica.

Van der Boor, M., Borst, S., Van Leeuwaarden, J., and Mukherjee, D. 2019. Scalable load balancing in networked systems: universality properties and stochastic coupling methods. Pages 3893–3923 of: *Proceedings of the International Congress of Mathematicians (ICM 2018),*. World Scientific.

Van Dijk, N.M. 1993. *Queueing Networks and Product Forms: a Systems Approach*. Wiley.

Van Houtum, G., Zijm, H., Adan, I., and Wessels, J. 1998. Bounds for performance characteristics: a systematic approach via cost structures. *Stochastic Models*, **14**(1-2), 205–224.

Van Leeuwaarden, J., Mathijsen, B., and Zwart, B. 2019. Economies-of-Scale in Many-Server Queueing Systems: Tutorial and Partial Review of the QED Halfin–Whitt Heavy-Traffic Regime. *SIAM Review*, **61**(3), 403–440.

Van Zant, P. 2014. *Microchip Fabrication: A Practical Guide to Semiconductor Processing, 6th Edition*. McGraw-Hill.

Visschers, J. 2000. *Random Walks with Geometric Jumps*. Ph.D. thesis, Technische Universiteit Eindhoven.

Visschers, J., Adan, I., and Weiss, G. 2012. A product form solution to a system with multi-type jobs and multi-type servers. *Queueing Systems*, **70**(3), 269–298.

Vvedenskaya, N., Dobrushin, R., and Karpelevich, F. 1996. Queueing system with selection of the shortest of two queues: An asymptotic approach. *Problemy Peredachi Informatsii*, **32**(1), 20–34.

Walrand, J. 1988. *An Introduction to Queueing Networks*. Prentice-Hall.

Ward, A.R. 2012. Asymptotic analysis of queueing systems with reneging: a survey of results for FIFO, single class models. *Surveys in Operations Research and Management Science*, **17**(1), 1–14.

Weber, R.R. 1978. On the optimal assignment of customers to parallel servers. *Journal of Applied Probability*, **15**(2), 406–413.

Wein, L.M. 1988. Scheduling semiconductor wafer fabrication. *IEEE Transactions on Semiconductor Manufacturing*, **1**(3), 115–130.

Wein, L.M. 1990a. Optimal control of a two-station Brownian network. *Mathematics of Operations Research*, **15**(2), 215–242.

Wein, L.M. 1990b. Scheduling networks of queues: heavy traffic analysis of a two-station network with controllable inputs. *Operations Research*, **38**(6), 1065–1078.

Wein, L.M. 1991. Brownian networks with discretionary routing. *Operations Research*, **39**(2), 322–340.

Wein, L.M. 1992. Scheduling networks of queues: heavy traffic analysis of a multistation network with controllable inputs. *Operations Research*, **40**(3, supplement 2), S312–S334.

Weiss, G. 1995. On optimal draining of fluid reentrant lines. Pages 91–103 of: Kelly, F.P., and Williams, R.J. (eds), *Stochastic Networks*. IMA series, Vol. 71, Springer.

Weiss, G. 2005. Jackson networks with unlimited supply of work. *Journal of Applied Probability*, **42**(3), 879–882.

Weiss, G. 2008. A simplex based algorithm to solve separated continuous linear programs. *Mathematical Programming*, **115**(1), 151–198.

Weiss, G. 2020. Directed FCFS infinite bipartite matching. *Queueing Systems*, **96**(3–4), 387–418.

Whitt, W. 1982. On the heavy-traffic limit theorem for GI/G/∞ queues. *Advances in Applied Probability*, **14**(1), 171–190.

Whitt, W. 1986. Deciding which queue to join: some counterexamples. *Operations Research*, **34**(1), 55–62.

Whitt, W. 2002. *Stochastic Process Limits*. Springer.

Whitt, W. 2006. Fluid models for multiserver queues with abandonments. *Operations Research*, **54**(1), 37–54.

Whittle, P. 1986. *Systems in Stochastic Equilibrium*. Wiley.

Williams, R.J. 1998. Diffusion approximations for open multiclass queueing networks: sufficient conditions involving state space collapse. *Queueing Systems*, **30**(1), 27–88.

Winston, W. 1977. Optimality of the shortest line discipline. *Journal of Applied Probability*, **14**(1), 181–189.

Wolff, R.W. 1982. Poisson arrivals see time averages. *Operations Research*, **30**(2), 223–231.

Wolff, R.W. 1989. *Stochastic Modeling and the Theory of Queues*. Prentice-Hall.

Yang, J.K. 2009. A Simple proof for the stability of global FIFO queueing networks. *Acta Mathematicae Applicatae Sinica, English Series*, **25**(4), 647–654.

Zeltyn, S., and Mandelbaum, A. 2005. Call centers with impatient customers: many-server asymptotics of the M/M/n+ G queue. *Queueing Systems*, **51**(3-4), 361–402.

Index

abandonment, 15, 348–360, 406
  abandonment rate, 350, 357
arrival process, 3
arrival theorem, 133

backwards recurrence time, 27
balance equations, 127, 132, 173, 237,
    262, 270, 355
  detailed balance equations, 8, 128
  global balance equations, 35, 38, 128
  partial balance equations, 129, 137,
      138, 390, 401, 412
bandit processes, 53–55
birth and death processes, 7–12,
    105–107, 141, 336, 353, 360
birth and death queues, 7–12
bottleneck workload process, 303
Bramson's FCFS network, 165
Brownian bridge, 326, 329, 331
Brownian control problem, 275–278,
    280, 285, 291, 298, 302–303, 309,
    312, 319
Brownian motion, 102–104, 112–115
  distribution of maximum, 113
  multivariate, 111
Burke's theorem, 128
busy period, 33–34
  exceptional first service bp, 34
busy time, 84

coefficient of variation, 16, 32, 68, 79,
    265, 338
compatibility graph, 387, 412
compensation method, 224, 262–264
complete resource pooling, 283, 302,
    394, 396, 407
complexity
  #P-hard, 391, 396
  NP-hard, 46, 146, 240

continuous mapping theorem, 72, 152,
    338
convergence, 76–77
  almost sure, 72
  in distribution, 72
  in probability, 76, 81
  law of the iterated logarithm, 76
  weak, 72
convex majorization, 364, 385
correlation, 96, 100, 237
covariance calculations, 153
Cox Process, *see* doubly stochastic
    Poisson process
criss-cross network, 278–283, 308
customer average, 17

density dependent Markov chains,
    366–369
departure process, 5, 33, 49, 84, 96, 128,
    237
diffusion limits, 90–95, 150
diffusion process
  Brownian bridge, 326
  Ito calculus representation, 102
  Kiefer process, 326, 329, 342
  multivariate reflected Brownian
      motion, 111, 152
  Ornstein–Uhlenbeck process, 99, 105,
      337, 354
  reflected Brownian motion, 104, 115
diffusion scaling, 86
distributional form of Little's law, 36,
    391, 412
Donsker's theorem, 75, 77
doubly exponential rate, 370
doubly stochastic Poisson process, 367

efficiency driven service, 352
elementary renewal theorem, 23, 183
embedded Markov chains, 35–39

equilibrium distribution, 27, 52, 67, 100, 139, 344, 356
ergodicity, 18, 19, 27, 128, 131, 170–175, 183
Erlang loss system, 4, 11, 21, 142
extreme allocation available, 204

FCFS-ALIS policy, 388–392
FCLT
    Donsker's theorem, 75, 77
    for random walks, 75
    for renewal processes, 80
flowtime, 43
fluid limit, 88, 148, 178
fluid limit model, 180
fluid model equations, 180, 349
fluid model solutions, 180
fluid scaling, 86
fluid stability, 183
    weak, 186
forward recurrence time, 27
Foster criterion, 174
free time, 273
FSLLN
    for random walks, 74
    for renewal processes, 80

general state space Markov processes, 170–175
generalized cut constraints, 318
Gittins index, 53–55
global FCFS, *see* time stamp policy
global stability region, 192–195

Halfin–Whitt regime, 333–347, 352–354, 358
Harris Markov process, 171
Harris positive recurrent, 171, 183
Harris recurrent, 171
hazard rate, 264, 349
head of the line policy, 84
head of the line proportional processor sharing, 192, 199

idle time, 84, 274
infinite bipartite matching, 392–403
    directed matching, 403–406
    link lengths, 397
    matching rates, 397
    stationary distribution, 395, 401
    time-reversal theorem, 395, 400
    uniqueness theorem, 394, 397
infinite sum of product forms, 224, 262
infinite virtual queues, 221–223, 249–252

input queued switch, 212–214
input-output matrix, 147, 162, 203, 216, 230, 242
insensitivity, 52, 100, 139
instability examples, 163–167
Israeli queue, 15
Ito integral, 116
Ito process, 116

Jackson networks, 125, 131
join shortest of d policy , 363–375
join the shortest queue, 261–264, 375–383
    with jockeying, 270

K machine M repairmen model, 14
Kelly's lemma, 131, 141, 401
Kelly-type networks, 140, 192, 198, 306
Kendall's three field notation, 4
Kiefer process, 326, 329, 342
Kigman's bound, 67
Kumar–Seidman Rybko–Stolyar network, 166, 176, 197, 226
Kurtz theorem, 368

length biasing, 25–27
    backwards recurrence time, 27
    forward recurrence time, 27
    length of the current interval, 27
likelihood ratio, *see* Radon–Nikodym derivative
Lindley's equation, 5, 13, 62, 67, 69
Lindley's integral equation, 66
linear complementarity problem, 146, 156
link lengths, 397
Little's law, 16–21
    applications, 20
    distributional form, 36, 391, 412
Loynes construction, 61–64, 69, 154, 398
Lu–Kumar network, 163–164, 192–195
Lyapunov functions, 168, 173, 187, 207
    piecewise linear, 192–195

manufacturing system, 109–111, 239–240
matching rates, 397
maximum pressure policy, 203–205
    application to MCQN, 211
    fluid stability, 209
    for input queued switch, 214
    for network of switches, 216
    non-preemptive, 210
    non-splitting, 210

parametrization, 208
rate stability, 206, 209, 210
maximum throughput, 199, 226, 275, 285, 305, 406
mean field representation, 364
multi-class queueing networks, 161–163
    diffusion limits, 268
    divergent, 186
    fluid stability proofs, 187–192
    global stability, 192
    in balanced heavy traffic, 272–274
    positive Harris recurrent, 183
    rate stable, 186
    stable under maximum pressure, 211
    state space collapse, 268
    unstable, 163
nominal allocation, 205, 273
NP-hard, 46, 146, 240

order statistics, 46
Ornstein–Uhlenbeck process, 99, 105, 337, 354

Palm measure, 27, 400
parallel skilled service system, 387
PASTA, 28–31, 36, 50, 67, 133, 135, 392
patience distribution, 348
Perron–Frobenius theorem, 126
petite set, 171
phase-type distributions, 35
piecewise deterministic Markov process, 26, 154, 170
piecewise linear Lyapunov functions, 192–195
policy
    $c\mu$, 53
    earliest due date, 192, 198, 199
    first come first served, 4, 32, 49, 165, 169, 176, 181, 196, 198, 304, 321, 388
    head of the line processor sharing, 198
    last come first served, 4, 49, 52, 169
    non-preemptive, 20
    processor sharing, 4, 52
    queue and idleness ratio, 408–411
    shortest expected processing time first, 45
    shortest processing time first, 43, 56
    shortest remaining processing time first, 47–49, 57
    Smith rule, 53
    time stamp, 192, 199

work conserving, 20
Pollaczek–Khinchine formula, 32, 36
positive matrices, 126
priority scheduling, 50–52
processing network, 200–203
product-form solution, 130, 131, 135, 137, 143, 390, 395, 405
push-pull network, 226–228, 232–235

quality and efficiency driven service, 352, 358–360
quality driven service, 352
quality of service, 407
quasi birth and death queues, 35
queue and idleness ratio policy, 408–411
queue dynamics, 84, 145, 161, 201, 390
queue with balking, 15
queue with vacations, 15
queueing system
    $\cdot/G/\infty$ queue, 325–331
    $\cdot/G/s$ queue, 338–347
    $G/D/\infty$ queue, 98
    $G/G/\infty$ queue, 98–100
    $G/G/n+G$ queue, 348–352
    $G/G/s$ queue, 344, 346
    $G/G/1$ queue, 61–68, 70, 83–97, 107
    $G/M/\infty$ queue, 99
    $G/M/s$ queue, 338
    $G/M/1$ queue, 37–39
    $M/G/\infty$ queue, 100
    $M/G/1$ queue, 16, 31–33, 35–37
    $M/M/\infty$ queue, 10, 99
    $M/M/n+G$ queue, 354–360
    $M/M/n+M$ queue, 352–354
    $M/M/s$ queue, 11, 333–337
    $M/M/1$ queue, 9, 52, 130

Radon–Nikodym derivative, 114
random walk, 64–67, 73–78
re-entrant line, 163, 190–192, 196, 225, 236, 239, 244, 254
reflected Brownian motion, 104, 115
    multivariate, 111, 152
reflection principle, 112
regeneration times, 34
reneging, 15, *see* abandonment
renewal function, 23, 40
renewal process, 23, 79–81
    delayed, 28
renewal reward theorem, 25–27, 33, 40, 109, 118
renewal theory, 23–25

resource pooling, 10, 41, 264–265, 283, 302, 394, 396, 406
reverse Leontief network, 210
ride-sharing, 403
routing matrix, 125, 218

#P-hard, 391, 396
scheduling, 42–47
  batch scheduling, 43
  stream scheduling, 46
service regime
  efficiency driven, 352
  quality and efficiency driven, 352, 358–360
  quality driven, 352
simulation, 6–7, 12, 258, 412
Skorohod embedding, 77, 121
Skorohod reflection, 86–88, 281, 310, 381
  oblique, 111–112, 147–148, 188
  two-sided, 107–109, 288
small set, 171
snapshot principle, 34
sojourn time, 6, 9, 18, 31, 33, 37, 50, 52, 133–134, 264, 321, 360, 371, 383
square root staffing, 335, 352, 376
stability, 19, 61, 130, 155, 170–175, 183, 187, 389
  rate stability, 18, 126, 146, 174, 186, 206
standby customer, 41
state space collapse, 266–268, 283, 302, 411
static planning problem, 205, 209, 230, 409
  dual, 303
stationary probabilities, 8, 130, 131, 135, 137, 224, 264, 352, 363, 390, 395, 405
stationary process, 18
stochastic integral, *see* Ito process
Stone's theorem, 106, 336, 354
stopping time, 24, 55, 78, 103, 118, 170
strict Leontief network, 204
strong approximation, 78–79, 97–98
strong Markov property, 103, 113

taxi stand model, 13
threshold policy, 268, 282, 310
tightness, 72
time average, 17
time change lemma, 72, 80, 152, 154, 338, 379, 382

time reversibility, 8, 27, 127–129
time-reversal, 104, 131, 395, 400
traffic equations, 126
two sided regulator, 107–109, 117–119

uniform integrability, 24, 40, 183, 185
uniformly small set, 171
utilization, 21, 205, 221, 255

virtual machine, 167

waiting time, 5, 6, 32, 51, 62, 67, 350, 356–358, 363
Wald's equation, 24, 40, 52, 133
Wiener process, *see* Brownian motion
Wiener–Hopf decomposition, 66
work conservation, 21–23, 32, 84, 147, 162, 189
workload formulation, 275, 280, 287, 291, 297

Printed in the United States
by Baker & Taylor Publisher Services